Detection of Signals in Noise

This is a volume in

ELECTRICAL SCIENCE

A Series of Monographs and Texts

Editors: Henry G. Booker and Nicholas DeClaris

A complete list of titles in this series appears at the end of this volume.

DETECTION OF SIGNALS IN NOISE

ANTHONY D. WHALEN

Bell Telephone Laboratories
Whippany, New Jersey

ACADEMIC PRESS, INC.
Harcourt Brace Jovanovich, Publishers
San Diego New York Berkeley Boston
London Sydney Tokyo Toronto

ACADEMIC PRESS, INC.
1250 Sixth Avenue, San Diego, California 92101

United Kingdom Edition published by
ACADEMIC PRESS, INC. (LONDON) LTD.
24/28 Oval Road, London NW1 7DX

LIBRARY OF CONGRESS CATALOG CARD NUMBER: 76-137630

PRINTED IN THE UNITED STATES OF AMERICA

87 9 8

Contents

Preface

This book is intended to serve as an introduction to the principles and applications of the statistical theory of signal detection. Emphasis is placed on those principles which have been found to be particularly useful in practice. These principles are applied to detection problems encountered in digital communications, radar, and sonar. In large part the book has drawn upon the open literature: texts, technical journals, and industrial and university reports. The author's intention was to digest this material and present a readable and pedagogically useful treatment of the subject.

The theory is presented in a manner tempered by practical experience, sometimes at a sacrifice of rigor. The reader is not generally cautioned about existence proofs, convergence, interchanging operators, etc. Normally, the material is not presented in all possible generality. While an all-encompassing general approach to a subject is mathematically elegant, insight and understanding may be sacrificed. An example of the approach taken is the treatment of additive noise. It is initially assumed that additive noise is white and Gaussian. Since much of the work in the field and many practical receiver designs are based on white Gaussian noise, the solutions based on this assumption are thoroughly treated first. The derivations are more straightforward and lead easily to understanding the fundamentals and deriving numerical results. Afterward, the more general approach which assumes nonwhite noise is introduced.

A summary of the book is as follows: Probability and random processes are briefly reviewed in detail sufficient to acquaint the casual student of probability with the tools necessary to read the text. Narrowband signals, their complex representation, and their properties are discussed with the aid of the Hilbert transform. Probability functions resulting from the Gaussian assumption are derived and graphically displayed. Hypothesis testing is applied to the detection of signals. Digital communications and detection applications (e.g., radar and sonar) are treated, optimum receivers are derived, and their performance graphically displayed. Estimation theory is discussed especially as it relates to the estimation of signal amplitude, phase, frequency, and time of arrival. Bounds on the variance of estimators are determined. Space–time or multichannel processing emphasizing the sampled approach is treated. Problem exercises, references, and a supplementary bibilography are included after each chapter.

Early versions of this manuscript were used for one-semester courses given at the Bell Telephone Laboratories. Thus, the book should be suitable for a graduate course in signal detection theory. It may also be suitable for an undergraduate course for special students who have established interests in the material and who have already had a course in probability and random processes. The author hopes the book can be used for self-teaching, and serve as well as a reference book for the practicing engineer.

Acknowledgments

I wish to acknowledge those people who directly or indirectly contributed to this book. I am indebted to R. M. Lauver for his continuous encouragement to teach and to publish; to J. K. Wolf for his instruction and allowing me to use his lecture notes which guided my presentation of material; to B. P. Bogert for his moral support during preparation of the final draft; and especially to G. H. Robertson for making available to me his repertoire of computer programs, which were extensively used for numerical computations; to J. D. Patterson for a critical reading of the final draft; to Miss N. Lockwood for faithfully typing many versions of the manuscript over a period of five years; to the manuscript reviewers from Bell Telephone Laboratories for their helpful comments. But most of all, I want to thank my wife and my children, who have been patient and understanding beyond measure. This book is dedicated to them.

Detection of Signals in Noise

Chapter 1

Probability†

1.1 Probability in Brief

It is intuitively appealing to associate the probability of an event (outcome) with the relative frequency of occurrence of that event. For example, suppose we conduct an experiment (such as a dice roll) having possible outcomes A, B, C, etc. We repeat the experiment N times and record the number of times that each event occurs. Note these by n_A, n_B, n_C respectively. Their relative frequencies of occurrence are given by the ratios n_A/N, n_B/N, n_C/N. In the limit as N approaches infinity we say that these ratios approach the true probability of occurrence of the event. That is,

$$\lim_{N\to\infty} n_A/N \to P(A), \qquad \lim_{N\to\infty} n_B/N \to P(B), \qquad \text{etc.}$$

Thus, the probability is a number between 0 and 1, inclusive. One cannot in practice determine such probabilities with absolute certainty.

The set of all possible outcomes of an experiment is called the sample space of the experiment. An event may be a single member of the sample space or it may be a collection of possible outcomes. In either case the occurrence or nonoccurrence of an event is determined by the outcome of the experiment. We shall use bracketed symbols to denote events. For example, $\{A\}$ is the subset of the sample space whose members have property A.

† Very readable treatments of probability may be found in Davenport and Root (*1*) and Papoulis (*2*). For additional references consult the bibliography at the end of this chapter.

For any event $\{A\}$ there exists the event $\{\text{not } A\}$ which we denote $\{\bar{A}\}$. The event† $\{A \text{ or } \bar{A}\}$ is the set of all possible outcomes (the certain event). The event $\{A \text{ and } \bar{A}\}$ is the set with no members (the null event). An immediate consequence of the relative frequency definition is that the probabilities of the certain and null events are one and zero respectively. If the events $\{A\}$ and $\{B\}$ are mutually exclusive (that is, if one occurs, the other cannot) then for the event $\{A \text{ or } B\}$ we have $P(A \text{ or } B) = P(A) + P(B)$.

Suppose two experiments are conducted whose possible outcomes are denoted by $A_i, i = 1, 2, \ldots,$ and $B_j, j = 1, 2, \ldots$ respectively. Then the joint event $\{A_i \text{ and } B_j\}$ is defined. Associated with such a joint event is a probability as in the case of single events. Denote the probability of this joint event by $P(A_i, B_j)$. If the A_i's and B_j's are exhaustive, that is, if no other events are possible, then‡

$$\sum_i \sum_j P(A_i, B_j) = 1$$

$$\sum_i P(A_i, B_j) = P(B_j), \qquad \sum_j P(A_i, B_j) = P(A_i)$$

1.2 Conditional Probability and Statistical Independence

Conditional probability concerns knowledge of one event given that another event has occurred. Denote the probability that event $\{B\}$ occurs given that event $\{A\}$ occurs by $P(B \mid A)$. The conditional probability of "$\{B\}$ given $\{A\}$" is defined as

$$P(B \mid A) = P(A, B)/P(A) \tag{1-1}$$

provided that $P(A) \neq 0$. Similarly, the probability that $\{A\}$ occurs given that $\{B\}$ occurs is

$$P(A \mid B) = P(A, B)/P(B) \qquad (P(B) \neq 0) \tag{1-2}$$

If the experiment consists of mutually exclusive and exhaustive outcomes $B_i, i = 1, 2, \ldots,$ then

$$\sum_i P(B_i \mid A) = 1$$

If it is determined for two events $\{A\}$ and $\{B\}$ that $P(A \mid B) = P(A)$, it then follows from the definition of conditional probability that

$$P(A, B) = P(A) \cdot P(B) \tag{1-3}$$

† The event $\{A \text{ or } B\}$ implies the occurrence of either $\{A\}$ or $\{B\}$ or both. The event $\{A \text{ and } B\}$ implies the occurrence of both $\{A\}$ and $\{B\}$.

‡ The summations are intended to be carried out over all values of the index.

It also follows that $P(B \mid A) = P(B)$. Thus, knowledge that one of these events occurs provides no information about the probability of occurrence of the other event. Such events are said to be *statistically independent*.

For three events $\{A_1\}$, $\{A_2\}$, and $\{A_3\}$ to be statistically independent, the following relations must be satisfied:

$$\begin{aligned} P(A_1, A_2) &= P(A_1)\,P(A_2), & P(A_1, A_3) &= P(A_1)\,P(A_3) \\ P(A_2, A_3) &= P(A_2)\,P(A_3), & P(A_1, A_2, A_3) &= P(A_1)\,P(A_2)\,P(A_3) \end{aligned} \qquad (1\text{-}4)$$

For more than three events to be independent, the probability of events taken two, three, four, etc. at a time must be equal to the product of the probability of these individual events.

1.3 Probability Distribution Functions

We shall define a *random variable* or *variate* to be a real-valued function of the sample space, that is, a real-valued function of the outcome of an experiment. For example, in the tossing of a die, the number of dots which appear is a random variable or variate. Any function of the number of dots is also a random variable. If the number of values which the random variable may assume (the sample space) is finite or countably infinite,† then it is said to be a discrete random variable. Otherwise, it is a continuous random variable.

Suppose we have a discrete random variable‡ which may take on any one of, say, six possible values x_i where $x_6 > x_5 > x_4 > x_3 > x_2 > x_1$. The corresponding probabilities, denoted $P(x_i)$ or $P(X = x_i)$, are represented in Fig. 1-1. For such an example it is reasonable to inquire into the probability that the random variable takes on values less than or equal to x_3, say. In this case, $P(X \leq x_3) = P(x_1) + P(x_2) + P(x_3)$. The function of x defined by $P(X \leq x)$ is called the *probability distribution function* of the random variable X. (It is also called the *distribution function* and *cumulative distribution function*.) Figure 1-1 also shows the cumulative distribution for the preceding example. The outcome $\{X \leq x\}$ is an event in the usual probability sense so that the cumulative distribution function must satisfy the properties previously discussed. In particular, $P(X < -\infty) = 0$ and $P(X < \infty) = 1$. Also, the probability that X falls in the interval $x_i < X \leq x_j$ is $P(X \leq x_j) - P(X \leq x_i) = P(x_i < X \leq x_j)$.

† A set of numbers is countable if it can be placed in a one-to-one correspondence with the set of positive integers.

‡ In this chapter, the random variable is denoted by an upper case letter, and elements of the sample space are denoted by the corresponding lower case letter. However, the use of different symbols to distinguish these becomes cumbersome so that in remaining chapters only one symbol is used provided its interpretation is clear when taken in context.

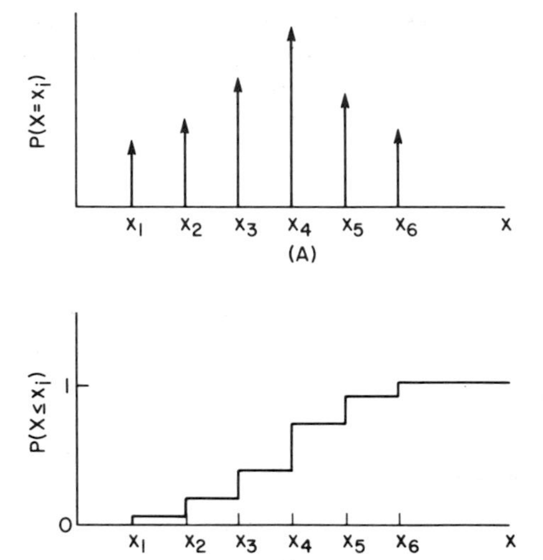

Fig. 1-1 Probability functions of a discrete random variable. (A) Impulse functions to represent discrete probabilities. (B) Cumulative distribution function.

The above is easily extrapolated to two random variables (*bivariate* distributions) or more (*multivariate* distributions). For two random variables, X and Y, which may be continuous or discrete the following are true:

$$P(X \leq -\infty,\ Y \leq y) = 0, \qquad P(X \leq x,\ Y \leq -\infty) = 0$$
$$P(X \leq \infty,\ Y \leq \infty) = 1 \tag{1-5}$$
$$P(X \leq x,\ Y \leq \infty) = P(X \leq x), \qquad P(X \leq \infty,\ Y \leq y) = P(Y \leq y)$$

1.4 Continuous Random Variables

Consider a random variable X which has a continuous cumulative distribution function as shown in Fig. 1-2. This is an example of a *continuous random variable*. Such a random variable can take on a noncountable number of values. For example, the sample space might be the entire real line. For a continuous random variable we define a *probability density function* (or simply *density function*) as the derivative of the cumulative distribution function, provided it exists. Denoting the probability density function of a random variable X by $p(x)$, we have

$$p(x) = dP(X \leq x)/dx \tag{1-6}$$

A specification of a density function must include the range over which it is applicable.

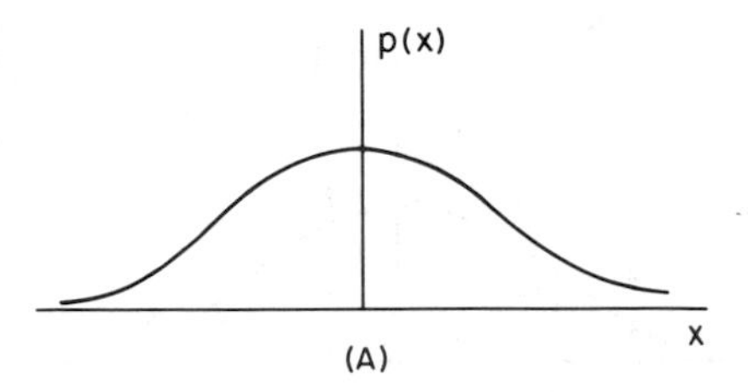

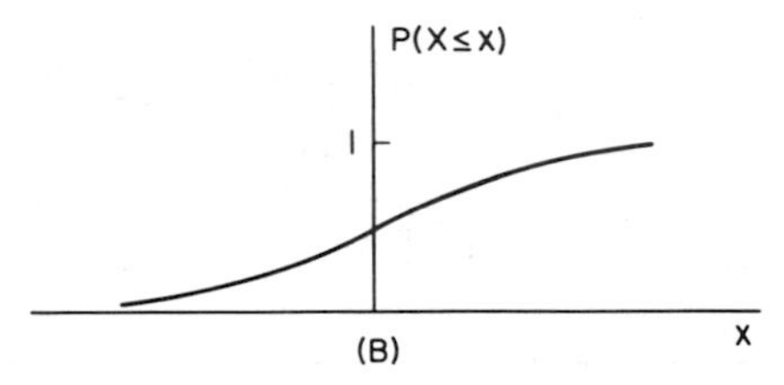

Fig. 1-2 Probability functions of a continuous random variable. (A) The density function. (B) The cumulative distribution function.

If the function $P(X \leq x)$ is in the form of a definite integral, the differentiation may be accomplished using Leibnitz' rule (*3*):

If $\phi(x) = \int_{a(x)}^{b(x)} f(t, x)\, dt$ where $a(x)$ and $b(x)$ are differentiable functions of x, and $f(t, x)$ and $\partial f(t, x)/\partial x$ are continuous in both x and t, then

$$\frac{d\phi}{dx} = \int_{a(x)}^{b(x)} \frac{\partial f(t, x)}{\partial x}\, dt + f[b(x), x]\, \frac{db(x)}{dx} - f[a(x), x]\, \frac{da(x)}{dx}$$

Since the cumulative distribution function is nondecreasing, $p(x) \geq 0$. An example of a probability density function is shown in Fig. 1-2. Using Dirac delta functions† (impulse functions) we may also define the probability density function for a discrete random variable as the derivative of the cumulative distribution function. Delta functions arise at the points of discontinuity as shown in Fig. 1-1. For the example in this figure, the density function may be expressed

$$p(x) = \sum_{i=1}^{6} P(x)\, \delta(x - x_i)$$

In general, a random variable may be of the mixed type and have a cumulative distribution function which consists of a part having jump discontinuities and a part which is everywhere continuous. An example of such a random variables is shown in Fig. 1-3.

† The Dirac delta function $\delta(x)$ is infinite when its argument is zero and is zero otherwise. It has unit area so $\int_{-\infty}^{\infty} \delta(x)\, dx = 1$. Also, $\int_{-\infty}^{\infty} f(x)\, \delta(x - a)\, dx = f(a)$. See Lighthill (*4*) for a discussion of generalized functions.

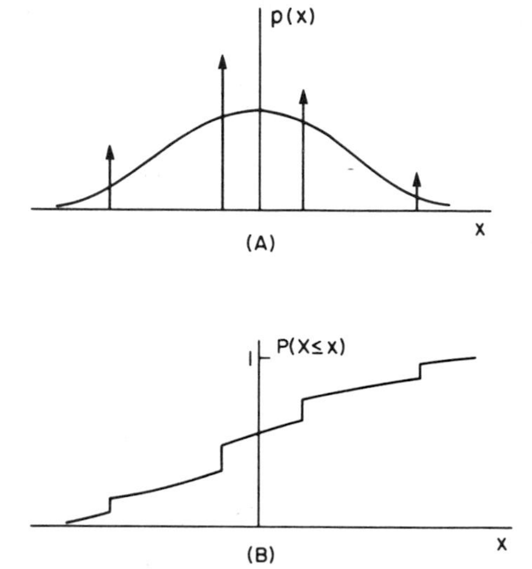

Fig. 1-3 Probability functions of a mixed random variable. (A) The density function. (B) The cumulative distribution function.

From the definition of the probability density function it follows immediately that

$$P(X \leq a) = \int_{-\infty}^{a} p(x)\,dx$$

The quantity $p(x)\,dx$ can be interpreted as the probability that the random variable falls between x and $x + dx$. The probability that the random variable falls within the interval $a < x \leq b$ is given by

$$P(a < X \leq b) = \int_{a}^{b} p(x)\,dx$$

For a continuous random variable, the probability that the variable falls within an interval approaches zero as the interval decreases. This is easily seen by replacing b by $a + \varepsilon$ in the above expression and allowing ε to approach zero.

For two random variables, X and Y, the *joint probability density function* denoted $p(x, y)$ is defined as

$$p(x, y) = \frac{\partial^2}{\partial x\,\partial y} P(X \leq x, Y \leq y) \tag{1-7}$$

It then follows that†

$$P(X \leq a, Y \leq b) = \int_{-\infty}^{b} dy \int_{-\infty}^{a} dx \, p(x, y)$$

$$P(X \leq a) = \int_{-\infty}^{\infty} dy \int_{-\infty}^{a} dx \, p(x, y)$$

$$P(Y \leq b) = \int_{-\infty}^{\infty} dx \int_{-\infty}^{b} dy \, p(x, y)$$

It may be similarly shown that

$$p(x) = \int_{-\infty}^{\infty} p(x, y) \, dy$$

and

$$p(y) = \int_{-\infty}^{\infty} p(x, y) \, dx$$

Probability density functions arising in this manner are sometimes referred to as *marginal density functions.*

The *conditional probability density function* for a variable Y given the random variable X is defined by

$$p(y \mid x) = p(x, y)/p(x) \qquad (p(x) \neq 0) \tag{1-8}$$

Thus, $P(Y \leq b \mid x) = \int_{-\infty}^{b} p(y \mid x) \, dy$ is interpreted as the probability that $\{Y \leq y\}$ given that $\{X = x\}$. The definition of conditional probability density function can be extended to a set of random variables $\{Y_i, i = 1, \ldots, n\}$ conditioned on yet another set of random variables $\{X_j, j = 1, \ldots, m\}$. Thus

$$p(y_1, \ldots, y_n \mid x_1, \ldots, x_m) = \frac{p(y_1, \ldots, y_n, x_1, \ldots, x_m)}{p(x_1, \ldots, x_m)}$$

The condition for *statistical independence* may also be stated in terms of probability density functions: random variables $X, Y, \ldots, Z$ are statistically independent if and only if

$$p(x, y, \ldots, z) = p(x)p(y) \cdots p(z) \tag{1-9}$$

for all values of $x, y, \ldots, z$.

† Following a common convention, the notation $p(\cdot)$ will be used to denote the probability density functions for X and for Y, as well as for the joint probability density function of X and Y even though they have different functional form. The arguments are meant to distinguish these functions. Whenever this convention may not be clear in context, subscripts will be used to distinguish different functional forms.

1.5 Functions of Random Variables

Functions of one or more random variables commonly occur in detection theory, as well as other areas involving probability and statistics. A function of one random variable, $Y = f(X)$, is specified by observing the real number x that results from performing the experiment and then carrying out the arithmetic operation defined by $y = f(x)$. Typical examples appear in Fig. 1-4. This may be generalized to functions of more than one random variable,

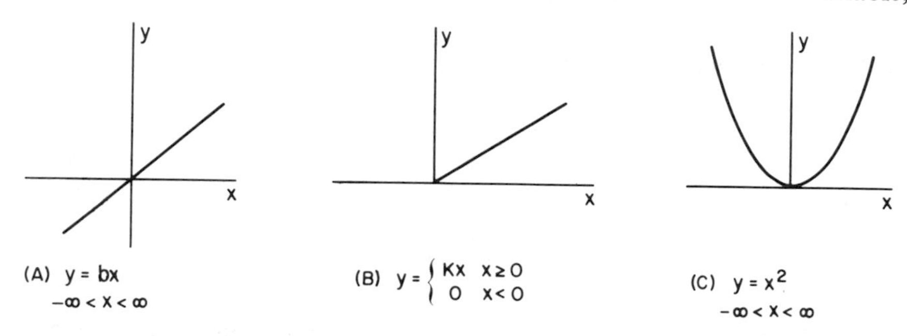

Fig. 1-4 Examples of functions of random variables. (A) Linear operator. (B) Half-wave rectification. (C) Square law.

for example, $Y = f(X, Z)$. This operation is specified by observing the pair of real numbers x and z and carrying out the arithmetic operation $y = f(x, z)$. An example is the sum, $y = x + z$.

To demonstrate a direct method of determining the statistics of a function of a random variable, consider the case shown in Fig. 1-4A. This is a linear function $y = bx$ involving only a scale change. Assume the probability density function of X is known and that the density function of Y is to be determined. The probability that $\{Y \leq y\}$ is equal to the probability that $\{X \leq y/b\}$. That is,† $P_Y\{Y \leq y\} = P_X\{X \leq y/b\}$. It follows directly from the definition of the probability density function that

$$p_Y(y) = \frac{d}{dy} P_X\{X \leq y/b\}$$

Since the distribution function is nondecreasing, the derivative cannot be negative, and it may be shown using Leibnitz' rule that

$$p_Y(y) = \frac{1}{|b|} p_X(x = y/b)$$

where $|\cdot|$ indicates absolute value. The range of Y is the range of X multiplied by b.

† To avoid possible confusion here, the distribution function of X and Y are denoted by $P_X(\cdot)$ and $P_Y(\cdot)$ respectively.

EXAMPLE 1.5-1 If for the above case the density function of X is exponential (in other words, X is exponentially distributed), then $p_X(x) = e^{-x}$ for $x \geq 0$. It follows immediately that

$$p_Y(y) = \frac{1}{|b|} e^{-y/b} \qquad \begin{cases} y \geq 0 & \text{if} \quad b > 0 \\ y \leq 0 & \text{if} \quad b < 0 \end{cases}$$

The density functions for X and Y for $b = 2$ are shown in Fig. 1-5.

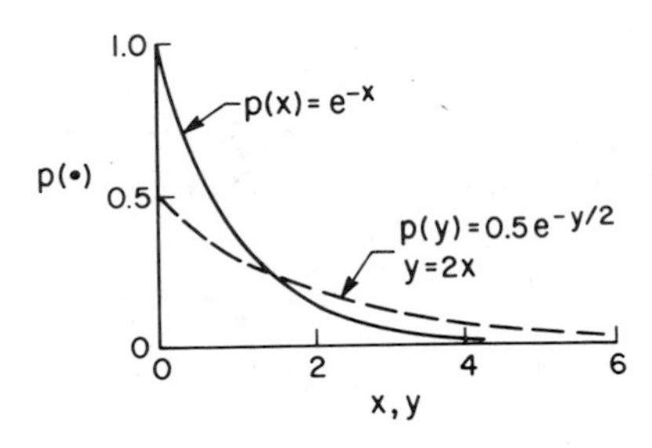

Fig. 1-5 Density functions for Example 1.5-1.

EXAMPLE 1.5-2 Let $p_X(x)$ have the same density function as in the preceding example. Determine the density function of Y where $Y = X + a$.

Using the direct method,

$$P_Y(Y \leq y) = P_X(X \leq y - a)$$

Now

$$p_Y(y) = \frac{d}{dy} P_Y(Y \leq y) = \frac{d}{dy} P_X(X \leq y - a)$$

so that

$$p_Y(y) = p_X(x = y - a) = e^{-(y-a)}, \qquad y > a$$

This is shown in Fig. 1-6. □ □

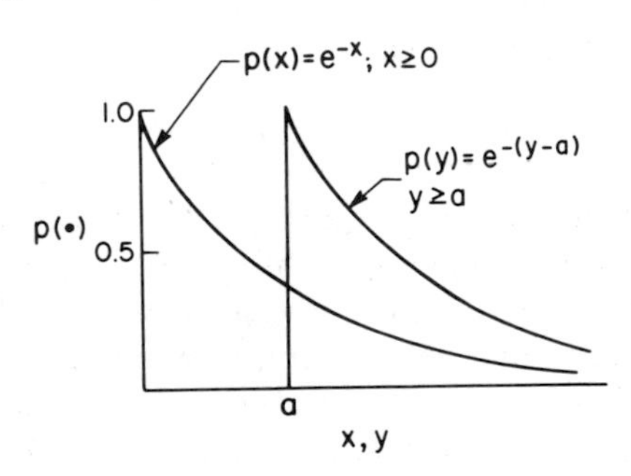

Fig. 1-6 Density functions for Example 1.5-2.

The cases discussed below, while more complicated, involve the same principle as the simple cases discussed above. One or more random variables are transformed or mapped into another set of random variables. The probability that the multidimensional random variable falls within a given region of the sample space is the same as the probability that the transformed multidimensional random variable falls within the transformed region of the new sample space.

Suppose we have a set of random variables, say $X_1, X_2, \ldots, X_N$, for which the joint probability density function denoted $p_X(x_1, x_2, \ldots, x_N)$ is known, and we wish to determine the joint probability density function denoted $p_Y(y_1, y_2, \ldots, y_N)$ of a new set of random variables, say $Y_1, Y_2, \ldots, Y_N$, which are functionally related to the X's. That is,

$$\begin{aligned} Y_1 &= g_1(X_1, X_2, \ldots, X_N) \\ &\vdots \\ Y_N &= g_N(X_1, X_2, \ldots, X_N) \end{aligned} \tag{1-10}$$

For example,

$$Y_1 = X_1 + X_2, \qquad Y_2 = X_1 - X_2$$

For the moment, the number (N) of new variables, the Y_i's, is assumed to be equal to the number of old variables, the X_i's. Assume that the new random variables are single-valued continuous functions of the old random variables (with continuous partial derivatives everywhere) and further that the old variables may be expressed as single-valued continuous functions of the new variables:

$$\begin{aligned} X_1 &= f_1(Y_1, Y_2, \ldots, Y_N) \\ &\vdots \\ X_N &= f_N(Y_1, Y_2, \ldots, Y_N) \end{aligned} \tag{1-11}$$

Continuing the preceding example,

$$X_1 = \tfrac{1}{2}(Y_1 + Y_2), \qquad X_2 = \tfrac{1}{2}(Y_1 - Y_2)$$

Then, to each point in the sample space of the X_i's there corresponds one and only one point in the sample space of the Y_i's, so that there is a one-to-one mapping of the old variables into the new variables.

Suppose that a particular set of sample points are contained within a region A of the X domain. Due to the functional relationship between the X_i's and Y_i's, the region A maps into a region B in the Y domain as depicted in Fig. 1-7. The probability that a sample point $x_1, \ldots, x_N$ falls in A is the

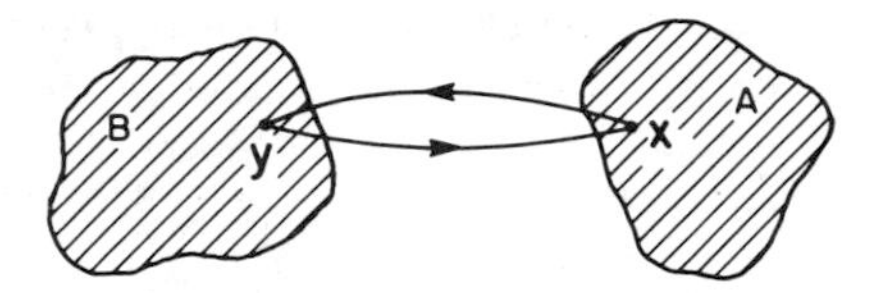

Fig. 1-7 Illustration of one-to-one mapping of space A into space B. **x** and **y** denote the multidimensional variables $x_1, x_2, \ldots, x_N$ and $y_1, y_2, \ldots, y_N$ respectively.

same as the probability that a sample point $y_1, \ldots, y_N$ falls into region B. Therefore

$$\int \underset{A}{\cdots} \int p_X(x_1, x_2, \ldots, x_N)\, dx_1\, dx_2 \cdots dx_N$$
$$= \int \underset{B}{\cdots} \int p_Y(y_1, y_2, \ldots, y_N)\, dy_1\, dy_2 \cdots dy_N \tag{1-12}$$

where $p_X(x_1, \ldots, x_N)$ is the joint probability density function of the X_i's, which is given, and $p_Y(y_1, \ldots, y_N)$ is the joint probability density function of the Y_i's which is to be determined.

By appropriately changing variables in the first integral we get (5)

$$\int \underset{A}{\cdots} \int p_X(x_1, \ldots, x_N)\, dx_1 \cdots dx_N$$
$$= \int \underset{B}{\cdots} \int p_X(x_1 = f_1(y_1, \ldots, y_N), \ldots, x_N = f_N(y_1, \ldots, y_N))|J|\, dy_1 \cdots dy_N$$

where $|J|$ is the absolute value of the Jacobian of the transformation.† The Jacobian is the determinant of the matrix defined by

$$J = \begin{vmatrix} \dfrac{\partial f_1}{\partial Y_1} & \cdots & \dfrac{\partial f_N}{\partial Y_1} \\ & \vdots & \\ \dfrac{\partial f_1}{\partial Y_N} & \cdots & \dfrac{\partial f_N}{\partial Y_N} \end{vmatrix} \tag{1-13}$$

The absolute value of the Jacobian is used since the limits of integration are assumed to be in ascending order. Continuing the example

$$J = \begin{vmatrix} \frac{1}{2} & \frac{1}{2} \\ \frac{1}{2} & -\frac{1}{2} \end{vmatrix}$$

and $|J| = \frac{1}{2}$.

† This method, of course, is the standard way of changing variables in an integral and is in no way unique to probability theory.

Comparing the integral with that of Eq. (1-12), we see that the old and new joint probability density functions are related by the equation

$$p_Y(y_1, \ldots, y_N) = p_X(x_1 = f_1(y_1, \ldots, y_N), \ldots, x_N = f_N(y_1, \ldots, y_N)) \, |J| \tag{1-14}$$

This equation, along with the proper region of definition of the y's, is the complete specification of the density function.

EXAMPLE 1.5-3 Assume that the independent random variables X_1 and X_2 are normally (Gaussian) distributed† with density functions

$$p(x_1) = \frac{1}{(2\pi)^{1/2}} e^{-x_1^2/2} \qquad \text{and} \qquad p(x_2) = \frac{1}{(2\pi)^{1/2}} e^{-x_2^2/2}$$

Determine the density functions of Y_1 and Y_2 where

$$Y_1 = X_1 + X_2 = g_1(X_1, X_2)$$
$$Y_2 = X_1 - X_2 = g_2(X_1, X_2)$$

Since the variables are independent, the joint density function is

$$p(x_1, x_2) = p(x_1)p(x_2) = \frac{1}{2\pi} e^{-(x_1^2 + x_2^2)/2}$$

The inverse transformations are

$$X_1 = \tfrac{1}{2}(Y_1 + Y_2) = f_1(Y_1, Y_2)$$

and

$$X_2 = \tfrac{1}{2}(Y_1 - Y_2) = f_2(Y_1, Y_2)$$

The Jacobian is

$$\begin{vmatrix} \dfrac{\partial f_1}{\partial Y_1} & \dfrac{\partial f_2}{\partial Y_1} \\[2ex] \dfrac{\partial f_1}{\partial Y_2} & \dfrac{\partial f_2}{\partial Y_2} \end{vmatrix} = \begin{vmatrix} \frac{1}{2} & \frac{1}{2} \\[2ex] \frac{1}{2} & -\frac{1}{2} \end{vmatrix} = -\tfrac{1}{2}$$

Therefore, $|J| = \frac{1}{2}$, and

$$p_Y(y_1, y_2) = \frac{1}{4\pi} e^{-(y_1^2 + y_2^2)/4} = \frac{1}{(2\pi)^{1/2} 2^{1/2}} e^{-y_1^2/4} \frac{1}{(2\pi)^{1/2} 2^{1/2}} e^{-y_2^2/4}$$

For this case Y_1 and Y_2 are statistically independent Gaussian random variables. □ □

† See Chap. 4.

The preceding covered the case where the number of new random variables was the same as the number of old random variables. The case for which the number of new random variables is less than the number of old random variables may be similarly treated. We demonstrated with an example.

EXAMPLE 1.5-4 Given two random variables X_1 and X_2 with a joint density function $p_X(x_1, x_2)$. Determine the density function $p_Y(y)$ of a single new random variable Y where

$$Y = X_1 + X_2 = g_1(X_1, X_2)$$

We temporarily keep the number of old and new variables equal by determining the joint density function of Y and, say, X_2. Denote this by $p_1(x_2, y)$. Having done this, carrying out the operation $p_Y(y) = \int p_1(x_2, y)\, dx_2$ produces the desired result. Specifically, let

$$p(x_1, x_2) = \frac{1}{2\pi} e^{-(x_1{}^2 + x_2{}^2)/2}, \qquad -\infty < x_1, x_2 < \infty$$

As before, we may write $X_1 = Y - X_2 = f_1(Y, X_2)$ and $X_2 = X_2$. Thus, $f_2(Y, X_2)$ is equal to X_2, and

$$\begin{vmatrix} \dfrac{\partial f_1}{\partial Y} & \dfrac{\partial f_2}{\partial Y} \\ \dfrac{\partial f_1}{\partial X_2} & \dfrac{\partial f_2}{\partial X_2} \end{vmatrix} = \begin{vmatrix} 1 & 0 \\ -1 & 1 \end{vmatrix} = 1$$

Therefore,

$$p_1(x_2, y) = \frac{1}{2\pi} e^{-(y^2 - 2x_2 y + 2x_2{}^2)/2}$$

and

$$p_Y(y) = \int_{-\infty}^{\infty} \frac{1}{2\pi} e^{-(y^2 - 2x_2 y + 2x_2{}^2)/2}\, dx_2$$

Carrying out the integration produces

$$p_Y(y) = \frac{1}{2(\pi)^{1/2}} e^{-y^2/4} \qquad \square\,\square$$

When the functional relationships between the old and new random variables are not single valued, special care is required. For example, the output of a square-law device as shown in Fig. 1-4C is given by $y = x^2$ where x is the input. We require the probability density function of Y given that

$p_X(x)$ is known (assume $-\infty < x < \infty$). Obviously, for $Y < 0$, $p(y) = 0$. For $Y > 0$,

$$\begin{aligned} P(Y \leq y) &= P(-y^{1/2} \leq X \leq y^{1/2}) \\ &= P(X \leq y^{1/2}) - P(X < -y^{1/2}) \\ &= \int_{-\infty}^{y^{1/2}} p_X(x)\,dx - \int_{-\infty}^{-y^{1/2}} p_X(x)\,dx \end{aligned}$$

Therefore, since

$$p_Y(y) = \frac{d}{dy} P(Y \leq y)$$

it follows that

$$p_Y(y) = \frac{p_X(x = y^{1/2}) + p_X(x = -y^{1/2})}{2y^{1/2}} \tag{1-15}$$

Example 1.5-5 Given

$$p_X(x) = \frac{1}{(2\pi)^{1/2}} e^{-x^2/2}, \qquad -\infty < x < \infty$$

Determine $p_Y(y)$ where $Y = X^2$.

From Eq. (1-15) it follows immediately that

$$p_Y(y) = \begin{cases} \dfrac{e^{-y/2}/(2\pi)^{1/2} + e^{-y/2}/(2\pi)^{1/2}}{2y^{1/2}}, & y \geq 0 \\ 0, & y < 0 \end{cases}$$

or

$$p_Y(y) = \begin{cases} e^{-y/2}/(2\pi y)^{1/2}, & y \geq 0 \\ 0, & 0 < y \end{cases}$$

Had this been carelessly done by the Jacobian method without accounting for the two-to-one mapping of x into y, the answer would be different (and incorrect) by a factor of 2.

1.6 Characteristic Functions

A function which is useful for determining the probability density function of a sum of independent random variables is the characteristic function. It is also useful for determining the moments of a random variable (discussed in the next chapter). The characteristic function† of a random variable X is defined as

$$C(j\omega) = \int_{-\infty}^{\infty} p(x)e^{j\omega x}\, dx \tag{1-16}$$

That is, $C(j\omega)$ is the Fourier transform of the probability density function.‡ (Actually it is the complex conjugate of the usual Fourier transform.) A similar function, called the moment generating function, is defined as the Laplace transform of the density function. A little later, we shall treat the characteristic function as an average.

Since

$$\left| \int_{-\infty}^{\infty} p(x)e^{j\omega x}\, dx \right| \le \int_{-\infty}^{\infty} p(x)\, dx = 1$$

it follows that the characteristic function always exists. For a discrete random variable X, the characteristic function is

$$C(j\omega) = \sum_i e^{j\omega x_i} P(X = x_i) \tag{1-17}$$

The density function may be found from the characteristic function by the Fourier inversion formula

$$p(x) = \frac{1}{2\pi} \int_{-\infty}^{\infty} C(j\omega)e^{-j\omega x}\, d\omega \tag{1-18}$$

EXAMPLE 1.6-1 Determine the characteristic function of

$$p(x) = \frac{1}{(2\pi)^{1/2}} e^{-x^2/2}, \qquad -\infty < x < \infty$$

(This is the density function for a normalized Gaussian variable.) From Eq. (1-16)

$$C(j\omega) = \int_{-\infty}^{\infty} \frac{1}{(2\pi)^{1/2}} e^{-x^2/2} e^{j\omega x}\, dx$$

After completing the square in the exponent

$$C(j\omega) = e^{-\omega^2/2} \frac{1}{(2\pi)^{1/2}} \int_{-\infty}^{\infty} e^{-(x-j\omega)^2/2}\, dx = e^{-\omega^2/2} \qquad \square\,\square \tag{1-19}$$

† A summary of the common probability density functions, moments, and their characteristic functions appears in Farison (*6*).

‡ Consult Campbell and Foster (*7*), for tables of Fourier transform pairs. Papoulis (*8*), has a good treatment of the Fourier integral.

In a manner similar to the single variate case, a joint characteristic function may be defined for a joint probability density function $p(x, y)$ as the two-dimensional Fourier transform

$$C(j\omega, jv) = \int_{-\infty}^{\infty} \int_{-\infty}^{\infty} p(x, y)e^{j\omega x}e^{jvy}\, dx\, dy \tag{1-20}$$

The inversion formula is

$$p(x, y) = \frac{1}{4\pi^2} \int_{-\infty}^{\infty} \int_{-\infty}^{\infty} C(j\omega, jv)e^{-j\omega x}e^{-jvy}\, d\omega\, dy \tag{1-21}$$

The definitions can be similarly extended to the multivariate case.

Sum of Random Variables†

Suppose we wish to determine the probability density function of $Z = X + Y$. The joint density function of X and Y is denoted $p_{XY}(x, y)$. In a manner similar to that of Example 1.5-4, we express the "old" variables in terms of the "new" variables. That is

$$Y = Z - X, \qquad X = X$$

The Jacobian of the transformation is 1. From Eq. (1-14), the density function for z and x is

$$p(z, x) = p_{XY}(x, y = z - x)$$

and finally

$$p(z) = \int_{-\infty}^{\infty} p_{XY}(x, y = z - x)\, dx \tag{1-22}$$

If the random variables X and Y are statistically independent, then

$$p_{XY}(x, y) = p_X(x) \cdot p_Y(y)$$

It follows that

$$p_Z(z) = \int_{-\infty}^{\infty} p_X(x)\, p_Y(z - x)\, dx \tag{1-23}$$

or

$$p_Z(z) = p_X(z) * p_Y(z) \tag{1-24}$$

where the symbol $*$ denotes convolution. This result can be extended to the sum of more than two random variables. The conclusion is that the probability density function of the sum of independently distributed random variables is given by the convolution of their respective probability density functions.

† For products of random variables, see Springer and Thompson (9).

This result can be related to the characteristic function. It is well known in Fourier transform theory that the transform of a convolution is the product of the respective Fourier transforms. Denote the characteristic functions of $p_Z(z)$, $p_X(x)$, and $p_Y(y)$ by $C_Z(j\omega)$, $C_X(j\omega)$, and $C_Y(j\omega)$ respectively. Then it follows from Eq. (1-24) that

$$C_Z(j\omega) = C_X(j\omega)C_Y(j\omega) \tag{1-25}$$

The probability density function of Z is determined by the Fourier inverse transform of $C_Z(j\omega)$.

EXAMPLE 1.6-2 Determine the probability density function of $Z = X + Y$ by means of characteristic functions. Assume that X and Y are statistically independent with

$$p(x) = \frac{1}{(2\pi)^{1/2}} e^{-x^2/2}, \qquad -\infty < x < \infty$$

$$p(y) = \frac{1}{(2\pi)^{1/2}} e^{-y^2/2}, \qquad -\infty < y < \infty$$

From Example 1.6-1 the characteristic functions are

$$C_X(j\omega) = e^{-\omega^2/2} \qquad \text{and} \qquad C_Y(j\omega) = e^{-\omega^2/2}$$

The characteristic function for Z is therefore

$$C_Z(\omega) = e^{-\omega^2}$$

and it follows that

$$p_Z(z) = \frac{1}{2\pi} \int_{-\infty}^{\infty} e^{-\omega^2} e^{-j\omega z}\, d\omega$$

The evaluation of the integral results in

$$p_Z(z) = \frac{1}{2\pi^{1/2}} e^{-z^2/4}$$

which is in agreement with the results of Example 1.5-4.

1.7 Averages

For a discrete random variable X with probabilities $P(x_i)$, $i = 1, \ldots, N$, we define the *statistical average* (also called the *expectation, mean, ensemble average*) of X to be

$$E\{X\} = \sum_{i=1}^{N} x_i P(x_i) \tag{1-26}$$

where $E\{\cdot\}$ denotes the expectation or averaging operation. This is a weighted sum of the values which the random variable can assume. The weights are, of course, the probability of occurrence of the values.

The definition may be extended to include the average of a function of X. For example, if $Y = g(X)$, then

$$E\{Y\} = \sum_{i=1}^{N} y_i P(y_i)$$

This is equivalent to

$$E\{g(X)\} = \sum_{i=1}^{N} g(x_i)P(x_i) \tag{1-27}$$

For continuous random variables, the corresponding averages are computed using integrals. For example, if X is a continuous random variable with density function $p(x)$, $-\infty < x < \infty$, then the expected value of X is defined as

$$E\{X\} = \int_{-\infty}^{\infty} x\, p(x)\, dx \tag{1-28}$$

In general, the average of a function $g(X)$ is

$$E\{g(X)\} = \int_{-\infty}^{\infty} g(x)p(x)\, dx \tag{1-29}$$

provided the integral exists. Of particular interest is the expected value of X^n which is called the nth moment of X.

$$E\{X^n\} = \int_{-\infty}^{\infty} x^n\, p(x)\, dx \tag{1-30}$$

The first moment, $E\{X\} \triangleq m$, is of course the mean. Also of interest are the nth *central moments* or nth moments about the mean, that is,

$$E\{(X - m)^n\} = \int_{-\infty}^{\infty} (x - m)^n\, p(x)\, dx \tag{1-31}$$

The first central moment is zero. The second central moment is of particular significance and is usually denoted by $\sigma^2 = E\{(X - m)^2\}$. It is called the *variance* and gives a measure of the spread of the random variable. The square root of the variance is called the *standard deviation* of X. Figure 1-8 shows two examples displaying mean and variance.

The case of two or more random variables can be treated similarly. For example, consider two continuous random variables X and Y with joint

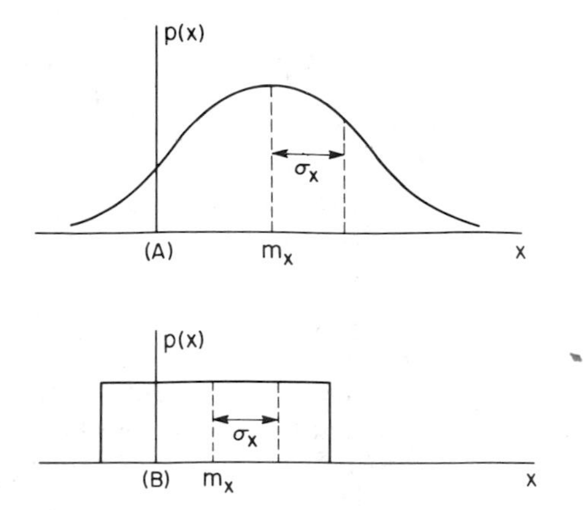

Fig. 1-8 Examples to show mean and standard deviation for (A) Gaussian distribution, and (B) uniform distribution.

density function $p(x, y)$. The expected value of a function of these variables $g(X, Y)$ is defined as

$$E\{g(X, Y)\} = \int_{-\infty}^{\infty} \int_{-\infty}^{\infty} g(x, y)p(x, y)\,dx\,dy \tag{1-32}$$

In particular, the $(n + k)$th order *joint moment* (or cross moment) of the random variables X and Y is†

$$E\{X^n Y^k\} = \int_{-\infty}^{\infty} \int_{-\infty}^{\infty} x^n y^k p(x, y)\,dx\,dy \tag{1-33}$$

The corresponding *joint central moment* is

$$\mu_{nk} = E\{(X - m_x)^n (Y - m_y)^k\} \tag{1-34}$$

where m_x and m_y are the means of X and Y respectively. Of particular significance is the joint moment

$$\sigma_{xy} = E\{(X - m_x)(Y - m_y)\} \tag{1-35}$$

which is called the *covariance* of the random variables X and Y.

The *normalized covariance* of X and Y is defined by

$$\rho_{xy} = \frac{E\{(X - m_x)(Y - m_y)\}}{[E\{(X - m_x)^2\}E\{(Y - m_y)^2\}]^{1/2}} = \frac{\sigma_{xy}}{\sigma_x \sigma_y} \tag{1-36}$$

where σ_x and σ_y are the standard deviations of X and Y respectively. This is also known as the *correlation coefficient*. It can be shown (*1*) that $|\rho_{xy}| \leq 1$.

† Note that if X and Y are statistically independent, then $E\{X^n Y^k\} = E\{X^n\}E\{Y^k\}$.

The random variables X and Y are said to be uncorrelated if

$$E\{XY\} = E\{X\}E\{Y\} \tag{1-37}$$

If such variables are uncorrelated, it is then easily shown that their normalized covariance or correlation coefficient is zero. Note that if two random variables are statistically independent, they are uncorrelated. However, the converse is not true except in special cases. The random variables X and Y are *orthogonal* if

$$E\{XY\} = 0 \tag{1-38}$$

A summary of statistical averages appears in Table 1-1.

It is now appropriate to reexamine the characteristic function, Eq. (1-16), repeated below:

$$C(j\omega) = \int_{-\infty}^{\infty} p(x)e^{j\omega x}\,dx$$

It should be clear that this is the expectation of $e^{j\omega X}$. That is,

$$C(j\omega) = E\{e^{j\omega X}\} \tag{1-39}$$

The characteristic function is useful for generating the moments of a random variable. Assuming that the derivative exists, we have

$$\frac{dC(j\omega)}{d\omega} = \int_{-\infty}^{\infty} jxp(x)e^{j\omega x}\,dx$$

Evaluating it at $\omega = 0$ produces

$$\left.\frac{dC(j\omega)}{d\omega}\right|_{\omega=0} = j\int_{-\infty}^{\infty} xp(x)\,dx$$

The integral is $E\{X\}$ from which it follows that

$$-j\left.\frac{dC(j\omega)}{d\omega}\right|_{\omega=0} = E\{X\} \tag{1-40}$$

The nth derivative of the characteristic function evaluated at $\omega = 0$ may be shown to be

$$\left.\frac{d^nC(j\omega)}{d\omega^n}\right|_{\omega=0} = j^n\int_{-\infty}^{\infty} x^n p(x)\,dx$$

It follows that the nth moment is

$$E\{X^n\} = (-j)^n \left.\frac{d^nC(j\omega)}{d\omega^n}\right|_{\omega=0} \tag{1-41}$$

TABLE 1-1 SUMMARY OF STATISTICAL AVERAGES[a]

NAME	DEFINITION	INDICATED OPERATION	
		Discrete case	Continuous case
Mean = m. Also called average, expectation, ensemble average	$E\{X\}$	$\sum_{i=1}^{N} x_i P(x_i)$	$\int_{-\infty}^{\infty} xp(x)\,dx$
Variance = σ^2 (standard deviation = σ)	$E\{(X-m_x)^2\}$	$\sum_{i=1}^{N} (x_i - m_x)^2 P(x_i)$	$\int_{-\infty}^{\infty} (x-m_x)^2 p(x)\,dx$
nth moment	$E\{X^n\}$	$\sum_{i=1}^{N} x_i^n P(x_i)$	$\int_{-\infty}^{\infty} x^n p(x)\,dx$
nth central moment or nth moment about the mean	$E\{(X-m_x)^n\}$ (=0 for $n=1$)	$\sum_{i=1}^{N} (x_i - m_x)^n P(x_i)$	$\int_{-\infty}^{\infty} (x-m_x)^n p(x)\,dx$
Mean of a function $g(x)$	$E\{g(X)\}$	$\sum_{i=1}^{N} g(x_i) P(x_i)$	$\int_{-\infty}^{\infty} g(x)p(x)\,dx$
$(n+k)$th joint moment	$E\{X^n Y^k\}$	$\sum_{i=1}^{N}\sum_{j=1}^{M} x_i^n y_j^k P(x_i, y_j)$	$\int_{-\infty}^{\infty}\int_{-\infty}^{\infty} x^n y^k p(x,y)\,dx\,dy$
$(n+k)$th joint central moment	$E\{(X-m_x)^n (Y-m_y)^k\}$	$\sum_{i=1}^{N}\sum_{j=1}^{M} (x_i - m_x)^n (y_j - m_y)^k P(x_i, y_j)$	$\int_{-\infty}^{\infty}\int_{-\infty}^{\infty} (x-m_x)^n (y-m_y)^k p(x,y)\,dx\,dy$
Covariance	$E\{(X-m_x)(Y-m_y)\}$	$\sum_{i=1}^{N}\sum_{j=1}^{M} (x_i - m_x)(y_j - m_y) P(x_i, y_j)$	$\int_{-\infty}^{\infty}\int_{-\infty}^{\infty} (x-m_x)(y-m_y) p(x,y)\,dx\,dy$
Mean of a function $g(x, y)$	$E\{g(X, Y)\}$	$\sum_{i=1}^{N}\sum_{j=1}^{M} g(x_i, y_j) P(x_i, y_j)$	$\int_{-\infty}^{\infty}\int_{-\infty}^{\infty} g(x,y)p(x,y)\,dx\,dy$

[a] For the discrete cases, outcomes are $x_i (i = 1, \ldots, N)$ and y_j $(j = 1, \ldots, M)$ with probability $P(x_i)$ and $P(y_j)$ respectively. For the continuous cases X and Y are defined over the interval $(-\infty < x, y < \infty)$ and have continuous density functions $p(x)$ and $p(y)$ respectively. The means are indicated by m_x and m_y.

Exercises

1.1 Consider a coin toss experiment where the probability of a head is p, and that of a tail is $q = 1 - p$.

(a) Show that the probability of exactly i heads in N independent trials is given by the binomial distribution

$$\binom{N}{i} p^i q^{N-i}, \qquad 0 \le i \le N$$

(b) Show that the mean and variance of i are Np and Npq respectively.

(c) Show that the characteristic function, $E\{e^{j\omega i}\}$, is equal to $(pe^{j\omega} + q)^N$.

(d) Find the first two moments of i by using the characteristic function.

1.2 The Poisson distribution is given by

$$P(k) = \frac{e^{-\mu}\mu^k}{k!}, \qquad k \text{ integer}, \quad k \ge 0$$

(a) Show that the mean and variance are equal to μ.

(b) Show that the characteristic function is $\exp[\mu(e^{j\omega} - 1)]$.

1.3 The exponential probability density function is

$$p(x) = \frac{1}{\sigma} \exp\left[-\left(\frac{x - \alpha}{\sigma}\right)\right] \qquad x \ge \alpha, \quad \sigma > 0$$

(a) Show that the mean and variance of x are $\alpha + \sigma$ and σ^2 respectively.

(b) Show that the characteristic function is

$$C(j\omega) = \frac{e^{j\omega\alpha}}{1 - j\omega\sigma}$$

1.4 The uniform probability density function with zero mean value may be expressed

$$p(x) = \begin{cases} 1/2a, & -a \le x \le a \\ 0, & \text{otherwise} \end{cases}$$

Show that the variance is $a^2/3$ and the characteristic function is

$$C(j\omega) = \frac{1}{a\omega} \sin \omega a$$

1.5 The Gaussian (also called normal) probability density function with mean and variance denoted by μ and σ^2 respectively is

$$p(x) = \frac{1}{(2\pi)^{1/2}\sigma} \exp \frac{(x - \mu)^2}{-2\sigma^2}, \qquad -\infty < x < \infty$$

(a) Show that the characteristic function is equal to

$$C(j\omega) = \exp\left(j\mu\omega - \frac{\omega^2\sigma^2}{2}\right)$$

(b) The odd central moments are obviously zero. By successive differentiation of $\int_{-\infty}^{\infty} e^{-\alpha x^2}\,dx$ with respect to α, show that the even order central moments of the Gaussian variable are

$$E\{X^m\} = 1 \cdot 3 \cdot 5 \cdots (m-1)\sigma^m, \qquad m \text{ even}$$

1.6 Assume Y is a Gaussian (normal) random variable with mean and variance μ and σ^2 respectively. We say that the random variable X is log-normally distributed if $Y = \ln X$, and Y is Gaussian.

(a) Show that the log normal density may be written

$$p(x) = \frac{1}{(2\pi)^{1/2}\sigma x} \exp \frac{(\ln x - \mu)^2}{-2\sigma^2}, \qquad x \geq 0$$

(b) Show that the first and second moments of x are

$$E\{X\} = \exp\left(\mu + \frac{\sigma^2}{2}\right)$$

$$E\{X^2\} = \exp(2\mu + 2\sigma^2)$$

1.7 A quantizer characteristic or analog-to-digital encoder is shown in Fig. 1-9. The input, denoted X, is usually a continuous variable. The output Y is a discrete variable.

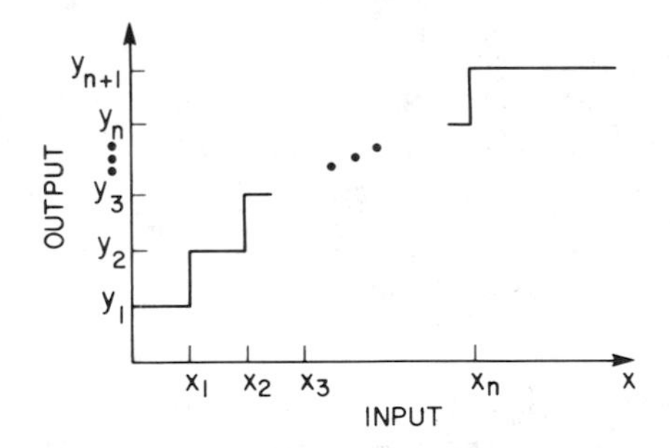

Fig. 1-9 Quantizer characteristic (analog-to-digital converter).

(a) Write an expression for the probability density of the output in terms of the density function $p(x)$ of the input.

(b) Assume $p(x) = e^{-x}$, $x \geq 0$. What is the probability density function of the sum of two statistically independent samples of Y.

1.8 Assume the random variables $X_1, X_2, \ldots, X_n$ are statistically independent with means μ_i and variances σ_i^2.

(a) Show that the mean and variance of the sample mean, defined as $\bar{x} = (1/n) \sum_{i=1}^{n} x_i$, are

$$\frac{1}{n} \sum_{i=1}^{n} \mu_i \qquad \text{and} \qquad \frac{1}{n^2} \sum_{i=1}^{n} \sigma_i^2$$

respectively.

(b) Suppose the X_i's are identically distributed Gaussian variables with zero mean values, what is the probability density function of $\bar{x}$. Is $\bar{x}$ a Gaussian random variable?

(c) Suppose the X_i's are identically distributed exponential variables, that is,

$$p(x) = \frac{1}{\sigma} \exp\left[-\left(\frac{x-\alpha}{\sigma}\right)\right], \qquad x \geq \alpha, \quad \sigma > 0$$

Is $\bar{x}$ exponentially distributed?

1.9 In the preceding problem assume the random variables X_i are statistically dependent. Define

$$\sigma_{|i-j|}^2 \triangleq E\{(X_i - \mu_i)(X_j - \mu_j)\}$$

Show that the variance of the sample mean may be expressed as

$$\frac{\sigma_0^2}{n} + \frac{2}{n} \sum_{i=1}^{n-1} \left(1 - \frac{i}{n}\right) \sigma_i^2$$

1.10 Derive an expression for the density function of a product of two continuous variables such as $Z = XY$ where $x, y \geq \varepsilon > 0$.

1.11 Assume the random variables X and Y are statistically independent with zero mean value and variance σ^2. Assume X and Y are Gaussian distributed. For the transformation $R = +(X^2 + Y^2)^{1/2}$ and $\theta = \tan^{-1} Y/X$, find the joint density function for R and θ, as well as the density function of R, and of θ. (Note: The inverse transformations are $X = R \cos \theta$ and $Y = R \sin \theta$.)

1.12 Denote n statistically independent identically distributed random variables as $X_1, X_2, \ldots, X_n$ and their density and distribution functions by $f(x)$ and $F(x)$ respectively. Define a random variable which is the maximum of the X_i's. That is $Y = \max\{X_1, X_2, \ldots, X_n\}$. Show that the probability density function of Y is

$$p(y) = nF^{n-1}(y)f(y)$$

What is the result if $f(x) = e^{-x}$, $x \geq 0$?

1.13 Assume for random variables W and X that

$$E\{W\} = E\{X\} = 0, \qquad E\{WX\} = \rho$$

and the variance of each is σ^2. Consider the transformation

$$Y = aW, \qquad Z = bW + cX$$

where a, b, and c are constants. Find a, b, and c such that $E\{Y^2\} = E\{Z^2\} = 1$ and $E\{YZ\} = 0$.

1.14 Assume we have the discrete random variables S, N, and R with the possible values

$$\begin{aligned} &s_i, && i = 1, \ldots, U \\ &n_i, && i = 1, \ldots, V \\ &r_i, && i = 1, \ldots, W \end{aligned}$$

(a) Show that

$$P(s_i \mid r_j) = \frac{P(r_j \mid s_i)P(s_i)}{\sum_{i=1}^{U} P(r_j \mid s_i)P(s_i)}$$

This is called Bayes theorem.

(b) Suppose further that $R = S + N$, and assume $P(s_1 = 1) = P(s_2 = -1) = \frac{1}{2}$ and $P(n_1 = 1) = P(n_2 = -1) = \frac{1}{2}$. Determine $P(s_i \mid r_j)$ for all i and j. (This problem can be cast as a detection or estimation problem. We want to guess the value of S given that we know only R. We would therefore find useful the probability that S takes on each value given that R is known.)

1.15 For continuous random variables W and X, show that

$$p(w \mid x) = \frac{p(x \mid w)p(w)}{\int p(x \mid w)p(w)\, dw}$$

This is Bayes theorem expressed in terms of probability density functions (see Exercise 1.14).

1.16 For the continuous random variable Y, show that

$$\int_{-\infty}^{\infty} p(y \mid x)\, dy = 1$$

1.17 For continuous random variables W, X, and Y show that

$$p(w \mid x, y)p(x \mid y) = p(w, x \mid y)$$

1.18 Suppose the random variables W, X, Y, and Z are such that $p(w, x \mid y, z) = p(w \mid y)p(x \mid z)$. Show that

(a) $p(w \mid y, z) = p(w \mid y)$ or $p(x \mid y, z) = p(x \mid z)$

(b) $p(w, z \mid y) = p(w \mid y)p(z \mid y)$ or $p(x, z \mid y) = p(x \mid z)p(z \mid y)$

(c) $p(z \mid y, w) = p(z \mid y)$ or $p(y \mid z, x) = p(y \mid z)$

References

1. Davenport, W. B. and Root, W. L., "An Introduction to the Theory of Random Signals and Noise." McGraw-Hill, New York, 1958.
2. Papoulis, A., "Probability, Random Variables, and Stochastic Processes." McGraw-Hill, New York, 1965.
3. Wylie, C. R. Jr., "Advanced Engineering Mathematics." McGraw-Hill, New York, 1951.
4. Lighthill, M. J., "Introduction to Fourier Analysis and Generalized Functions." Cambridge Univ. Press, London and New York, 1958.
5. Apostol, T. M., "Mathematical Analysis." Addison-Wesley, Reading, Massachusetts, 1957.
6. Farison, J. B., On calculating moments for some common probability laws, *IEEE Trans. Inform. Theory* **IT-11**, No. 4, 586–589 (1965).
7. Campbell, G. A., and Foster, R. M., "Fourier Integrals for Practical Applications." Van Nostrand, Princeton, New Jersey, 1948.
8. Papoulis, A., "The Fourier Integral and Its Applications." McGraw-Hill, New York, 1962.
9. Springer, M. D., and Thompson, W. E., The distribution of products of independent random variables, *J. Soc. Ind. Appl. Math.* **14**, No. 3 (1966).

SUPPLEMENTARY BIBLIOGRAPHY

Cramér, H., "Mathematical Methods of Statistics." Princeton Univ. Press, Princeton, New Jersey, 1958.

Feller, W., "An Introduction to Probability Theory and Its Applications." Vol. I, 2nd Ed., 1957, Vol. II, Wiley, New York, 1966.

Fry, T. C., "Probability and Its Engineering Uses." 2nd Ed., Van Nostrand, Princeton, New Jersey, 1965.

Gnedenko, B. V., "The Theory of Probability" (Transl. by B. D. Seckler). Chelsea, New York, 1962.

Hald, A., "Statistical Theory with Engineering Applications." Wiley, New York, 1952.

Kendall, M. G., and Stuart, A., "The Advanced Theory of Statistics," Vol. 1, Distribution Theory. Hafner, New York, 1958.

Loève, M., "Probability Theory." 3rd Ed. Van Nostrand, Princeton, New Jersey, 1963.

Mood, A. M., and Graybill, F. A., "Introduction to the Theory of Statistics," 2nd Ed. McGraw-Hill, New York, 1963.

Parzen, E., "Modern Probability Theory and Its Applications." Wiley, New York, 1960.

Uspensky, J. V., "Introduction to Mathematical Probability." McGraw-Hill, New York, 1937.

Wilks, S. S., "Mathematical Statistics." Wiley, New York, 1962.

Chapter 2

Random Processes

2.1 Introduction

In the last chapter a random variable was defined as a real-valued function of the outcome of an experiment. If the experiment continues in time, we call the outcome a *random process* or *stochastic process*.† The value which the random variable assumes is a function of time. A typical example is sequential temperature or wind velocity recordings. Another example much closer to our interests is the noise encountered in communication and radar receivers. If a recording instrument is connected to the output of a radar receiver, for example, the recording might look like that shown in Fig. 2-1A. If the same measurement is performed on a second "identical" radar receiver operating under "identical" conditions, the recording might look like that shown in Fig. 2-1B. The signal could be a signal picked up by the receiver antenna or could have been internally generated circuit noise. In any event such stochastic processes exhibit unpredictable changes with time and defy precise prediction.

However, such processes often exhibit average properties that do not change with time. For example, in the radar case, if recordings were taken from an "infinite" number of "identical" radars operating under "identical" conditions, we might find that the average value and average deviation of each recording were the same. Histograms which show the amplitude distribution of the records might look the same, the average number of axis crossings

† Davenport and Root (*1*) and Papoulis (*2*) provide very readable treatments of this material.

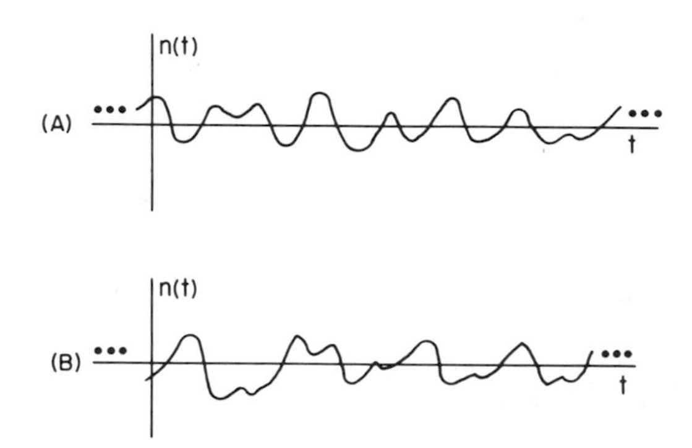

Fig. 2-1 Recordings of "typical" outputs of a radar receiver.

might also be identical. These and many other properties might be identical. It is upon such properties that a statistical description can be founded.

In the above discussion the collection of all possible recordings is called an *ensemble*. The individual members of the ensemble are called *sample functions*. The relationship between ensemble averages and sample function averages is discussed in Section 2.4.

2.2 Relation to Probability

In the last chapter probability and random variables were discussed with no mention of any time dependence. Yet, all that was said also applies when time is introduced. Before pursuing this, a discussion of notation will be worthwhile.

If we are to remain consistent with the notation of the preceding chapter, then for a random process $X(t)$ the members of the ensemble of sample functions would be denoted by $x(t)$. Maintaining the distinction between the process and its realizations as functions of time is not only cumbersome but can lead to superfluous notation. For example, the random process derived from $X(t)$ by the operation

$$y(t) = \int_0^t x(\tau)\, d\tau$$

is defined by its sample functions and nothing is gained by pointing out that $y(t)$ is a sample function for some process $Y(t)$. For these reasons, when no possibility of confusion exists we will ignore the formality and simply identify a process with its sample functions. For example, the expected value of the process $x(t)$ at time t_1 is written $E\{x_1\}$ which, of course, is to be interpreted as $E\{X(t_1)\}$.

Suppose we have a random process denoted $x(t)$. Typical sample functions might look like those shown in Fig. 2-2, where the superscripts are used to distinguish different possible members of the ensemble. Denote by x_1 a sample taken at time t_1. It is a random variable and has a probability density function $p(x_1)$ which in theory may be obtained by observing all members of the ensemble at time t_1. We may similarly take samples at instants of time $t_1, t_2, \ldots, t_N$ and denote these as $x_1, x_2, \ldots, x_N$ respectively. These are

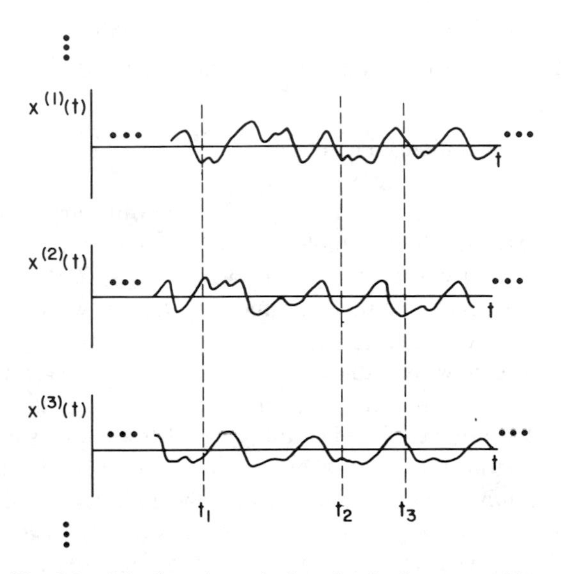

Fig. 2-2 Members (sample functions) of an ensemble.

associated with a joint event and have a joint probability density function $p(x_1, x_2, \ldots, x_N)$ which in theory may also be obtained by observing all members of the ensemble. The relations for joint density functions discussed in Chap. 1 also hold for these functions. In particular,

$$p(x_1, \ldots, x_N) \geq 0 \tag{2-1}$$

$$\int_{-\infty}^{\infty} \cdots \int_{-\infty}^{\infty} p(x_1, x_2, \ldots, x_N)\, dx_1\, dx_2 \cdots dx_N = 1 \tag{2-2}$$

$$\int_{-\infty}^{\infty} \cdots \int_{-\infty}^{\infty} p(x_1, \ldots, x_k, x_{k+1}, \ldots, x_N)\, dx_{k+1} \cdots dx_N = p(x_1, \ldots, x_k) \qquad k < N \tag{2-3}$$

$$p(x_1, \ldots, x_k \mid x_{k+1}, \ldots, x_N) = \frac{p(x_1, \ldots, x_N)}{p(x_{k+1}, \ldots, x_N)} \tag{2-4}$$

$$\int_{-\infty}^{\infty} p(x_N \mid x_1, \ldots, x_{N-1})\, dx_N = 1 \tag{2-5}$$

Suppose we have another random process, say $y(t)$. Denote its samples at times $t_1, \ldots, t_M$ as $y_1, \ldots, y_M$. As in the preceding case, we assume that the joint probability density functions $p(y_1, \ldots, y_M)$ and $p(x_1, \ldots, x_N, y_1, \ldots, y_M)$ exist. The two random processes are *statistically independent random processes* if

$$p(x_1, \ldots, x_N, y_1, \ldots, y_M) = p(x_1, \ldots, x_N)p(y_1, \ldots, y_M) \qquad (2\text{-}6)$$

for all sets of samples of $x(t)$ and $y(t)$ and all N and M.

2.3 Ensemble Correlation Functions

Again suppose that $x_{t_1}, x_{t_2}, \ldots, x_{t_N}$ are random variables pertaining to a random process $x(t)$. The samples are chosen at times $t_1, t_2, \ldots, t_N$ and have a joint probability density function $p(x_{t_1}, \ldots, x_{t_N})$. If we chose to select the samples at different instants of time, say $t_i - \tau$, we have a new set of random variables $x_{t_1-\tau}, x_{t_2-\tau}, \ldots, x_{t_N-\tau}$. The joint probability density function for these new variables is $p(x_{t_1-\tau}, \ldots, x_{t_N-\tau})$. If the probability density functions corresponding to time instants t_n are the same as those functions corresponding to time instants $t_n - \tau$ for all values of N and τ, then these density functions do not change with a shift of the time origin. A random process having such time-invariant probability density functions is said to be a *strictly stationary* or *stationary* random process. Other random processes are either *wide-sense stationary* (defined below), or *nonstationary*.

Let x_1 and x_2 be the random variables that refer to time samples at t_1 and t_2 of a random process $x(t)$. The *autocorrelation function*† $R_x(t_1, t_2)$ for a real random process is defined as

$$R_x(t_1, t_2) = E\{x_1, x_2\} = \iint x_1 x_2 p(x_1, x_2)\, dx_1\, dx_2 \qquad (2\text{-}7)$$

(The limits of integration are omitted for convenience. It is to be understood that they extend over the entire sample space.) If we represent t_1 as t, t_2 as $t - \tau$, then

$$R_x(t_1, t_2) = R_x(t, t - \tau) = \iint x_t x_{t-\tau} p(x_t, x_{t-\tau})\, dx_t\, dx_{t-\tau} \qquad (2\text{-}8)$$

If the process is stationary, then

$$p(x_t, x_{t-\tau}) = p(x_{t+t'}, x_{t+t'-\tau}) \qquad (2\text{-}9)$$

for all t', so that the autocorrelation function is independent of the time origin, and depends only on the time difference τ. That is

$$R_x(t, t-\tau) = R_x(\tau) \qquad (2\text{-}10)$$

† These are *ensemble* autocorrelation functions. *Time* autocorrelation functions are defined in Section 2.5.

for a stationary process. In analogy to Eq. (1-36), a correlation coefficient, called the *normalized autocovariance function*, of the random process may be defined as

$$\rho_x(t_1, t_2) = \frac{E\{(x_1 - m_1)(x_2 - m_2)\}}{\sigma_1 \sigma_2} \tag{2-11}$$

$$= \frac{R_x(t_1, t_2) - m_1 m_2}{\sigma_1 \sigma_2} \tag{2-12}$$

where

$$m_i = E\{x_i\} \qquad \text{and} \qquad \sigma_i^2 = E\{(x_i - m_i)^2\}$$

If the process is stationary, and $t_1 - t_2 = \tau$, then $m_1 = m_2 = m$, $\sigma_1 = \sigma_2 = \sigma$, and

$$\rho_x(\tau) = \frac{R_x(\tau) - m^2}{\sigma^2} \tag{2-13}$$

A process which is not strictly stationary, but has a time invariant mean value and an autocorrelation function for which $R_x(t, t - \tau) = R_x(\tau)$, is said to be *wide-sense stationary*. Obviously a process which is strictly stationary is also stationary in the wide sense. The converse is not necessarily true.†

If the random process in question is complex, the autocorrelation function is defined as

$$R_x(t_1, t_2) = E\{x_1 x_2^*\} \tag{2-14}$$

where the asterisk denotes complex conjugate. In the discussion which follows the random process may be complex. The probability density function of a complex random variable is two-dimensional and real. The two dimensions are for the real and imaginary parts.

Consider two random processes $x(t)$ and $y(t)$. Their autocorrelation functions are $R_x(t_1, t_2)$ and $R_y(t_1, t_2)$. We define the *crosscorrelation functions*

$$R_{xy}(t_1, t_2) = E\{x_1 y_2^*\} = \iint x_1 y_2^* p(x_1, y_2)\, dx_1\, dy_2 \tag{2-15}$$

and

$$R_{yx}(t_1, t_2) = E\{y_1 x_2^*\} = \iint y_1 x_2^* p(y_1, x_2)\, dy_1\, dx_2 \tag{2-16}$$

If the crosscorrelation functions do not depend on the time origin, then the processes are said to be *jointly wide-sense stationary*. Thus, for such processes

$$E\{x_t\, y_{t-\tau}^*\} = R_{xy}(\tau) \qquad \text{and} \qquad E\{y_t\, x_{t-\tau}^*\} = R_{yx}(\tau)$$

† An exception is the Gaussian process which is discussed in Chap. 4.

Properties of Correlation Functions

Assume that the random processes $x(t)$ and $y(t)$ are individually and jointly wide-sense stationary. Express the autocorrelation function as

$$R_x(\tau) = \iint x_t x_{t-\tau}^* p(x_t, x_{t-\tau})\, dx_t\, dx_{t-\tau}$$

Then

$$R_x(-\tau) = \iint x_t x_{t+\tau}^* p(x_t, x_{t+\tau})\, dx_t\, dx_{t+\tau}$$

and

$$R_x^*(-\tau) = \iint x_t^* x_{t+\tau} p(x_t, x_{t+\tau})\, dx_t\, dx_{t+\tau}$$

Substituting $t' = t + \tau$ and rearranging terms slightly yields

$$R_x^*(-\tau) = \iint x_{t'} x_{t'-\tau} p(x_{t'-\tau}, x_{t'})\, dx_{t'}\, dx_{t'-\tau}$$

or

$$R_x^*(-\tau) = R_x(\tau) \tag{2-17}$$

If the random process is *real*, then

$$R_x(-\tau) = R_x(\tau)$$

The autocorrelation function of a real random process is therefore an even function. Similarly, for the crosscorrelation function

$$R_{xy}(\tau) = \iint x_t y_{t-\tau}^* p(x_t, y_{t-\tau})\, dx_t\, dy_{t-\tau}$$

and

$$R_{xy}(-\tau) = \iint x_t y_{t+\tau}^* p(x_t, y_{t+\tau})\, dx_t\, dy_{t+\tau}$$

Substituting $t' = t + \tau$ in the integrals and taking the complex conjugate of each side of the equation yields

$$R_{xy}^*(-\tau) = \iint y_{t'} x_{t'-\tau}^* p(x_{t'-\tau}, y_{t'})\, dy_{t'}\, dx_{t'-\tau} = R_{yx}(\tau)$$

or

$$R_{xy}(\tau) = R_{yx}^*(-\tau) \tag{2-18}$$

It is worth noting that for a zero mean wide-sense stationary process

$$R_x(0) = E\{x_t x_t^*\} = E\{|x_t|^2\} = \sigma^2 \tag{2-19}$$

It is also true that (*1*)

$$|R_x(\tau)| \le R_x(0) \tag{2-20}$$

and

$$|R_{xy}(\tau)| \le [R_x(0)R_y(0)]^{1/2} \tag{2-21}$$

Therefore, the autocorrelation function has a peak at its origin. This is not in general true for the crosscorrelation function.

EXAMPLE 2.3-1 Determine the autocorrelation function of a sinewave with random phase, $x(t) = \sin(\omega_o t + \theta)$, where ω_o is a constant and θ is a random variable uniformly distributed over the interval $(0, 2\pi)$,

$$p(\theta) = \begin{cases} 1/2\pi, & 0 \le \theta \le 2\pi \\ 0, & \text{otherwise} \end{cases}$$

This is an example of a *periodic random process*. A periodic random process $x(t)$ is one for which $x(t) = x(t + T)$ where T is the period.

From previous definitions,

$$R_x(t, t - \tau) = E\{x_t x_{t-\tau}\}$$

This is equal to

$$\begin{aligned} R_x(t, t - \tau) &= E\{\sin(\omega_o t + \theta)\sin[\omega_o(t - \tau) + \theta]\} \\ &= E\{\sin(\omega_o t + \theta)[\sin(\omega_o t + \theta)\cos \omega_o\tau - \cos(\omega_o t + \theta)\sin \omega_o\tau]\} \\ &= \cos \omega_o\tau \; E\{\sin^2(\omega_o t + \theta)\} \\ &\quad \sin \omega_o\tau \; E\{\sin(\omega_o t + \theta)\cos(\omega_o t + \theta)\} \end{aligned}$$

We must determine the expectation of functions of θ, $g_1(\theta) = \sin^2(\omega_o t + \theta)$ and $g_2(\theta) = \sin(\omega_o t + \theta)\cos(\omega_o t + \theta)$. The expectation is given by

$$E\{g_1(\theta)\} = \int_0^{2\pi} g_1(\theta)p(\theta)\, d\theta = \int_0^{2\pi} \sin^2(\omega_o t + \theta)\frac{d\theta}{2\pi} = \frac{1}{2}$$

and

$$E\{g_2(\theta)\} = \int_0^{2\pi} g_2(\theta)p(\theta)\, d\theta = \int_0^{2\pi} \sin(\omega_o t + \theta)\cos(\omega_o t + \theta)\frac{d\theta}{2\pi} = 0$$

Therefore

$$R_x(t, t - \tau) = R_x(\tau) = \tfrac{1}{2}\cos \omega_o\tau$$

(The correlation function of a periodic random process is itself periodic.) Since the autocorrelation is independent of time and dependent only on the time difference τ, the process is (at least) wide-sense stationary. (To be wide-sense stationary, the mean must also be independent of time; this is easily verified.)

EXAMPLE 2.3-2 Determine the autocorrelation function of the binary waveform (3) shown in Fig. 2-3. In each interval of unit length the waveform may take on values ± 1 each with probability $\frac{1}{2}$. The value of the waveform in any one interval is assumed independent of the value in any other interval. In order for the process to be wide-sense stationary, no specific time origin has been chosen.

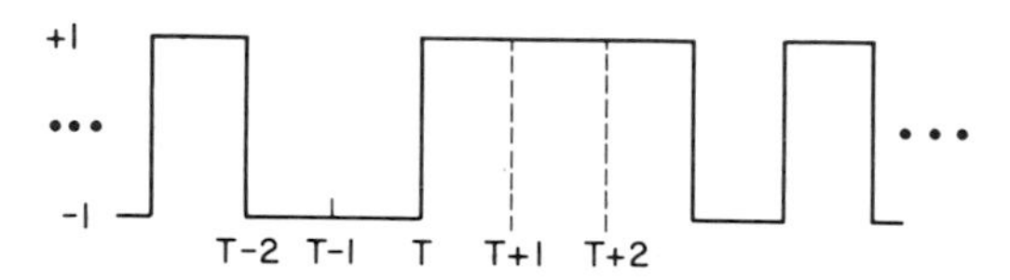

Fig. 2-3 A binary waveform.

We form

$$\begin{aligned} R_x(t, t-\tau) = E\{x_t x_{t-\tau}\} &= \sum_{\text{all } x_t} \sum_{\text{all } x_{t-\tau}} x_t x_{t-\tau} P(x_t, x_{t-\tau}) \\ &= (1)(1)P(x_t = 1, x_{t-\tau} = 1) \\ &\quad + (1)(-1)P(x_t = 1, x_{t-\tau} = -1) \\ &\quad + (-1)(1)P(x_t = -1, x_{t-\tau} = 1) \\ &\quad + (-1)(-1)P(x_t = -1, x_{t-\tau} = -1) \\ &= P(x_t = 1 \mid x_{t-\tau} = 1)P(x_{t-\tau} = 1) \\ &\quad - P(x_t = 1 \mid x_{t-\tau} = -1)P(x_{t-\tau} = -1) \\ &\quad - P(x_t = -1 \mid x_{t-\tau} = 1)P(x_{t-\tau} = 1) \\ &\quad + P(x_t = -1 \mid x_{t-\tau} = -1)P(x_{t-\tau} = -1) \end{aligned}$$

Now

$$P(x_{t-\tau} = 1) = P(x_{t-\tau} = -1) = \tfrac{1}{2}$$

Also

$$P(x_t = -1 \mid x_{t-\tau} = -1) = 1 - P(x_t = 1 \mid x_{t-\tau} = -1)$$

and

$$P(x_t = -1 \mid x_{t-\tau} = +1) = 1 - P(x_t = 1 \mid x_{t-\tau} = 1)$$

Therefore,

$$\begin{aligned} R_x(t, t-\tau) &= \tfrac{1}{2}P(x_t = 1 \mid x_{t-\tau} = 1) \\ &\quad - \tfrac{1}{2}P(x_t = 1 \mid x_{t-\tau} = -1) \\ &\quad - \tfrac{1}{2}[1 - P(x_t = 1 \mid x_{t-\tau} = 1)] \\ &\quad + \tfrac{1}{2}[1 - P(x_t = 1 \mid x_{t-\tau} = -1)] \\ &= P(x_t = 1 \mid x_{t-\tau} = 1) - P(x_t = 1 \mid x_{t-\tau} = -1) \end{aligned}$$

If $|\tau| > 1$, then t and $t - \tau$ are in different intervals and the corresponding values of x_t and $x_{t-\tau}$ are independent. Then

$$P(x_t = 1 \mid x_{t-\tau} = 1) = P(x_t = 1) = P(x_t = 1 \mid x_{t-\tau} = -1) = \tfrac{1}{2}$$

so for $|\tau| > 1$, $R_x(t, t - \tau) = 0$.

For $|\tau| < 1$, $P(x_t = 1 \mid x_{t-\tau} = -1)$ is the probability that t and $t - \tau$ are in different intervals and that there is a change in the value of the waveform in the two intervals. This is a statement of a joint event. The events however are independent and so the joint probability is a product of the probability of each event. The probability that t and $t - \tau$ are in different intervals is zero if $\tau = 0$, and is one if $\tau = \pm 1$. The probability will vary linearly with τ between these extremes. That is, for $|\tau| \leq 1$ the probability that t and $t - \tau$ are in different intervals is equal to $|\tau|$. The other portion of the joint probability, the probability that the waveform changes value, is equal to $\frac{1}{2}$ since the values are independent from pulse to pulse. Therefore,

$$P(x_t = 1 \mid x_{t-\tau} = -1) = |\tau|/2, \qquad |\tau| \leq 1$$

Now

$$P(x_t = 1 \mid x_{t-\tau} = 1) = 1 - P(x_t = -1 \mid x_{t-\tau} = 1)$$

So that for $|\tau| < 1$

$$R_x(t, t - \tau) = 1 - P(x_t = -1 \mid x_{t-\tau} = 1) - P(x_t = 1 \mid x_{t-\tau} = -1)$$

The latter two probabilities using the previous arguments are seen to be equal so that

$$R_x(t, t - \tau) = 1 - 2P(x_t = +1 \mid x_{t-\tau} = -1)$$

and finally

$$R_x(\tau) = \begin{cases} 1 - |\tau|, & |\tau| \leq 1 \\ 0, & \text{otherwise} \end{cases}$$

This is shown in Fig. 2-4.

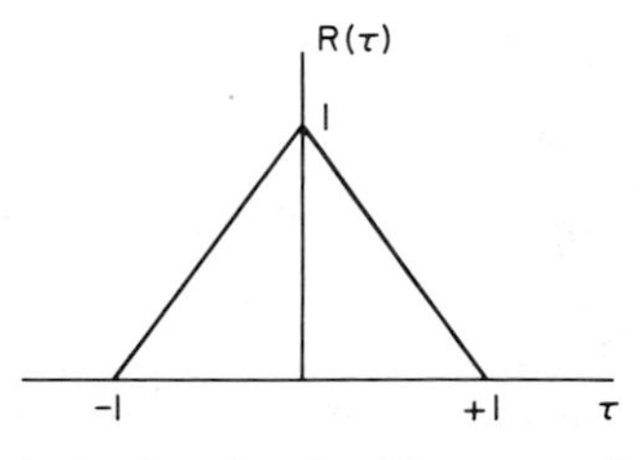

Fig. 2-4 Correlation function for binary waveform of Fig. 2-3.

EXAMPLE 2.3-3 Determine the autocorrelation function of the product $z(t) = x(t)y(t)$ where $x(t)$ and $y(t)$ are statistically independent and stationary random processes.

By definition

$$R_z(t, t-\tau) = E\{z(t)z(t-\tau)\} = E\{x(t)x(t-\tau)y(t)y(t-\tau)\}$$

Since the processes are statistically independent

$$R_z(t, t-\tau) = E\{x(t)x(t-\tau)\}E\{y(t)y(t-\tau)\} = R_x(\tau)R_y(\tau)$$

Thus the autocorrelation of the product is dependent only on the time difference τ

$$R_z(\tau) = R_x(\tau)R_y(\tau)$$

In summary, the autocorrelation function of the product of statistically independent stationary random processes is the product of the respective autocorrelation functions. The process formed by the product is therefore also stationary.

Suppose further that $y(t) = \cos(\omega_0 t + \theta)$ as in Example 2.3-1. Then the autocorrelation function of $z(t)$ is

$$R_z(\tau) = \tfrac{1}{2}R_x(\tau)\cos \omega_0\tau$$

An example of this is shown in Fig. 2-5 for an $R_x(\tau)$ which changes slowly.

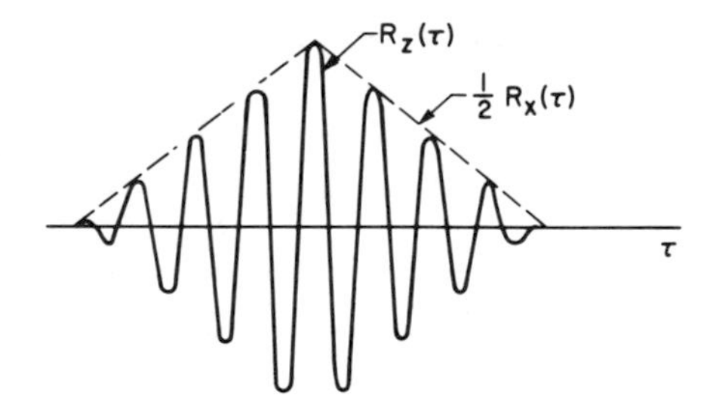

Fig. 2-5 Autocorrelation function for Example 2.3-3.

2.4 Time Averages

The averaging operations discussed previously have all been ensemble averages. Conceptually, at any given instant of time each member of the ensemble is sampled and these samples are averaged accordingly. In this section we briefly discuss averages formed from "typical" sample functions or realizations of the random process.

Time averages are important. Indeed, in practice only sample functions are available, and inferences about ensemble characteristics must be drawn from these.

Define the time average, denoted by $\langle \cdot \rangle$, of a sample function $x(t)$ by

$$\langle x(t) \rangle = \lim_{T \to \infty} (1/2T) \int_{-T}^{T} x(t)\, dt \tag{2-22}$$

if this limit exists. It is possible that this average may not exist for some or all sample functions and that the average may be different for different sample functions. An example of the latter is shown in Fig. 2-6 in which each sample

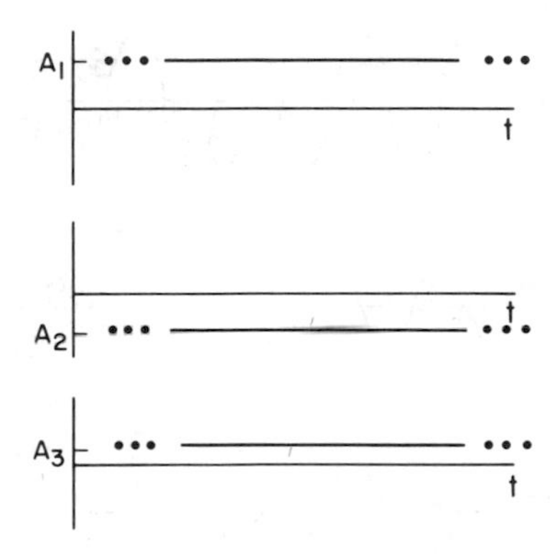

Fig. 2-6 Sample functions with unequal time and ensemble averages.

function is a constant with time. The time average of each sample function is equal to the value A_i which is implied to be different from sample function to sample function. An ensemble average, if it exists, would be a constant. Hence, the time averages are not in general equal to the ensemble average.†

An important property of random processes is *ergodicity* (*1*). If a random process is ergodic, then ensemble averages and the corresponding time averages are equal with probability one. In simplified terms, if a process is ergodic, each sample function in a statistical sense looks very much like any other sample function. An example of a process which is not ergodic is shown in Fig. 2-6 where each sample function is a constant with time and so does not assume those values which the other sample functions assume. Generally, ergodicity is difficult to prove so it is frequently assumed to be

† A good treatment of the relation of ensemble averages to time averages, and ergodicity appears in Brown (*4*).

true. For a zero mean stationary Gaussian process† with autocorrelation function $R(\tau)$, a condition which implies ergodicity is (*1*)

$$\int_{-\infty}^{\infty} |R(\tau)|\, d\tau < \infty \tag{2-23}$$

EXAMPLE 2.4-1 Consider the function

$$f(t) = A_i \cos(\omega_o t + \theta_i)$$

where A_i and θ_i are statistically independent, and θ_i is uniformly distributed over the interval $(0, 2\pi)$. Typical sample functions are shown in Fig. 2-7. Is

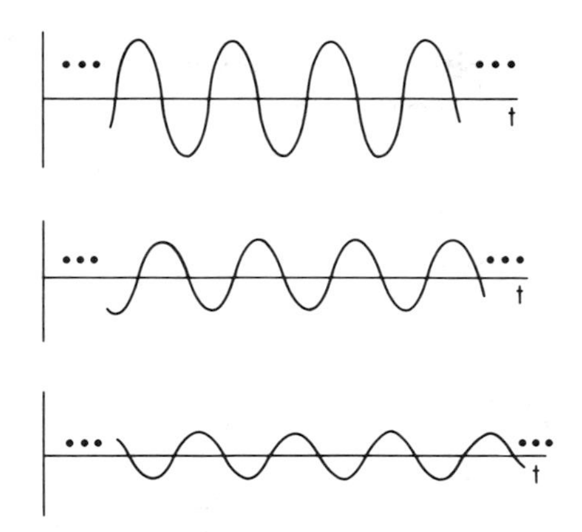

Fig. 2-7 Typical sample functions for Example 2.4-1.

the process ergodic? It is easily found that

$$E\{f(t)\} = 0 \qquad \text{and} \qquad \langle f(t) \rangle = 0$$

so that the ensemble mean and time average means are equal. On the other hand

$$E\{f^2(t)\} = E\{A_i^2\}E\{\cos^2(\omega_o t + \theta_i)\} = E\{A_i^2/2\}$$

whereas

$$\langle f^2(t) \rangle = A_i^2/2$$

which is not in general equal to $E\{A_i^2/2\}$. The process is therefore not ergodic. □ □

† See Chap. 4.

A time average is itself a random variable and one could investigate its statistical properties. For the above example

$$E\{\langle f^2(t)\rangle\} = E\{A_i^2/2\}$$

so that the ensemble average of the time average is equal to the original ensemble average.

2.5 Time Correlation Functions

In analogy to ensemble autocorrelation functions, a *time autocorrelation function* may be defined as

$$\mathscr{R}_x(\tau) = \lim_{T\to\infty} (1/2T) \int_{-T}^{T} x(t)x^*(t-\tau)\,dt \tag{2-24}$$

It is inherent in this definition that the time autocorrelation is a function of only the difference τ. If the process is ergodic, then the ensemble and time autocorrelation functions are equal with probability one.

The *time crosscorrelation* function of two sample functions $x(t)$ and $y(t)$ is defined as

$$\mathscr{R}_{xy}(\tau) = \lim_{T\to\infty} (1/2T) \int_{-T}^{T} x(t)y^*(t-\tau)\,dt \tag{2-25}$$

Again, the argument is only a function of the time difference τ. If the processes are jointly ergodic, the ensemble and time crosscorrelation functions are equal with probability one.

2.6 Power Spectral Density

The concept of "frequency" has long proven useful in engineering studies. One way in which such a concept is introduced is the Fourier series in which a periodic signal is decomposed into "frequency" components.† The frequency implies a rate at which the signal changes. That is, if a signal has "high" frequency components, we expect to see "rapid" variations. For signals which are not periodic, the Fourier integral is useful. For random processes, frequency can be introduced through the autocorrelation function.

† The Fourier series of a periodic function $r(t)$ is given by $r(t) = (1/T^{1/2}) \sum_{k=-\infty}^{\infty} r_k e^{jk\omega_0 T}$ where T is the period, $k\omega_o$ denotes the frequency of the components ($\omega_o = 2\pi/T$), and the amplitude of the Fourier components are

$$r_k = (1/T^{1/2}) \int_0^T r(t)e^{-jk\omega_0 t}\,dt$$

We define the *power spectral density* (also called spectral density) of a wide-sense stationary random process $x(t)$ to be the Fourier transform of the autocorrelation function (the Wiener–Kinchine theorem)

$$S_x(\omega) = \int_{-\infty}^{\infty} R_x(\tau)e^{-j\omega\tau}\,d\tau \tag{2-26}$$

The inverse relation is

$$R_x(\tau) = \frac{1}{2\pi}\int_{-\infty}^{\infty} S_x(\omega)e^{j\omega\tau}\,d\omega \tag{2-27}$$

From Eq. (2-19) we see that $R_x(0) = \sigma^2 = E\{x^2(t)\}$. This is the average power of a zero mean process. Now, from Eq. (2-27)

$$R_x(0) = \frac{1}{2\pi}\int_{-\infty}^{\infty} S_x(\omega)\,d\omega = E\{x^2(t)\} \tag{2-28}$$

so that the average power of a random process may be obtained by integrating the power spectral density over its entire range of definition. The power contributed by any particular frequency band Ω is given by $(1/2\pi)\int_\Omega S_x(\omega)\,d\omega$.

Without regard to stationarity we can also define a power spectral density function for a sample function as

$$\mathscr{S}_x(\omega) = \int_{-\infty}^{\infty} \mathscr{R}_x(\tau)e^{-j\omega\tau}\,d\tau \tag{2-29}$$

provided it exits. If the process is ergodic then the time autocorrelation function $\mathscr{R}_x(\tau)$ produces the same spectrum as $R_x(\tau)$ with probability one.†

There are two common cases for which the integrals in Eqs. (2-26) and (2-29) do not exist in the usual sense. If the random process in question has a nonzero mean, then the Fourier transform will not exist in the usual sense. However a nonzero mean can be represented by a delta function at the origin of the frequency domain. The weight of the delta function is the amount of power in the steady or dc (direct current) component. A second case occurs when $x(t)$ has periodic components which correspond to discrete frequencies. Such components will produce delta functions in the frequency domain.

In Eq. (2-17) it was shown that $R_x^*(-\tau) = R_x(\tau)$. Using this relation and the complex conjugate of Eq. (2-26) produces

$$S_x^*(\omega) = \int_{-\infty}^{\infty} R_x^*(\tau)e^{j\omega\tau}\,d\tau = \int_{-\infty}^{\infty} R_x(\tau)e^{-j\omega\tau}\,d\tau$$

† For references on spectrum analysis in practice, consult the bibliographies of Chaps. 2 and 3.

Therefore,

$$S_x^*(\omega) = S_x(\omega) \tag{2-30}$$

so that the power spectral density is a real function. Furthermore, if $x(t)$ is also a real function, then $R_x(\tau)$ is real and

$$S_x^*(\omega) = \int_{-\infty}^{\infty} R_x(\tau)e^{j\omega\tau}\, d\tau = S_x(-\omega)$$

Combining this result with Eq. (2-30) we get $S_x(\omega) = S_x(-\omega)$, so that the spectral density of a real process is an even function.

EXAMPLE 2.6-1 For the sinewave with random phase in Example 2.3-1, determine the power spectral density.

The autocorrelation function is given by

$$R_x(\tau) = \tfrac{1}{2}\cos \omega_o\tau = \tfrac{1}{4}(e^{j\omega_o\tau} + e^{-j\omega_o\tau})$$

The spectral density using Eq. (2-26) is

$$\begin{aligned} S_x(\omega) &= \tfrac{1}{4}\int_{-\infty}^{\infty} e^{-j(\omega-\omega_o)\tau}\, d\tau + \tfrac{1}{4}\int_{-\infty}^{\infty} e^{-j(\omega+\omega_o)\tau}\, d\tau \\ &= \tfrac{1}{2}\pi\, \delta(\omega - \omega_o) + \tfrac{1}{2}\pi\, \delta(\omega + \omega_o) \end{aligned}$$

This is shown in Fig. 2-8. The total power is $\frac{1}{2}$ which is in agreement with $R_x(0)$ and $E\{x^2(t)\}$.

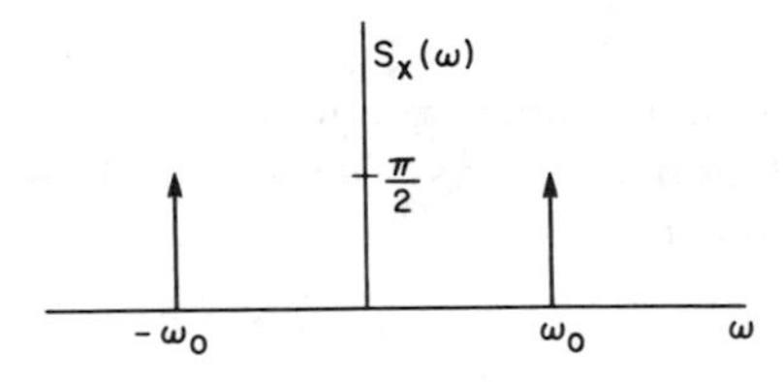

Fig. 2-8 Power spectral density for a randomly phased sinewave of radian frequency ω_o.

EXAMPLE 2.6-2 Determine the spectral density for the binary waveform of Example 2.3-2.

The autocorrelation function is

$$R_x(\tau) = \begin{cases} 1 - |\tau|, & |\tau| \le 1 \\ 0, & \text{otherwise} \end{cases}$$

The spectral density is

$$S_x(\omega) = \int_{-1}^{1} (1 - |\tau|)e^{-j\omega\tau}\, d\tau = 2\int_{0}^{1} (1 - \tau) \cos \omega\tau\, d\tau$$

or

$$S_x(\omega) = \frac{\sin^2(\omega/2)}{(\omega/2)^2}$$

This is shown in Fig. 2-9.

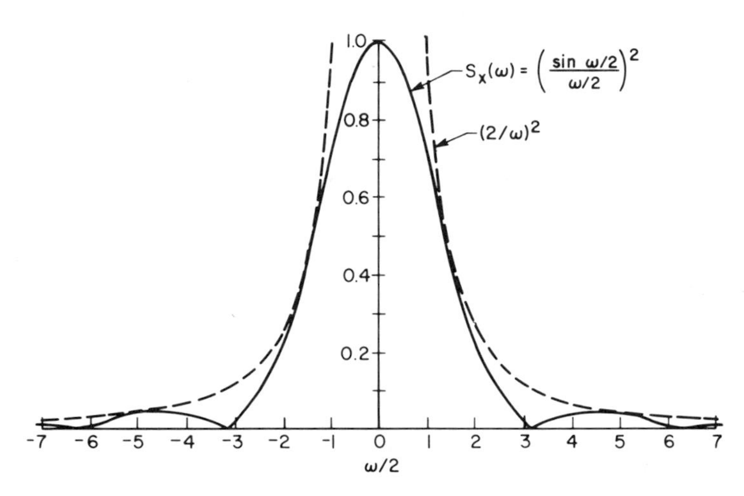

Fig. 2-9 Power spectral density for binary waveform of Fig. 2-3.

EXAMPLE 2.6-3 Determine the spectral density for the product of the binary waveform of Example 2.3-2 and the randomly phased sine wave of Example 2.3-1.

In Example 2.3-3 it was shown that the autocorrelation function of a product of statistically independent random processes is the product of the individual autocorrelation functions. Denote the binary waveform by $x(t)$ and the product by $y(t)$. That is

$$y(t) = x(t)\cos(\omega_o t + \theta)$$

Then

$$R_y(\tau) = \tfrac{1}{2}R_x(\tau)\cos\omega_o\tau = \tfrac{1}{4}R_x(\tau)e^{j\omega_o\tau} + \tfrac{1}{4}R_x(\tau)e^{-j\omega_o\tau}$$

The power spectral density of $y(t)$ may then be determined to be

$$S_y(\omega) = \tfrac{1}{4}\int_{-1}^{1} R_x(\tau)e^{-j(\omega-\omega_o)\tau}\,d\tau + \tfrac{1}{4}\int_{-1}^{1} R_x(\tau)e^{-j(\omega+\omega_o)\tau}\,d\tau$$

$$= \tfrac{1}{2}\pi S_x(\omega - \omega_o) + \tfrac{1}{2}\pi S_x(\omega + \omega_o)$$

where $S_x(\omega)$ is the spectral density of the binary waveform, Fig. 2-9. Then, if the "center frequency" ω_o is sufficiently high, the spectrum of $y(t)$ will be as shown in Fig. 2-10.

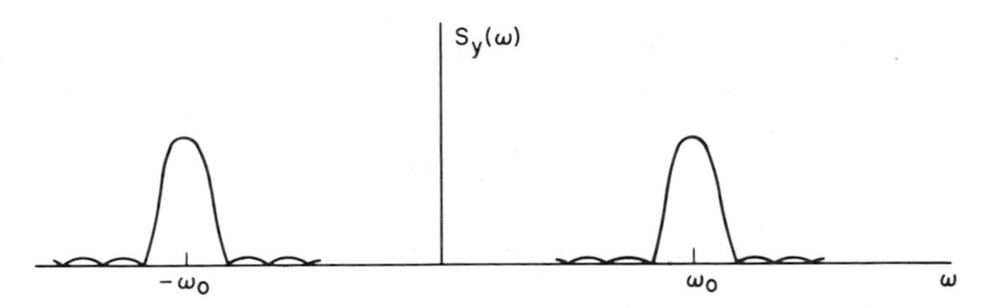

Fig. 2-10 Power spectral density of sinewave amplitude modulated by a binary waveform.

Cross-Spectral Densities

For two random processes $x(t)$ and $y(t)$ which are jointly wide-sense stationary and have a crosscorrelation function $R_{xy}(\tau)$, we define their *power cross-spectral density* (or simply cross-spectral density) to be

$$S_{xy}(\omega) = \int_{-\infty}^{\infty} R_{xy}(\tau)e^{-j\omega\tau}\, d\tau \tag{2-31}$$

The inverse transformation is

$$R_{xy}(\tau) = \frac{1}{2\pi}\int_{-\infty}^{\infty} S_{xy}(\omega)e^{j\omega\tau}\, d\omega \tag{2-32}$$

From Eq. (2-18), $R_{xy}(\tau) = R^*_{yx}(-\tau)$, so the complex conjugate of Eq. (2-31) is

$$\begin{aligned} S^*_{xy}(\omega) &= \int_{-\infty}^{\infty} R^*_{xy}(\tau)e^{+j\omega\tau}\, d\tau \\ &= \int_{-\infty}^{\infty} R_{yx}(-\tau)e^{+j\omega\tau}\, d\tau \\ &= \int_{-\infty}^{\infty} R_{yx}(\tau)e^{-j\omega\tau}\, d\tau \end{aligned}$$

Therefore

$$S^*_{xy}(\omega) = S_{yx}(\omega) \tag{2-33}$$

If the processes $x(t)$ and $y(t)$ are real, then

$$S^*_{xy}(\omega) = \int_{-\infty}^{\infty} R_{xy}(\tau)e^{+j\omega\tau}\, d\tau = S_{xy}(-\omega) \tag{2-34}$$

Combining this result with Eq. (2-33), we determine for real processes that

$$S_{yx}(\omega) = S_{xy}(-\omega) \tag{2-35}$$

For sample functions of random processes $x(t)$ and $y(t)$, the time cross-spectral density is defined as

$$\mathscr{S}_{xy}(\omega) = \int_{-\infty}^{\infty} \mathscr{R}_{xy}(\tau) e^{-j\omega\tau}\, d\tau \tag{2-36}$$

For jointly ergodic processes, this will be equal to $S_{xy}(\omega)$ with probability one.

2.7 Response of Linear Filters

One of the basic operations involved in signal processing is that of filtering. If a time-varying signal $f_i(t)$ is applied to the input of a filter, a signal $f_o(t)$ will be observed at its output. The filter may be defined by all possible pairs of input–output signals. Symbolically, we represent these pairs as $f_i(t) \to f_o(t)$. A restriction on all physically realizable (causal) filters is that they cannot have an output before an input occurs. For example, if $f_i(t) = 0$ for $t < t_o$, then for a passive† physically realizable filter $f_o(t) = 0$ for $t < t_o$. A filter whose components do not change with time is called a time-invariant or stationary filter. For example, if $f_i(t) \to f_o(t)$, then for a time-invariant filter $f_i(t + t_0) \to f_o(t + t_0)$.

A particularly important class of filters is the linear filter. If $f_i(t) \to f_o(t)$, and $g_i(t) \to g_o(t)$, then for a linear filter

$$Af_i(t) + Bg_i(t) \to Af_o(t) + Bg_o(t)$$

for all values of A and B.

For a linear, time-invariant filter, the input and output are related through the convolution operation. That is,

$$f_o(t) = \int_{-\infty}^{\infty} h(\tau) f_i(t - \tau)\, d\tau = h(t) * f_i(t) \tag{2-37}$$

where $h(\tau)$ is called the impulse response or weighting function of the filter, and $*$ denotes convolution. If the filter is physically realizable, $h(t) = 0$ for $t < 0$. If $f_i(t) = \delta(t)$, a Dirac delta impulse function, the output $f_o(t)$ is equal to $h(t)$.

The above integral relationship is the convolution of the input function with the weighting function. Therefore if we denote the Fourier transforms of $f_o(t)$, $h(t)$, $f_i(t)$ by $F_o(j\omega)$, $H(j\omega)$, $F_i(j\omega)$, then

$$F_o(j\omega) = H(j\omega) F_i(j\omega) \tag{2-38}$$

We define $H(j\omega)$ as the transfer function of the filter.

† We assume throughout that all filters are passive in the sense that they contain no energy sources.

The Fourier transform of the impulse function is

$$H(j\omega) = \int_{-\infty}^{\infty} h(t)e^{-j\omega\tau}\, dt$$

If the impulse function is real, then

$$H^*(j\omega) = \int_{-\infty}^{\infty} h(t)e^{j\omega\tau}\, dt = H(-j\omega) \tag{2-39}$$

The equations for linear filters can be generalized to filters having multiple inputs and outputs.

Throughout this section it is assumed that all inputs are bounded, and the linear system is stable. A signal $f(t)$ is bounded if there exists some positive constant M such that $|f(t)| \leq M < \infty$. A linear system is stable if

$$\int_{-\infty}^{\infty} |h(\tau)|\, d\tau < \infty$$

With these assumptions, if the system is stable and the input is bounded, then the output is also bounded.

Unless stated otherwise, all filters considered will be stable, time-invariant linear filters.

Random Inputs

Suppose that the input to a linear system $h(t)$ is a sample function $x(t)$ of a random process. Then the output of the filter, obtained by convolving the input with the impulse response, is a sample function of another random process. In this section we shall investigate the statistical properties of the output random process given those of the input random process.

Because of the assumptions of bounded inputs and stable filters, it follows that for each sample function the convolution integral relating the input to the output will converge. In particular, assume that the bounded sample function $x(t)$ of a wide-sense stationary process is applied to a time-invariant linear stable filter. The output is given by

$$y(t) = \int_{0}^{\infty} h(\tau)x(t-\tau)\, d\tau \tag{2-40}$$

Denoting the expected value of the input by m_x, the expected value of the output m_y is

$$m_y = E\{y(t)\} = E\left\{\int_{0}^{\infty} h(\tau)x(t-\tau)\, d\tau\right\}$$

(See Sect. 2.2 for a discussion of notation especially as it relates to random processes and sample functions of random processes.) Interchanging the order of the expectation and integration operators we find

$$m_y = \int_0^\infty h(\tau)E\{x(t-\tau)\}\,d\tau$$

and finally

$$m_y = m_x \int_0^\infty h(\tau)\,d\tau$$

From the definition of the transfer function,

$$H(j\omega) = \int_0^\infty h(t)e^{-j\omega t}\,dt$$

we find

$$H(0) = \int_0^\infty h(t)\,dt \tag{2-41}$$

Therefore, $m_y = m_x H(0)$, a constant independent of time.

We may similarly determine the autocorrelation function of the output $R_y(t, t-\tau)$

$$\begin{aligned} R_y(t, t-\tau) &= E\{y_t\, y_{t-\tau}\} \\ &= E\left\{\int_0^\infty h(\eta)x(t-\eta)\,d\eta \int_0^\infty h(\alpha)x(t-\tau-\alpha)\,d\alpha\right\} \\ &= E\left\{\int_0^\infty \int_0^\infty h(\eta)h(\alpha)x(t-\eta)x(t-\tau-\alpha)\,d\eta\,d\alpha\right\} \\ &= \int_0^\infty \int_0^\infty h(\eta)h(\alpha)E\{x(t-n)x(t-\tau-\alpha)\}\,d\eta\,d\alpha \end{aligned}$$

The input process is assumed to be wide-sense stationary so that

$$E\{x(t-\eta)x(t-\tau-\alpha)\} = R_x(\tau+\alpha-\eta)$$

Therefore

$$R_y(t, t-\tau) = \int_0^\infty \int_0^\infty h(\eta)h(\alpha)R_x(\tau+\alpha-\eta)\,d\eta\,d\alpha = R_y(\tau)$$

Since this depends only on the time difference τ and not the value of the time origin t, it follows that the output is wide-sense stationary. Consequently the response of a time-invariant linear system to a wide-sense stationary input is itself wide-sense stationary. Therefore, the spectral density is defined and is

$$S_y(\omega) = \int_{-\infty}^\infty R_y(\tau)e^{-j\omega\tau}\,d\tau$$

If the expression for $R_y(\tau)$ is substituted into this equation we get

$$S_y(\omega) = \int_{-\infty}^{\infty} d\tau \int_0^{\infty} d\eta \int_0^{\infty} d\alpha\, h(\eta)h(\alpha)R_x(\tau + \alpha - \eta)e^{-j\omega\tau}$$

By including $e^{-j\omega\eta}e^{j\omega\eta}$ and $e^{-j\omega\alpha}e^{j\omega\alpha}$, each of which is unity, under the integrals and properly grouping the expressions, the following equation results

$$S_y(\omega) = \int_0^{\infty} h(\eta)e^{-j\omega\eta}\, d\eta \int_0^{\infty} h(\alpha)e^{+j\omega\alpha}\, d\alpha \int_{-\infty}^{\infty} R_x(\tau + \alpha - \eta)e^{-j\omega(\tau+\alpha-\eta)}\, d\tau$$

The first and second integrals are

$$H(j\omega) = \int_0^{\infty} h(\eta)e^{-j\omega\eta}\, d\eta$$

and

$$H^*(j\omega) = \int_0^{\infty} h(\alpha)e^{j\omega\alpha}\, d\alpha$$

Their product is then $|H(j\omega)|^2$. The last integral, after a change of variable ($t' = \tau + \alpha - \eta$), is the spectral density of the input, namely

$$S_x(\omega) = \int_{-\infty}^{\infty} R_x(t')e^{-j\omega t'}\, dt'$$

It then follows that the output spectral density of a time-invariant linear filter having a wide-sense stationary input is given by

$$S_y(\omega) = |H(j\omega)|^2 S_x(\omega) \tag{2-42}$$

We can also use this expression to determine the output autocorrelation function

$$R_y(\tau) = \frac{1}{2\pi}\int_{-\infty}^{\infty} |H(j\omega)|^2 S_x(\omega)e^{j\omega\tau}\, d\omega$$

EXAMPLE 2.7-1 Assume the input to the low-pass† filter shown in Fig. 2-11 is "white noise;" that is, $S_x(\omega) = N_o/2(-\infty < \omega < \infty)$. The corresponding autocorrelation function is $(N_o/2)\delta(\tau)$. Determine the output spectral density and autocorrelation function.

† The filter attenuates high frequency components much more than the low frequency components of the input.

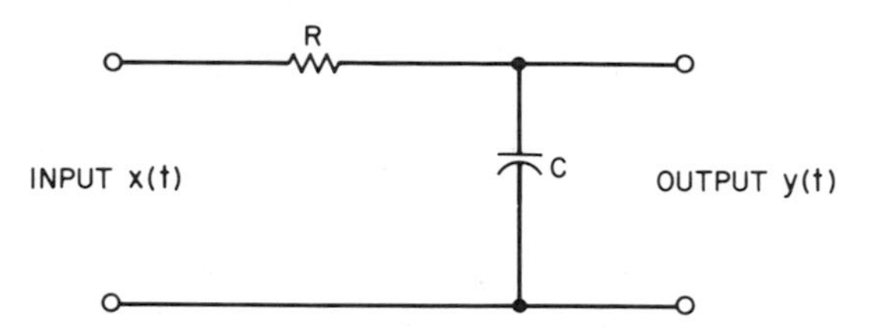

Fig. 2-11 A simple lowpass filter.

The transfer function of the filter is†

$$H(j\omega) = \frac{1/j\omega c}{R + 1/j\omega c} = \frac{1}{1 + j\omega RC}$$

and its magnitude squared is

$$|H(j\omega)|^2 = \frac{1}{1 + \omega^2(RC)^2}$$

The output spectral density is, from Eq. (2-42),

$$\frac{N_o}{2} \frac{1}{1 + \omega^2(RC)^2}$$

and the autocorrelation function is

$$R_y(\tau) = \frac{N_o}{4\pi} \int_{-\infty}^{\infty} \frac{e^{j\omega\tau}}{1 + \omega^2(RC)^2} \, d\omega = \frac{N_o}{2\pi} \int_0^{\infty} \frac{\cos \omega\tau}{1 + \omega^2(RC)^2} \, d\omega$$

$$= \frac{N_o}{4RC} e^{-|\tau/RC|}$$

EXAMPLE 2.7-2 Determine the output spectral density of a running integrator. Denoting the input by $x(t)$, the output $y(t)$ of such an integrator is given by

$$y(t) = \int_{t-T}^{t} x(u) \, du$$

where T, the integration time, is fixed.

We start by determining an expression for the output autocorrelation function, $R_y(\tau)$. The output is stationary since the running integrator is a linear system. The output at time $t - \tau$ is

$$y(t - \tau) = \int_{t-\tau-T}^{t-\tau} x(v) \, dv$$

† Consult any text on circuit or network theory for methods to determine transfer functions.

so that $R_y(\tau)$ is

$$R_y(\tau) = \int_{t-T}^{t} du \int_{t-\tau-T}^{t-\tau} dv\, R_x(u - v)$$

Substituting

$$R_x(u - v) = \frac{1}{2\pi} \int_{-\infty}^{\infty} S_x(\omega) e^{j\omega(u-v)}\, d\omega$$

and changing the order of integration we get

$$R_y(\tau) = \int_{-\infty}^{\infty} \frac{d\omega}{2\pi} S_x(\omega) \int_{t-T}^{t} du\, e^{j\omega u} \int_{t-\tau-T}^{t-\tau} dv\, e^{-j\omega v}$$

The integration with respect to u produces

$$\frac{e^{j\omega t}(1 - e^{-j\omega T})}{j\omega}$$

and the integration with respect to v produces

$$\frac{e^{-j\omega(t-\tau)}(1 - e^{j\omega T})}{-j\omega}$$

Their product is

$$e^{j\omega\tau} \frac{\sin^2 \omega T/2}{(\omega/2)^2}$$

Finally

$$R_y(\tau) = \frac{1}{2\pi} \int_{-\infty}^{\infty} S_x(\omega) \frac{\sin^2 \omega T/2}{(\omega/2)^2} e^{j\omega\tau}\, d\omega$$

By noting the integrand we see that the spectral density of $y(t)$ is

$$S_y(\omega) = S_x(\omega) \frac{\sin^2 \omega T/2}{(\omega/2)^2} \qquad \square\,\square$$

Another useful relationship which may be derived is the crosscorrelation between the input and output of a linear system. Denote the wide-sense stationary input by $x(t)$ and the output by $y(t)$. Then

$$y(t) = \int_0^{\infty} h(\eta) x(t - \eta)\, d\eta$$

and

$$y(t)x(t - \tau) = \int_0^{\infty} h(\eta) x(t - \eta) x(t - \tau)\, d\eta$$

Then

$$R_{yx}(\tau) = \int_0^\infty h(\eta)R_x(\tau - \eta)\, d\eta \tag{2-43}$$

This is the convolution of $h(\tau)$ and $R_x(\tau)$. Consequently,

$$S_{yx}(\omega) = H(j\omega)S_x(\omega) \tag{2-44}$$

EXAMPLE 2.7-3 For the single-input, multiple-output linear filter, Fig. 2-12, determine the cross-spectral density of $y_1(t)$ and $y_2(t)$.

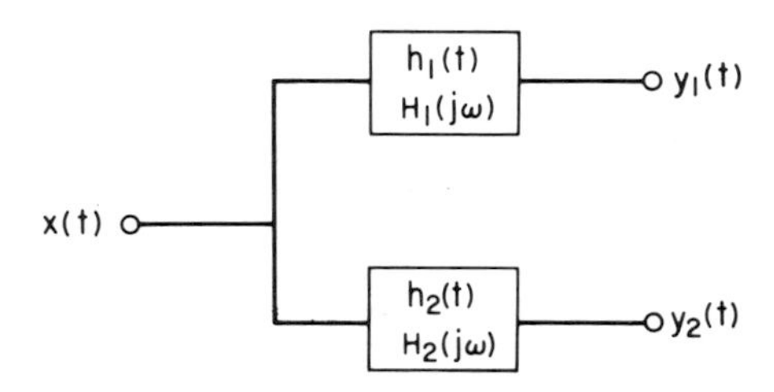

Fig. 2-12 A single-input, multiple-output linear system.

We write the expressions for $y_1(t)$ and $y_2(t - \tau)$.

$$y_1(t) = \int_0^\infty h_1(u)x(t - u)\, du$$

and

$$y_2(t - \tau) = \int_0^\infty h_2(v)x(t - \tau - v)\, dv$$

From these it is easily shown that

$$R_{y_1y_2}(\tau) = \int_0^\infty \int_0^\infty h_1(u)h_2(v)R_x(\tau + v - u)\, du\, dv$$

Replacing

$$R_x(\tau + v - u) = \frac{1}{2\pi}\int_{-\infty}^{\infty} S_x(\omega)e^{j\omega(\tau + v - u)}\, d\omega$$

and regrouping yields

$$R_{y_1y_2}(\tau) = \int_{-\infty}^{\infty} \frac{d\omega}{2\pi} S_x(\omega)e^{j\omega\tau} \int_0^\infty du\, h_1(u)e^{-j\omega u} \int_0^\infty dv\, h_2(v)e^{j\omega v}$$

The integration with respect to u and v produces $H_1(j\omega)$ and $H_2{}^*(j\omega)$ respectively. Then

$$R_{y_1y_2}(\tau) = \frac{1}{2\pi}\int_{-\infty}^{\infty} H_1(j\omega)H_2{}^*(j\omega)S_x(\omega)e^{j\omega\tau}\, d\omega$$

By inspection, we see that

$$S_{y_1y_2}(\omega) = H_1(j\omega)H_2^*(j\omega)S_x(\omega)$$

From this expression it is clear that if $H_1(j\omega)$ and $H_2(j\omega)$ do not overlap, the processes $y_1(t)$ and $y_2(t)$ are uncorrelated.

Exercises

2.1 Assume the random processes $x(t)$ and $y(t)$ are individually and jointly stationary.

(a) Find the autocorrelation function of $z(t) = x(t) + y(t)$.
(b) Repeat for the case when $x(t)$ and $y(t)$ are uncorrelated.
(c) Repeat for uncorrelated signals with zero means.

2.2 Assume the stationary random process $x(t)$ is periodic with period T, that is, $x(t) = x(t - T)$. Show that $R_x(\tau) = R_x(\tau + T)$.

2.3 For the complex function $z(t) = e^{j(\omega t + \theta)}$, assume θ is uniformly distributed $(0, 2\pi)$. Find

$$E\{z(t)z^*(t - \tau)\} \qquad \text{and} \qquad E\{z(t)z(t - \tau)\}$$

2.4 A first-order Markov process has the property that

$$p(x_n \mid x_{n-1}, x_{n-2}, \ldots, x_1) = p(x_n \mid x_{n-1})$$

For such a process show that

$$p(x_n, x_{n-1}, \ldots, x_1) = p(x_n \mid x_{n-1})p(x_{n-1} \mid x_{n-2}) \cdots p(x_2 \mid x_1)p(x_1)$$

2.5 Assume that $x(t)$ is a stationary input to a time invariant linear filter with transfer function $H(j\omega)$. Denote its output by $y(t)$. Show that

$$S_y(\omega) = H^*(j\omega)S_{yx}(\omega)$$

2.6 White noise with autocorrelation function $(N_0/2)\delta(\tau)$ is inserted into a filter with $|H(j\omega)|^2$ as shown in Fig. 2-13. What is the total noise power measured at the filter output?

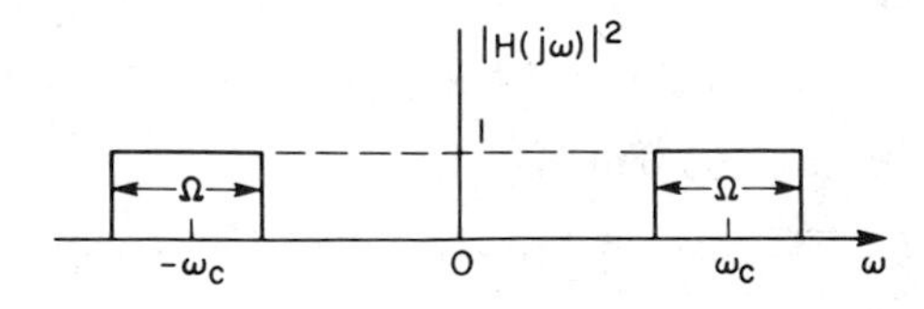

Fig. 2-13 Absolute value (squared) of transfer function.

2.7 For white noise with $R(\tau) = (N_0/2)\delta(\tau)$ as an input to the low-pass RC filter (Example 2.7-1), show that the autocorrelation function of the output can be expressed as

$$R(t_3 - t_1) = \frac{R(t_3 - t_2)R(t_2 - t_1)}{R(0)}$$

for $t_3 > t_2 > t_1$.

2.8 Consider a zero mean stationary process $x(t)$ inserted into a linear filter

$$h(t) = ae^{-at}, \qquad t \geq 0$$

(a) Show that the power spectral density of the filter output is

$$\frac{a^2}{a^2 + \omega^2} S_x(\omega)$$

(b) If the filter is a truncated exponential

$$h(t) = \begin{cases} ae^{-at}, & 0 \leq t \leq T \\ 0, & \text{otherwise} \end{cases}$$

show that the power spectral density of the output is

$$\frac{a^2}{a^2 + \omega^2}(1 - 2e^{-aT}\cos \omega T + e^{-2aT})S_x(\omega)$$

2.9 For the system of Fig. 2-14, assume $x(t)$ is stationary. Show that the

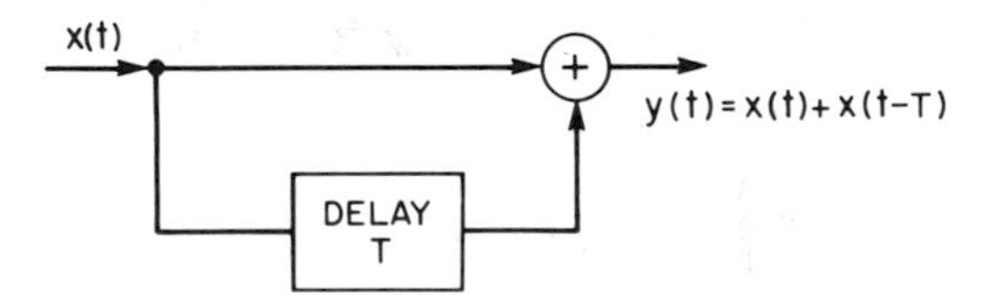

Fig. 2-14 A linear system.

power spectral density of $y(t)$ is

$$S_y(\omega) = 2S_x(\omega)(1 + \cos \omega T)$$

2.10 Let $x(t)$ be a zero mean stationary input to a filter $h(t)$, and denote its output by $y(t)$. The output correlation function is denoted by $R_y(\tau)$. Let x_1, x_2, x_3, and x_4 be samples of the process $x(t)$ and assume that

$$E\{x_1x_2x_3x_4\} = E\{x_1x_2\}E\{x_3x_4\} + E\{x_1x_3\}E\{x_2x_4\} + E\{x_1x_4\}E\{x_2x_3\}$$

Show for the output process that

$$E\{y(t_1)y(t_2)y(t_3)y(t_4)\} = R_y(\Delta_{12})R_y(\Delta_{34}) + R_y(\Delta_{13})R_y(\Delta_{24}) + R_y(\Delta_{14})R_y(\Delta_{23})$$

where $\Delta_{ij} = t_i - t_j$.

2.11 Assume a signal

$$f(t) = a\cos(t + \phi)$$

where a may or may not be a random variable. Assume ϕ is uniformly distributed in the range $(0, 2\pi)$. Find both the time autocorrelation function and ensemble autocorrelation function. Under what conditions on the amplitude a are the autocorrelation functions equal?

2.12 Consider the random process

$$x(t) = a\cos(\omega t + \theta)$$

where a is a constant, θ is uniformly distributed $(0, 2\pi)$, and ω is a random variable with a probability density which is an even function of its argument. That is $p(\omega) = p(-\omega)$. Show that the power spectral density of $x(t)$ is $a^2\pi p(\omega)$.

2.13 An amplitude modulated signal is given by

$$y(t) = a(t)\cos(\omega_c t + \theta)$$

where ω_c is a constant, $a(t)$ is a random process, and θ is uniformly distributed $(0, 2\pi)$. Determine the autocorrelation function and power spectral density of $y(t)$.

2.14 For a signal

$$z(t) = x(t)\cos\omega t - y(t)\sin\omega t$$

assume $x(t)$ and $y(t)$ are stationary.

(a) Find the ensemble correlation function of $z(t)$.
(b) Repeat the above but with $R_x(\tau) = R_y(\tau)$, and $R_{xy}(\tau) = 0$, thus showing that $R_z(\tau) = R_x(\tau)\cos\omega\tau$.

2.15 For the stationary random process $x(t)$, show that

$$\lim_{\varepsilon\to 0} E\{[x(t+\varepsilon) - x(t)]^2\} = 0$$

if $R_x(t, s)$ is continuous at $t = s$.

2.16 Assume that the derivative of a random process $x(t)$ exists; that is

$$\frac{dx(t)}{dt} \triangleq \lim_{\varepsilon\to 0} \frac{x(t+\varepsilon) - x(t)}{\varepsilon}$$

exists. In the following problems, assume that the indicated operations exist.

(a) Find $E\{dx(t)/dt\}$.
(b) Show that the autocorrelation function of the derivative is $-d^2R_x(\tau)/d\tau^2$.
(c) Show $E\{x(t)\, dx(t)/dt\} = dR_x(\tau)/d\tau\,|_{\tau=0}$.
(d) Show $E\{x(t)\, dx(t-\tau)/dt\} = -dR_x(\tau)/d\tau$.

2.17 Consider the process $dx(t)/dt$, if it exists, to be generated by passing the process $x(t)$ through a filter with transfer function $H(j\omega) = j\omega$. What are the power cross spectral density of $x(t)$ and $dx(t)/dt$, and the power spectral density of $dx(t)/dt$?

References

1. Davenport, W. B. and Root, W. L., "An Introduction to the Theory of Random Signals and Noise." McGraw-Hill, New York, 1958.
2. Papoulis, A., "Probability, Random Variables, and Stochastic Processes." McGraw-Hill, New York, 1965.
3. Rice, S. O., Mathematical Analysis of Random Noise, reprinted in "Selected Papers on Noise and Stochastic Processes" (N. Wax, ed.). Dover, New York, 1954.
4. Brown, W. M., Time statistics of noise, *IRE Trans. Inform. Theory*, December 1958.

SUPPLEMENTARY BIBLIOGRAPHY

Bartlett, M. S., "An Introduction to Stochastic Processes." 2nd Ed. Cambridge Univ. Press, London and New York, 1966.

Bendat, J. S., "Principles and Applications of Random Noise Theory." Wiley, New York, 1958.

Deutsch, R., "Nonlinear Transformations of Random Processes." Prentice-Hall, Englewood Cliffs, New Jersey, 1962.

Doob, J. L. "Stochastic Processes." Wiley, New York, 1953.

Feller, W., "An Introduction to Probability Theory and Its Applications," Vol. II. Wiley, New York, 1966.

Golomb, S. W., ed., "Digital Communications." Prentice Hall, Englewood Cliffs, New Jersey, 1964.

Grenander, U. and Rosenblatt, M., "Statistical Analysis of Stationary Time Series." Wiley, New York, 1957.

Laning, J. H. Jr., and Battin, R. H., "Random Processes in Automatic Control." McGraw-Hill, New York, 1956.

Lee, Y. W., "Statistical Theory of Communication." Wiley, New York, 1960.

Loève, M., "Probability Theory." 3rd Ed. Van Nostrand, Princeton, New Jersey, 1963.

Middleton, D., "An Introduction to Statistical Communication Theory." McGraw-Hill, New York, 1960.

Schwartz, M., Bennett, W. R., and Stein, S., "Communications Systems and Techniques." McGraw-Hill, New York, 1966.

Yaglom, A. M., "An Introduction to the Theory of Stationary Random Functions" (Transl. by R. A. Silverman). Prentice Hall, Englewood Cliffs, New Jersey, 1962.

Chapter 3

Narrowband Signals

3.1 Introduction

In practice, many of the signals, systems, and filters of interest satisfy the narrowband assumption. This means that the frequency content for signals, or frequency response for filters is confined to a relatively narrow region around a "carrier" or center frequency. This center frequency, in turn, is relatively distant from zero frequency or dc. For example, it is not uncommon for a radar to operate at 10^9 cycles per second (hertz, abbreviated Hz) and have a frequency span (bandwidth) of interest equal to 10^5 Hz. Voice communications, which require approximately 3000 Hz bandwidth, are commonly modulated onto a carrier in the megahertz range for transmission over wire or microwave facilities. An example of a signal which approximately satisfies the narrowband properties is shown in Fig. 3-1. For such a signal, the period of the carrier frequency is much less than the duration of the signal. This particular signal, or any signal of finite duration, can only satisfy approximately the narrowband assumption because a signal limited in time cannot be simultaneously limited in frequency. However, in most cases the energy falling outside a narrow band is negligible, and a narrowband approximation is very reasonable.

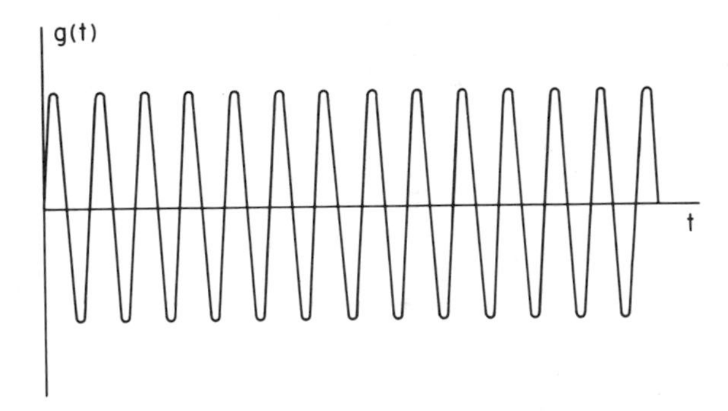

Fig. 3-1 A narrowband signal $g(t) = A \sin 2\pi f_0 t, 0 \leq t \leq T, T \gg 1/f_0$.

3.2 Deterministic Signal

We shall confine our attention here to signals which are completely known, and Fourier transformable. Let $f(t)$ represent a signal and assume it is quadratically integrable. That is

$$\int_{-\infty}^{\infty} |f(t)|^2 \, dt < \infty$$

Then its Fourier transform exists and is given by†

$$F(j\omega) = \int_{-\infty}^{\infty} f(t) e^{-j\omega t} \, dt$$

The inversion formula is

$$f(t) = \frac{1}{2\pi} \int_{-\infty}^{\infty} F(j\omega) e^{j\omega t} \, d\omega$$

When $f(t)$ is real,

$$F^*(j\omega) = \int_{-\infty}^{\infty} f(t) e^{j\omega t} \, d\omega$$

so that

$$F^*(j\omega) = F(-j\omega) \tag{3-1}$$

† The Fourier transform of a signal is often called its spectrum. However, the power spectral density introduced in Chap. 2 is also referred to as a spectrum. To distinguish these, the Fourier transform is sometimes called the amplitude spectrum.

It will be instructive here, and useful for later purposes, to compute the Fourier transform of a signal such as that shown in Fig. 3-1. The signal is $g(t) = A \cos \omega_o t$, where $0 \le t \le T$ and $2\pi/\omega_o \ll T$. The Fourier transform is

$$G(j\omega) = \int_0^T A \cos \omega_o t \, e^{-j\omega t} \, dt = \frac{-A}{2j} \left[\frac{e^{-j(\omega+\omega_o)t}}{(\omega + \omega_o)} + \frac{e^{j(\omega-\omega_o)t}}{(\omega - \omega_o)} \right]_0^T$$

Neglecting the mathematical details, this may be expressed as

$$G(j\omega) = \frac{AT}{2} e^{-j(\omega+\omega_o)T/2} \frac{\sin(\omega + \omega_o)T/2}{(\omega + \omega_o)T/2} + \frac{AT}{2} e^{-j(\omega-\omega_o)T/2} \frac{\sin(\omega - \omega_o)T/2}{(\omega - \omega_o)T/2} \tag{3-2}$$

Note the symmetry about the frequencies $\pm\omega_o$. The absolute value squared is

$$\left(\frac{2}{AT}\right)^2 |G(j\omega)|^2 = \frac{\sin^2(\omega + \omega_o)T/2}{[(\omega + \omega_o)T/2]^2} + \frac{\sin^2(\omega - \omega_o)T/2}{[(\omega - \omega_o)T/2]^2} + \frac{2 \sin(\omega + \omega_o)T/2}{(\omega + \omega_o)T/2} \cdot \frac{\sin(\omega - \omega_o)T/2}{(\omega - \omega_o)T/2} \cos \omega_o T \tag{3-3}$$

If the carrier frequency ω_o is sufficiently large, then the last term is small for all values of ω and may be neglected. With the same assumption, it also follows that the first term dominates when ω is near $-\omega_o$ and the second term dominates when ω is near ω_o. Then

$$|G(j\omega)|^2 \approx |G_+(j\omega)|^2 + |G_-(j\omega)|^2$$

where

$$|G_+(j\omega)|^2 \triangleq \left(\frac{AT}{2}\right)^2 \frac{\sin^2(\omega - \omega_o)T/2}{[(\omega - \omega_o)T/2]^2} \tag{3-4}$$

is the term which dominates for positive frequencies and

$$|G_-(j\omega)|^2 \triangleq \left(\frac{AT}{2}\right)^2 \frac{\sin^2(\omega + \omega_o)T/2}{[(\omega + \omega_o)T/2]^2} \tag{3-5}$$

dominates for negative frequencies. These are shown in Fig. 3-2.

For the above illustration, the real signal contained both positive and negative frequency components. We now define a new signal, which will be complex, that contains only positive frequencies (*1*, *2*). That is,

$$f_+(t) \triangleq \frac{1}{2\pi} \int_0^\infty F(j\omega) e^{j\omega t} \, d\omega \tag{3-6}$$

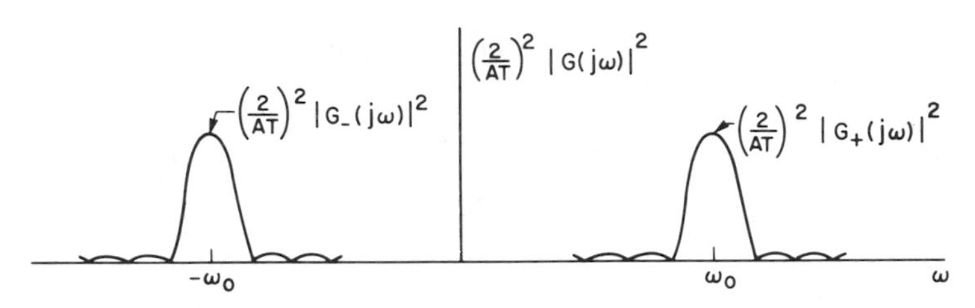

Fig. 3-2 Magnitude squared of transform of signal $g(t) = A \cos \omega_0 t, 0 \leq t \leq T$, $2\pi/\omega_0 \ll T$.

Taking the complex conjugate

$$f_+{}^*(t) = \frac{1}{2\pi} \int_0^\infty F^*(j\omega) e^{-j\omega t}\, d\omega \tag{3-7}$$

Using Eq. (3-1),

$$f_+{}^*(t) = \frac{1}{2\pi} \int_0^\infty F(-j\omega) e^{-j\omega t}\, d\omega = \frac{1}{2\pi} \int_{-\infty}^0 F(j\omega) e^{j\omega t}\, d\omega \tag{3-8}$$

The last form shows that $f_+{}^*(t)$ contains only negative frequencies and we denote it as $f_-(t)$. Then

$$f(t) = f_+(t) + f_-(t) = f_+(t) + f_+{}^*(t) \tag{3-9}$$

from which it follows that

$$f(t) = 2\, \mathrm{Re}\, f_+(t) \tag{3-10}$$

where Re denotes "real part of."

Now let $f(t)$ be a narrowband signal whose power is concentrated in the vicinity of $\omega = \Omega$, and $\omega = -\Omega$. Define a new function which has a Fourier transform given by

$$\tilde{F}(\omega) = \begin{cases} 2F(\omega + \Omega), & -\Omega < \omega \\ 0, & \omega < -\Omega \end{cases} \tag{3-11}$$

This translates $F(\omega)$ to the left by an amount Ω, doubles the transform, and cuts off all frequencies below $\omega = -\Omega$.† An illustration of this is shown in Fig. 3-3.

Consider the expression for $f_+(t)$, Eq. (3-6)

$$f_+(t) = \frac{1}{2\pi} \int_0^\infty F(v) e^{jvt}\, dv$$

† The choice of Ω is arbitrary. In most problems, however, there is a value of Ω which clearly simplifies the results. See Example 3.2-1.

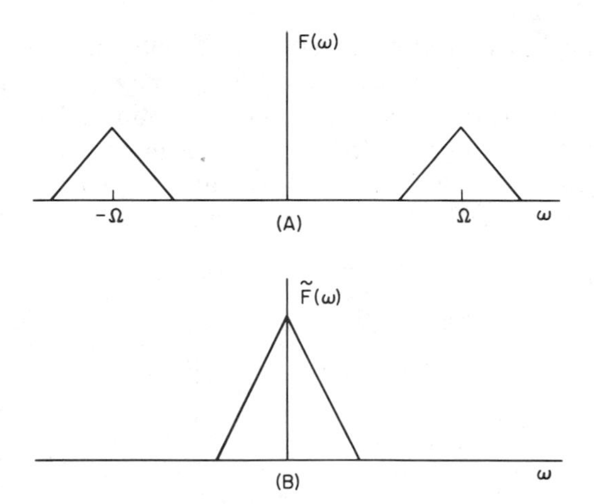

Fig. 3-3 A narrowband spectrum (A), and the low frequency translation (B).

Making a change of variable, $v = \omega + \Omega$, this becomes

$$f_+(t) = \frac{1}{2\pi}\int_{-\Omega}^{\infty} F(\omega + \Omega)e^{j(\omega+\Omega)t}\,d\omega \tag{3-12}$$

Substituting for $F(\omega + \Omega)$, Eq. (3-11),

$$f_+(t) = \frac{1}{2}\int_{-\Omega}^{\infty} \frac{\tilde{F}(\omega)}{2\pi}\,e^{j(\omega+\Omega)t}\,d\omega$$

But $\tilde{F}(\omega) = 0$ for $\omega < -\Omega$, so that the lower limit of integration may be taken to be $-\infty$. Then

$$f_+(t) = \frac{e^{j\Omega t}}{2}\int_{-\infty}^{\infty} \frac{\tilde{F}(\omega)}{2\pi}\,e^{j\omega t}\,d\omega = \tilde{f}(t)\frac{e^{j\Omega t}}{2} \tag{3-13}$$

or

$$\tilde{f}(t) = 2f_+(t)e^{-j\Omega t}$$

where we defined

$$\tilde{f}(t) = \int_{-\infty}^{\infty} \frac{\tilde{F}(\omega)}{2\pi}\,e^{j\omega t}\,d\omega \tag{3-14}$$

Since $f(t) = 2\,\mathrm{Re}\,f_+(t)$, the original signal is

$$f(t) = \mathrm{Re}\,\tilde{f}(t)e^{j\Omega t}$$

Since $\tilde{F}(\omega)$ is nonzero only at low frequencies as shown in Fig. 3-3, it follows that $\tilde{f}(t)$ is a slowly changing function of time. We call $\tilde{f}(t)$ the complex envelope of the narrowband process $f(t)$. The absolute value of the complex envelope is called the envelope or amplitude modulation of the narrowband process. (A complex representation may also be used for functions which are not narrowband; however, the notion of envelope for such cases has little physical significance.) The argument or phase of $\tilde{f}(t)$ is called the phase modulation, and the derivative of the phase is the frequency modulation. Breaking $\tilde{f}(t)$ into real and imaginary components

$$\tilde{f}(t) = x(t) + jy(t) \tag{3-15}$$

where $x(t)$ and $y(t)$ are real functions, we have for the envelope

$$\mathscr{E}(t) = [x^2(t) + y^2(t)]^{1/2} \tag{3-16}$$

The phase is given by

$$\phi(t) = \tan^{-1} y(t)/x(t) \tag{3-17}$$

The complex envelope can now be represented in polar form by

$$\tilde{f}(t) = \mathscr{E}(t)\exp[j\phi(t)] \tag{3-18}$$

Further discussion of narrowband signals and filters will be deferred until the Hilbert transform is introduced in Sect. 3.3.

EXAMPLE 3.2-1 For the signal in Fig. 3-1, $g(t) = A \cos \omega_o t$, $(0 \leq t \leq T$, $2\pi/\omega_o \ll T)$ determine $g_+(t)$ and $\tilde{g}(t)$ using suitable approximations.

From the definition of $g_+(t)$, Eq. (3-6),

$$g_+(t) = \frac{1}{2\pi}\int_0^\infty G(j\omega)e^{j\omega t}\,d\omega$$

where $G(j\omega)$ is given in Eq. (3-2). Using the assumption that $2\pi/\omega_o \ll T$, the first term of $G(j\omega)$ is negligible for positive frequencies. Then

$$g_+(t) = \frac{1}{2\pi}\int_0^\infty \frac{AT}{2}\, e^{-j(\omega-\omega_o)T/2}\,\frac{\sin(\omega-\omega_o)T/2}{(\omega-\omega_o)T/2}\, e^{j\omega t}\,d\omega$$

Making a change of variable $(\nu = \omega - \omega_o)$ this becomes

$$g_+(t) = \left(\frac{AT}{2}\right)\frac{e^{j\omega_o t}}{2\pi}\int_{-\omega_o}^{\infty} \frac{\sin \nu T/2}{\nu T/2}\, e^{j\nu(t-T/2)}\,d\nu$$

For the assumption that $2\pi/\omega_o \ll T$, the $(\sin \nu T/2)/(\nu T/2)$ term in the integrand is negligible for $\nu < -\omega_o$. Therefore

$$\begin{aligned} g_+(t) &= \left(\frac{AT}{2}\right)\frac{e^{j\omega_o t}}{2\pi}\int_{-\infty}^{\infty} \frac{\sin \nu T/2}{\nu T/2}\, e^{j\nu(t-T/2)}\,d\nu \\ &= \left(\frac{AT}{2}\right)\frac{e^{j\omega_o t}}{2\pi}\, q\left(t-\frac{T}{2}\right) \end{aligned}$$

where we defined

$$q(t) = \int_{-\infty}^{\infty} \frac{\sin \nu T/2}{\nu T/2} e^{j\nu t} \, d\nu = 2 \int_{0}^{\infty} \frac{\sin(\nu T/2) \cos \nu t}{\nu T/2} \, d\nu$$

From tables of integrals (Peirce (*3*), for example)

$$q(t) = \begin{cases} 0, & t < -T/2, \quad t > T/2 \\ \pi/T, & t = -T/2, T/2 \\ 2\pi/T, & -T/2 < t < T/2 \end{cases}$$

It follows that

$$g_+(t) = \begin{cases} 0, & t < 0, \quad t > T \\ (A/4)e^{j\omega_o t}, & t = 0, T \\ (A/2)e^{j\omega_o t}, & 0 < t < T \end{cases}$$

To find $\tilde{g}(t)$ we translate $G_+(j\omega)$ [see Fig. 3-2 and Eq. (3-4)] to the origin. We need not shift $G_+(j\omega)$ by ω_o since $\tilde{g}(t)$ is arbitrary to within a phase factor. It should be clear from Eq. (3-4) that the computation for $\tilde{g}(t)$ will be simplified if the shift is ω_o. Now

$$\tilde{G}(\omega) = 2G_+(\omega + \omega_o)$$

Using the positive frequency term of Eq. (3-2)

$$\tilde{g}(t) = \frac{AT}{2\pi} \int_{-\infty}^{\infty} \frac{\sin \omega T/2}{\omega T/2} e^{j\omega(t - T/2)} \, d\omega$$

From previous results in this example, this is seen to be

$$\tilde{g}(t) = 2g_+(t)e^{-j\omega_o t} = \begin{cases} 0, & t < 0, \quad t > T \\ A/2, & t = 0, T \\ A, & 0 < t < T \end{cases}$$

3.3 Hilbert Transform

In dealing with narrowband signals and complex representations it is often convenient to employ the Hilbert transform (*4*). Given a real-valued function $x(t)$ in the interval $-\infty < t < \infty$, its Hilbert transform, denoted by $\hat{x}(t)$ or $H \cdot x(t)$, is defined by

$$H \cdot x(t) = \hat{x}(t) = \frac{1}{\pi} P \int_{-\infty}^{\infty} \frac{x(\tau)}{t - \tau} \, d\tau \tag{3-19}$$

where the symbol P denotes the principal value† of the integral. For notational simplicity the symbol P will be omitted. The inverse transform is given by

$$x(t) = H^{-1} \cdot \hat{x}(t) = -\frac{1}{\pi} \int_{-\infty}^{\infty} \frac{\hat{x}(\tau)}{t - \tau} \, d\tau \tag{3-20}$$

By a change of variable we get the following equivalent forms for the Hilbert transform and its inverse:

$$\hat{x}(t) = -\frac{1}{\pi} \int_{-\infty}^{\infty} \frac{x(t + \tau)}{\tau} \, d\tau = \frac{1}{\pi} \int_{-\infty}^{\infty} \frac{x(t - \tau)}{\tau} \, d\tau \tag{3-21}$$

and

$$x(t) = \frac{1}{\pi} \int_{-\infty}^{\infty} \frac{\hat{x}(t + \tau)}{\tau} \, d\tau = -\frac{1}{\pi} \int_{-\infty}^{\infty} \frac{\hat{x}(t - \tau)}{\tau} \, d\tau \tag{3-22}$$

***Properties of the Hilbert Transform* (4)**

(1) From the first definition of the Hilbert transform, Eq. (3-19), it may be seen that the Hilbert transform is the convolution of $x(t)$ with $1/\pi t$, that is,

$$\hat{x}(t) = x(t) * \frac{1}{\pi t} = \frac{1}{\pi} \int_{-\infty}^{\infty} \frac{x(\tau)}{t - \tau} \, d\tau \tag{3-23}$$

(1A) If $X(f)$ represents the Fourier transform of $x(t)$, that is, $\mathscr{F} \cdot x(t) = X(f)$, it follows that the Fourier transform of the Hilbert transform is given by

$$\mathscr{F} \cdot \hat{x}(t) = -jX(f) \operatorname{sgn}(f) \tag{3-24}$$

where $\operatorname{sgn}(f)$, the signum function, is defined by

$$\operatorname{sgn}(f) = \begin{cases} 1 & \text{for } f > 0 \\ 0 & \text{for } f = 0 \\ -1 & \text{for } f < 0 \end{cases}$$

Proof: Since $x(t)$ is convolved with $1/\pi t$, it follows that the Fourier transform of the Hilbert transform is the product of the Fourier transforms of $x(t)$ and $1/\pi t$. The Fourier transform for the latter is $-j \operatorname{sgn}(f)$.

† By the principal value of an integral $P\int_{-\infty}^{\infty} g(t)\, dt$ we mean

$$\lim_{\varepsilon \to 0} \left[\int_{-\infty}^{-\varepsilon} g(t)\, dt + \int_{\varepsilon}^{\infty} g(t)\, dt \right]$$

as opposed to the definition

$$\lim_{R \to 0} \int_{-\infty}^{R} g(t)\, dt + \lim_{S \to 0} \int_{S}^{\infty} g(t)\, dt$$

where R and S approach zero independently of each other. See, for example, Wylie (5).

(2) The Hilbert transform of $x(t) = \cos(\omega t + \phi)$ is $\sin(\omega t + \phi)$, and the Hilbert transform of $\sin(\omega t + \phi)$ is $-\cos(\omega t + \phi)$.

Proof:

$$H \cdot x(t) = \hat{x}(t) = -\frac{1}{\pi}\int_{-\infty}^{\infty} \frac{\cos[\omega(t+\tau)+\phi]}{\tau}\, d\tau$$

or

$$\hat{x}(t) = -\frac{1}{\pi}\int_{-\infty}^{\infty} \frac{\cos(\omega t+\phi)\cos\omega\tau}{\tau}\, d\tau + \frac{1}{\pi}\int_{-\infty}^{\infty} \frac{\sin(\omega t+\phi)\sin\omega\tau}{\tau}\, d\tau$$

The principal value of the first integral is zero since the integrand is odd. Therefore

$$\hat{x}(t) = \frac{\sin(\omega t+\phi)}{\pi}\int_{-\infty}^{\infty} \frac{\sin\omega\tau}{\tau}\, d\tau$$

This integral is equal to π for $\omega > 0$, so that

$$H \cdot \cos(\omega t+\phi) = \sin(\omega t+\phi) \tag{3-25}$$

We shall leave the proof of the second half of this property until property (5). From this we see that the Hilbert transform behaves as a 90° phase shifter.

(3) Both $x(t)$ and $\hat{x}(t)$ have the same power in the interval $(-\infty < t < \infty)$; that is,

$$\lim_{T\to\infty} \frac{1}{2T}\int_{-T}^{T} x^2(t)\, dt = \lim_{T\to\infty} \frac{1}{2T}\int_{-T}^{T} \hat{x}^2(t)\, dt$$

Proof: This may easily be shown with the aid of property (8). The proof is left as an exercise.

(4) The functions $x(t)$ and $\hat{x}(t)$ are orthogonal in the interval $-\infty < t < \infty$, that is,

$$\lim_{T\to\infty} \frac{1}{2T}\int_{-T}^{T} x(t)\hat{x}(t)\, dt = 0 \tag{3-26}$$

Proof: This may easily be shown with the aid of property (9). The proof is left as an exercise.

(5) The Hilbert transform of the Hilbert transform of a function is the negative of the original function. That is,

$$H \cdot \hat{x}(t) = -x(t) \tag{3-27}$$

Proof: By the definition given in Eq. (3-19) we have

$$H \cdot \hat{x}(t) = \frac{1}{\pi} \int_{-\infty}^{\infty} \frac{\hat{x}(\tau)}{t - \tau} \, d\tau = -x(t)$$

This last equality follows from the definition of the inverse transform given in Eq. (3-20). It follows from this property and property (2) that $H \cdot \sin(\omega t + \phi) = -\cos(\omega t + \phi)$.

(6) If $y(t) = v(t)*x(t)$, then the Hilbert transform of $y(t)$ is given by

$$\hat{y}(t) = v(t)*\hat{x}(t) \tag{3 28}$$

Proof:

$$\hat{y}(t) = \frac{1}{\pi} \int_{-\infty}^{\infty} \frac{y(\tau)}{t - \tau} \, d\tau = \frac{1}{\pi} \int_{-\infty}^{\infty} \int_{-\infty}^{\infty} \frac{v(\eta)x(\tau - \eta)}{t - \tau} \, d\tau \, d\eta$$

Rearranging terms,

$$\hat{y}(t) = \int_{-\infty}^{\infty} d\eta \; v(\eta) \int_{-\infty}^{\infty} \frac{d\tau}{\pi} \frac{x(\tau - \eta)}{t - \tau}$$

By a change of variable ($\xi = \tau - \eta$) in the inner integral, we get

$$\hat{y}(t) = \int_{-\infty}^{\infty} d\eta \; v(\eta) \int_{-\infty}^{\infty} \frac{d\xi}{\pi} \frac{x(\xi)}{t - \eta - \xi}$$

The inner integral is by definition, $\hat{x}(t - \eta)$, so

$$\hat{y}(t) = \int_{-\infty}^{\infty} d\eta \; v(\eta)\hat{x}(t - \eta)$$

which is seen to be $v(t)*\hat{x}(t)$. Consequently,

$$\hat{y}(t) = v(t)*\hat{x}(t)$$

It may similarly be shown that $\hat{y}(t)$ can also be written

$$\hat{y}(t) = x(t)*\hat{v}(t)$$

(7) If a signal $a(t)$ has a Fourier transform $A(\omega)$ which is band limited, that is,

$$A(\omega) = \begin{cases} A(\omega), & |\omega| < W \\ 0, & \text{otherwise} \end{cases}$$

then

$$H \cdot [a(t)\cos \omega_c t] = a(t)\sin \omega_c t \tag{3-29}$$

and

$$H \cdot [a(t)\sin \omega_c t] = -a(t)\cos \omega_c t \tag{3-30}$$

provided $\omega_c > W$.

Proof: We shall first determine the Fourier transform of $a(t)\cos\omega_c t$. The signal may be written

$$x(t) = a(t)\cos\omega_c t = \frac{a(t)e^{j\omega_c t}}{2} + \frac{a(t)e^{-j\omega_c t}}{2}$$

Then the Fourier transform is

$$X(\omega) = \int_{-\infty}^{\infty} \frac{a(t)}{2} e^{-j(\omega-\omega_c)t}\, dt + \int_{-\infty}^{\infty} \frac{a(t)}{2} e^{-j(\omega+\omega_c)t}\, dt$$

so that

$$X(\omega) = \frac{A(\omega - \omega_c)}{2} + \frac{A(\omega + \omega_c)}{2}$$

This is illustrated in Fig. 3-4.

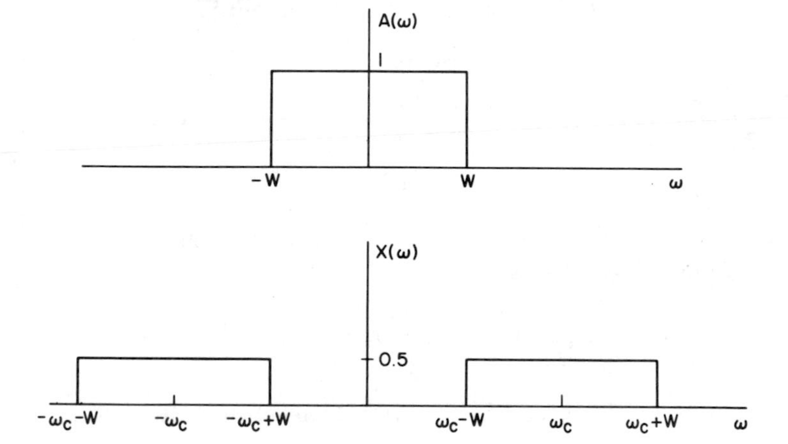

Fig. 3-4 The Fourier transform of $x(t) = a(t)\cos\omega_c t$, $W \leq \omega_c$.

By property (1) the Fourier transform of $\hat{x}(t)$ is given by

$$\mathscr{F}\cdot\hat{x}(t) = -jX(\omega)\,\text{sgn}(\omega)$$

It can be seen from Fig. 3-4 that

$$X(\omega) = \begin{cases} \dfrac{A(\omega - \omega_c)}{2}, & \omega > 0 \\[2ex] \dfrac{A(\omega + \omega_c)}{2}, & \omega < 0 \end{cases}$$

so that

$$\mathscr{F} \cdot \hat{x}(t) = -jX(\omega)\,\mathrm{sgn}(\omega) = \begin{cases} -\dfrac{j}{2} A(\omega - \omega_c), & \omega > 0 \\ +\dfrac{j}{2} A(\omega + \omega_c), & \omega < 0 \end{cases}$$

We now determine the inverse Fourier transform of the above:

$$\hat{x}(t) = \frac{1}{2\pi}\int_{-\infty}^{\infty} -jX(\omega)\,\mathrm{sgn}(\omega) e^{j\omega t} d\omega$$

$$= \frac{1}{2\pi}\int_{0}^{\infty} -\frac{j}{2} A(\omega - \omega_c) e^{j\omega t}\, d\omega + \frac{1}{2\pi}\int_{-\infty}^{0} \frac{j}{2} A(\omega + \omega_c) e^{j\omega t}\, d\omega$$

By noting Fig. 3-4 we see that $A(\omega - \omega_c)$ is zero for $\omega_c + W < \omega < \omega_c - W$, and $A(\omega + \omega_c)$ is zero for $-\omega_c + W < \omega < -\omega_c - W$. Therefore,

$$\hat{x}(t) = \frac{1}{2\pi}\int_{\omega_c - W}^{\omega_c + W} -\frac{j}{2} A(\omega - \omega_c) e^{j\omega t}\, d\omega$$

$$+ \frac{1}{2\pi}\int_{-\omega_c - W}^{-\omega_c + W} \frac{j}{2} A(\omega + \omega_c) e^{j\omega t}\, d\omega$$

By a change of variables ($v = \omega - \omega_c$) in the first integral and ($v = \omega + \omega_c$) in the second integral we get

$$\hat{x}(t) = -\frac{j}{2}\frac{e^{j\omega_c t}}{2\pi}\int_{-W}^{W} A(v) e^{jvt}\, dv + \frac{j}{2}\frac{e^{-j\omega_c t}}{2\pi}\int_{-W}^{W} A(v) e^{jvt}\, dv$$

The integrals are the inverse Fourier transform of $A(\omega)$. Therefore

$$\hat{x}(t) = \frac{j}{2} a(t)(e^{-j\omega_c t} - e^{j\omega_c t}) \qquad \text{or} \qquad \hat{x}(t) = a(t) \sin \omega_c t$$

which we set out to determine.

The second part, $H \cdot [a(t)\sin \omega_c t] = -a(t)\cos \omega_c t$, may easily be shown using property (5).

(8) Let $\mathscr{R}_x(\tau)$ be the time autocorrelation function of $x(t)$, defined as (see Sects. 2.5 and 2.6)

$$\mathscr{R}_x(\tau) = \lim_{T\to\infty} \frac{1}{2T}\int_{-T}^{T} x(t)x(t - \tau)\, dt$$

and let

$$\mathscr{S}_x(\omega) = \int_{-\infty}^{\infty} \mathscr{R}_x(\tau) e^{-j\omega\tau}\, d\tau$$

be its spectral density. If $\mathscr{R}_{\hat{x}}(\tau)$ is the autocorrelation function of $\hat{x}(t)$ and $\mathscr{S}_{\hat{x}}(\omega)$ its spectral density, then we have the following relationships

$$\mathscr{R}_{\hat{x}}(\tau) = \mathscr{R}_x(\tau) \tag{3-31}$$

and

$$\mathscr{S}_{\hat{x}}(\omega) = \mathscr{S}_x(\omega) \tag{3-32}$$

Proof: Using the second definition of the Hilbert transform

$$\mathscr{R}_{\hat{x}}(\tau) = \lim_{T\to\infty} \frac{1}{2T} \int_{-T}^{T} dt \int_{-\infty}^{\infty} \frac{d\eta}{\pi} \int_{-\infty}^{\infty} \frac{d\xi}{\pi} \frac{x(t+\eta)}{\eta} \frac{x(t-\tau+\xi)}{\xi}$$

Interchanging the order of integration and limiting

$$\mathscr{R}_{\hat{x}}(\tau) = \int_{-\infty}^{\infty} \frac{d\eta}{\pi\eta} \int_{-\infty}^{\infty} \frac{d\xi}{\pi\xi} \lim_{T\to\infty} \frac{1}{2T} \int_{-T}^{T} dt\, x(t+\eta)x(t-\tau+\xi)$$

The latter limit and integral is $\mathscr{R}_x(\eta + \tau - \xi)$. Therefore

$$\mathscr{R}_{\hat{x}}(\tau) = \int_{-\infty}^{\infty} \frac{d\xi}{\pi\xi} \int_{-\infty}^{\infty} \frac{d\eta}{\pi} \frac{\mathscr{R}_x(\eta+\tau-\xi)}{\eta}$$

The inner integral is $-\hat{\mathscr{R}}_x(\tau - \xi)$, so that

$$\mathscr{R}_{\hat{x}}(\tau) = \frac{-1}{\pi} \int_{-\infty}^{\infty} \frac{\hat{\mathscr{R}}_x(\tau-\xi)}{\xi}\, d\xi$$

Using Eq. (3-22) this is equal to $\mathscr{R}_x(\tau)$. Thus

$$\mathscr{R}_{\hat{x}}(\tau) = \mathscr{R}_x(\tau)$$

This being true, it follows directly that

$$\mathscr{S}_{\hat{x}}(\omega) = \mathscr{S}_x(\omega)$$

Therefore, a function and its Hilbert transform have the same time correlation function and spectral density.

It may also be shown that if $x(t)$ is a sample function of a wide-sense stationary process having an ensemble autocorrelation function $R_x(\tau) = E\{x(t)x(t-\tau)\}$, then

$$R_{\hat{x}}(\tau) = R_x(\tau) \tag{3-33}$$

The proof is quite similar to that above and is left as an exercise.

(9) The time crosscorrelation of the functions $\hat{x}(t)$ and $x(t)$, defined by

$$\mathscr{R}_{\hat{x}x}(\tau) = \lim_{T\to\infty} \frac{1}{2T} \int_{-T}^{T} \hat{x}(t)x(t-\tau)\,dt$$

is equal to the Hilbert transform of the time autocorrelation† $\mathscr{R}_x(\tau)$ of $x(t)$, that is,

$$\mathscr{R}_{\hat{x}x}(\tau) = \hat{\mathscr{R}}_x(\tau) \tag{3-34}$$

Proof: Writing $\hat{x}(t) = -(1/\pi)\int_{-\infty}^{\infty} x(t+\eta)/\eta\,d\eta$, we have

$$\mathscr{R}_{\hat{x}x}(\tau) = \lim_{T\to\infty} \frac{1}{2T} \int_{-T}^{T} dt \int_{-\infty}^{\infty} \frac{x(t+\eta)x(t-\tau)}{-\eta\pi}\,d\eta$$

Changing the order of integration,

$$\mathscr{R}_{\hat{x}x}(\tau) = \int_{-\infty}^{\infty} \frac{d\eta}{-\pi\eta} \lim_{T\to\infty} \frac{1}{2T} \int_{-T}^{T} dt\, x(t+\eta)x(t-\tau)$$

The integral on the right is $\mathscr{R}_x(\tau+\eta)$, so that

$$\mathscr{R}_{\hat{x}x}(\tau) = \frac{-1}{\pi} \int_{-\infty}^{\infty} \frac{\mathscr{R}_x(\eta+\tau)}{\eta}\,d\eta = H \cdot \mathscr{R}_x(\tau)$$

Therefore,

$$\mathscr{R}_{\hat{x}x}(\tau) = \hat{\mathscr{R}}_x(\tau)$$

It can be similarly shown that

$$\mathscr{R}_{x\hat{x}}(\tau) = -\hat{\mathscr{R}}_x(\tau) \tag{3-35}$$

(9A) A corollary to this property is that

$$\mathscr{R}_{\hat{x}x}(\tau) = -\mathscr{R}_{\hat{x}x}(-\tau) \tag{3-36}$$

and

$$\mathscr{R}_{x\hat{x}}(-\tau) = -\mathscr{R}_{x\hat{x}}(\tau) \tag{3-37}$$

That is, $\mathscr{R}_{\hat{x}x}(\tau)$ and $\mathscr{R}_{x\hat{x}}(\tau)$ are odd functions, and in particular $x(t)$ and $\hat{x}(t)$ are uncorrelated at the same time. That is

$$\mathscr{R}_{x\hat{x}}(0) = 0 \tag{3-38}$$

† Note, if the sign in the definition of crosscorrelation is reversed, that is if

$$\mathscr{R}^{+}_{\hat{x}x}(\tau) \triangleq \lim_{T\to\infty} \frac{1}{2T} \int_{-T}^{T} \hat{x}(t)x(t+\tau)\,dt$$

then $\mathscr{R}^{+}_{\hat{x}x}(\tau) = -\hat{\mathscr{R}}_x(\tau)$. We shall however conform to our original notation which uses the minus sign, that is, $x(t-\tau)$.

Proof: In Sect. 2.3 it was shown that for *real* functions $R_{x\hat{x}}(\tau) = R_{\hat{x}x}(-\tau)$. A similar relation can be shown to hold for time autocorrelation functions as well. Therefore using property (9)

$$\hat{\mathscr{R}}_x(\tau) = \mathscr{R}_{x\hat{x}}(-\tau) \qquad \text{and} \qquad \hat{\mathscr{R}}_x(\tau) = -\mathscr{R}_{x\hat{x}}(\tau)$$

Hence $\mathscr{R}_{x\hat{x}}(-\tau) = -\mathscr{R}_{x\hat{x}}(\tau)$ and $\mathscr{R}_{\hat{x}x}(\tau) = -\mathscr{R}_{\hat{x}x}(-\tau)$, which were to be shown. Since these are odd functions it follows that $\mathscr{R}_{x\hat{x}}(0) = \mathscr{R}_{\hat{x}x}(0) = 0$.

Similarly, it can be shown for a wide-sense stationary random process with sample function $x(t)$, and ensemble autocorrelation function $R_x(\tau)$, where $R_x(\tau) = E\{x(t)x(t-\tau)\}$, that

$$R_{x\hat{x}}(\tau) = -\hat{R}_x(\tau) \tag{3-39}$$

$$R_{\hat{x}x}(\tau) = \hat{R}_x(\tau) \tag{3-40}$$

$$R_{\hat{x}}(\tau) = R_x(\tau) \tag{3-41}$$

3.4 Signal Preenvelope

It was shown in Sect. 3.2 that a real signal $f(t)$ can be represented as

$$f(t) = \operatorname{Re} \tilde{f}(t)e^{j\Omega t}$$

For a narrowband signal, the complex envelope $\tilde{f}(t)$ is a slowly changing function of time. By definition, its Fourier transform is truncated at $\omega = -\Omega$. That is, $\tilde{F}(\omega)$ is identically zero for $\omega < -\Omega$. Multiplying $\tilde{f}(t)$ by $e^{j\Omega t}$ shifts the spectrum to the right by an amount Ω. The resulting complex signal has no negative frequency components.

We now define the *preenvelope* of a real signal $f(t)$. The preenvelope of a real signal $f(t)$, is the complex-valued function (6)

$$f_p(t) = f(t) + j\hat{f}(t) \tag{3-42}$$

The real signal is, of course, the real part of the preenvelope $f_p(t)$. The preenvelope is also called the *analytic signal*. The envelope of $f(t)$ is the absolute value $|f_p(t)|$ of its preenvelope.

From property (1) of the Hilbert transform, it is an easy matter to show that the Fourier transform of $f_p(t)$ is

$$\mathscr{F} \cdot f_p(t) = \begin{cases} 2F(j\omega), & \omega > 0 \\ F(j\omega), & \omega = 0 \\ 0, & \omega < 0 \end{cases}$$

From Sect. 3.2, we therefore see that $f_p(t) = 2f_+(t) = \tilde{f}(t)e^{j\Omega t}$.

EXAMPLE 3.4-1 Suppose we have a function

$$f(t) = \begin{cases} \cos \omega_c t, & 0 \le t \le T, \quad 2\pi/\omega_c \ll T \\ 0, & \text{otherwise} \end{cases}$$

Determine its envelope.

The preenvelope is

$$f_p(t) = f(t) + j\hat{f}(t)$$

If $f(t)$ is $\cos \omega_c t$ then $\hat{f}(t)$ is $\sin \omega_c t$, so that

$$f_p(t) = \cos \omega_c t + j \sin \omega_c t = e^{j\omega_c t}$$

The envelope is then

$$|f_p(t)| = 1, \qquad 0 \leq t \leq T$$

EXAMPLE 3.4-2 Let the narrowband signal $f(t)$ be defined as

$$f(t) = x(t)\cos \omega_c t - y(t)\sin \omega_c t, \qquad -\infty < t < \infty$$

where $x(t)$ and $y(t)$ have Fourier transforms which are nonzero only over a narrow interval around $\omega = 0$. Determine the preenvelope and envelope of $f(t)$.

The Hilbert transform of $f(t)$ is

$$\hat{f}(t) = x(t)\sin \omega_c t + y(t)\cos \omega_c t$$

The preenvelope is therefore

$$\begin{aligned} f_p(t) &= x(t)(\cos \omega_c t + j \sin \omega_c t) + y(t)(\cos \omega_c t + j \sin \omega_c t) \\ &= |x(t) + jy(t)|e^{j\omega_c t} \end{aligned}$$

The envelope is therefore

$$|f_p(t)| = [x^2(t) + y^2(t)]^{1/2}$$

3.5 Narrowband Filters

The complex representation for signals may be extended to filters as well (*7*). A narrowband filter is one whose response in the frequency domain is confined to a narrow band around some center frequency and for which the ratio of center frequency to filter bandwidth is quite large. A narrowband filter might have a weighting function and transfer function much like those shown in Figs. 3-1 and 3-2 respectively.

Before introducing the narrowband filter, we shall prove a theorem for general time-invariant linear filters (*6*). Let $s(t)$ be a real input to a time-invariant linear filter having a real impulse response function $h(t)$. The pre-envelope of the output is the response of $h(t)$ to the preenvelope of $s(t)$ as the input.

Proof: The output of the filter is given by

$$y(t) = h(t) * s(t)$$

The preenvelope of $y(t)$ is

$$y_p(t) = h(t)*s(t) + jH \cdot [h(t)*s(t)]$$
$$= h(t)*s(t) + jh(t)*\hat{s}(t)$$

The latter equality follows from property (6) of the Hilbert transform. Thus

$$y_p(t) = h(t)*[s(t) + j\hat{s}(t)]$$

But the complex envelope of the input is $s_p(t) = s(t) + j\hat{s}(t)$, and it follows immediately that

$$y_p(t) = h(t)*s_p(t)$$

which completes the proof.

EXAMPLE 3.5-1 Let the real signal input to a filter be

$$s(t) = A \cos \omega_c t, \qquad 0 \le t \le T, \quad 2\pi/\omega_c \ll T$$

Let the filter impulse response be

$$h(t) = \cos \omega_c t, \qquad 0 \le t \le T_F$$

Determine the preenvelope of the output. (The advantage of complex signal representation will not be brought out in this example. Indeed, this problem is complicated a bit more by going to a complex signal representation. However, the method and approximations will be found useful shortly.)

The preenvelope of the input signal is immediately seen to be $s_p(t) = Ae^{j\omega_c t}$, $0 \le t \le T$. The preenvelope of the output is

$$y_p(t) = \int_L Ae^{j\omega_c \tau} \cos \omega_c(t - \tau)\, d\tau$$

The limits of integration, denoted L, are those regions of τ which satisfies the constraints

$$0 \le \tau \le T, \qquad 0 \le t - \tau \le T_F$$

and are given in Table 3-1.

TABLE 3-1 LIMITS OF INTEGRATION[a]

$T_F < T$		$T_F = T$		$T_F > T$	
Time	Limits	Time	Limits	Time	Limits
$0 < t < T_F$	$0, t$	$0 < t < T$	$0, t$	$0 < t < T$	$0, t$
$T_F < t < T$	$t - T_F, t$	$T < t < 2T$	$t - T, t$	$T < t < T_F$	$0, T$
$T < t < T + T_F$	$t - T_F, T$			$T_F < t < T + T_F$	$t - T_F, T$

[a] Integral is zero outside these regions.

The preenvelope may be expressed as

$$y_p(t) = e^{j\omega_c t} \int_L \tfrac{1}{2}A \, d\tau + e^{-j\omega_c t} \int_L \tfrac{1}{2}Ae^{j2\omega_c\tau} \, d\tau$$

Except for the region near $t = 0$, the second term may be neglected since it is small compared to the first. To be more specific, once $t \gg 2\pi/\omega_c$ the second integral may be neglected. The preenvelope may then be expressed

$$y_p(t) \approx e^{j\omega_c t} \int_L \tfrac{1}{2}A \, d\tau$$

The real output signal is

$$y(t) = \cos \omega_c t \int_L \tfrac{1}{2}A \, d\tau$$

The envelope of the output is

$$|y_p(t)| = \int_L \tfrac{1}{2}A \, d\tau$$

and is shown in Fig. 3-5 for several values of T_F.

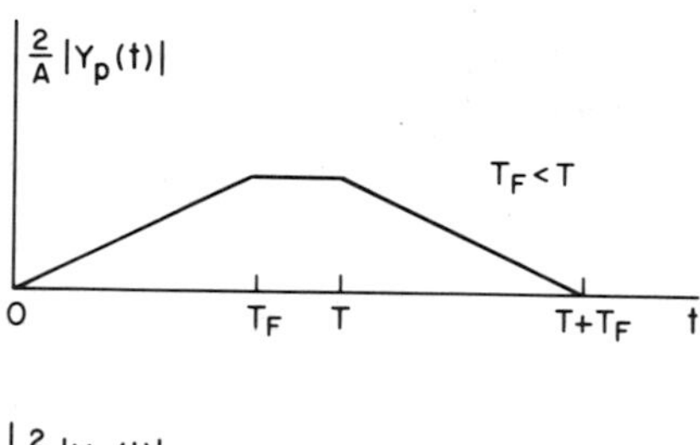

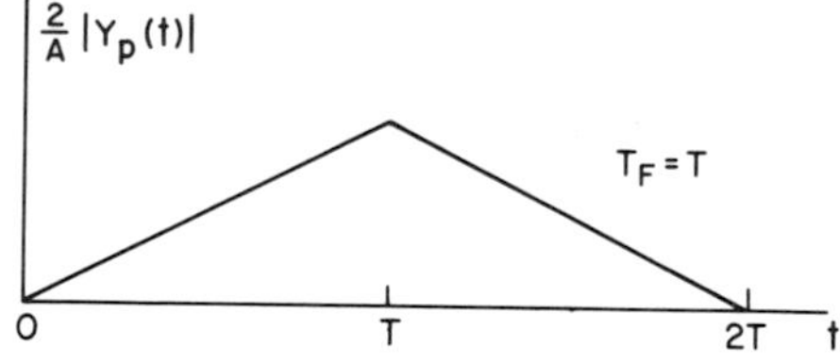

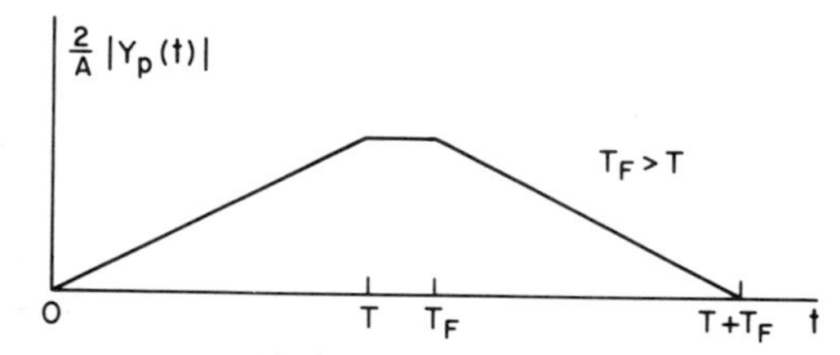

Fig. 3-5 Envelope of output signal for Example 3.5-1.

Narrowband Filter Representation

In analogy to the representation of narrowband signals, we may represent the impulse response or weighting function of a narrowband filter whose response is centered around a frequency Ω by (*8*)

$$h(t) = \text{Re}\, 2\tilde{h}(t)e^{j\Omega t} \tag{3-43}$$

The factor 2 is introduced for later convenience. We shall call $\tilde{h}(t)$ the complex impulse response (*7*). A method for obtaining it is discussed later.

We next determine an input–output relation for narrowband filters in terms of the complex envelopes and complex impulse response. Let $x(t)$ be a real input to a narrowband filter with a real impulse response $h(t)$. The output is denoted by $y(t)$. Denote the preenvelope of the input by $x_p(t)$

$$x_p(t) = x(t) + j\hat{x}(t) = \tilde{x}(t)e^{j\Omega t}$$

and the preenvelope of the output by $y_p(t)$

$$y_p(t) = y(t) + j\hat{y}(t) = \tilde{y}(t)e^{j\Omega t}$$

From the previous section we know that

$$y_p(t) = \int h(\tau)x_p(t-\tau)\, d\tau$$

or

$$\tilde{y}(t)e^{j\Omega t} = \int h(\tau)\tilde{x}(t-\tau)e^{j\Omega(t-\tau)}\, d\tau$$

Therefore

$$\tilde{y}(t) = \int h(\tau)e^{-j\Omega\tau}\tilde{x}(t-\tau)\, d\tau$$

For a complex function $k(t)$ we may write its real part as

$$\text{Re}\, k(t) = k(t)/2 + k^*(t)/2$$

Consequently, the real part of $2\tilde{h}(\tau)e^{j\Omega\tau}$ which by Eq. (3-43) is equal to $h(\tau)$ may be written

$$h(\tau) = \tilde{h}(\tau)e^{j\Omega\tau} + \tilde{h}^*(\tau)e^{-j\Omega\tau}$$

It then follows that

$$h(\tau)e^{-j\Omega\tau} = \tilde{h}(\tau) + \tilde{h}^*(\tau)e^{-j2\Omega\tau}$$

Substituting this in the expression for $\tilde{y}(t)$:

$$\tilde{y}(t) = \int \tilde{h}(\tau)\tilde{x}(t-\tau)\, d\tau + \int \tilde{h}^*(\tau)\tilde{x}(t-\tau)e^{-j2\Omega\tau}\, d\tau$$

As we saw in Example 3.5-1 for narrowband signals and filters, the second integral may be neglected since it is small compared to the first integral. Thus

$$\tilde{y}(t) \approx \int \tilde{h}(\tau)\tilde{x}(t-\tau)\, d\tau \tag{3-44}$$

Therefore, the complex envelope of the output is obtained by convolving the complex envelope of the input with the complex impulse response of the filter. This is one of the advantages of using complex notation for narrowband signals and filters, since the real signals and real impulse responses contain high frequency terms which are cumbersome to handle.

If we are given the transfer function of a narrowband filter, the complex impulse response $\tilde{h}(t)$ may be determined in the following way (*7*). Write the transfer function as

$$H(j\omega) = \tilde{H}(\omega - \Omega) + \tilde{H}^*(-\omega - \Omega) \tag{3-45}$$

This satisfies the requirement that $H^*(j\omega) = H(-j\omega)$ for a real impulse response. For a filter with lumped circuit elements, the poles of $H(j\omega)$ are located near $\omega = \pm\Omega$. By including the poles near $+\Omega$ into $\tilde{H}(\omega - \Omega)$, and the poles near $-\Omega$ into $\tilde{H}^*(-\omega - \Omega)$ it may be shown that

$$h(t) = \text{Re}\, 2\tilde{h}(t)e^{j\Omega t}$$

and $\tilde{h}(t)$ is given by

$$\tilde{h}(t) = \frac{1}{2\pi}\int_{-\infty}^{\infty} \tilde{H}(j\omega)e^{j\omega t}\, d\omega$$

Consequently, we take the positive frequency portion of $H(j\omega)$, shift it to the origin, and take the inverse Fourier transform to determine $\tilde{h}(t)$.

EXAMPLE 3.5-2 Determine the output complex envelope for the filter and signal of Example 3.5-1. The complex envelope and complex impulse response are respectively

$$\tilde{s}(t) = A, \qquad 0 \le t \le T$$

and

$$\tilde{h}(t) = \tfrac{1}{2}, \qquad 0 \le t \le T_F$$

The complex envelope of the output is

$$\tilde{y}(t) = \int_L \tfrac{1}{2}A\, d\tau$$

where the limits of integration are as shown in Table 3-1. Since the function is real, this is also the envelope.

3.6 Narrowband Processes

In analogy to deterministic narrowband signals, a random process is said to be narrowband if its spectral density is zero except for a narrow region around a high carrier frequency. If $n(t)$ is a sample function of such a real random process with zero mean, it may be expressed as (9)

$$n(t) = x(t)\cos \omega_c t - y(t)\sin \omega_c t$$

where the spectral densities of $x(t)$ and $y(t)$ are zero except for a narrow region around $\omega = 0$. The signals $x(t)$ and $y(t)$ are referred to as quadrature components. From Example 3.4-2, the envelope of the narrowband process is

$$\mathscr{E}(t) = [x^2(t) + y^2(t)]^{1/2}$$

In this section we shall derive certain important relationships for the correlation functions associated with narrowband random processes. These will be particularly useful for later work involving Gaussian processes (Chap. 4).

We shall now derive expressions for the autocorrelation function of $x(t)$ and $y(t)$. Assume that $n(t)$ is a wide-sense stationary process with autocorrelation function $R_n(\tau)$. Using properties (8) and (9) of the Hilbert transform Sect. 3.3 we get the following relationships:

$$R_{n\hat{n}}(\tau) = -\hat{R}_n(\tau) \qquad (3\text{-}39)$$

$$R_{\hat{n}n}(\tau) = \hat{R}_n(\tau) \qquad (3\text{-}40)$$

$$R_{n\hat{n}}(0) = R_{\hat{n}n}(0) = 0 \qquad (3\text{-}38)$$

$$R_{\hat{n}}(\tau) = R_n(\tau) \qquad (3\text{-}33)$$

From property (7) of the Hilbert transform,

$$\hat{n}(t) = x(t)\sin \omega_c t + y(t)\cos \omega_c t \qquad (3\text{-}46)$$

Combining this with $n(t)$ and solving for $x(t)$ and $y(t)$,

$$x(t) = n(t)\cos \omega_c t + \hat{n}(t)\sin \omega_c t \qquad (3\text{-}47)$$

$$y(t) = \hat{n}(t)\cos \omega_c t - n(t)\sin \omega_c t \qquad (3\text{-}48)$$

We are now in a position to determine $R_x(\tau)$ and $R_y(\tau)$ in terms of $R_n(\tau)$. Thus,

$$\begin{aligned} R_x(\tau) = E\{x(t)x(t-\tau)\} = {} & E\{n(t)n(t-\tau)\}\cos \omega_c t \cos \omega_c(t-\tau) \\ & + E\{\hat{n}(t)n(t-\tau)\}\sin \omega_c t \cos \omega_c(t-\tau) \\ & + E\{n(t)\hat{n}(t-\tau)\}\cos \omega_c t \sin \omega_c(t-\tau) \\ & + E\{\hat{n}(t)\hat{n}(t-\tau)\}\sin \omega_c t \sin \omega_c(t-\tau) \end{aligned}$$

Replacing the expectations by the appropriate correlation functions

$$R_x(\tau) = R_n(\tau)\cos \omega_c t \cos \omega_c(t-\tau) + R_{\hat{n}n}(\tau)\sin \omega_c t \cos \omega_c(t-\tau)$$
$$+ R_{n\hat{n}}(\tau)\cos \omega_c t \sin \omega_c(t-\tau) + R_{\hat{n}}(\tau)\sin \omega_c t \sin \omega_c(t-\tau)$$

Using Eqs. (3-33), (3-39), and (3-40)

$$\begin{aligned} R_x(\tau) &= R_n(\tau)\cos \omega_c t \cos \omega_c(t-\tau) + \hat{R}_n(\tau)\sin \omega_c t \cos \omega_c(t-\tau) \\ &\quad - \hat{R}_n(\tau)\cos \omega_c t \sin \omega_c(t-\tau) + R_n(\tau)\sin \omega_c t \sin \omega_c(t-\tau) \\ &= R_n(\tau)[\cos \omega_c t \cos \omega_c(t-\tau) + \sin \omega_c t \sin \omega_c(t-\tau)] \\ &\quad - \hat{R}_n(\tau)[\cos \omega_c t \sin \omega_c(t-\tau) - \sin \omega_c t \cos \omega_c(t-\tau)] \end{aligned}$$

Finally,

$$R_x(\tau) = R_n(\tau)\cos \omega_c \tau + \hat{R}_n(\tau)\sin \omega_c \tau \tag{3-49}$$

It is left as an exercise to show that

$$R_y(\tau) = R_n(\tau)\cos \omega_c \tau + \hat{R}_n(\tau)\sin \omega_c \tau \tag{3-50}$$

Therefore,

$$R_x(\tau) = R_y(\tau) \tag{3-51}$$

Three points are noteworthy:

1. If $n(t)$ is wide-sense stationary, then $x(t)$ and $y(t)$ are also wide-sense stationary.
2. The autocorrelation functions of $x(t)$ and $y(t)$ are identical. In particular, $R_x(0) = R_y(0)$ so that their variances are equal. Since the autocorrelation functions are the same, so are the power spectral density functions.
3. From Eq. (3-49) or (3-50) with $\tau = 0$ we see that $R_n(0) = R_x(0) = R_y(0)$, so that the power in either quadrature component is equal to the power of the process $n(t)$.

We next determine $R_{xy}(\tau)$. Formally,

$$\begin{aligned} R_{xy}(\tau) &= E\{x(t)y(t-\tau)\} \\ &= E\{n(t)\hat{n}(t-\tau)\}\cos \omega_c t \cos \omega_c(t-\tau) \\ &\quad + E\{\hat{n}(t)\hat{n}(t-\tau)\}\sin \omega_c t \cos \omega_c(t-\tau) \\ &\quad - E\{n(t)n(t-\tau)\}\cos \omega_c t \sin \omega_c(t-\tau) \\ &\quad - E\{\hat{n}(t)n(t-\tau)\}\sin \omega_c t \sin \omega_c(t-\tau) \end{aligned}$$

or

$$R_{xy}(\tau) = R_{n\hat{n}}(\tau)\cos \omega_c t \cos \omega_c(t-\tau) + R_{\hat{n}}(\tau)\sin \omega_c t \cos \omega_c(t-\tau)$$
$$- R_n(\tau)\cos \omega_c t \sin \omega_c(t-\tau) - R_{\hat{n}n}(\tau)\sin \omega_c t \sin \omega_c(t-\tau)$$

Again using Eqs. (3-33), (3-39), and (3-40),

$$R_{xy}(\tau) = R_n(\tau)[\sin \omega_c t \cos \omega_c(t-\tau) - \cos \omega_c t \sin \omega_c(t-\tau)]$$
$$- \hat{R}_n(\tau)[\cos \omega_c t \cos \omega_c(t-\tau) + \sin \omega_c t \sin \omega_c(t-\tau)]$$

and finally

$$R_{xy}(\tau) = R_n(\tau)\sin \omega_c\tau - \hat{R}_n(\tau)\cos \omega_c\tau \tag{3-52}$$

It may similarly be shown that

$$R_{yx}(\tau) = -R_n(\tau)\sin \omega_c\tau + \hat{R}_n(\tau)\cos \omega_c\tau \tag{3-53}$$

Comparing Eqs. (3-52) and (3-53)

$$R_{xy}(\tau) = -R_{yx}(\tau) \tag{3-54}$$

Since the random processes $x(t)$ and $y(t)$ are real, it follows from Eq. (2-18) that $R_{yx}(\tau) = R_{xy}(-\tau)$. Therefore,

$$R_{xy}(\tau) = -R_{xy}(-\tau) \tag{3-55}$$

and so the crosscorrelation of $x(t)$ and $y(t)$ is an odd function. In particular

$$R_{xy}(0) = 0 \tag{3-56}$$

Therefore samples of $x(t)$ and $y(t)$ taken at the same instant of time are uncorrelated.

We point out in passing that we may express

$$R_n(\tau) = R_x(\tau)\cos \omega\tau + R_{xy}(\tau)\sin \omega\tau \tag{3-57}$$

We next show that $R_{xy}(\tau) = 0$ for all τ if the following conditions are satisfied:

$$S_n(\omega) = 0, \qquad \omega \geq 2\omega_c \tag{3-58a}$$

and

$$S_n(\omega_c + \omega) = S_n(\omega_c - \omega), \qquad 0 < \omega < \omega_c \tag{3-58b}$$

Equation (3-58) states that the spectral density is an even function about the carrier frequency for positive frequencies.

Using property (1), Eq. (3-24), of the Hilbert transform, the Fourier transform of the crosscorrelation function, Eq. (3-52), is

$$S_x(\omega)_y = \frac{S_n(\omega-\omega_c)}{2j}[1 - \operatorname{sgn}(\omega-\omega_c)] - \frac{S_n(\omega+\omega_c)}{2j}[1 + \operatorname{sgn}(\omega+\omega_c)]$$
$$- \frac{S_n(\omega+\omega_c)}{j} \tag{3-59}$$

Using assumptions (3-58) it follows that

$$S_{xy}(\omega) = \begin{cases} 0, & \omega > \omega_c \\ -j[S_n(\omega - \omega_c) - S_n(\omega + \omega_c)], & -\omega_c < \omega < \omega_c \\ 0, & \omega < -\omega_c \end{cases} \quad (3\text{-}60)$$

It is easily verified that $S_{xy}(\omega)$ is an odd function of frequency. Its inverse transform may be expressed as

$$R_{xy}(\tau) = \frac{j}{\pi}\int_0^{\omega_c} S_{xy}(\omega)\sin \omega\tau \, d\omega$$

$$= \frac{1}{\pi}\int_0^{\omega_c} [S_n(\omega - \omega_c) - S_n(\omega + \omega_c)]\sin \omega\tau \, d\omega \quad (3\text{-}61)$$

After several changes of variables and some manipulation this may be shown to be

$$R_{xy}(\tau) = \frac{-1}{\pi}\int_{-\omega_c}^{\omega_c} S_n(\omega + \omega_c)\sin \omega\tau \, d\omega \quad (3\text{-}62)$$

Since the spectral density $S_n(\omega + \omega_c)$ is an even function for $-\omega_c < \omega < \omega_c$ and $\sin \omega\tau$ is odd, the integral is zero. Therefore

$$R_{xy}(\tau) = 0, \qquad \text{all } \tau \quad (3\text{-}63)$$

Thus, the random processes are uncorrelated under assumptions (3-58). For these assumptions, we also see from Eq. (3-57) that the autocorrelation function of the process may be expressed as

$$R_n(\tau) = R_x(\tau)\cos \omega_c\tau$$

and

$$S_n(\omega) = \tfrac{1}{2}S_x(\omega - \omega_c) + \tfrac{1}{2}S_x(\omega + \omega_c)$$

3.7 Fourier Series Representation

A common representation of periodic signals is a Fourier series of sines and cosines, or complex exponentials. If the signal is not periodic, the Fourier series may still be used to represent the signal over any fixed interval of time. Outside that interval a new set of Fourier coefficients would have to be generated.

An integrable square function $f(t)$ can be represented over the interval $(0, T)$ by the Fourier series

$$f(t) = \frac{1}{T^{1/2}}\sum_k F(k)e^{jk\omega_0 t}, \qquad 0 \le t \le T \quad (3\text{-}64)$$

where

$$F(k) = \frac{1}{T^{1/2}} \int_0^T f(t) e^{-jk\omega_0 t}\, dt \tag{3-65}$$

and $2\pi/\omega_0 = T$. The series representation converges as a limit in the mean. The particular form of normalization $(1/T^{1/2})$ used here is for convenience in working with the variance or covariance of the Fourier components. It is an easy matter to show that

$$\int_0^T f^2(t)\, dt = \sum_k |F(k)|^2$$

The Fourier coefficients, since they are complex, may be expressed as†

$$F(k) = F_c(k) + jF_s(k)$$

where

$$F_c(k) = \frac{1}{T^{1/2}} \int_0^T f(t) \cos k\omega_0 t\, dt$$

$$F_s(k) = \frac{-1}{T^{1/2}} \int_0^T f(t) \sin k\omega_0 t\, dt \tag{3-66}$$

Consider the Fourier series of a narrowband process in the interval $0 \le t \le T$

$$f(t) = x(t) \cos \omega_c t - y(t) \sin \omega_c t$$

For convenience, we express $\omega_c = m2\pi/T$. Then the Fourier coefficients will be approximately zero except for k in the vicinity of m, and extending over a range corresponding to the bandwidth of $f(t)$.

Applying the usual techniques and approximations for narrowband processes, the real and imaginary parts of $F(k)$ may be expressed as

$$F_c(k) = \frac{1}{2T^{1/2}} \int_0^T x(t) \cos (\omega_c - k\omega_0)t\, dt - \frac{1}{2T^{1/2}} \int_0^T y(t) \sin(\omega_c - k\omega_0)t\, dt \tag{3-67}$$

and

$$F_s(k) = \frac{1}{2T^{1/2}} \int_0^T x(t) \sin(\omega_c - k\omega_0)t\, dt + \frac{1}{2T^{1/2}} \int_0^T y(t) \cos(\omega_c - k\omega_0)t\, dt \tag{3-68}$$

† The terms $F_c(k)$ and $F_s(k)$ are sometimes called quadrature pairs. At times, $F_c(k)$ is also called the "inphase" component, and $F_s(k)$ the "quadrature" component.

Then,

$$F(k) = \frac{1}{2T^{1/2}} \int_0^T [x(t) + jy(t)]e^{j\omega_k t}\, dt \tag{3-69}$$

where $\omega_k = \omega_c - k\omega_o$. Therefore, for narrowband processes, the Fourier coefficients can be determined from the complex envelope of the signal.

We now briefly examine the statistical properties of the Fourier coefficients. Assume that the basic process is zero mean with an autocorrelation function $R_f(\tau)$ and a power spectral density $S_f(\omega)$.

The average of the magnitude squared of each coefficient is (*10*)

$$E\{F(k)F^*(k)\} = \frac{1}{T}\int_0^T dt \int_0^T du\, R_f(t-u)e^{-j\omega_k(t-u)}$$

$$= \int_{-T}^{T} \left(1 - \frac{|v|}{T}\right) R_f(v)e^{-j\omega_k v}\, dv$$

Now, if the value of T is much greater than the interval over which $R_f(v)$ is significant, then the expression above can be approximated by

$$E\{F(k)F^*(k)\} \cong \int_{-\infty}^{\infty} R_f(v)e^{-j\omega_k v}\, dv = S_f(\omega_k) \tag{3-70}$$

Thus for large values of T, the expected value of $|F(k)|^2$ is equal to the power spectral density at frequency ω_k. The terms $|F(k)|^2$ make up the periodogram (*11*) which can be used as an estimate of the power spectral density. It is well known (*10*), however, that for a Gaussian process the variance of this estimator does not decrease as $T \to \infty$.†

The covariance of the coefficients with different frequency indices may be expressed as (*10*)

$$E\{F(k)F^*(l)\} = \frac{1}{T}\int_0^T dt\, e^{-j\omega_o(k-l)t} \int_{t-T}^{t} dv\, R_f(v)e^{-j\omega_o l v}$$

As $T \to \infty$, $\omega_o \to 0$ and the inner integral approaches a constant $\int_{-\infty}^{\infty} R_f(v)\, dv$ so that the covariance vanishes. It is therefore only in the limit of large T that the coefficients become uncorrelated. That is

$$\lim_{T\to\infty} E\{F(k)F^*(l)\} = 0 \tag{3-71}$$

† For a discussion of spectrum analysis in practice, consult Blackman & Turkey (*11*), and the references in the supplementary bibliography at the end of this chapter and that of Chap. 2.

(It is often assumed in practice, even for modest value of T, that the coefficients are uncorrelated.) The fact that the coefficients are correlated for finite T will lead us, in later chapters, to consider a generalized Fourier series (Karhunen–Loeve expansion) which has uncorrelated coefficients.

For the real part of the coefficient, it can be shown that

$$E\{F_c^2(k)\} = \tfrac{1}{2} \int_{-T}^{T} \left(1 - \frac{|v|}{T}\right) R_f(v) \cos \omega_k v \, dv$$

which for large values of T becomes

$$E\{F_c^2(k)\} \cong \tfrac{1}{2} S_f(\omega_k) \tag{3-72}$$

It can be similarly shown for the imaginary part that

$$E\{F_s^2(k)\} \cong \tfrac{1}{2} S_f(\omega_k) \tag{3-73}$$

Now consider two narrowband processes $f_1(t)$ and $f_2(t)$, $0 \leq t \leq T$,

$$f_1(t) = x_1(t) \cos \omega_c t - y_1(t) \sin \omega_c t$$

and

$$f_2(t) = x_2(t) \cos \omega_c t - y_2(t) \sin \omega_c t$$

The correlation and power spectral densities are $R_{f_1}(\tau)$, $S_{f_1}(\omega)$ and $R_{f_2}(\tau)$, $S_{f_2}(\omega)$. The crosscorrelation function is $R_{f_1 f_2}(\tau)$ and the cross-spectral density is $S_{f_1 f_2}(\omega)$. The Fourier coefficients are

$$F_1(k) = F_{1c}(k) + jF_{1s}(k) = \frac{1}{T^{1/2}} \int_0^T f_1(t) e^{-jk\omega_0 t} \, dt$$

$$F_2(k) = F_{2c}(k) + jF_{2s}(k) = \frac{1}{T^{1/2}} \int_0^T f_2(t) e^{-jk\omega_0 t} \, dt$$

The correlation between the coefficients may be expressed as

$$E\{F_1(k) F_2^*(k)\} = \int_{-T}^{T} \left(1 - \frac{|v|}{T}\right) R_{f_1 f_2}(v) e^{-jk\omega_0 v} \, dv$$

which for large T becomes

$$\lim_{T \to \infty} E\{F_1(k) F_2^*(k)\} = S_{f_1 f_2}(\omega_k) \tag{3-74}$$

which is the cross-spectral density at the frequency corresponding to the particular Fourier coefficient.

For the narrowband case, the following additional relations may also be shown:

$$\lim_{T\to\infty} E\{F_1(k)F_2{}^*(l)\} = 0, \qquad k \neq l \tag{3-75a}$$

$$\lim_{T\to\infty} E\{F_{1c}(k)F_{2c}(k)\} = \tfrac{1}{2}\,\mathrm{Re}\, S_{f_1f_2}(\omega_k) \tag{3-75b}$$

$$\lim_{T\to\infty} E\{F_{1s}(k)F_{2s}(k)\} = \tfrac{1}{2}\,\mathrm{Re}\, S_{f_1f_2}(\omega_k) \tag{3-75c}$$

$$\lim_{T\to\infty} E\{F_{1s}(k)F_{2c}(k)\} = \tfrac{1}{2}\,\mathrm{Im}\, S_{f_1f_2}(\omega_k) \tag{3-75d}$$

$$\lim_{T\to\infty} E\{F_{1c}(k)F_{2s}(k)\} = -\tfrac{1}{2}\,\mathrm{Im}\, S_{f_1f_2}(\omega_k) \tag{3-75e}$$

All of the above relations can be written as covariance matrices. Thus

$$\lim_{T\to\infty} E\begin{bmatrix} F_{1c}^2(k) & F_{1c}(k)F_{1s}(k) & F_{1c}(k)_{2c}(k) & F_{1c}(k)F_{2s}(k) \\ F_{1s}(k)F_{1c}(k) & F_{1s}^2(k) & F_{1s}(k)F_{2c}(k) & F_{1s}(k)F_{2s}(k) \\ F_{2c}(k)F_{1c}(k) & F_{2c}(k)F_{1s}(k) & F_{2c}^2(k) & F_{2c}(k)F_{2s}(k) \\ F_{2s}(k)F_{1c}(k) & F_{2s}(k)F_{1s}(k) & F_{2s}(k)F_{2c}(k) & F_{2s}^2(k) \end{bmatrix}$$

$$= \left[\begin{array}{cc|cc} \tfrac{1}{2}S_{f_1}(\omega_k) & 0 & \tfrac{1}{2}\,\mathrm{Re}\, S_{f_1f_2}(\omega_k) & -\tfrac{1}{2}\,\mathrm{Im}\, S_{f_1f_2}(\omega_k) \\ 0 & \tfrac{1}{2}S_{f_1}(\omega_k) & \tfrac{1}{2}\,\mathrm{Im}\, S_{f_1f_2}(\omega_k) & \tfrac{1}{2}\,\mathrm{Re}\, S_{f_1f_2}(\omega_k) \\ \hline \tfrac{1}{2}\,\mathrm{Re}\, S_{f_1f_2}(\omega_k) & \tfrac{1}{2}\,\mathrm{Im}\, S_{f_1f_2}(\omega_k) & \tfrac{1}{2}S_{f_2}(\omega_k) & 0 \\ -\tfrac{1}{2}\,\mathrm{Im}\, S_{f_1f_2}(\omega_k) & \tfrac{1}{2}\,\mathrm{Re}\, S_{f_1f_2}(\omega_k) & 0 & \tfrac{1}{2}S_{f_2}(\omega_k) \end{array}\right] \tag{3-76}$$

The dashed lines are put in to delineate four particular parts of the matrix. Such a matrix partition is of value for discussing complex samples and isomorphic matrices (Chap. 11). As a preview of that, consider the covariance matrix of the complex Fourier components.

$$\lim_{T\to\infty} E\begin{bmatrix} F_1(k)F_1{}^*(k) & F_1(k)F_2{}^*(k) \\ F_2(k)F_1{}^*(k) & F_2(k)F_2{}^*(k) \end{bmatrix} = \begin{bmatrix} S_{f_1}(\omega_k) & S_{f_1f_2}(\omega_k) \\ S_{f_1f_2}^*(\omega_k) & S_{f_2}(\omega_k) \end{bmatrix} \tag{3-77}$$

The matrices (3-76) and (3-77) are isomorphic. For example, a matrix of the form $\left[\begin{smallmatrix} \alpha & -\beta \\ \beta & \alpha \end{smallmatrix}\right]$ is isomorphic to the complex number $\alpha + j\beta$. Therefore, the matrix

$$\begin{bmatrix} S_{f_1}(\omega_k) & 0 \\ 0 & S_{f_1}(\omega_k) \end{bmatrix}$$

is isomorphic to $S_{f_1}(\omega_k)$, and the matrix

$$\begin{bmatrix} \text{Re } S_{f_1 f_2}(\omega_k) & -\text{Im } S_{f_1 f_2}(\omega_k) \\ \text{Im } S_{f_1 f_2}(\omega_k) & \text{Re } S_{f_1 f_2}(\omega_k) \end{bmatrix}$$

is isomorphic to $S_{f_1 f_2}(\omega_k)$. The utility of this will become apparent in Chap. 11.

Exercises

3.1 Let $a(t)$, $-\infty < t < \infty$, be a known function with Fourier transform $A(\omega)$. Assume $A(\omega) = 0$ for $\omega > \omega_B$ where $\omega_B \ll \omega_c$.

(a) Determine and relate the Fourier transforms of $a(t)\cos \omega_c t$ and $\frac{1}{2}a(t)e^{j\omega_c t}$.

(b) Determine and relate the Fourier transforms of $a(t)\sin \omega_c t$ and $-j\frac{1}{2}a(t)e^{j\omega_c t}$.

(c) Relate the Fourier transforms of $a(t)\cos \omega_c t$ and $a(t)\sin \omega_c t$.

3.2 Show that (a) the Hilbert transform of an even function is odd and (b) the Hilbert transform of an odd function is even.

3.3 Prove the following relations concerning ensemble correlation functions of a wide sense stationary process $x(t)$ and its Hilbert transform $\hat{x}(t)$.

(a) $R_{x\hat{x}}(\tau) = -\hat{R}_x(\tau)$

(b) $R_{\hat{x}x}(\tau) = \hat{R}_x(\tau)$

(c) $R_{\hat{x}}(\tau) = R_x(\tau)$

(d) Show that $R_{x\hat{x}}(\tau)$ is an odd function.

3.4 Prove Properties (3) and (4) for Hilbert transforms. That is

$$\lim_{T\to\infty} \frac{1}{2T} \int_{-T}^{T} x^2(t)\, dt = \lim_{T\to\infty} \frac{1}{2T} \int_{-T}^{T} \hat{x}^2(t)\, dt$$

and

$$\lim_{T\to\infty} \frac{1}{2T} \int_{-T}^{T} x(t)\hat{x}(t)\, dt = 0$$

3.5 The analytic function or preenvelope of $x(t)$ is

$$z(t) = x(t) + j\hat{x}(t)$$

Show that

$$E\{z(t)z^*(t-\tau)\} = 2[R_x(\tau) + j\hat{R}_x(\tau)]$$

and

$$E\{z(t)z(t-\tau)\} = 0$$

Determine the power spectral density of $z(t)$.

3.6 If a signal $x(t)$ is bandlimited to Ω, show that the magnitude squared of its signal preenvelope is bandlimited to 2Ω.

3.7 For a frequency modulated signal

$$x(t) = \cos(\omega_c t + m(t))$$

Assume $dm(t)/dt \ll \omega_c$ so that the narrowband assumption is reasonable. What is the complex envelope of the signal? What is the envelope of the signal?

3.8 For the narrowband stationary process

$$n(t) = x(t)\cos \omega_c t - y(t)\sin \omega_c t$$

derive the relation shown in Eq. (3-50). That is,

$$R_y(\tau) = R_n(\tau)\cos \omega_c \tau + \hat{R}_n(\tau)\sin \omega_c \tau$$

3.9 For the narrowband stationary process

$$n(t) = x(t)\cos \omega_c t - y(t)\sin \omega_c t$$

(a) Show that the autocorrelation function of $n(t)$ may be expressed

$$R_n(\tau) = R_x(\tau)\cos \omega_c \tau + R_{xy}(\tau)\sin \omega_c \tau$$

(b) If the process is such that Eq. (3-58) is satisfied, what is the envelope of the autocorrelation function of $n(t)$, and show that the power spectral density of $x(t)$ may be expressed

$$S_x(\omega) = S_n(\omega - \omega_c) + S_n(\omega + \omega_c), \qquad -\omega_c < \omega < \omega_c$$

3.10 Consider the RLC filter shown in Fig. 3-6. Define $\omega_0^2 = 1/LC$, $\omega_1 = 1/RC$, $\omega_1 \ll \omega_0$. Show that the complex impulse response is given approximately by

$$\tilde{h}(t) = \frac{\omega_1}{2} \exp\left(-\frac{\omega_1}{2} t\right), \qquad t \geqslant 0$$

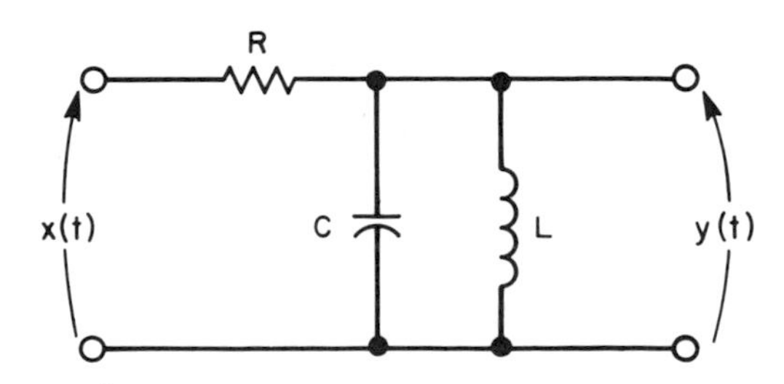

Fig. 3-6 An RLC filter.

3.11 For the RLC circuit in Fig. 3-7, assume the input is a voltage, and the output is the current in the circuit. Define $\omega_0{}^2 = 1/LC$, $\omega_1 = 1/RC$, $\omega_1 \ll \omega_0$. Show that the complex impulse response is given approximately by $e^{-\omega_1 t}/-2L$, $t \geqslant 0$.

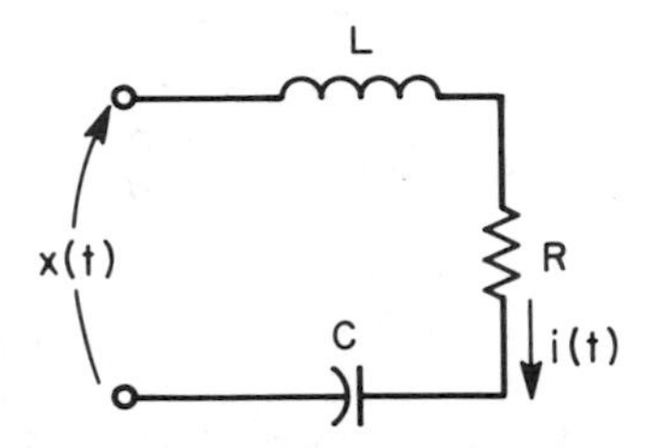

Fig. 3-7 An RLC filter.

References

1. Gabor, D., Theory of communication, *J. Inst. Elect. Eng.*, **93 (III)**, (1946).
2. Kelly, E. J., Reed, I. S., and Root, W. L., The detection of radar echoes in noise, I, *J. Soc. Ind. Appl. Math.* **8,** No. 2, 309–341 (1960).
3. Peirce, B. O., and Foster, R. M., "A Short Table of Integrals." Ginn, Boston, 1957.
4. Kuo, F. F., and Freeney, S. L., Hilbert transforms and modulation theory, *Proc. Nat. Electron. Conf.* Vol. XVIII (October 1962).
5. Wylie, C. R. Jr., "Advanced Engineering Mathematics." McGraw-Hill, New York, 1951.
6. Dugundji, J., Envelopes and pre-envelopes of real waveforms, *IRE Trans. Inform. Theory* March (1958).
7. Helstrom, C. W., "Statistical Theory of Signal Detection." Pergamon Press, Oxford, 1960.
8. Middleton, D., "An Introduction to Statistical Communication Theory." McGraw-Hill, New York, 1960.
9. Rice, S. O., "Mathematical Analysis of Random Noise," Reprinted in Selected Papers on Noise and Stochastic Processes (N. Wax, ed.). Dover, New York, 1954.
10. Davenport, W. B., and Root, W. L., An Introduction to the Theory of Random Signals and Noise, McGraw-Hill Book Co., 1958.
11. Blackman, R. B., and Tukey, J. W., "The Measurement of Power Spectra." Dover, New York, 1958.

SUPPLEMENTARY BIBLIOGRAPHY

On Complex Signal Representation:

Arens, R. A., Complex processes for envelopes of normal noise, *IRE Trans. Inform. Theory* **IT-3,** 204–207 (1957).

Bedrosian, E., The analytic signal representation of modulated waveforms, *Proc. IRE* (October 1962).

Grettenberg, T. L., A representation theorem for complex normal processes, correspondence, *IFEE Trans. Inform. Theory* **IT-11,** No. 2, 305–306 (1965).

Nuttall, A. H., High-order covariance functions for complex gaussian processes, correspondence, *IRE Trans. Inform. Theory* **IT-8,** 255–256 (1962).
Reed, I. S., On a moment theorem for complex gaussian processes, correspondence, *IRE Trans. Inform. Theory* **IT-8,** No. 3, 194–195 (1962).
Voelcker, H. B., Toward a unified theory of modulation, Part I, phase envelope relationships, *Proc. IEEE* **54,** No. 3 (1966).
Voelcker, H. B., Part II, Zero Manipulation (May 1966).
Zakai, M., Second order properties of the pre-envelope and envelope processes, correspondence, *IRE Trans. Inform. Theory* **IT-6,** 556–557 (1960).

On Spectrum Analysis: (See also Chap. 2 Bibliography)

Bendat, J. S., and Piersol, A. G., "Measurement and Analysis of Random Data." Wiley, New York, (1966).
Bogert, B. P., "On Practicing Spectrum Analysis, Human Communication: A Unified View" (P. Denes, ed.). McGraw-Hill, New York, to be published.
Kharkevich, A. A., "Spectra and Analyses." Consultants Bureau, New York, 1960.

The following appear in *Technometrics* **3,** No. 2, (1961):

Jenkins, G. M., General considerations in the analysis of spectra.
Parzen, E., Mathematical considerations in the estimation of spectra.
Tukey, J. W., Discussion emphasizing the connection between analysis of variance and spectrum analysis.
Goodman, N. R., Some comments on spectral analysis of time series.

Chapter 4

Gaussian Derived Processes

4.1 Gaussian Properties

The most important probability density function in detection theory is the Gaussian or normal density function, and the most often encountered random process is the Gaussian random process. There are sound technical reasons for this which include the important and far reaching central limit theorem which is stated below. In some cases, the Gaussian assumption is employed whether justified or not. Both the Gaussian distribution and the Gaussian process have properties which make analytical solutions tractable. Since the Gaussian assumption is so common, other related distributions are also frequently encountered. These distributions, their derivations, and properties are discussed in later sections.

Before presenting the central limit theorem several definitions are required. Let $x_1, x_2, \ldots, x_n$ be a sequence of independent random variables. Define (*1*)

$$\begin{aligned} S_1 &= x_1 \\ S_2 &= x_1 + x_2 \\ &\vdots \\ S_n &= x_1 + x_2 + \cdots + x_n \end{aligned}$$

Define the normalized random variable

$$z_n = \frac{S_n - E\{S_n\}}{\sigma_{S_n}} \tag{4-1}$$

where σ_{S_n} is the standard deviation of S_n. Two forms for the central limit theorem will be given.

The *central limit theorem* (*1–5*) for identically distributed, statistically independent random variables with finite means and variances states that the random variable z_n [Eq. (4-1)] converges as $n \to \infty$ to a normally distributed random variable with zero mean and unit variance.

The second form of the theorem does not depend on the variables being identically distributed but requires the existence of more than the mean and variance.

Define the $(2 + \delta)$th central moment (*1*),

$$\mu(2 + \delta; n) = E\{[x_n - E\{x_n\}]^{2+\delta}\} \tag{4-2}$$

The central limit theorem for independent random variables with finite means and finite $(2 + \delta)$th central moments, for some $\delta > 0$, states that the random variable z_n will converge to a normally distributed random variable provided

$$\lim_{n\to\infty} \frac{1}{(\sigma_{S_n})^{2+\delta}} \sum_{k=1}^{n} \mu(2 + \delta; k) = 0 \tag{4-3}$$

This form of the theorem is also called Laplace–Liapounoff's theorem (*3*). Equation (4-3) is Liapounoff's condition for the validity of this form of the theorem (*1*).

The consequence of these theorems is that if a physical process, for example circuit noise, is the sum of many independent actions, then if the conditions of one of these theorems is satisfied, the process will tend to be Gaussian.† This is the basis for applying the Gaussian assumption to many physical phenomena called noise.

A real random variable y defined over the interval $(-\infty < y < \infty)$ having the probability density function

$$p(y) = \frac{1}{(2\pi)^{1/2}\sigma_y} \exp\left[-\frac{(y - m_y)^2}{2\sigma_y{}^2}\right] \tag{4-4}$$

is a Gaussian random variable. The mean and variance of y are given by m_y and $\sigma_y{}^2$ respectively. It follows that a normalized Gaussian variable $z = (y - m_y)/\sigma_y$ has the density function

$$p(z) = \frac{1}{(2\pi)^{1/2}} e^{-z^2/2}, \qquad -\infty < z < \infty \tag{4-5}$$

† There is also a related distribution, called the log-normal distribution (Exercise 1.6) which is obtained by an inverse log operation on a Gaussian variable. It is a useful distribution when dealing with the product of many independent random variables [see Aitchinson and Brown (*6*)].

This is illustrated in Fig. 4-1. The probability distribution function is

$$P(z \leq a) = \frac{1}{(2\pi)^{1/2}} \int_{-\infty}^{a} e^{-z^2/2}\, dz \tag{4-6}$$

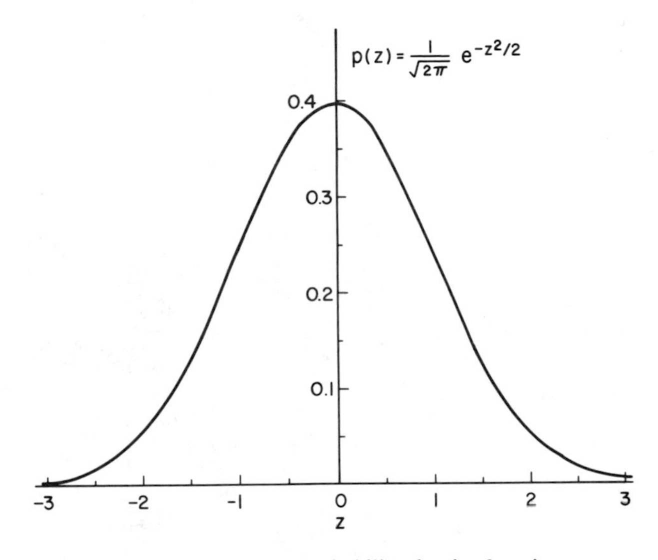

Fig. 4-1 Gaussian probability density function.

Although this integral cannot be put in closed form, it is tabulated in many sources (*7, 8*). The cumulative distribution function is shown in Fig. 4-2. The abscissa of this figure is the "normal probability" scale and is such that a graph of a Gaussian distribution will be a straight line.

The characteristic function for a Gaussian variable with mean m_y and variance $\sigma_y{}^2$ is easily shown to be

$$C(j\omega) = \exp\left[j\omega m_y - \frac{\omega^2 \sigma_y{}^2}{2}\right] \tag{4-7}$$

Two zero-mean random variables y_1 and y_2 are jointly Gaussian distributed if the joint density function can be expressed as

$$p(y_1, y_2) = \frac{1}{2\pi(\sigma_{y_1}^2 \sigma_{y_2}^2 - \sigma_{y_1 y_2}^2)^{1/2}} \exp\left[\frac{-\sigma_{y_2}^2 y_1{}^2 + 2\sigma_{y_1 y_2} y_1 y_2 - \sigma_{y_1}^2 y_2{}^2}{2(\sigma_{y_1}^2 \sigma_{y_2}^2 - \sigma_{y_1 y_2}^2)}\right] \tag{4-8}$$

where σ_{y_1}, and σ_{y_2} are the variances of y_1 and y_2 respectively, and the covariance is

$$\sigma_{y_1 y_2} = E\{y_1 y_2\}$$

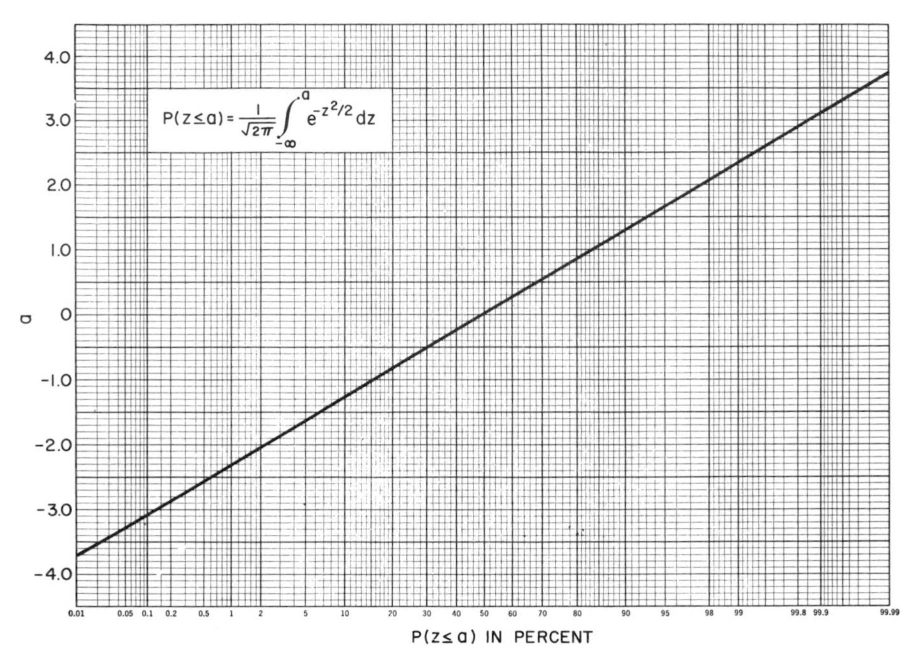

Fig. 4-2 Gaussian cumulative distribution function.

Using the normalized covariance

$$\rho_{12} = \frac{\sigma_{y_1 y_2}}{\sigma_{y_1}\sigma_{y_2}}$$

the bivariate density function may be expressed

$$p(y_1, y_2) = \frac{1}{2\pi\sigma_{y_1}\sigma_{y_2}(1-\rho_{12}^2)^{1/2}} \exp\left[\frac{-\sigma_{y_2}^2 {y_1}^2 + 2\rho_{12}\sigma_{y_1}\sigma_{y_2} y_1 y_2 - \sigma_{y_1}^2 {y_2}^2}{2\sigma_{y_1}^2\sigma_{y_2}^2(1-\rho_{12}^2)}\right] \tag{4-9}$$

Note that if $\rho = 0$, then

$$p(y_1, y_2) = p(y_1) \cdot p(y_2)$$

implying that uncorrelated Gaussian variables are statistically independent.

The joint characteristic function of two zero-mean Gaussian variables is (*9*)

$$C(j\omega_1, j\omega_2) = \exp[-\tfrac{1}{2}({\omega_1}^2 + 2\rho_{12}\omega_1\omega_2 + {\omega_2}^2)] \tag{4-10}$$

The multivariate characteristic function is given in Eq. (4-23).

Matrix Form

The bivariate and multivariate Gaussian density functions may be written more compactly if use is made of matrix notation (*10*). For the above case of two variables y_1 and y_2 having zero mean, we define a *covariance matrix*†

$$\mathbf{R} = \begin{bmatrix} E\{y_1{}^2\} & E\{y_1 y_2\} \\ E\{y_1 y_2\} & E\{y_2{}^2\} \end{bmatrix} = \begin{bmatrix} \sigma_{y1}^2 & \rho\sigma_{y_1}\sigma_{y_2} \\ \rho\sigma_{y_1}\sigma_{y_2} & \sigma_{y2}^2 \end{bmatrix} \tag{4-11}$$

It is easily verified that the determinant of $\mathbf{R}$, denoted by $|\mathbf{R}|$, is

$$|\mathbf{R}| = \sigma_{y_1}^2 \sigma_{y_2}^2 (1 - \rho^2) \tag{4-12}$$

The inverse of the matrix $\mathbf{R}$, denoted as $\mathbf{R}^{-1}$, is

$$\mathbf{R}^{-1} = \frac{1}{|\mathbf{R}|} \begin{bmatrix} \sigma_{y2}^2 & -\rho\sigma_{y_1}\sigma_{y_2} \\ -\rho\sigma_{y_1}\sigma_{y_2} & \sigma_{y1}^2 \end{bmatrix} \tag{4-13}$$

Define the column matrix, or vector

$$\mathbf{y} = \begin{bmatrix} y_1 \\ y_2 \end{bmatrix} \tag{4-14}$$

Its transpose, denoted as $\mathbf{y}'$, is

$$\mathbf{y}' = [y_1 \quad y_2] \tag{4-15}$$

The matrix product

$$\mathbf{y}'\mathbf{R}^{-1}\mathbf{y} = \frac{[y_1 \quad y_2]}{|\mathbf{R}|} \begin{bmatrix} \sigma_{y2}^2 & -\rho\sigma_{y_1}\sigma_{y_2} \\ -\rho\sigma_{y_1}\sigma_{y_2} & \sigma_{y1}^2 \end{bmatrix} \begin{bmatrix} y_1 \\ y_2 \end{bmatrix} \tag{4-16}$$

may be verified by direct calculation to be

$$\mathbf{y}'\mathbf{R}^{-1}\mathbf{y} = \frac{\sigma_{y_2}^2 y_1{}^2 - 2\rho\sigma_{y_1}\sigma_{y_2} y_1 y_2 + \sigma_{y_1}^2 y_2{}^2}{\sigma_{y_1}^2 \sigma_{y_2}^2 (1 - \rho^2)} \tag{4-17}$$

If these expressions are substituted into the bivariate Gaussian density function, Eq. (4-9), we find that

$$p(\mathbf{y}) \triangleq p(y_1, y_2) = \frac{1}{2\pi|\mathbf{R}|^{1/2}} e^{-\mathbf{y}'\mathbf{R}^{-1}\mathbf{y}/2} \tag{4-18}$$

If the variables have nonzero means m_1 and m_2 respectively, the bivariate density is written

$$p(\mathbf{y}) = \frac{1}{2\pi|\mathbf{R}|^{1/2}} e^{-(\mathbf{y}-\mathbf{m})'\mathbf{R}^{-1}(\mathbf{y}-\mathbf{m})/2} \tag{4-19}$$

† Upper case boldface letters, such as **R**, will normally denote matrices of arbitrary size, and lower case boldface letters, such as **y**, will normally denote column vectors, or what is the same, a matrix of one column.

where

$$\mathbf{m} = \begin{bmatrix} m_1 \\ m_2 \end{bmatrix}$$

and the covariance matrix is

$$\mathbf{R} = \begin{bmatrix} E\{(y_1 - m_1)^2\} & E\{(y_1 - m_1)(y_2 - m_2)\} \\ E\{(y_1 - m_1)(y_2 - m_2)\} & E\{(y_2 - m_2)^2\} \end{bmatrix} \tag{4-20}$$

The above formulation extends easily to the multivariate case of n Gaussian variables. Define the column vectors

$$\mathbf{y} = \begin{bmatrix} y_1 \\ y_2 \\ \vdots \\ y_n \end{bmatrix} \quad \text{and} \quad \mathbf{m} = \begin{bmatrix} E\{y_1\} \\ E\{y_2\} \\ \vdots \\ E\{y_n\} \end{bmatrix} = \begin{bmatrix} m_1 \\ m_2 \\ \vdots \\ m_n \end{bmatrix}$$

and the covariance matrix

$$\mathbf{R} = \begin{bmatrix} E\{(y_1 - m_1)^2\} & \ldots & E\{(y_1 - m_1)(y_n - m_n)\} \\ \vdots & & \vdots \\ E\{(y_n - m_n)(y_1 - m_1)\} & \ldots & E\{(y_n - m_n)^2 \end{bmatrix} \tag{4-21}$$

It should be obvious that the ijth term $E\{(y_i - m_i)(y_j - m_j)\}$ is equal to the jith term so that the covariance matrix is symmetric. The multivariate Gaussian density function is given by

$$p(\mathbf{y}) = \frac{1}{(2\pi)^{n/2}|\mathbf{R}|^{1/2}} e^{-(\mathbf{y}-\mathbf{m})'\mathbf{R}^{-1}(\mathbf{y}-\mathbf{m})/2} \tag{4-22}$$

Except for the exponent of the 2π term, this is identical in form to that given in Eq. (4-19).

The multivariate characteristic function is (see Exercise 4.2)

$$C(j\boldsymbol{\omega}) = e^{j\mathbf{m}'\boldsymbol{\omega} - \boldsymbol{\omega}'\mathbf{R}\boldsymbol{\omega}/2} \tag{4-23}$$

where $\boldsymbol{\omega}$ is the column vector

$$\boldsymbol{\omega} = \begin{bmatrix} \omega_1 \\ \omega_2 \\ \vdots \\ \omega_n \end{bmatrix}$$

Consider the covariance matrix of Eq. (4-21). If the variables are uncorrelated, that is,

$$E\{(y_i - m_i)(y_j - m_j)\} = 0, \qquad i \neq j$$

the covariance matrix is diagonal,

$$\mathbf{R} = \begin{bmatrix} \sigma_1^2 & 0 \cdots 0 \\ 0 & \vdots \\ \vdots & \\ 0 & \cdots \quad \sigma_n^2 \end{bmatrix}$$

Then $|\mathbf{R}|^{1/2}$ is $\prod_{i=1}^{n} \sigma_i$ where $\prod_{i=1}^{n}$ denotes the n-fold product. It may be verified directly that the exponential term in Eq. (4-22) becomes

$$\exp\left[\sum_{i=1}^{n} -\frac{(y_i - m_i)^2}{2\sigma_i^2}\right] = \prod_{i=1}^{n} \exp\left[-\frac{(y_i - m_i)^2}{2\sigma_i^2}\right]$$

It follows that

$$p(\mathbf{y}) = \prod_{i=1}^{n} \frac{1}{(2\pi)^{1/2}\sigma_i} \exp\left[-\frac{(y_i - m_i)^2}{2\sigma_i^2}\right] = \prod_{i=1}^{n} p(y_i)$$

which is the product of n single variate Gaussian density functions. Therefore, if n Gaussian variables are uncorrelated, they are also statistically independent.

Linear Transformations

We next discuss a linear transformation of a multivariate Gaussian vector. For convenience, assume that the vector mean is zero. The linear transformation is given by

$$\begin{aligned} x_1 &= l_{11}y_1 + l_{12}y_2 + \cdots + l_{1n}y_n \\ x_2 &= l_{21}y_1 + l_{22}y_2 + \cdots + l_{2n}y_n \\ &\vdots \\ x_n &= l_{n1}y_1 + l_{n2}y_2 + \cdots + l_{nn}y_n \end{aligned} \tag{4-24}$$

In matrix form

$$\mathbf{x} = \mathbf{L}\mathbf{y} \tag{4-25}$$

where $\mathbf{y}$ is distributed as in Eq. (4-22) with $\mathbf{m} = \mathbf{0}$. Denote the inverse of $\mathbf{L}$ by $\mathbf{\Gamma}$. ($\mathbf{L\Gamma} = \mathbf{I}$ where $\mathbf{I}$ is the unit matrix.) Then

$$\mathbf{y} = \mathbf{\Gamma x} \tag{4-26}$$

relates the old variables to the new variables. It is not difficult to show that the Jacobian of the transformation is equal to the determinant of $\mathbf{\Gamma}$ and $|\mathbf{\Gamma}| = 1/|\mathbf{L}|$. Then, the distribution for $\mathbf{x}$ is

$$p(\mathbf{x}) = \frac{1}{(2\pi)^{n/2}(|\mathbf{L}|^2 \cdot |\mathbf{R}|)^{1/2}} e^{-(\mathbf{\Gamma x})'\mathbf{R}^{-1}(\mathbf{\Gamma x})/2}$$

Noting that

$$(\mathbf{\Gamma x})' = \mathbf{x}'\mathbf{\Gamma}' \qquad \text{and} \qquad \mathbf{\Gamma}' = (\mathbf{L}^{-1})' = (\mathbf{L}')^{-1}$$

the exponential term becomes

$$\mathbf{x}'\mathbf{\Gamma}'\mathbf{R}^{-1}\mathbf{\Gamma x}$$

Define

$$\mathbf{F}^{-1} = \mathbf{\Gamma}'\mathbf{R}^{-1}\mathbf{\Gamma} = (\mathbf{L}')^{-1}\mathbf{R}^{-1}\mathbf{L}^{-1} = (\mathbf{LRL}')^{-1}$$

Then it follows that

$$\mathbf{F} = \mathbf{LRL}' \tag{4-27}$$

and its determinant is

$$|\mathbf{F}| = |\mathbf{L}| \cdot |\mathbf{R}| \cdot |\mathbf{L}'| = |\mathbf{L}|^2 \cdot |\mathbf{R}|$$

The density function for $\mathbf{x}$ may now be expressed

$$p(\mathbf{x}) = \frac{1}{(2\pi)^{n/2}|\mathbf{F}|^{1/2}} e^{-\mathbf{x}'\mathbf{F}^{-1}\mathbf{x}/2} \tag{4-28}$$

This will be recognized as a multivariate Gaussian density function. We have shown that a linear transformation of a multivariate Gaussian vector produces a new multivariate Gaussian vector. The covariance matrix of the transformed variables is given by Eq. (4-27). (This covariance matrix could have been directly determined, see Chap. 11.)

Linear Transformation to Decorrelate Random Variables

Before continuing, we shall need several definitions and theorems concerning matrices. These are stated without proof. (See Guillemin (*11*) for a treatment. Chapter 11 contains other matrix relations.)

If $\mathbf{x}$ is a column vector with n elements, and $\mathbf{R}$ is a square matrix having n rows and columns (order n), then

$$Q = \mathbf{x}'\mathbf{R}\mathbf{x}$$

is called the quadratic form associated with the matrix $\mathbf{R}$. The quadratic form is a scalar. For example, define

$$\mathbf{x} = \begin{pmatrix} x_1 \\ x_2 \end{pmatrix} \qquad \text{and} \qquad \mathbf{R} = \begin{pmatrix} r_{11} & r_{12} \\ r_{21} & r_{22} \end{pmatrix}$$

Then the quadratic form is

$$Q = r_{11}x_1^2 + r_{12}x_1x_2 + r_{21}x_1x_2 + r_{22}x_2^2$$

The matrix $\mathbf{R}$ is said to be *positive definite* if its quadratic form is positive for all nonzero vectors $\mathbf{x}$. Similarly the matrix $\mathbf{R}$ is said to be *positive semi-definite* if its associated quadratic form is never negative for any vector $\mathbf{x}$ and is zero for some nonzero vector. If a matrix is positive definite, its determinant is not zero and the matrix has an inverse which is also positive definite. A matrix whose determinant is nonzero is called *nonsingular*.

A square nonsingular matrix $\mathbf{A}$ is *orthogonal* if its transpose is equal to its inverse ($\mathbf{A}' = \mathbf{A}^{-1}$). If this is so then $\mathbf{AA}' = \mathbf{I}$, the unit matrix. An example of such a matrix is one which describes a rotation of a Cartesian coordinate system. Such a transformation preserves the magnitude of a vector as well as the angles between arbitrary vectors.

Now consider the matrix equation

$$\mathbf{Rx} = \lambda \mathbf{x} \tag{4-29}$$

where $\mathbf{R}$ is a square symmetric matrix of order n, and $\mathbf{x}$ is a column vector, and λ is a scalar. For this equation to hold for nonzero $\mathbf{x}$ it is necessary that

$$|\mathbf{R} - \lambda \mathbf{I}| = 0 \tag{4-30}$$

The determinant in (4-30), which is called the characteristic polynomial of the matrix $\mathbf{R}$, is a polynomial of degree n in λ and there are n roots. If the matrix is symmetric and positive definite, the roots are real and positive. These roots, denoted λ_i, are called the eigenvalues of $\mathbf{R}$. The vectors $\mathbf{x}$ which correspond to the eigenvalues are called the eigenvectors. Denote the eigenvector corresponding to the eigenvalue λ_i by $\mathbf{x}^{(i)}$. Then the following equations are satisfied

$$\mathbf{Rx}^{(i)} = \lambda_i \mathbf{x}^{(i)}, \qquad i = 1, \ldots, n \tag{4-31}$$

Assuming the eigenvalues are distinct, the eigenfunctions are orthogonal. Since the eigenvectors are arbitrary to within a multiplicative constant, the constant may be chosen to normalize the eigenvectors to form an orthonormal set. That is,

$$\mathbf{x}^{(i)\prime}\mathbf{x}^{(j)} = \delta_{ij}$$

where the Kronecker delta function

$$\delta_{ij} = \begin{cases} 1, & i = j \\ 0, & i \neq j \end{cases}$$

We now form a matrix of the eigenvectors

$$\mathbf{A} = \begin{bmatrix} x_1^{(1)} & \cdots & x_1^{(n)} \\ x_2^{(1)} & \cdots & x_2^{(n)} \\ \vdots & & \\ x_n^{(1)} & \cdots & x_n^{(n)} \end{bmatrix} \tag{4-32}$$

and a diagonal matrix consisting of the eigenvalues

$$\boldsymbol{\lambda} = \begin{bmatrix} \lambda_1 & 0 & \cdots 0 \\ 0 & \lambda_2 & \cdots 0 \\ \vdots & & \\ 0 & 0 & \cdots \lambda_n \end{bmatrix} \tag{4-33}$$

The matrix $\mathbf{A}$ can be shown to be an orthogonal matrix. The n matrix equations, Eq. (4-31), can be expressed as

$$\mathbf{RA} = \mathbf{A}\boldsymbol{\lambda}$$

Premultiplying this equation by $\mathbf{A}'$ we get

$$\mathbf{A}'\mathbf{RA} = \boldsymbol{\lambda} \tag{4-34}$$

(Recall $\mathbf{A}'\mathbf{A} = \mathbf{I}$ for an orthogonal matrix $\mathbf{A}$.) Comparing this to Eq. (4-27), it is seen that we have found a transformation $\mathbf{A}'$ which transforms $\mathbf{R}$ into a diagonal matrix. The matrix $\boldsymbol{\lambda}$ is positive definite if $\mathbf{R}$ is positive definite.

The preceding is now applied to a multivariate or vector random variable. If such a vector random variable has a covariance matrix $\mathbf{R}$ which is positive definite, the linear transformation ($\mathbf{A}'$) of the variables is such that the covariance matrix of the transformed variables is diagonal. This implies that the transformed variables are uncorrelated. Furthermore, if the original variables were Gaussian the transformed variables will be statistically independent.

Mixed Central Moments

We shall now derive a particularly useful relationship for the mixed central moments of a Gaussian distribution. For zero mean Gaussian variables $x_1, x_2, \ldots, x_n$ we desire the mixed moment $E\{x_1^{b_1} x_2^{b_2} \cdots x_n^{b_n}\}$ where the b_i are positive integers. Denote $B = \sum_{i=1}^n b_i$. Then

$$E\{x_1^{b_1} \cdots x_n^{b_n}\} = (-j)^B \frac{\partial^B}{\partial\omega_1^{b_1} \cdots \partial\omega_n^{b_n}} C(j\omega_1, \ldots, j\omega_n)\bigg|_{\text{all } \omega_i = 0} \tag{4-35}$$

The characteristic function is

$$C(j\boldsymbol{\omega}) = e^{-\boldsymbol{\omega}'\mathbf{R}\boldsymbol{\omega}/2}$$

which can also be expressed as

$$C(j\boldsymbol{\omega}) = \exp\left\{-\tfrac{1}{2}\sum_{i=1}^{n}\sum_{k=1}^{n} \omega_k r_{ki} \omega_i\right\} \tag{4-36}$$

where r_{ki} is the kith element of the covariance matrix $\mathbf{R}$. Using the exponential expansion

$$e^{-x} = 1 - x + \frac{x^2}{2!} - \cdots = \sum_{p=0}^{\infty} (-1)^p \frac{x^p}{p!}$$

the characteristic function may be expressed as

$$C(j\boldsymbol{\omega}) = \sum_{p=0}^{\infty} \frac{(-1)^p}{2^p p!} \left[\sum_{i=1}^{n} \sum_{k=1}^{n} \omega_k r_{ki} \omega_i \right]^p \tag{4-37}$$

Therefore $C(j\boldsymbol{\omega})$ may be expanded in a series, each term of which will contain the ω's in different combinations and in different powers. If an ω_i appears with a power less than b_i, then the derivative operation of Eq. (4-35) will cause the term containing the ω_i to be zero. Similarly, if ω_i appears with an exponent greater than b_i, the operation of evaluating the derivative at $\omega_i = 0$ will also cause the term to be zero. Therefore the only terms of $C(j\boldsymbol{\omega})$ which contribute to the mixed moment are terms which contain $\omega_1^{b_1}\omega_2^{b_2} \cdots \omega_n^{b_n}$. From Eq. (4-37) we see that this can occur only when $p = B/2$. (Even for $p = B/2$ there are terms which do not contribute to the moment computation.) It also follows that if B is odd, the mixed moment is zero. Taking only the $p = B/2$ term we may express the mixed moment as

$$E\{x_1^{b_1} \cdots x_n^{b_n}\} = \frac{1}{2^{B/2}(B/2)!} \frac{\partial^B}{\partial\omega_1^{b_1} \cdots \partial\omega_n^{b_n}} \left[\sum_{i=1}^{n} \sum_{k=1}^{n} \omega_k r_{ki} \omega_i \right]^{B/2} \Bigg|_{\text{all } \omega_i = 0} \tag{4-38}$$

An important case of practical interest occurs when $n = 4$ and $b_i = 1$. Then

$$E\{x_1 x_2 x_3 x_4\} = \frac{1}{8} \frac{\partial^4}{\partial\omega_1\, \partial\omega_2\, \partial\omega_3\, \partial\omega_4} \sum_{i=1}^{4} \sum_{k=1}^{4} \sum_{l=1}^{4} \sum_{j=1}^{4} \omega_k \omega_i \omega_l \omega_j r_{ki} r_{lj} \Bigg|_{\text{all } \omega_i = 0} \tag{4-39}$$

The only terms which will be nonzero are those for which $i \neq k \neq l \neq j$. There are 24 such terms. The contribution of each will be $r_{ki} r_{lj}/8$. Since $r_{ki} = r_{ik}$, it may be shown that

$$E\{x_1 x_2 x_3 x_4\} = r_{12} r_{34} + r_{13} r_{24} + r_{14} r_{23}$$

or in more familiar form

$$E\{x_1 x_2 x_3 x_4\} = E\{x_1 x_2\} E\{x_3 x_4\} + E\{x_1 x_3\} E\{x_2 x_4\} + E\{x_1 x_4\} E\{x_2 x_3\} \tag{4-40}$$

EXAMPLE 4.1-1 A common application of Eq. (4-40) is the determination of the autocorrelation function of $y(t) = x^2(t)$ where $x(t)$ is a zero mean Gaussian process.

Denoting the autocorrelation functions of $x(t)$ and $y(t)$ by $R_x(\tau)$ and $R_y(\tau)$ respectively, we have

$$R_y(\tau) = E\{x^2(t)x^2(t-\tau)\} = E\{x(t)x(t)x(t-\tau)x(t-\tau)\}$$

Applying Eq. (4-40)

$$R_y(\tau) = R_x^{\,2}(0) + 2R_x^{\,2}(\tau) \tag{4-41}$$

The Gaussian Random Process

The earlier discussion of multivariate Gaussian vectors can be used to define the Gaussian random process by first interpreting the variables x_i as being samples of a random process taken at time instants t_i. Then, a random process is defined to be Gaussian if at all times t_i $(1 \leq i \leq n)$ and for all values of n the joint density function of these samples is given by the multivariate Gaussian density function.

In terms of time samples each element of the covariance matrix Eq. (4-21) may be written

$$r_{ij} = E\{(x_i - m_i)(x_j - m_j)\} = E\{x_i x_j\} - m_i m_j$$

In terms of the autocorrelation function, this is

$$r_{ij} = R_x(t_i, t_j) - m_i m_j$$

Therefore, any joint density function of the process depends on only the autocorrelation function and the mean as a function of time. If the process is wide-sense stationary, the mean value is constant and the covariance function will depend only on time differences such as $t_i - t_j$ and not on t_i and t_j individually. Since this is true for all time instants and any number of samples, it follows that a wide-sense stationary Gaussian process is also stationary in the strict sense.

Summary of Gaussian Properties

The Gaussian properties previously discussed are summarized below:

1. A linear transformation of a multivariate Gaussian vector produces another Gaussian vector. This includes as a special case the integral of a Gaussian process, which can be shown by considering the integral as a limiting Riemann sum (*9*). As another special case, the sum of dependent *jointly* Gaussian variables is Gaussian. It of course follows that the sum of statistically independent Gaussian variables is Gaussian.
2. If Gaussian random variables are uncorrelated, they are also statistically independent.

3. For variables which have a multivariate Gaussian distribution and a covariance matrix which is positive definite, a linear transformation can be found which transforms the dependent variables into a new set of statistically independent Gaussian variables.
4. For zero mean Gaussian variables x_1, x_2, x_3, and x_4, the following mixed moment relationship holds

$$E\{x_1x_2x_3x_4\} = E\{x_1x_2\}E\{x_3x_4\} + E\{x_1x_3\}E\{x_2x_4\} + E\{x_1x_4\}E\{x_2x_3\}$$

5. If a Gaussian process is stationary in the wide sense, it is also stationary in the strict sense.

4.2 Sum of a Sine Wave and a Gaussian Process

We shall derive the distribution of a single sample of the sum of a randomly phased sine wave and a Gaussian process (*12*, *13*),

$$r(t) = A \sin(\omega t + \theta) + n(t)$$

where θ is uniform $(0, 2\pi)$ and $n(t)$ is zero mean Gaussian with variance σ^2. The phase θ and the noise are assumed statistically independent. For a fixed value of θ the function $r(t)$ is a Gaussian random process but is not stationary. This follows from the fact that, for a given value of θ, the average value of $r(t)$ is $A \cos(\omega t + \theta)$ which is time dependent. To put it in other words, if we select at random a single sine wave of unknown but constant phase and add it to each member of an ensemble of Gaussian sample functions, the resulting ensemble will be a nonstationary but nevertheless Gaussian random process. On the other hand, if we combined the entire ensemble of randomly phased sine waves with the ensemble of Gaussian functions, the resulting ensemble will not be Gaussian but will be at least wide-sense stationary. We now derive the first-order density function of $r(t)$.

The characteristic function of a randomly phased sine wave $x = A \sin(\omega t + \theta)$ is (*12*)

$$C_x(j\omega) = \frac{1}{2\pi}\int_0^{2\pi} \exp\,[j\omega A \sin(\omega t + \theta)]\,d\theta = J_0(A\omega) \tag{4-42}$$

The inverse transformation would show that $p(x) = (1/\pi)(A^2 - x^2)^{-1/2}$ for $-A < x < A$.

For a sample of the Gaussian noise process, the characteristic function is

$$C_n(j\omega) = e^{-\sigma^2\omega^2/2} \tag{4-43}$$

Since a sample of $r(t)$ is the sum of two statistically independent samples, its characteristic function is

$$C_r(j\omega) = J_0(A\omega)e^{-\sigma^2\omega^2/2} \tag{4-44}$$

The inverse transformation produces (*13*) the density function

$$p(r) = \frac{1}{(2\pi\sigma^2)^{1/2}} \sum_{k=0}^{\infty} \frac{(-r^2/2\sigma^2)^k}{k!} \, {}_1F_1\left(k + \frac{1}{2}; 1; -\frac{A^2}{2\sigma^2}\right)$$

where ${}_1F_1(a; b; x)$ is the confluent hypergeometric function (*8*, *14*) defined by

$${}_1F_1(a; b; x) = 1 + \frac{a}{b}\frac{x}{1!} + \frac{a(a+1)}{b(b+1)}\frac{x^2}{2!} + \frac{a(a+1)(a+2)}{b(b+1)(b+2)}\frac{x^3}{3!} + \cdots \tag{4-45}$$

To simplify the expression somewhat, define the normalized variable $v = r/\sigma$. Also define $\alpha^2 = A^2/\sigma^2$, twice the ratio of signal power to noise power. The density function for the normalized variable is

$$p(v) = \frac{1}{(2\pi)^{1/2}} \sum_{k=0}^{\infty} \frac{(-v^2)^k}{2^k k!} \, {}_1F_1\left(k + \frac{1}{2}; 1; -\frac{\alpha^2}{2}\right)$$

The density function is shown plotted in Fig. 4-3. The departure from Gaussian is very evident for high signal amplitudes. Obviously, as the sine wave amplitude decreases, the density function approaches Gaussian.

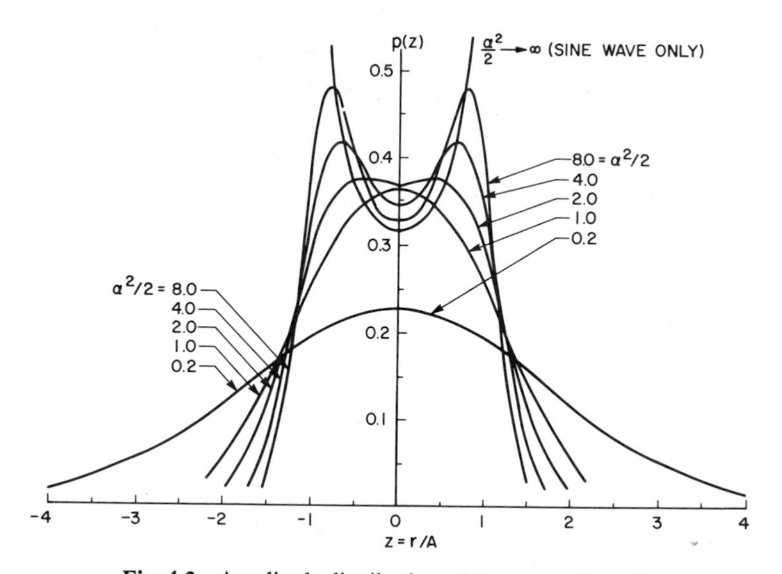

Fig. 4-3 Amplitude distribution of sine wave plus noise.

4.3 Distribution of the Envelope of a Narrowband Gaussian Process

We show that the envelope of a narrowband Gaussian process† is Rayleigh distributed.

The zero mean narrowband process is expressible as

$$n(t) = x(t)\cos \omega_c t - y(t)\sin \omega_c t$$

Denote its autocorrelation function by $R_n(\tau)$. The power or variance of the process is $R_n(0) = \sigma^2$. Since both $x(t)$ and $y(t)$ may be obtained by a linear operation, see Eqs. (3-47) and (3-48), on the Gaussian process $n(t)$, it follows that they are also Gaussian processes. They also have zero mean and variance σ^2. The envelope is given by (see Sect. 3.4)

$$z(t) = [x^2(t) + y^2(t)]^{1/2}$$

In Sect. 3.6 it was shown that $x(t)$ and $y(t)$ were uncorrelated at the same time instants. It follows that these Gaussian variables are statistically independent. Their joint density function is

$$p(x, y) = \frac{1}{2\pi\sigma^2} \exp\left[\frac{(x^2 + y^2)}{-2\sigma^2}\right]$$

where we have dropped the explicit time dependence. Define the "new" variables

$$z = [x^2 + y^2]^{1/2}, \qquad z \geq 0$$

and

$$\theta = \tan^{-1} y/x, \qquad 0 \leq \theta \leq 2\pi$$

(Note that the narrowband signal could be expressed as $n(t) = z(t)\cos(\omega_c t + \theta)$.) The inverse relations are

$$x = z \cos \theta, \qquad y = z \sin \theta$$

and the Jacobian of the transformation is z (which is positive). The joint density for z and θ is

$$p(z, \theta) = ze^{-z^2/2\sigma^2}/2\pi\sigma^2 \tag{4-46}$$

The density function for z can be obtained by integrating over θ

$$p(z) = \int_0^{2\pi} p(z, \theta)\, d\theta$$

† Unless specified otherwise all processes will be assumed to be wide-sense stationary.

This is easily shown to produce

$$p(z) = \frac{z}{\sigma^2} \exp\left[-\frac{z^2}{2\sigma^2}\right], \qquad z \geq 0 \tag{4-47}$$

which is the well-known Rayleigh distribution. The normalized distribution for $v = z/\sigma$ is shown in Fig. 4-4 along with other curves to be discussed shortly.

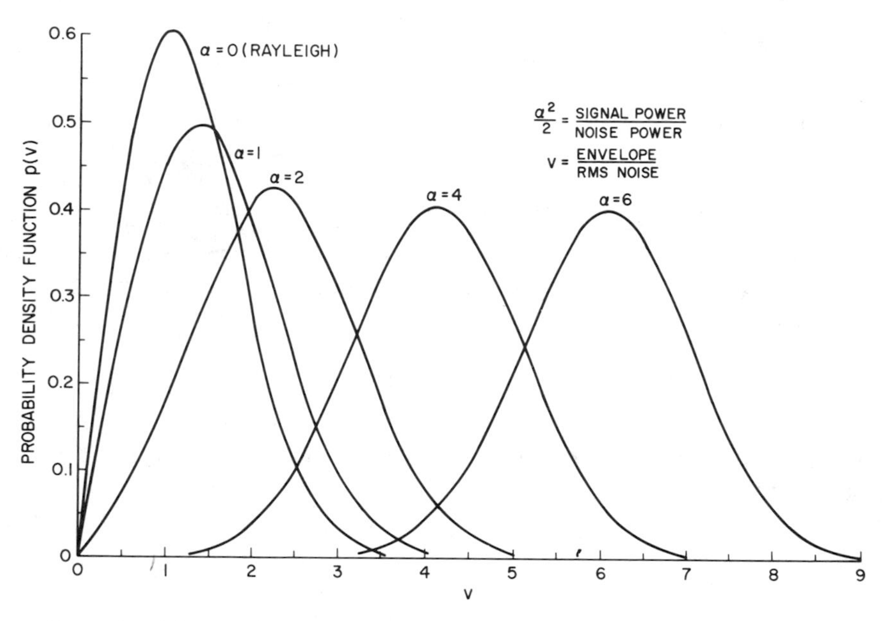

Fig. 4-4 Rician density function. $\alpha^2/2$ = signal power/noise power; v = envelope/rms noise.

The mean of the Rayleigh distribution is

$$E\{z\} = (\pi/2)^{1/2}\sigma$$

and the second moment is

$$E\{z^2\} = 2\sigma^2$$

By performing the integration $p(\theta) = \int_0^\infty p(z, \theta)\, dz$ it may be shown that the phase is uniformly distributed, that is, $p(\theta) = 1/2\pi$, $0 \leq \theta \leq 2\pi$.

From Eqs. (4-46) and (4-47) it follows that

$$p(z, \theta) = p(z)p(\theta) \tag{4-48}$$

so the envelope and phase samples at the same instant of time are statistically independent random *variables*. However, for envelope and phase samples at times t_1 and t_2, and denoted $z_1, \theta_1, z_2, \theta_2$, it may be shown that (*9*)

$$p(z_1, \theta_1, z_2, \theta_2) \neq p(z_1, \theta_1)\, p(z_2, \theta_2)$$

so the envelope and phase are not statistically independent random *processes*.

4.4 Envelope of a Sine Wave Plus Narrowband Noise

We now determine the density function for the envelope of a sine wave plus a narrowband Gaussian process (*12, 15, 16*). In this case the signal is of the form

$$\begin{aligned} r(t) &= A\cos(\omega_c t + \theta) + n(t) \\ &= [A\cos\theta + x(t)]\cos\omega_c t - [A\sin\theta + y(t)]\sin\omega_c t \end{aligned}$$

where θ is uniformly distributed $(0, 2\pi)$. The amplitude A and frequency ω_c are assumed known. The envelope is

$$z(t) = \{[A\cos\theta + x(t)]^2 + [A\sin\theta + y(t)]^2\}^{1/2}$$

Denote

$$z_c(t) = A\cos\theta + x(t), \qquad z_s(t) = A\sin\theta + y(t)$$

The envelope can then be expressed as

$$z(t) = [z_c^2(t) + z_s^2(t)]^{1/2}$$

For any given value of θ, both $z_c(t)$ and $z_s(t)$ are Gaussian variables and may be shown to be uncorrelated, hence statistically independent. Furthermore, for a given value of θ,

$$E\{z_c(t)\} = A\cos\theta \qquad \text{and} \qquad E\{z_s(t)\} = A\sin\theta$$
$$V\{z_c(t)\} = V\{z_s(t)\} = \sigma^2$$

where $V\{\cdot\}$ denotes variance. Therefore the density function of z_c and z_s conditioned on the phase θ is (dropping the explicit time dependence)

$$p(z_c, z_s \mid \theta) = \frac{1}{2\pi\sigma^2}\exp\left\{-\frac{1}{2\sigma^2}[(z_c - A\cos\theta)^2 + (z_s - A\sin\theta)^2]\right\}$$

The envelope is

$$z = [z_c^2 + z_s^2]^{1/2}, \qquad z \geq 0$$

Define a new variable ϕ

$$\phi = \tan^{-1} z_s/z_c, \qquad 0 \leq \phi \leq 2\pi$$

Then the expressions for the old variables in terms of the new variables is

$$z_s = z \sin \phi, \qquad z_c = z \cos \phi$$

and the Jacobian is z. Therefore

$$p(z, \phi \mid \theta) = \frac{z}{2\pi\sigma^2} \exp\left\{\frac{-1}{2\sigma^2} [z^2 + A^2 - 2Az \cos(\theta - \phi)]\right\} \tag{4-49}$$

and

$$p(z \mid \theta) = \int_0^{2\pi} p(z, \phi \mid \theta)\, d\phi$$

$$= \frac{z}{2\pi\sigma^2} \exp\left\{-\frac{1}{2\sigma^2} [z^2 + A^2]\right\} \int_0^{2\pi} \exp\left[\frac{Az}{\sigma^2} \cos(\theta - \phi)\right] d\phi$$

The integral is (*12*)

$$\int_0^{2\pi} \exp\left[\frac{Az}{\sigma^2} \cos(\theta - \phi)\right] d\phi = 2\pi I_0\left(\frac{Az}{\sigma^2}\right) \tag{4-50}$$

where $I_0(x)$ is the modified Bessel function† of zero order and is shown in Fig. 4-5. The conditional distribution of the envelope becomes

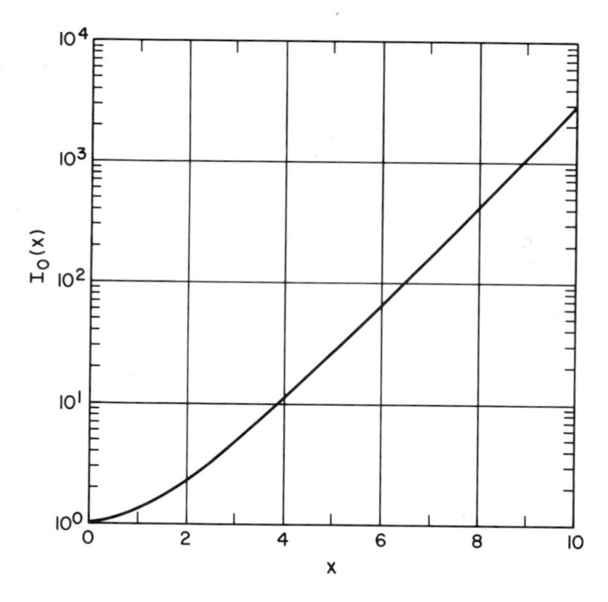

Fig. 4-5 Modified Bessel function.

† $I_0(x) = \sum_{n=0}^{\infty} \frac{x^{2n}}{2^{2n}(n!)^2}$ See (*8*).

$$p(z \mid \theta) = \frac{z}{\sigma^2} \exp\left[-\frac{1}{2\sigma^2}(z^2 + A^2)\right] I_0\left(\frac{Az}{\sigma^2}\right) \tag{4-51}$$

This is independent of ϕ so that the density function of the envelope of a sine wave plus narrowband Gaussian noise is

$$p(z) = \frac{z}{\sigma^2} \exp\left[-\frac{1}{2\sigma^2}(z^2 + A^2)\right] I_0\left(\frac{Az}{\sigma^2}\right) \tag{4-52}$$

This density will be referred to as Rician. It is sometimes called the generalized Rayleigh. As the amplitude A approaches zero, this function approaches the Rayleigh function as we would expect.

For convenience, define a normalized variable $v = z/\sigma$ and the constant $\alpha = A/\sigma$. The Rician density function may then be expressed as

$$p(v) = v \exp\left[\frac{v^2 + \alpha^2}{-2}\right] I_0(\alpha v) \tag{4-53}$$

This is shown in Fig. 4-4 for different values of α. The moments of the envelope are given by (*12*)

$$E\{z^n\} = (2\sigma^2)^{n/2} \Gamma\left(\frac{n}{2} + 1\right) {}_1F_1\left(-\frac{n}{2}; 1; -\frac{\alpha^2}{2}\right) \tag{4-54}$$

where $\alpha^2/2 = A^2/2\sigma^2$ is the input signal-to-noise ratio. The hypergeometric function is given in Eq. (4-45). The mean and standard deviation of z/σ are shown in Fig. 4-6. For large values of $\alpha^2/2$, the density function becomes

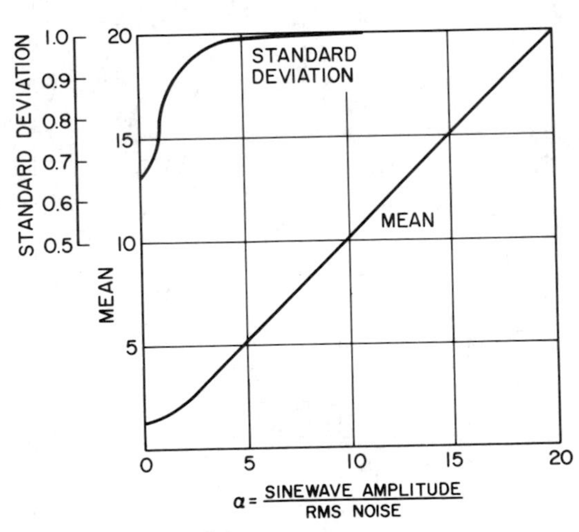

Fig. 4-6 Mean and standard deviation for Rician distribution (mean and standard deviation normalized to rms noise).

approximately Gaussian in the region around its mean value. See Exercise 4.8.

We shall find it convenient to plot one minus the cumulative probability distribution of a Rician variable; that is,

$$Q(\alpha, \beta) = 1 - \int_0^\beta v \exp\left[\frac{v^2 + \alpha^2}{-2}\right] I_0(\alpha v)\, dv$$

$$= \int_\beta^\infty v \exp\left[\frac{v^2 + \alpha^2}{-2}\right] I_0(\alpha v)\, dv \tag{4-55}$$

This is also called the Marcum Q-function (*17*) and is discussed further in Sect. 7.4. This function is shown in Figs. 4-7 and 4-8. Because the curves of Fig. 4-8 are very nearly straight lines for large values of $\alpha^2/2$, it follows that the Gaussian function is a good approximation to the Rician density function in the region about the mean.

We shall also derive the probability density function of the phase ϕ. As discussed in Sect. 4.2, it is sometimes important to distinguish whether one sample function of the phase ensemble or the entire phase ensemble is being considered. We shall derive the distribution of phase ϕ for a given value of θ. We therefore have conceptually the ensemble formed by adding a sine wave of phase θ to each member of the noise ensemble.

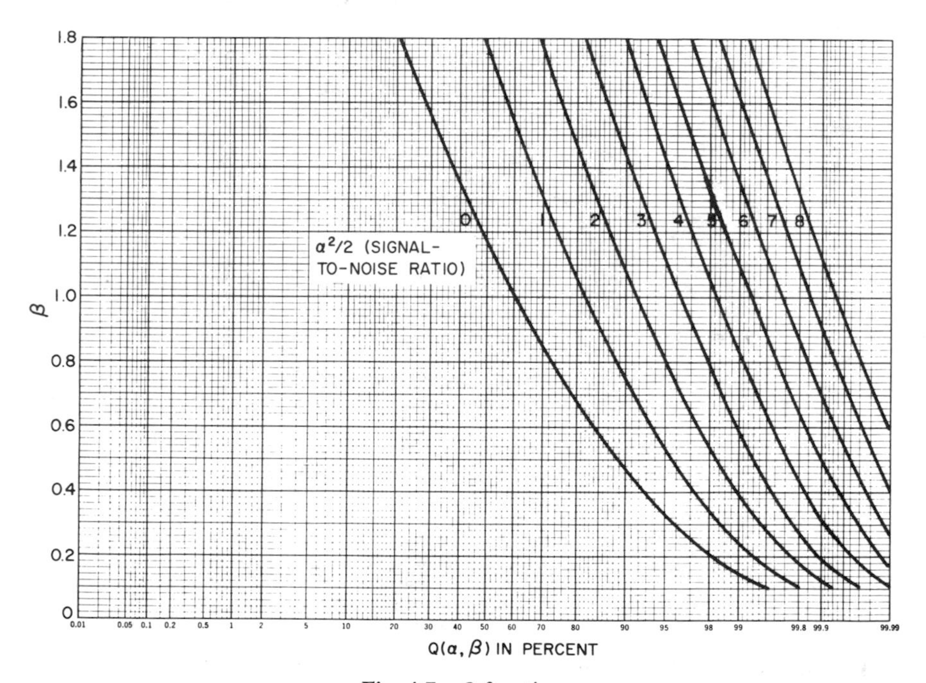

Fig. 4-7 Q-function.

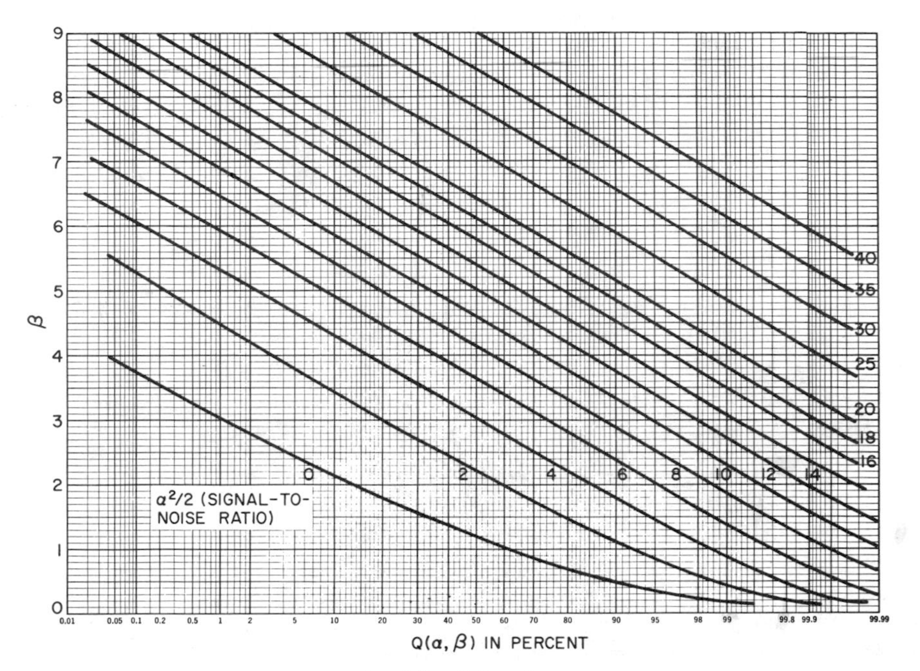

Fig. 4-8 *Q*-function.

Equation (4-49) is a logical starting point for such a determination. We perform the operation

$$p(\phi \mid \theta) = \int_0^\infty p(z, \phi \mid \theta)\, dz$$

and obtain

$$p(\phi \mid \theta) = \exp\left[-\frac{A^2}{2\sigma^2}\sin^2(\theta - \phi)\right]\left(\frac{1}{2\pi\sigma^2}\right)\int_0^\infty z \exp\left\{\frac{[z - A\cos(\theta - \phi)]^2}{-2\sigma^2}\right\} dz$$

After some manipulation we can show that

$$p(\phi \mid \theta) = \frac{\exp(-A^2/2\sigma^2)}{2\pi} + \frac{A\cos(\theta - \phi)}{2(2\pi)^{1/2}\sigma}\exp\left[-\frac{A^2}{2\sigma^2}\sin^2(\theta - \phi)\right] \times \left\{1 + \operatorname{erf}\left[\frac{A\cos(\theta - \phi)}{2^{1/2}\sigma}\right]\right\} \tag{4-56}$$

where the error function is defined as

$$\operatorname{erf}(x) = \frac{2}{\pi^{1/2}}\int_0^x e^{-z^2}\, dz \tag{4-57}$$

The phase function is shown in Fig. 4-9 for several values of the parameter $A^2/2\sigma^2$. This is the ratio of signal power to noise power.

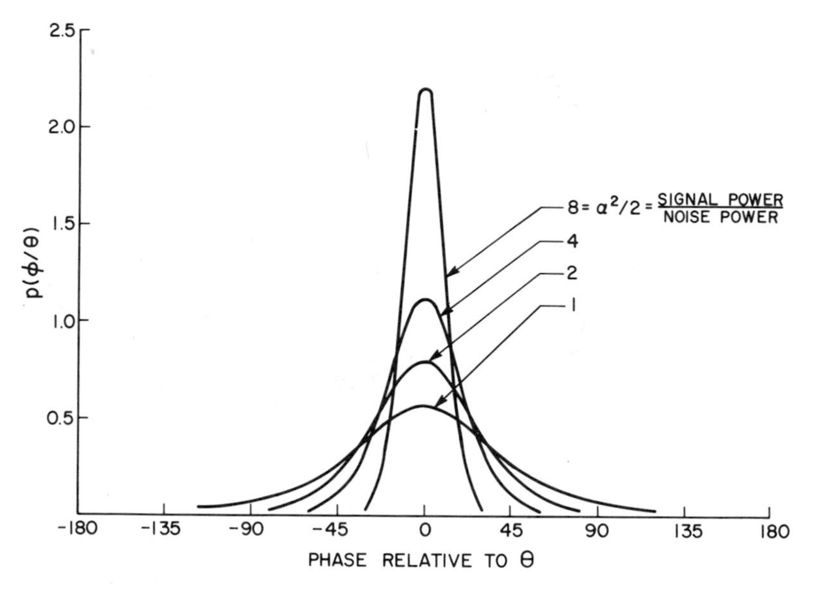

Fig. 4-9 Density function for phase of sine wave plus noise.

4.5 Envelope Squared of Narrowband Process

The envelope detector of the previous section is often used in practice since it can be implemented easily. However it is difficult to work analytically with the envelope detector primarily because of the square root operation. For this reason it is often convenient to do the analysis for a quadratic detector which is much easier. The quadratic detector obtains the square of the envelope. In many applications the difference in performance between the envelope and quadratic detectors is slight (see Chap. 8) so this substitution is justified.

From Sect. 4.3 the envelope of the narrowband Gaussian signal is

$$z = [x^2 + y^2]^{1/2}$$

where the time dependence is suppressed. The density function, from Eq. (4-47), is

$$p(z) = \frac{z}{\sigma^2} \exp\left[-\frac{z^2}{2\sigma^2}\right], \qquad z \geq 0$$

The envelope squared is obviously

$$u = z^2, \qquad z, u \geq 0 \tag{4-58}$$

The density function can be determined immediately by a change of variable in Eq. (4-47). This produces

$$p(u) = \frac{1}{2\sigma^2} \exp\left[-\frac{u}{2\sigma^2}\right], \qquad u \geq 0 \tag{4-59}$$

which is an exponential distribution.

A special case of interest occurs for $\sigma^2 = 1$. For this we obviously have

$$p(u) = \tfrac{1}{2}e^{-u/2} \tag{4-60}$$

This is a special case of the chi-squared distribution with two degrees of freedom. (See Sect. 4.6.)

The characteristic function is easily verified to be

$$C_u(j\omega) = \frac{1}{1 - 2j\omega} \tag{4-61}$$

and the mean and variance are respectively

$$E\{u\} = 2, \qquad V\{u\} = 4$$

4.6 Chi-Squared Distribution

A frequent operation in detection of signals is "postdetection integration" (see Chap. 8) in which many independent samples of the output of a detector are summed to improve detection performance. If the input to a quadratic detector is a narrowband Gaussian process, the output will be chi-squared distributed. This variable is often denoted as χ^2. However, since the variable is χ^2 and not just χ, there is sometimes mild confusion and we shall generally avoid that notation in equations although we shall refer to the distribution as being "χ^2."

The derivation of the χ^2 distribution may be stated independently of detectors, and we do so as follows: let $x_1, x_2, \ldots, x_n$ be statistically independent Gaussian variables with zero mean and unit variance. Then the distribution of s where

$$s = x_1^2 + x_2^2 + \cdots + x_n^2 \tag{4-62}$$

is χ^2 distributed with n degrees of freedom (*18*).

We shall derive the χ^2 density function by first determining the characteristic function of s. From Eq. (1-15), the distribution of $y = x_i^2$ is

$$p(y) = \frac{e^{-y/2}}{(2\pi y)^{1/2}}, \qquad y \geq 0$$

and it may be verified that the characteristic function is

$$C_y(j\omega) = \frac{1}{(1 - 2j\omega)^{1/2}} \tag{4-63}$$

It follows that the characteristic function for the sum, Eq. (4-62), is

$$C_s(j\omega) = \frac{1}{(1 - 2j\omega)^{n/2}} \tag{4-64}$$

The probability density function, obtained by the inverse Fourier transformation, is (Campbell and Foster (*19*), pair 524.2)

$$p(s) = \frac{1}{2^{n/2}\Gamma(n/2)} s^{n/2-1} e^{-s/2}, \qquad s \geq 0 \tag{4-65}$$

This is defined as the χ^2 distribution with n degrees of freedom. The function is shown in Fig. 4-10 for several degrees of freedom. The cumulative distribu-

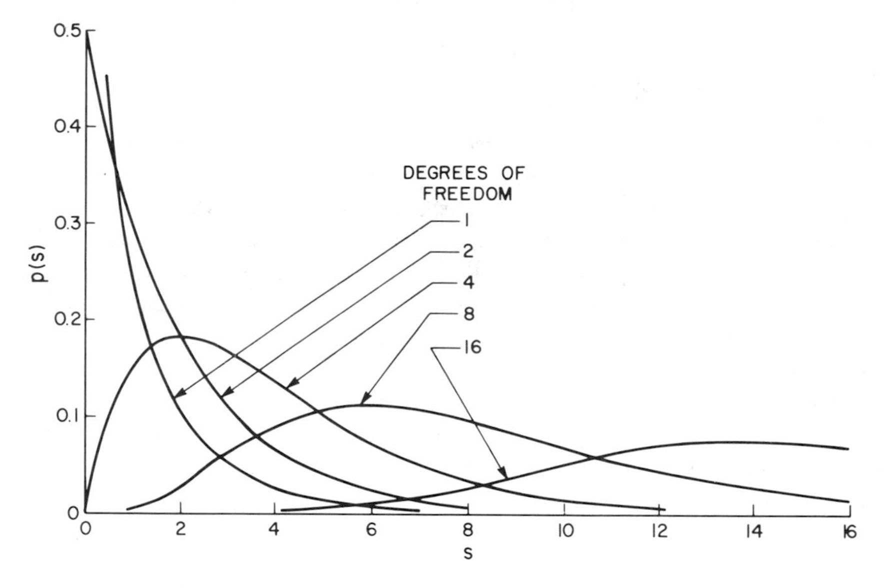

Fig. 4-10 Probability density functions for χ^2 variables.

tion in integral form is

$$P\{s \leq S\} = \int_0^S \frac{s^{n/2-1} e^{-s/2}}{2^{n/2}\Gamma(n/2)} \, ds \tag{4-66}$$

and is shown in Fig. 4-11. Making a change of variable ($t = s/2$) and substituting $p = n/2 - 1$ and $u = S/[2(p + 1)^{1/2}]$, this function can be put in the form

$$I(u, p) = \frac{1}{\Gamma(p + 1)} \int_0^{u(p+1)^{1/2}} t^p e^{-t} \, dt \tag{4-67}$$

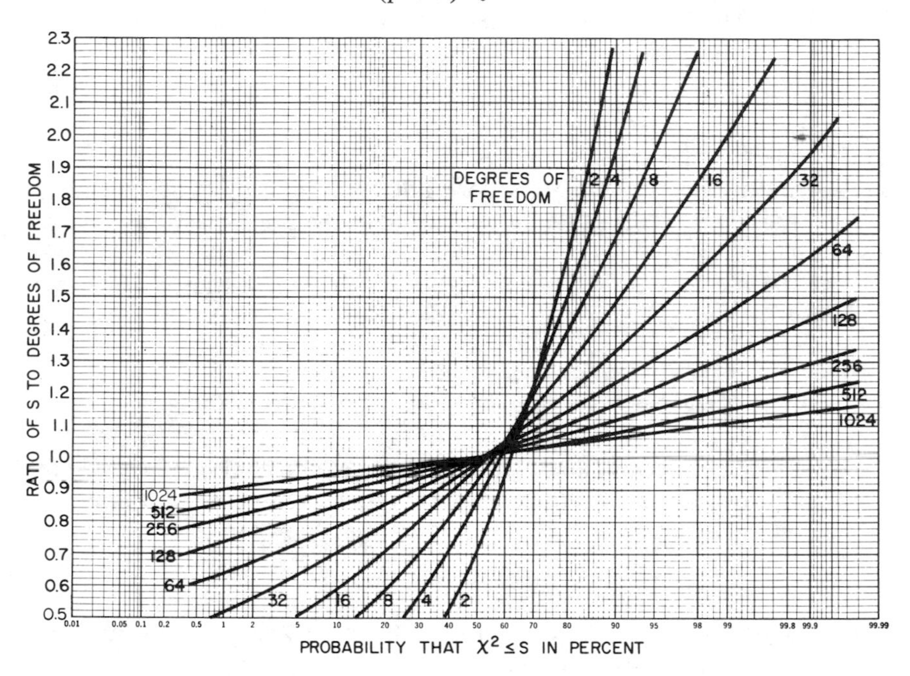

Fig. 4-11 Cumulative χ^2 distribution.

This is Pearson's incomplete gamma function, which is tabulated (*20*). Tabulated values of the χ^2 distribution also appear in Pearson and Hartley (*21*) and Fisher and Yates (*22*).

It should be clear that the sum of two statistically independent chi-squared variables is itself chi-squared. In particular, if s_N and s_M are statistically independent chi-squared variates with N and M degrees of freedom respectively, their sum is chi-squared distributed with $N + M$ degrees of freedom.

From Eq. (4-62) the mean and variance of s are easily determined. They are

$$E\{s\} = n \qquad \text{and} \qquad V\{s\} = 2n$$

Knowing just these moments, the number of degrees of freedom may be found using

$$n = 2E^2\{s\}/V\{s\}$$

As a final note in this section we determine the density function of

$$s' = y_1^2 + y_2^2 + \cdots + y_n^2$$

where the y_i are statistically independent Gaussian random variables with zero mean and variance σ^2. The distribution for s' may be obtained directly from Eq. (4-65) with the change of variable $s' = \sigma^2 s$. The result is

$$p(s') = \frac{1}{(\sigma^2)^{n/2} 2^{n/2} \Gamma(n/2)} (s')^{n/2-1} \exp\left[-\frac{s'}{2\sigma^2}\right] \tag{4-68}$$

This is called the gamma distribution. The previous comment relating to the sum of independent χ^2 variables does not apply for independent gamma variables *unless* σ^2 is the same for each. If σ^2 is the same, then the sum will also be gamma distributed.

For the special case when $n = 2$, the exponential distribution given in Eq. (4-59) is obtained.

4.7 Envelope Squared of a Sine Wave Plus a Narrowband Process

We consider here the square of the envelope of a randomly phased sine wave and a narrowband Gaussian process. To be consistent with Sect. 4.4, assume the signal to be

$$r(t) = A \cos(\omega_c t + \theta) + n(t)$$

where θ is uniformly distributed over the interval $(0, 2\pi)$. Using the notation of Sect. 4.4, the envelope squared is

$$q(t) = z^2(t) = [A \cos\theta + x(t)]^2 + [A \sin\theta + y(t)]^2$$

We can obtain the conditional density function for q by making the transformation $q = z^2$ in Eq. (4-51). The result is easily seen to be

$$p(q \mid \theta) = \frac{1}{2\sigma^2} \exp\left[-\frac{1}{2\sigma^2}(q + A^2)\right] I_0\left(\frac{Aq^{1/2}}{\sigma^2}\right)$$

Carrying out the operation

$$p(q) = \int_0^{2\pi} p(q \mid \theta) p(\theta)\, d\theta$$

where $p(\theta) = 1/2\pi$ will obviously produce

$$p(q) = \frac{1}{2\sigma^2} \exp\left[-\frac{1}{2\sigma^2}(q + A^2)\right] I_0\left(\frac{Aq^{1/2}}{\sigma^2}\right) \tag{4-69}$$

This is a special case of the noncentral χ^2 distribution discussed further in Sect. 4.8. The characteristic function, the derivation of which is discussed in Sect. 4.8, is

$$C_q(j\omega) = \exp\left(-\frac{A^2}{\sigma^2}\right)\left(\frac{1}{1-j2\sigma^2\omega}\right)\exp\left(\frac{A^2/\sigma^2}{1-j2\sigma^2\omega}\right) \tag{4-70}$$

4.8 Noncentral Chi-Squared Distribution

It was pointed out in Sect. 4.6 that a common operation in detection of signals is summing output samples of a quadratic detector. If the input is a narrowband Gaussian process, the sum of the statistically independent samples will be χ^2 distributed. This corresponds to the noise only case. If the input is a sine wave signal plus a narrowband Gaussian process, the sum will be noncentral χ^2 distributed (*23*). We now derive this distribution.

We are interested in the sum of random variables

$$q' = \sum_{i=1}^{N}(A + x_i)^2 \tag{4-71}$$

where the x_i are identically distributed and statistically independent Gaussian variables with zero mean and variance σ^2. (The noncentral χ^2 distribution will be defined with $\sigma^2 = 1$.) We proceed by first finding the probability density and characteristic functions of $q_1 = (A + x_1)^2$. The probability density function of $y_1 = A + x_1$ is obviously

$$p(y_1) = \frac{1}{(2\pi)^{1/2}\sigma}\exp\left[\frac{-(y_1 - A)^2}{2\sigma^2}\right]$$

Using Eq. (1-15), the density function for $q_1 = y_1{}^2$ is

$$p(q_1) = \left(\frac{1}{8\pi\sigma^2 q_1}\right)^{1/2}\left\{\exp\left[-\frac{(q_1^{1/2} - A)^2}{2\sigma^2}\right] + \exp\left[-\frac{(-q_1^{1/2} - A)^2}{2\sigma^2}\right]\right\}$$

Carrying out the square in the exponential terms and making use of the fact that $e^b + e^{-b} = 2\cosh b$, we get for the density function

$$p(q_1) = \left(\frac{1}{2\pi\sigma^2 q_1}\right)^{1/2}\exp\left(\frac{q_1 + A^2}{-2\sigma^2}\right)\cosh\left(\frac{Aq_1^{1/2}}{\sigma^2}\right) \tag{4-72}$$

The characteristic function (using Campbell and Foster, pair 651) is

$$C_{q_1}(j\omega) = \exp\left(\frac{-A^2}{2\sigma^2}\right)\left(\frac{1}{1-j2\sigma^2\omega}\right)^{1/2}\exp\left(\frac{A^2/2\sigma^2}{1-j2\sigma^2\omega}\right) \tag{4-73}$$

The characteristic function for q' is therefore

$$C_{q'}(j\omega) = \exp\left(-\frac{NA^2}{2\sigma^2}\right)\left(\frac{1}{1 - j2\sigma^2\omega}\right)^{N/2} \exp\left(\frac{NA^2/2\sigma^2}{1 - j2\sigma^2\omega}\right) \tag{4-74}$$

After some manipulation, and using Campbell and Foster, pair 650, the inversion of $C_{q'}(j\omega)$ produces the density function†

$$p(q') = \frac{1}{2\sigma^2}\left(\frac{q'}{\lambda'}\right)^{(N-2)/4} \exp\left(\frac{\lambda' + q'}{-2\sigma^2}\right) I_{N/2-1}\left[\frac{(q'\lambda')^{1/2}}{\sigma^2}\right] \tag{4-75}$$

The range is $q' > 0$, $\lambda' \triangleq A^2N$ is defined as the noncentral parameter, and $I_n(\cdot)$ is the modified Bessel Function (*24*) of the first kind and order m. To put this in more compact form, define a normalized variable $q = q'/\sigma^2$. Then, from Eq. (4-71),

$$q = \sum_{i=1}^{N}\left(\frac{A}{\sigma} + \frac{x_i}{\sigma}\right)^2 \tag{4-76}$$

The variable x_i/σ is Gaussian with zero mean and unit variance. The density function for q is then easily shown to be

$$p(q) = \frac{1}{2}\left(\frac{q}{\lambda}\right)^{(N-2)/4} \exp\left(-\frac{\lambda}{2} - \frac{q}{2}\right) I_{N/2-1}((q\lambda)^{1/2}) \tag{4-77}$$

where $\lambda = NA^2/\sigma^2$ is defined as the noncentral parameter. This is the noncentral χ^2 distribution with N degrees of freedom. Plots of the noncentral χ^2 function are shown in Figs. 4-12 and 4-13 for selected values of λ and N.

We now derive the probability distribution of

$$Q' = \sum_{i=1}^{N}(A_i + x_i)^2 \tag{4-78}$$

where the x_i are statistically independent zero mean Gaussian variables with variance σ^2. In particular, we shall relate the distribution of Q' to that of q', Eq. (4-71).

For a single sample $Q_i' = (A_i + x_i)^2$, the characteristic function follows immediately from Eq. (4-73).

$$C_{Q_i'}(j\omega) = \exp\left(-\frac{A_i^2}{2\sigma^2}\right)\left(\frac{1}{1 - j2\sigma^2\omega}\right)^{1/2} \exp\left(\frac{A_i^2/2\sigma^2}{1 - j2\sigma^2\omega}\right)$$

Consequently, the characteristic function for Q' is

$$C_{Q'}(j\omega) = \left(\frac{1}{1 - j2\sigma^2\omega}\right)^{N/2} \exp\left(\frac{\sum A_i^2}{-2\sigma^2}\right) \exp\left(\frac{\sum A_i^2/2\sigma^2}{1 - j2\sigma^2\omega}\right) \tag{4-79}$$

† In analogy to Sect. 4.6, we could call this the noncentral gamma density function. It is not standard usage however.

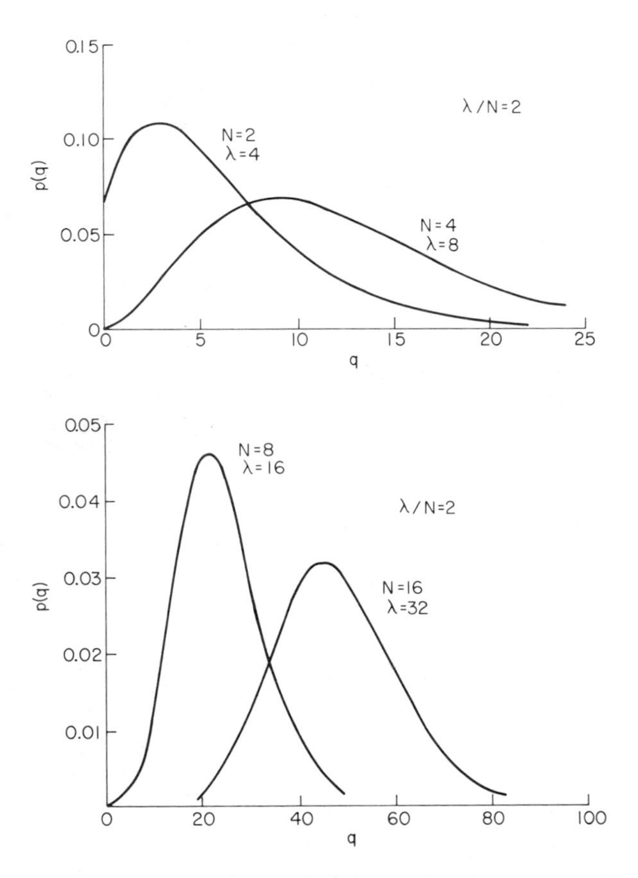

Fig. 4-12 Noncentral χ^2 density functions.

(The summations extend from $i = 1$ to N.) In analogy to Eq. (4-74), define $NA^2 = \sum_{i=1}^{N} A_i^2$. Then, Eq. (4-79) is identical to Eq. (4-74). Thus Q' has the same probability density function as q', and it is given by Eq. (4-75) except that the noncentral parameter is $\lambda' = \sum_{i=1}^{N} A_i^2$.

We are now in a position to generate the normalized noncentral χ^2 distribution in a manner which is more general than that given in Eq. (4-76). If the y_i's are statistically independent Gaussian random variables with mean m_i and unit variance, the sum given by

$$q = \sum_{i=1}^{N} y_i^2 \tag{4-80}$$

has a noncentral χ^2 distribution with N degrees of freedom and a noncentral parameter $\lambda = \sum_{i=1}^{N} m_i^2$. The noncentral χ^2 density function is given by Eq. (4-77).

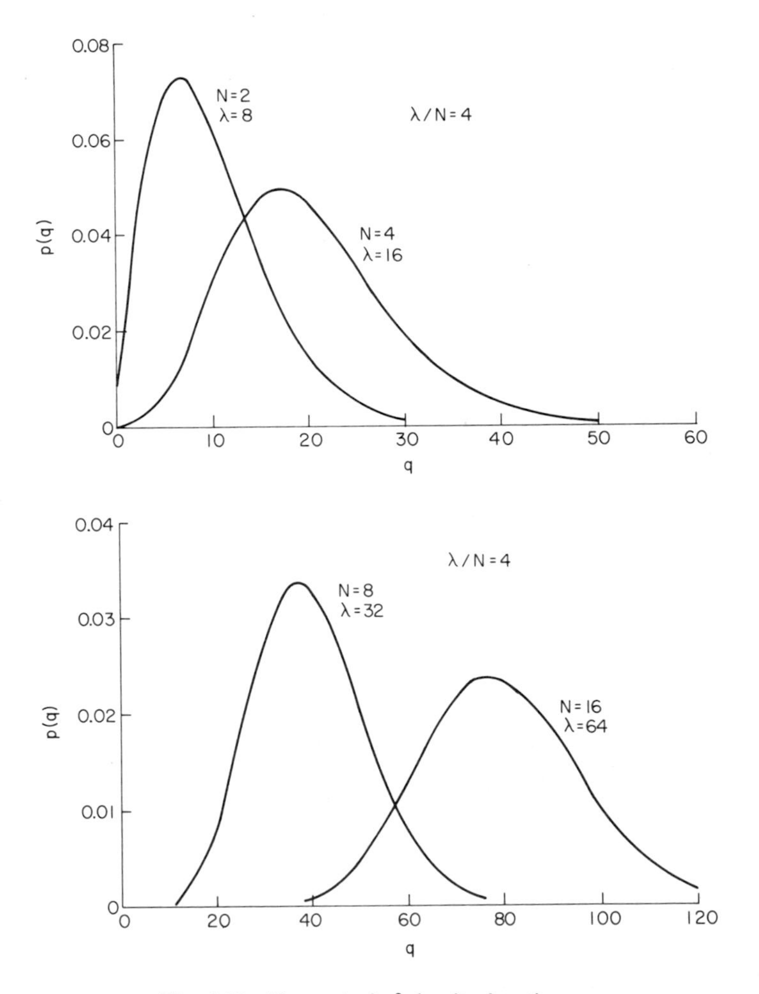

Fig. 4-13 Noncentral χ^2 density functions.

The mean and variance of q are easily shown to be

$$E\{q\} = \lambda + N \qquad \text{and} \qquad V\{q\} = 4\lambda + 2N$$

It should be clear that the sum of two statistically independent noncentral χ^2 variables is also a noncentral χ^2 variable. In particular if q_N and q_M are statistically independent noncentral χ^2 variables with N and M degrees of freedom and noncentral parameters λ_1 and λ_2 respectively, their sum is a noncentral χ^2 variable with $N + M$ degrees of freedom with a noncentral parameter given by $\lambda_1 + \lambda_2$.

We may now apply these results to the example of a sine wave plus a narrowband Gaussian process as inputs to a quadratic detector with post detection summation as shown in Fig. 4-14. The input to the quadratic detector is

$$r(t) = A\cos(\omega_o t + \theta) + n(t)$$

where the noise has variance σ^2. We represent the noise by

$$n(t) = x(t)\cos\omega_o t - y(t)\sin\omega_o t$$

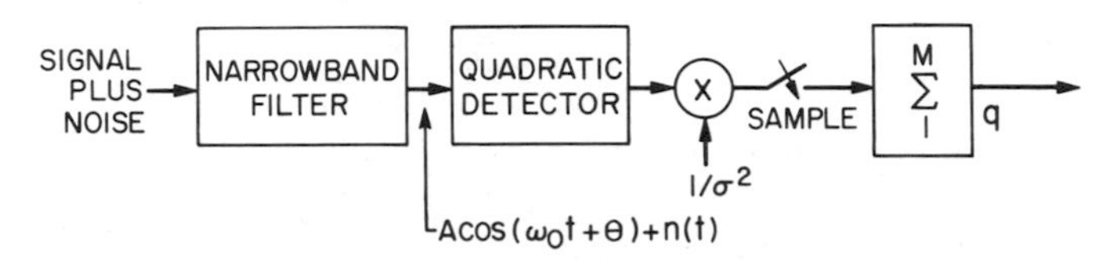

Fig. 4-14 Postdetection summation.

The samples at the output of the quadratic detector are assumed statistically independent. For normalization, divide the output of the detector by σ^2, the noise power.

The output of the quadratic detector is the envelope squared of the input signal

$$[A\cos\theta + x]^2 + [A\sin\theta + y]^2$$

The explicit time dependence has been suppressed. The output q of the post-detection summer is given by

$$q = \sum_{i=1}^{M}\left(\frac{A\cos\theta}{\sigma} + \frac{x_i}{\sigma}\right)^2 + \sum_{i=1}^{M}\left(\frac{A\sin\theta}{\sigma} + \frac{y_i}{\sigma}\right)^2 \qquad (4\text{-}81)$$

We immediately see that the first summation is noncentral χ^2 with M degrees of freedom and noncentral parameter

$$\lambda_1 = \sum_{i=1}^{M}\left(\frac{A}{\sigma}\cos\theta\right)^2 = \frac{MA^2}{\sigma^2}\cos^2\theta$$

The second summation is statistically independent of the first and has a noncentral χ^2 distribution with M degrees of freedom and noncentral parameter

$$\lambda_2 = \sum_{i=1}^{M}\left(\frac{A\sin\theta}{\sigma}\right)^2 = \frac{MA^2}{\sigma^2}\sin^2\theta$$

It therefore follows that q is noncentral χ^2 with $2M$ degrees of freedom and noncentral parameter $\lambda = MA^2/\sigma^2$. Note that the distribution is independent of θ, and that each sample of the detector output has two degrees of freedom. Using Eq. (4-77) the distribution of q is

$$p(q) = \frac{1}{2}\left(\frac{q}{\lambda}\right)^{(M-1)/2} \exp\left(-\frac{\lambda}{2} - \frac{q}{2}\right) I_{M-1}((q\lambda)^{1/2}) \tag{4-82}$$

for $q \geq 0$, and $\lambda = MA^2/\sigma^2$.

The ratio of the noncentral parameter and the degrees of freedom is equal to

$$\frac{\lambda}{2M} = \frac{A^2}{2\sigma^2}$$

which is the ratio of signal power to noise power at the input to the quadratic detector. Thus Figs. 4-12 and 4-13 for the noncentral density function are for signal-to-noise ratios of 2 and 4 respectively.

Exercises

4.1 Assume the Gaussian vector $\mathbf{x}$ is zero mean and has a covariance matrix $\mathbf{R}_x$. Show that the characteristic function of the random variable $q = \mathbf{x}'\mathbf{Q}\mathbf{x}$ is equal to

$$\frac{1}{|\mathbf{I} - j2\omega\mathbf{R}_x\mathbf{Q}|^{1/2}}$$

Hint: It is obvious if $\mathbf{x}$ is n-dimensional that

$$\int_{\text{all } \mathbf{x}} \frac{1}{(2\pi)^{n/2}|\mathbf{R}_x|^{1/2}} \exp(-\tfrac{1}{2}\mathbf{x}'\mathbf{R}_x^{-1}\mathbf{x})\, d\mathbf{x} = 1$$

4.2 For $\mathbf{x}$ an $n+1$ column vector prove the integral relation

$$\int_{-\infty}^{\infty} \frac{1}{(2\pi)^{n/2}|\mathbf{R}|^{1/2}} \exp(-\tfrac{1}{2}\mathbf{x}'\mathbf{R}^{-1}\mathbf{x} - 2\mathbf{A}\mathbf{x})\, d\mathbf{x} = \exp(\tfrac{1}{2}\mathbf{A}'\mathbf{R}'\mathbf{A})$$

Then derive the characteristic function of the multivariate Gaussian vector. That is, show

$$C(j\boldsymbol{\omega}) = \exp(j\mathbf{m}'\boldsymbol{\omega} - \tfrac{1}{2}\boldsymbol{\omega}'\mathbf{R}\boldsymbol{\omega})$$

where $\mathbf{m}$ and $\mathbf{R}$ are the mean and covariance matrices.

4.3 Assume white Gaussian noise of power spectral density $N_o/2$ is inserted into a filter with transfer function

$$H(j\omega) = \frac{1}{1 + j\omega/\omega_1}$$

Find the probability density functions of the output and the envelope of the output.

4.4 What is the probability density function of $z(t) = A\cos\omega_c t + n(t)$? Assume A and ω_c constant, and $n(t)$ zero mean Gaussian with power σ^2.

4.5 Consider the following orthogonal transformation of the correlated, zero mean variables y_1 and y_2

$$\begin{bmatrix}\cos\theta & -\sin\theta \\ \sin\theta & \cos\theta\end{bmatrix}\begin{bmatrix}y_1 \\ y_2\end{bmatrix}$$

Define $E\{y_1{}^2\} = \sigma_{y_1}^2$, $E\{y_2{}^2\} = \sigma_{y_2}^2$, and $E\{y_1 y_2\} = \rho\sigma_{y_1}\sigma_{y_2}$, with $\sigma_{y_1} = \sigma_{y_2}$. Show that to make the transformed variables uncorrelated θ must be chosen such that

$$\tan 2\theta = \frac{2\rho\sigma_{y_1}\sigma_{y_2}}{\sigma_{y_2}^2 - \sigma_{y_1}^2}$$

4.6 Consider the narrowband Gaussian process

$$n(t) = x(t)\cos\omega_c t - y(t)\sin\omega_c t$$

and assume that the spectral density is symmetric about the carrier frequency ω_c. Determine the multivariate probability density function $p(x_t, x_{t-\tau}, y_t, y_{t-\tau})$.

4.7 Consider the narrowband signal

$$z(t) = A\cos(\omega_c t + \theta) + n(t)$$

with $n(t)$ Gaussian and

$$n(t) = x(t)\cos\omega_c t - y(t)\sin\omega_c t$$

(a) Show that the correlation function of the envelope squared of $z(t)$ is

$$R(\tau) = A^4 + 4A^2\sigma^2 + 4\sigma^4 + 4\left(A^2 R_x(\tau) + R_x{}^2(\tau) + R_{xy}^2(\tau)\right)$$

(b) What is the power spectral density for the noise only case if the noise spectral density is symmetric about ω_c for $\omega > 0$.

4.8 Assume that the variables x and y are jointly Gaussian with zero mean. Denote $E\{x^2\} = \sigma_x{}^2$, $E\{y^2\} = \sigma_y{}^2$, $E\{xy\} = \rho\sigma_x\sigma_y$. Show that the conditional density function of y given x is

$$p(y\,|\,x) = \frac{1}{(2\pi)^{1/2}\sigma_y(1-\rho^2)^{1/2}}\exp\left[\frac{(y - \rho(\sigma_y/\sigma_x)x)^2}{-\sigma_y{}^2(1-\rho^2)}\right]$$

4.9 This exercise demonstrates that RC filtered white Gaussian noise (cf Example 2.7-1) is a Markov process (Exercise 2.4). Such a process, denote it $y(t)$, will have an exponential correlation function such as $e^{-\omega_1|\tau|}$. Define $t_3 - t_2 = t_2 - t_1 = \Delta$, $e^{-\omega_1|\Delta|} \triangleq \rho$, $y(t_3) = y_3$, $y(t_2) = y_2$, and $y(t_1) = y_1$ where $t_3 > t_2 > t_1$. Show that

$$p(y_1 \mid y_2, y_3) = \frac{1}{(2\pi)^{1/2}(1-\rho^2)^{1/2}} \exp\left[\frac{(y_1 - \rho y_2)^2}{-2(1-\rho^2)}\right]$$

and therefore that

$$p(y_1 \mid y_2, y_3) = p(y_1 \mid y_2)$$

4.10 Show that

(a) The eigenvalues of the matrix

$$\mathbf{R} = \begin{bmatrix} 2 & 3^{1/2}/2 \\ 3^{1/2}/2 & 3 \end{bmatrix}$$

are 1.5 and 3.5, and that the matrix of eigenvectors is

$$\mathbf{A} = \begin{bmatrix} 3^{1/2}/2 & 1/2 \\ -1/2 & 3^{1/2}/2 \end{bmatrix}$$

(b) Verify that the matrix $\mathbf{A}$ is orthogonal.

(c) Verify by direct calculation that $\mathbf{RA} = \mathbf{A}\boldsymbol{\lambda}$.

(d) If samples x_1 and x_2 have the covariance matrix $\mathbf{R}$, find a transformation on x_1 and x_2 such that the transformed variables have the unit covariance matrix $\mathbf{I}$. Relate the answer to both $\mathbf{A}$ and $\boldsymbol{\lambda}$.

4.11 What are the eigenvalues of the Markov covariance matrix

$$\begin{bmatrix} 1 & \rho & \rho^2 \\ \rho & 1 & \rho \\ \rho^2 & \rho & 1 \end{bmatrix}$$

4.12 From Fig. 4-8 it is clear that the Rician density function is approximately normal over a reasonable range of β for large values of signal-to-noise. Show that this is to be expected using the expression for the density function.

4.13 The chi variable is the square root of the χ^2 variable.

(a) Show that the probability density function for the chi variable with n degrees of freedom is

$$p(\chi) = \frac{\chi^{n-1} e^{-\chi^2/2}}{2^{(n-2)/2}\Gamma(n/2)}$$

(b) Do the same for the noncentral chi variable and show that

$$p(\chi) = \chi\left(\frac{\chi^2}{\lambda}\right)^{(n-2)/4} \exp\left(-\frac{\lambda}{2} - \frac{\chi^2}{2}\right) I_{n/2-1}((\chi^2\lambda)^{1/2})$$

4.14 Show that the mth order central moment of the central χ^2 variable with n degrees of freedom is

$$2^m\left(\frac{n}{2}\right)\left(\frac{n}{2}+1\right)\cdots\left(\frac{n}{2}+m-1\right)$$

4.15 Suppose x is central chi-square with N degrees of freedom, and y is noncentral with M degrees of freedom and noncentral parameter λ. Assume x and y statistically independent. What is the probability distribution of $z = x + y$? (We assume that x and y were generated by Gaussian variables of unit variance.)

4.16 The average power defined as $y = (1/T)\int_0^T x^2(t)\,dt$ of a zero mean Gaussian process is often approximated by a central chi-square probability distribution.

(a) Using the mean and variance of y, show that the approximate number of degrees of freedom of y is

$$\nu \approx R_x^2(0)\left[\frac{1}{T}\int_{-T}^{T}\left(1-\frac{|v|}{T}\right)R_x^2(v)\,dv\right]^{-1}$$

and that for large T

$$\nu \approx R_x^2(0)\left[\frac{1}{T}\int_{-\infty}^{\infty}R_x^2(v)\,dv\right]^{-1}$$

where $R_x(\tau)$ is the autocorrelation function of $x(t)$.

(b) Suppose the power spectral density of $x(t)$ is bandlimited in frequency to $|\omega| \le 2\pi B$. Within this range assume $S_x(\omega)$ is constant, that is, bandlimited white noise. For large T, show that $\nu \approx 2BT$.

4.17 Suppose we were to approximate, by matching the means and variances, a probability distribution of mean μ and variance σ^2 to a noncentral chi-square distribution with ν degrees of freedom and a noncentral parameter λ. Show that we would choose

$$\lambda = \frac{\sigma^2}{2} - \mu, \qquad \nu = 2\mu - \frac{\sigma^2}{2}$$

4.18 When chi-squared variables are added the result is chi-squared because of the normalization. This exercise shows that the same result does not necessarily hold for the gamma distribution. Assume that

$$z_1 = x_1^2 + x_2^2, \qquad z_2 = y_1^2 + y_2^2, \qquad \text{and} \qquad q = z_1 + z_2$$

where x_1, x_2 are statistically independent zero mean Gaussian with variance σ_1^2 and y_1, y_2 are similar except with variance σ_2^2, and $\sigma_1 \ne \sigma_2$.

(a) Show that the distribution of q is

$$p(q) = \frac{1}{2(\sigma_1^2 - \sigma_2^2)} \left[\exp\left(-\frac{q}{2\sigma_1^2}\right) - \exp\left(-\frac{q}{2\sigma_2^2}\right) \right]$$

Is this the gamma distribution?

(b) What is the distribution if $\sigma_1 = \sigma_2$?

References

1. Parzen, E., "Modern Probability Theory and Its Applications." Wiley, New York, 1960.
2. Gnedenko, B. V. and Kolmogorov, A. N.; "Limit Distributions for Sums of Independent Random Variables." Addison-Wesley, Reading, Massachusetts, 1954.
3. Uspensky, J. V. "Introduction to Mathematical Probability." McGraw-Hill, New York, 1937.
4. Doob, J. L., "Stochastic Processes." Wiley, New York, 1953.
5. Loève, M., "Probability Theory." 3rd Ed. Van Nostrand, Princeton, New Jersey, 1963.
6. Aitchinson, J. and Brown, J. A. C., "The Lognormal Distribution." Monograph 5. Cambridge Univ. Press, London and New York, 1966.
7. Tables of Normal Probability Functions, Table 23, Nat. Bur. Std., Appl. Math. Ser., Washington, D.C. (1953).
8. Abramowitz, M., and Stegun, I. A., Handbook of Mathematical Functions, Nat. Bur. of Std., Appl. Math. Ser. 55 (1966).
9. Davenport, W. B. Jr., and Root, W. L., "An Introduction to the Theory of Random Signals and Noise." McGraw-Hill, New York, 1958.
10. Miller, K. S., "Multidimensional Gaussian Distribution." Wiley, New York, 1964.
11. Guillemin, E. A., "The Mathematics of Circuit Analysis." Wiley, New York, 1949.
12. Rice, S. O., "Mathematical Analysis of Random Noise." Reprinted in Selected Papers on Noise and Stochastic Processes (N. Wax, ed.). Dover, New York, 1954.
13. Rice, S. O., Statistical properties of a sinewave plus random noise, *Bell System Tech. J.* (January 1948).
14. Middleton, D., "An Introduction to Statistical Communication Theory." McGraw-Hill, New York, 1960.
15. Bennett, W. R., Methods of solving noise problems, *Proc. IRE*, 609–638 (May 1956).
16. Lawson, J. L., and Uhlenbeck, G. E., "Threshold Signals, Radiation Laboratory Series," Vol. 24. McGraw-Hill, New York, 1950.
17. Marcum, J. I., and Swerling, P., Studies of target detection by pulsed radar, special monograph issue, *IRE Trans. Inform. Theory* **IT-6** (April 1960).
18. Hald, A., "Statistical Theory With Engineering Applications." Wiley, New York, 1952
19. Campbell, G. A., and Foster, R. M., "Fourier Integrals for Practical Applications." Van Nostrand, Princeton, New Jersey, 1954.
20. Pearson, K., "Tables of the Incomplete Γ-Function." Cambridge Univ. Press, London and New York, 1965.
21. Pearson, E. S., and Hartley, H. O., "Biometrika Tables for Statisticians," Vol. I, 3rd Ed. Cambridge Univ. Press, London and New York, 1966.
22. Fisher, R. A., and Yates, F., "Statistical Tables," 5th Ed. Hafner, New York, 1957.

23. Patnaik, P. B., The non-central χ^2 and F-distributions and their applications, *Biometrika* **36,** 202–232 (1949).
24. McLachlan, N. W., "Bessel Functions for Engineers," 2nd ed. Oxford Univ. Press, London and New York, 1955.

SUPPLEMENTARY BIBLIOGRAPHY

Anderson, T. W., "An Introduction to Multivariate Statistical Analysis." Wiley, New York, 1958.

Archer, C. O., Density Functions for Some Random Sums of Normal, Sinusoidal, and Rayleigh Variables, Systems Evaluation Division, Naval Missile Center, Point Mugu, California, Tech. Memo. TM–67–35, AD 657121 (9 August 1967).

Bennett, W. R., Distribution of the sum of randomly phased components, *Quart. Appl. Math.* **5** (1947).

Blötekjaer, K., An experimental investigation of some properties of band-pass limited gaussian noise, *IRE Trans. Inform. Theory* (September 1958).

Brennan, L. E. and Reed, I. S., A recursive method of computing the Q function, correspondence, *IEEE Trans. Inform. Theory* **IT-11,** No. 2, 312–313 (1965).

Burgess, R. E., The rectification and observation of signals in the presence of noise, *Phil. Mag. Ser.* 7 **42** (328) (May 1951).

Clarke, G. F., The Statistical Distribution of Sums of Mixed Pairs of Products of Partly-Correlated Gaussian Variates of Any Variance, Royal Aircraft Establishment, Tech. Rep. 67233, AD 663587 (September 1967).

Owen, D. B. and Wiesen, J. M., A method of computing bivariate normal probabilities, *Bell System Tech. J.* (March 1959).

Pierce, J. N., A markoff envelope process. *IRE Trans. Inform. Theory* (December 1958).

Price, R., A note of the envelope and phase-modulated components of narrow-band gaussian noise. *IRE Trans. Inform. Theory* (September 1955).

Rice, S. O., Distribution of the duration of fades in radio transmission; gaussian noise model, *Bell System Tech. J.* **37,** 581–635 (1958).

On the Energy of a Gaussian Process, and Quadratic Forms:

Grenander, U., Pollak, H. O., and Slepian, D., The distribution of quadratic forms in normal variates: A small sample theory with applications to spectral analysis, *J. Soc. Ind. Appl. Math.* **7,** No. 4 (December 1959).

Kac, M., and Siegert, A. J. F., On the theory of noise in radio receivers with square law detectors, *J. Appl. Phys.* **18** (April 1947).

Rice, S. O., Communication in the presence of noise, *Bell Syst. Tech. J.* **29,** (January 1950).

Ruben, H., Probability content of region under spherical normal distributions, IV: the distribution of homogenous and non-homogenous quadratic functions of normal variables, *Ann. Math. Statistics* **33,** No. 2 (June 1962).

Ruben, H., A new result on the distribution of quadratic forms, notes, *Ann. Math. Statistics* **34,** No. 4, 1582–1584 (1963).

Schwartz, Morton I., Distribution of the energy of a gaussian process, *IEEE Trans. Inform. Theory* (January 1970).

Slepian, D., Fluctuations of random noise power, *Bell System Tech. J.* **37,** No. 1 (January 1958).

On the Lognormal Distribution:

Jones, R. M., and Miller, K. S., On the multivariate lognormal distribution, *J. Ind. Appl. Math. Soc.* **16,** Pt. 2 (1966).

Marlow, N. A., A normal limit theorem for power sums of independent random variables, *Bell Syst. Tech. J.* **46,** No. 9, 2081–2089 (1967).

Nasell, I., Some properties of power sums of truncated normal random variables, *Bell Syst. Tech. J.* **46,** No. 9, 2091–2110 (1967).

Mitchell, R. L., Permanence of the log-normal distribution, *J. Opt. Soc. Amer.* **58,** No. 9 (September 1968).

Chapter 5

Hypothesis Testing

5.1 Introduction

In a digital communication system one of several possible signal waveforms is selected for transmission to the receiver. In the transmission medium connecting the transmitter to the receiver, the signal waveforms may undergo some distortion. In many cases this distortion is caused by processes which, because of their complexity or randomness, may not be known precisely. At the receiver itself, noise is unavoidably added to the signal to cause further distortion. The consequence is that to an observer at the receiver it is no longer certain which of the possible transmitted signals was recieved. A similar event occurs in radar. A known signal is transmitted into space and the receiver stands by to receive this signal after it is reflected by a target. Again the signal may be distorted and mixed with noise so that we may no longer be sure if in fact the signal returned at all. Thus, we are uncertain whether a target was illuminated by the signal or not. In other cases a receiver alone is used as a passive listening device. In this event, the source signal itself may not be precisely known. This signal of unknown form as in the previous cases is also mixed with receiver noise, thus increasing our uncertainty of its presence.

The above discussion employed uncertainty in making decisions. There is another aspect of signal processing, namely, estimation of signal parameters. Typical of such parameters are amplitude, phase, frequency, and time of arrival. The random nature of the signal and the noise accompanying such a signal do not allow us to determine with arbitrary precision the values of these parameters.

In all of these cases a signal, known precisely or imperfectly, is distorted by another random process so that the observable signal is itself a random process. It is not surprising then that statistical methods are used to guide the decision and estimation processes.

In this chapter the primary concern will be with the decision or detection process. Estimation of signal parameters is treated in Chap. 10.

5.2 Hypothesis Testing

One of the most important statistical tools for making decisions is hypothesis testing. The hypotheses are statements of the possible decisions that are being considered. For example, in a radar detection problem we might select two hypotheses, a target is present or no target is present. Corresponding to each hypothesis there is a probabilistic description of the possible outcomes. Coupling this probabilistic description with a criterion or measure of goodness that the decision will satisfy on the average dictates a dichotomy (for two hypotheses) of the sample space over which the outcome of the experiment is defined. This dichotomy represents the best (optimum) decision rule subject to the criterion of goodness.

We shall illustrate this with a simple but instructive example. Assume a binary communication system which transmits in a time interval, T sec long, a pulse $s(t)$ of unit amplitude or no signal at all. At the receiver noise $n(t)$ is unavoidably added to the signal. The problem is to decide, based on a single observation of $r(t) = s(t) + n(t)$, whether or not the signal is present. The situation is represented in Fig. 5-1. The decision would be more reliable

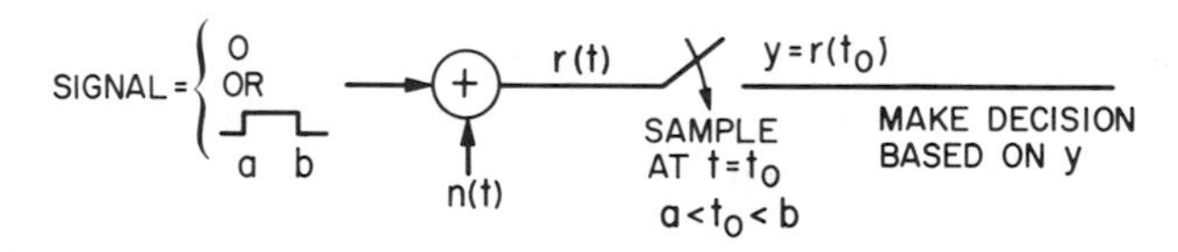

Fig. 5-1 Single observation receiver.

if we chose more than one sample out of the interval (a, b), but this will be considered later. We call the null hypothesis (H_0) the event that no signal is present, and the alternative hypothesis (H_1) the event that a one-volt signal is present. Symbolically:

H_0: no signal present $[r(t) = n(t)]$
H_1: signal present $[r(t) = 1 + n(t)]$

Based on the single observation, denoted $y = r(t_o)$, we must choose between these hypotheses. A criterion for making the decision must be selected. One reasonable criterion is to choose that hypothesis which is most likely to have occurred based on our single observation. That is, given a sample value y, which hypothesis is most probably true? The two conditional probabilities are denoted

$$P(H_0 \mid y) \qquad \text{and} \qquad P(H_1 \mid y)$$

These are the probability that H_0 is true given that $r(t_o) = y$; and the probability that H_1 is true given that $r(t_o) = y$. These probabilities are called *a posteriori probabilities*, so this is called the *maximum a posteriori probability criterion*. The decision rule is to choose H_0 if

$$P(H_0 \mid y) > P(H_1 \mid y) \qquad \text{or} \qquad \frac{P(H_0 \mid y)}{P(H_1 \mid y)} > 1$$

and choose H_1 otherwise.

The decision rule may also be expressed in terms of probability density functions, often a more convenient form. To accomplish this, the decision rule may be expressed as: choose H_0 if

$$P(H_0 \mid y \le Y \le y + dy) > P(H_1 \mid y \le Y \le y + dy)$$

and choose H_1 otherwise. Using the definition of conditional probability:

$$P(H_0 \mid y \le Y \le y + dy) = \frac{P(y \le Y \le y + dy \mid H_0)P(H_0)}{P(y \le Y \le y + dy)}$$

where $P(H_0)$ is the probability that hypothesis H_0 is true. The probability $P(H_1)$ that hypothesis H_1 is true is then $1 - P(H_0)$. These are called *a priori* probabilities. We may assign to the random variable Y a probability density function $p(y)$. Thus, $P(y \le Y \le y + dy) = p(y)\,dy$. Similarly, $P(y \le Y \le y + dy \mid H_0)$ may be replaced by $p_0(y)dy$. The subscript is used in the latter case since the probability is conditioned on H_0 being true.

Then

$$P(H_0 \mid y \le Y \le y + dy) = \frac{p_0(y)\,dyP(H_0)}{p(y)\,dy}$$

In the limit of arbitrarily small dy,

$$P(H_0 \mid y) = \frac{p_0(y)P(H_0)}{p(y)}$$

Similarly,

$$P(H_1 \mid y) = \frac{p_1(y)[1 - P(H_0)]}{p(y)}$$

The decision rule may then be written: choose H_0 if

$$\frac{p_0(y)P(H_0)}{p_1(y)[1 - P(H_0)]} > 1$$

or equivalently

$$\frac{p_0(y)}{p_1(y)} > \frac{1 - P(H_0)}{P(H_0)}$$

The reciprocal of this expression may also be used and the decision rule stated as: choose H_1 if

$$\frac{p_1(y)}{p_0(y)} \geq \frac{P(H_0)}{1 - P(H_0)} \tag{5-1}$$

and choose H_0 otherwise. The ratio $p_1(y)/p_0(y)$ is of particular importance, and is given the name *likelihood ratio*. A test based on this ratio is called a *likelihood ratio test*. The functions $p_1(y)$ and $p_0(y)$ are each called *likelihood functions*. We shall see later that for other criteria of goodness of interest to us, the decision process will involve this ratio and only the quantity on the right-hand side of the inequality will change.

For the above example, assume the noise is Gaussian distributed with zero mean and unit variance. This information is sufficient to compute the likelihood ratio. If H_0 is true, then with no signal present, y has the same density function as the noise. That is

$$p_0(y) = \frac{1}{(2\pi)^{1/2}} e^{-y^2/2}$$

If on the other hand H_1 is true so that a one-volt signal is present, y is still Gaussian but with a mean of one. Thus

$$p_1(y) = \frac{1}{(2\pi)^{1/2}} e^{-(y-1)^2/2}$$

The likelihood ratio, which we shall denote as $\lambda(y)$, is

$$\frac{p_1(y)}{p_0(y)} = \lambda(y) = \frac{e^{-(y-1)^2/2}}{e^{-y^2/2}} = e^{y-1/2}$$

Therefore the decision rule becomes, choose H_1 if

$$e^{y-1/2} \geq \frac{P(H_0)}{1 - P(H_0)}$$

It is frequently convenient to define a *log-likelihood ratio* which is the natural logarithm of the likelihood function. Since $\ln x$ is a monotonically increasing function of x, the inequality will still hold.

Then, using the log-likelihood, choose H_1 if

$$y - \tfrac{1}{2} \geq \ln\left[\frac{P(H_0)}{1 - P(H_0)}\right] \qquad \text{or} \qquad y \geq \tfrac{1}{2} + \ln\left[\frac{P(H_0)}{1 - P(H_0)}\right]$$

In this case, the sample y is called the *test statistic*. Thus the receiver which is optimum in the maximum a posteriori sense samples the signal and compares it to a threshold. If the sample is greater than $\frac{1}{2} + \ln P(H_0)/[1 - P(H_0)]$ it is assumed that a "1" was sent. Otherwise it is assumed that a "0" was sent.

Since y may take on any value in the interval $(-\infty, \infty)$, the sample space is the entire real line. The results above suggests a dichotomy of the real line with $\frac{1}{2} + \ln P(H_0)/[1 - P(H_0)]$ as the dividing point as shown in Fig. 5-2. A sample point which falls in a region R_0 results in H_0 being chosen, and a point which falls in region R_1 results in H_1 being chosen. To produce an easily associated notation, we shall denote the choice (decision) of hypotheses H_0 and H_1 as D_0 and D_1 respectively. We are now in a position to determine quantitatively the performance of the receiver.

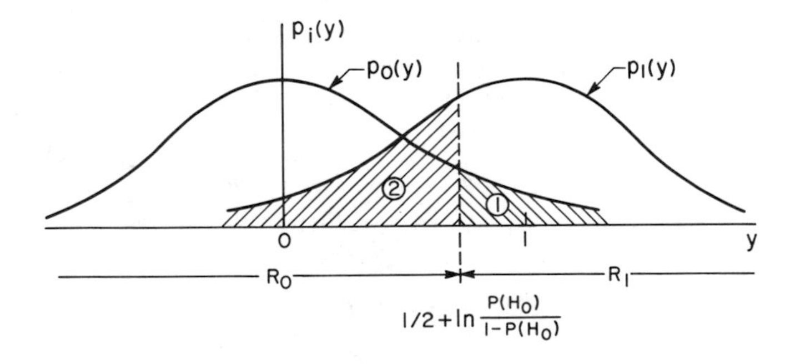

Fig. 5-2 Conditional probability density functions and dichotomy of sample space.

There are two types of errors which can be made. If we say a signal is present when in fact it is not, an *error of the first kind* is made. That is, we choose H_1 given that H_0 is true. Denote the probability of this as $P(D_1 \mid H_0)$ This is represented by area 1 in Fig. 5-2. In radar terminology this is called a *false alarm* error. It is equivalent to saying a target is present when there is none.

On the other hand, if H_0 is chosen when in fact H_1 is true, an *error of the second kind* is made. Denote the probability of this as $P(D_0 \mid H_1)$. This is represented by area 2 in Fig. 5-2. We shall frequently use $P(D_1 \mid H_1)$ which is the probability of choosing H_1 when H_1 is true, a correct decision. This is equal to $1 - P(D_0 \mid H_1)$, and in radar terminology is called the *probability of detection*. In radar terms it is the probability of correctly deciding that a target is present. In statistics, the false alarm probability is called the *size* of the test, and the probability of detection is called the *power* of the test.

With the aid of Fig. 5-2, the error probabilities may easily be determined. The false alarm probability is the probability that $y \geq \frac{1}{2} + \ln P(H_0)/[1 - P(H_0)]$ given that H_0 is true. If H_0 is true, then the density function $p_0(y)$ applies so that

$$P(D_1 \mid H_0) = \int_{R_1} p_0(y)\, dy = \frac{1}{(2\pi)^{1/2}} \int_{Y_T}^{\infty} e^{-y^2/2}\, dy$$

where

$$Y_T = \tfrac{1}{2} + \ln \frac{P(H_0)}{1 - P(H_0)}$$

Similarly, when H_1 is true

$$P(D_0 \mid H_1) = \int_{R_0} p_1(y)\, dy = \frac{1}{(2\pi)^{1/2}} \int_{-\infty}^{Y_T} e^{-(y-1)^2/2}\, dy$$

Finally, the average probability of error, denoted P_e, is

$$P_e = P(D_1 \mid H_0)P(H_0) + P(D_0 \mid H_1)[1 - P(H_0)]$$

In communications problems it is frequently assumed that each signal is sent with equal probability, that is $P(H_1) = 1 - P(H_0) = \frac{1}{2}$. Then, for the above example, R_1 is the region of the real line greater than one-half. It then follows for this example that

$$P(D_1 \mid H_0) = P(D_0 \mid H_1)$$

5.3 Bayes Criterion

The above example had for a criterion the maximum a posteriori probability. Throughout the example there was no particular weighting given to the two types of errors. As such, we more or less assumed that each type of error was equally harmful.

In many cases the consequences of each type of error are not equally important. For example, in a radar system, the consequence of saying a target is present when in fact there is none is quite different from saying that no target is present when in fact there is. To reflect these differences, costs may be assigned to each type of error.

Although cost functions are discussed here, in practice they may be extremely difficult if not impossible to generate. For example, in the above radar case what are the quantitative costs associated with each type of error? Nevertheless, their theoretical consideration does lead to some useful results. For those cases where the costs are impossible to evaluate, other criteria may be used.

Define C_{ij} as the cost associated with choosing hypothesis H_i when actually hypothesis H_j is true. In general, a cost can also be assigned to a correct decision. (A positive cost indicates a penalty.) Normally C_{00} and C_{11} can be assumed to be zero with no loss of generality. In any event, it is assumed that the cost of making an error is greater than the cost of a correct decision. Then $C_{10} - C_{00} > 0$ and $C_{01} - C_{11} > 0$.

We now determine the average cost or risk. Let hypothesis H_0 be chosen whenever the outcome of an experiment lies in a region R_0; let H_1 be chosen when the outcome lies in region R_1. Let the a priori probabilities of H_0 and H_1 be $P(H_0)$ and $1 - P(H_0)$ respectively. The average risk or cost for the decision procedure is given by

$$\begin{aligned}\bar{C} &= P(H_0)[P(D_0 \mid H_0)C_{00} + P(D_1 \mid H_0)C_{10}] \\ &\quad + [1 - P(H_0)][P(D_0 \mid H_1)C_{01} + P(D_1 \mid H_1)C_{11}]\end{aligned} \tag{5-2}$$

A reasonable criterion is to minimize the average cost. Thus the regions R_0 and R_1 are chosen to minimize $\bar{C}$. Substituting

$$P(D_1 \mid H_1) = 1 - P(D_0 \mid H_1) \qquad \text{and} \qquad P(D_1 \mid H_0) = 1 - P(D_0 \mid H_0)$$

into Eq. (5-2) we obtain

$$\begin{aligned}\bar{C} &= P(H_0)C_{10} + [1 - P(H_0)]C_{11} - P(H_0)(C_{10} - C_{00})P(D_0 \mid H_0) \\ &\quad + [1 - P(H_0)](C_{01} - C_{11})P(D_0 \mid H_1)\end{aligned} \tag{5-3}$$

In terms of the likelihood functions

$$P(D_0 \mid H_0) = \int_{R_0} p_0(y)\,dy \qquad \text{and} \qquad P(D_0 \mid H_1) = \int_{R_0} p_1(y)\,dy$$

The average cost may then be expressed as

$$\begin{aligned}\bar{C} &= P(H_0)C_{10} + [1 - P(H_0)]C_{11} \\ &\quad + \int_{R_0} \{[1 - P(H_0)](C_{01} - C_{11})p_1(y) - P(H_0)(C_{10} - C_{00})p_0(y)\}\,dy\end{aligned} \tag{5-4}$$

The only variable quantity is the region R_0. The first two terms are constants leaving the integral to be minimized with respect to R_0. This is easily done by including in the region R_0 only that portion of the y domain for which the integrand is negative. Thus the region R_0, where H_0 is chosen, is the region for which

$$P(H_0)(C_{10} - C_{00})p_0(y) \geq [1 - P(H_0)](C_{01} - C_{11})p_1(y) \tag{5-5}$$

In terms of the likelihood function, the decision rule is to choose H_0 when

$$\frac{p_1(y)}{p_0(y)} < \frac{P(H_0)(C_{10} - C_{00})}{[1 - P(H_0)](C_{01} - C_{11})}$$

In terms of the hypothesis H_1, the decision rule is, choose H_1 if

$$\lambda(y) \triangleq \frac{p_1(y)}{p_0(y)} \geq \frac{P(H_0)(C_{10} - C_{00})}{[1 - P(H_0)](C_{01} - C_{11})} \triangleq \lambda_o \tag{5-6}$$

We have found the region R_1 which minimizes the average cost given by Eq. (5-2). The resulting minimum cost is called the *Bayes risk*, and this criterion is called the *Bayes criterion.* By comparing Eqs. (5-6) and (5-1) for the case when $C_{10} - C_{00} = C_{01} - C_{11}$, it follows that the maximum a posteriori probability criterion is a special case of the Bayes criterion.

5.4 Minimum Error Probability Criterion

In communication systems it is usual to minimize the average error probability. No cost is associated with a correct decision, and the errors of each kind are assigned equal cost. Therefore, assume that

$$C_{00} = C_{11} = 0 \qquad \text{and} \qquad C_{01} = C_{10} = 1 \tag{5-7}$$

Using these costs, the average cost, Eq. (5-2), is

$$\bar{C} = P(H_0)P(D_1 \mid H_0) + [1 - P(H_0)]P(D_0 \mid H_1) \tag{5-8}$$

Thus, for the assumptions in Eq. (5-7), the average cost is the average error probability P_e, and minimizing the average error probability is equivalent to minimizing the Bayes risk. The decision rule becomes: choose H_1 if

$$\frac{p_1(y)}{p_0(y)} \geq \frac{P(H_0)}{1 - P(H_0)} \tag{5-9}$$

Note that this test is identical to that of Eq. (5-1) for the maximum a posteriori probability criterion. This test is also referred to as the *ideal observer* test (*1*, *2*).

5.5 Neyman–Pearson Criterion

In most communication systems where the errors are assumed to be of equal importance and the a priori probabilities are known, the criterion of minimum error probability is generally used. In a radar system, however, the a priori probabilities and the cost of each kind of error are difficult to determine. For such cases there is another criterion which involves neither a priori probabilities nor cost estimates. It is the Neyman–Pearson criterion. In radar terminology its objective is to maximize the probability of detection for a given probability of false alarm. This objective can be accomplished by using a likelihood ratio test. Specifically, there exists some nonnegative number η such that if hypothesis H_1 is chosen when

$$\lambda(y) = \frac{p_1(y)}{p_0(y)} \geq \eta \tag{5-10}$$

and hypothesis H_0 is chosen otherwise, then this rule yields the maximum $P(D_1 \mid H_1)$ for all tests subject to the constraint that $P(D_1 \mid H_0)$ is less than some predetermined constant. This can be proven by using the Bayes criterion.

We wish to maximize $P(D_1 \mid H_1)$ subject to the constraint $P(D_1 \mid H_0) = \alpha$. Since $P(D_1 \mid H_1) = 1 - P(D_0 \mid H_1)$, we may equivalently minimize $P(D_0 \mid H_1)$. Since $P(D_1 \mid H_0)$ is constant, adding it to $P(D_0 \mid H_1)$ will not influence the minimization. Consequently, maximizing $P(D_1 \mid H_1)$ is equivalent to minimizing

$$Q = P(D_0 \mid H_1) + \mu P(D_1 \mid H_0) \tag{5-11}$$

where μ is an arbitrary constant.† Substituting $C_{00} = C_{11} = 0$, $[1 - P(H_0)]$ $C_{01} = 1$ and $P(H_0)C_{10} = \mu$ in Eq. (5-2), the average cost becomes

$$\bar{C} = P(D_0 \mid H_1) + \mu P(D_1 \mid H_0) \tag{5-12}$$

This is the same as Eq. (5-11) and is the quantity to be minimized. Thus, the Neyman–Pearson criterion is a special case of the Bayes criterion. It has already been determined that $\bar{C}$ is a minimum when hypothesis H_1 is chosen to satisfy Eq. (5-6). Substituting the assumed values for C_{00}, C_{01}, C_{11}, C_{10} into this equation yields the rule: choose H_1 when

$$\frac{p_1(y)}{p_0(y)} \geq \mu$$

Thus a likelihood ratio test will maximize $P(D_1 \mid H_1)$ for a given $P(D_1 \mid H_0)$.

EXAMPLE 5.5-1 Define the random variable y as $y = s + n$, where n is a Gaussian random variable having zero mean and variance $\sigma^2 = 2$, and s is a constant equal to either 0 or 1. On the basis of a single sample y, determine an optimum decision rule to choose between the hypotheses

$$H_0\colon \quad s = 0, \qquad\qquad H_1\colon \quad s = 1$$

using a Neyman–Pearson test with $P(D_1 \mid H_0) = 0.1$.

† The constant μ is a Lagrange undetermined multiplier [Goldstein (*3*)].

The likelihood functions $p_0(y)$ and $p_1(y)$ are Gaussian density functions since n is a Gaussian variable. In particular

$$p_0(y) = \frac{1}{(2\pi)^{1/2}2^{1/2}} e^{-y^2/4} \qquad \text{and} \qquad p_1(y) = \frac{1}{(2\pi)^{1/2}2^{1/2}} e^{-(y-1)^2/4}$$

The likelihood ratio is

$$\lambda(y) = \frac{p_1(y)}{p_0(y)} = e^{(y/2)-1}$$

and the Neyman–Pearson test is to choose H_1 if

$$e^{(y/2)-1} \geq \lambda_o$$

The threshold λ_0 is chosen to satisfy the false alarm probability constraint. Since the exponential term is monotonically increasing with y, an equivalent test is to choose H_1 if $y \geq \gamma$. To determine the threshold, the false alarm probability is

$$P(D_1 \mid H_0) = \int_{R_1} p_0(y)\, dy = \int_\gamma^\infty \frac{1}{(2\pi)^{1/2}2^{1/2}} e^{-y^2/4}\, dy = 0.1$$

With a change of variable ($x = y/2^{1/2}$) this becomes

$$P(D_1 \mid H_0) = 0.1 = \int_{\gamma/2^{1/2}}^\infty \frac{1}{(2\pi)^{1/2}} e^{-x^2/2}\, dx$$

Therefore $\gamma = 1.8$, and the decision rule is to choose H_1 if $y \geq 1.8$; choose H_0 otherwise. The probability of detection based on the single observation y is

$$P(D_1 \mid H_1) = \int_{R_1} p_1(y)\, dy = \int_{1.8}^\infty \frac{1}{(2\pi)^{1/2}2^{1/2}} e^{-(y-1)^2/4}\, dy = 0.285$$

To express the decision rule in terms of the likelihood ratio $\lambda(y)$ and λ_o observe that $\lambda(\gamma) = \lambda_o$. Therefore, since $\gamma = 1.8$,

$$\lambda(\gamma) = \frac{p_1(\gamma)}{p_0(\gamma)} = \frac{(2\pi)^{-1/2}2^{-1/2}e^{-(.8)^2/4}}{(2\pi)^{-1/2}2^{-1/2}e^{-(1.8)^2/4}} = \lambda_o \simeq 1.9$$

The decision rule is then: choose H_1 if $\lambda(y) \geq 1.9$ and choose H_0 otherwise.

Example 5.5-2 It is of some interest to take the previous example and assume a priori probabilities for the signal "s." This is then analogous to a binary communication problem. The criterion shall be minimum error probability. Assume $P(H_0) = \frac{1}{2}$. Then from Eq. (5-9), $\lambda_o = 1$ so the decision is: choose H_1 if

$$\lambda(y) = \frac{p_1(y)}{p_0(y)} \geq 1$$

and choose H_0 otherwise. In terms of the likelihood functions, choose H_1 if

$$\frac{(2\pi)^{-1/2}2^{-1/2}e^{-(y-1)^2/4}}{(2\pi)^{-1/2}2^{-1/2}e^{-y^2/4}} \geq 1$$

The decision rule may then be shown to be: choose H_1 if $y \geq \frac{1}{2}$; choose H_0 otherwise. For this, the false alarm probability is

$$P(D_1 \mid H_0) = \int_{R_1} p_0(y)\, dy = \int_{1/2}^{\infty} \frac{1}{(2\pi)^{1/2}2^{1/2}} e^{-y^2/4}\, dy = 0.362$$

It may also be shown that $P(D_0 \mid H_1) = P(D_1 \mid H_0) = 0.362$ so that $P(D_1 \mid H_1) = 0.638$. Note that the probability of detection is higher than that determined for the previous example. However, the false alarm probability is also greater than the 0.1 constraint of the previous example.

5.6 Minimax Criterion

The Bayes criterion required both costs and a priori probabilities. For the minimum error probability criterion or the maximum a posteriori probability criterion, only the a priori probabilities were required. For the Neyman–Pearson criterion neither costs nor a priori probabilities were required. We shall now discuss the minimax criterion which requires cost functions but not a priori probabilities. It is a criterion for which the maximum possible cost is minimized. The reason for doing this will become apparent. We will demonstrate with an example.

Consider the example of Sect. 5.2 in which we had to decide on the basis of a single sample whether the sample was drawn from a Gaussian distribution with zero mean and unit variance or from a Gaussian distribution with unit mean and variance. Assume costs $C_{00} = C_{11} = 0$, $C_{10} = 1$, and $C_{01} = C$. The likelihood functions are

$$p_0(y) = \frac{1}{(2\pi)^{1/2}} e^{-y^2/2} \qquad \text{and} \qquad p_1(y) = \frac{1}{(2\pi)^{1/2}} e^{-(y-1)^2/2}$$

For the moment, assume that the a priori probability $P(H_0)$ is known. In this case the Bayes criterion would produce the decision rule, choose H_1 if

$$y \geq \tfrac{1}{2} + \ln\left[\frac{P(H_0)/C}{1 - P(H_0)}\right]$$

choose H_0 otherwise. For given $P(H_0)$ and C, the probability of each error may be determined. The resulting Bayes risk from Eq. (5-3) is

$$\bar{C} = P(H_0)P(D_1 \mid H_0) + C[1 - P(H_0)]P(D_0 \mid H_1) \tag{5-13}$$

This risk is shown in Fig. 5-3 as a function of the true a priori probability $P(H_0)$ for a cost $C = 2$.

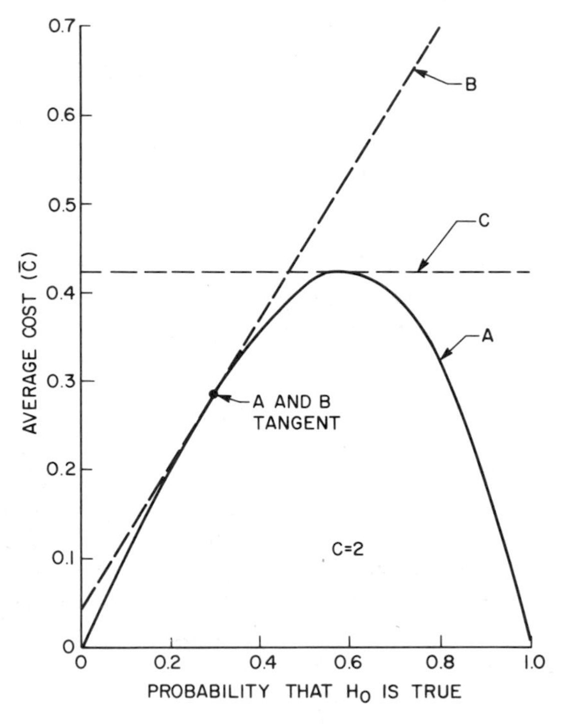

Fig. 5-3 Example of minimax solution. (A) Bayes minimum cost. (B) Cost for particular decision threshold. (C) Minimax cost.

Consider next the average cost if the a priori probability is unknown. In such an event a possibility is to guess at the a priori probability and set the decision threshold accordingly. Once the threshold is set, the error probabilities $P(D_0 \mid H_1)$ and $P(D_1 \mid H_0)$ are fixed and the average risk given by Eq. (5-13) is linear in $P(H_0)$. For a particular threshold, the average cost will be as shown by curve B in Fig. 5-3. For this case the a priori probability guess was 0.3. The resulting cost as a function of the true, but unknown, a priori probability is

$$\bar{C} = 0.809\, P(H_0) + 0.042$$

If in fact the a priori probability guess was correct, the resulting average cost would be the minimum. Otherwise, the cost would be greater than the Bayes cost. As shown in Fig. 5-3, if the true a priori probability is less than 0.5, there is little difference between the actual cost and the minimum cost. For true a priori probabilities beyond approximately 0.47 the average cost is greater than the *maximum* Bayes cost. If, in fact, the correct a priori probability was greater than, say, 0.8 the average cost would be excessive. Such high average costs can be avoided if the threshold is chosen to produce the operating curve labeled C in Fig. 5-3. This curve is tangent to the peak of the Bayes cost curve and results in the maximum Bayes cost. But, independent of the true a priori probability, the average cost is limited to this value. Therefore the maximum possible risk is minimized. This is the essence of the minimax criterion. In this example, the minimax solution corresponds to assuming that the a priori probability is 0.6. The resulting average cost is 0.42.

As a partial summary, the purpose of the minimax criterion is to minimize the maximum possible cost when the a priori probability of each hypothesis is unknown. Therefore, the solution involves finding the least favorable a priori probability among the Bayes solutions. This least favorable, and hence the minimax, solution may be determined by finding the maximum of the average cost as given by Eq. (5-2). Differentiating $\bar{C}$ with respect to $P(H_0)$ and setting the result to zero produces the condition for the minimax solution. It is

$$C_{10}P(D_1 \mid H_0) + C_{00}P(D_0 \mid H_0) = C_{01}P(D_0 \mid H_1) + C_{11}P(D_1 \mid H_1) \tag{5-14}$$

The left-hand side is the average cost if hypothesis H_0 is true, and the right-hand side is the average cost when hypothesis H_1 is true. The condition for the minimax solution is therefore to make these costs equal. The minimax cost may be found by substituting the condition of Eq. (5-14) back into Eq. (5-2). The result is

$$\bar{C} = C_{10}P(D_1 \mid H_0) + C_{00}P(D_0 \mid H_0) = C_{01}P(D_0 \mid H_1) + C_{11}P(D_1 \mid H_1) \tag{5-15}$$

which is the same as either side of Eq. (5-14).

EXAMPLE 5.6-1 In most binary communication systems the following cost functions are applicable:

$$C_{00} = C_{11} = 0 \qquad \text{and} \qquad C_{10} = C_{01} = 1$$

For such systems what is the condition for the minimax cost?

From Eq. (5-14) the condition for the minimax cost is

$$P(D_1 \mid H_0) = P(D_0 \mid H_1)$$

That is, the probabilities of each kind of error are made equal. It then follows that the minimax cost is the average probability of error.

5.7 Multiple Measurements

The preceding work on hypothesis testing was stated in terms of only one sample. In practice the decision process is usually based on many samples to improve the quality of the test, that is, to reduce the probability of error. Assuming that n samples $y_1, y_2, \ldots, y_n$ are taken, they may be represented by the vector $\mathbf{y}$. The joint probability density functions of the sample vector conditioned on each hypothesis are the likelihood functions, and the likelihood ratio is

$$\lambda(\mathbf{y}) = \frac{p_1(y_1, y_2, \ldots, y_n)}{p_0(y_1, y_2, \ldots, y_n)} = \frac{p_1(\mathbf{y})}{p_0(\mathbf{y})} \tag{5-16}$$

Although we shall not do so, it is an easy matter to show that the previous results for a single sample can be generalized to the multisample case. Therefore for the criteria of interest to us, the likelihood ratio test is optimum and the decision rule is to choose H_1 if $\lambda(\mathbf{y}) \geq \lambda_o$; choose H_0 if $\lambda(\mathbf{y}) < \lambda_o$. As before λ_o is a threshold which depends on the criterion and the a priori knowledge available.

The procedure for the case of multiple observations will be demonstrated by an example.

EXAMPLE 5.7-1 Based on n statistically independent Gaussian samples with variance σ^2, determine a likelihood ratio test to choose between the hypotheses that their mean is zero (H_0) or that their mean is one (H_1). The hypotheses are

$$H_0: \quad \text{mean} = 0, \qquad H_1: \quad \text{mean} = 1$$

and the decision rule is, choose H_1 if $\lambda(y_1, \ldots, y_n) \geq \lambda_o$; choose H_0 otherwise.

For hypothesis H_1, the random variables y_i are Gaussian with unity mean and variance σ^2. For H_0, the variables are Gaussian with zero mean and variance σ^2. Since the samples are independent, the likelihood functions are

$$p_1(\mathbf{y}) = \frac{1}{(2\pi)^{1/2}\sigma} \exp\left[-\frac{(y_1 - 1)^2}{2\sigma^2}\right] \cdots \frac{1}{(2\pi)^{1/2}\sigma} \exp\left[-\frac{(y_n - 1)^2}{2\sigma^2}\right]$$

and

$$p_0(\mathbf{y}) = \frac{1}{(2\pi)^{1/2}\sigma} \exp\left[-\frac{{y_1}^2}{2\sigma^2}\right] \cdots \frac{1}{(2\pi)^{1/2}\sigma} \exp\left[-\frac{{y_n}^2}{2\sigma^2}\right]$$

or

$$p_1(\mathbf{y}) = \left(\frac{1}{2\pi\sigma^2}\right)^{n/2} \exp\left[-\tfrac{1}{2}\sum_{i=1}^{n}\left(\frac{y_i - 1}{\sigma}\right)^2\right]$$

and

$$p_0(\mathbf{y}) = \left(\frac{1}{2\pi\sigma^2}\right)^{n/2} \exp\left[-\tfrac{1}{2}\sum_{i=1}^{n}\left(\frac{y_i}{\sigma}\right)^2\right]$$

The likelihood ratio is

$$\lambda(\mathbf{y}) = \exp\left[-\tfrac{1}{2}\sum_{i=1}^{n}\left(\frac{y_i - 1}{\sigma}\right)^2 + \tfrac{1}{2}\sum_{i=1}^{n}\left(\frac{y_i}{\sigma}\right)^2\right] = \exp\left[\tfrac{1}{2}\sum_{i=1}^{n}\frac{2y_i - 1}{\sigma^2}\right]$$

Therefore the decision rule is: choose H_1 if

$$\exp\left[\tfrac{1}{2}\sum_{i=1}^{n}\frac{2y_i - 1}{\sigma^2}\right] \geq \lambda_o$$

Using the log-likelihood ratio and manipulating the result, the decision rule becomes: choose H_1 if

$$\frac{1}{n}\sum_{i=1}^{n} y_i \geq \lambda_o'$$

where

$$\lambda_o' = \frac{1}{2} + \frac{\sigma^2}{n}\ln \lambda_o$$

The term $(1/n)\sum_{i=1}^{n} y_i$ will be recognized as the sample mean of the measurements. Denote it by $\hat{m}_y$. The receiver operation is therefore to take the n independent samples, calculate the average value of these samples, and compare the result to a threshold λ_o'. If the result is above this threshold, it is decided that the mean of the Gaussian distribution is one.

The probability of incorrectly choosing H_1 is

$$P(D_1 \mid H_0) = \int_{R_1} p_0(y_1, \ldots, y_n)\, dy_1 \cdots dy_n$$

where R_1 is the n-dimensional volume for which $(1/n)\sum_{i=1}^{n} y_i$ is greater than some constant, and in which hypothesis H_1 is chosen. However, the error probability can be determined more directly from the probability density functions for the test statistic $\hat{m}_y$.

Since $\hat{m}_y$ is the sum of independent Gaussian random variables, it too is Gaussian. Its variance is σ^2/n independent of the hypothesis. The mean is dependent on which hypothesis is assumed true. If H_0 is true, the mean of

$\hat{m}_y$ is zero. If H_1 is true, the mean is unity. Denote the density function of $\hat{m}_y$ for H_0 and H_1 as $p_0(\hat{m}_y)$ and $p_1(\hat{m}_y)$ respectively. (Their argument is meant to distinguish them from the likelihood functions.)

The probability for each kind of error may then be expressed as

$$P(D_1 \mid H_0) = \int_{\lambda_0'}^{\infty} p_0(\hat{m}_y)\, d\hat{m}_y = \int_{\lambda_0'}^{\infty} \left(\frac{n}{2\pi\sigma^2}\right)^{1/2} e^{-nz^2/2\sigma^2}\, dz$$

and

$$P(D_0 \mid H_1) = \int_{-\infty}^{\lambda_0'} p_1(\hat{m}_y)\, d\hat{m}_y = \int_{-\infty}^{\lambda_0'} \left(\frac{n}{2\pi\sigma^2}\right)^{1/2} e^{-n(z-1)^2/2\sigma^2}\, dz$$

5.8 Multiple Alternative Hypothesis Testing

In all the previous cases only binary type decisions were considered. In some problems there may be many other decisions possible. For example, many communication systems use alphabets of higher order than binary. With a quaternary alphabet one of four possible signals is selected for transmission. In such a case there would be four hypotheses from which to choose. Multiple alternative hypotheses (*5*) may also be used for estimation of signal parameters; more will be said about this in Chap. 10.

We shall proceed as for the Bayes criterion and assume that cost functions and a priori probabilities are known. Assume that one of n hypotheses is to be chosen and assign a cost, C_{ij}, to each decision. This is the cost of choosing hypothesis H_i when hypothesis H_j is true. The average cost is then

$$\bar{C} = \sum_{i=1}^{n} \sum_{j=1}^{n} C_{ij} P(D_i \mid H_j) P(H_j) \tag{5-17}$$

where $P(D_i \mid H_j)$ denotes the probability of choosing hypothesis H_i given that H_j is true. The Bayes criterion is to minimize the average cost. This implies that for any received set of samples, the hypothesis H_j which yields the lowest average cost is chosen. Given the set of samples $\mathbf{y}$, the cost associated with hypothesis H_j is

$$C_j = \sum_{i=1}^{n} C_{ji} P(H_i \mid \mathbf{y}) \tag{5-18}$$

where $P(H_i \mid \mathbf{y})$ is the probability that hypothesis H_i is true given $\mathbf{y}$. As in Sect. 5.2, this probability may be expressed

$$P(H_i \mid \mathbf{y}) = \frac{p_i(\mathbf{y}) P(H_i)}{p(\mathbf{y})} \tag{5-19}$$

where $p_i(\mathbf{y})$ is the probability density function for hypothesis H_i, and $p(\mathbf{y})$ is the marginal density function for $\mathbf{y}$. The cost associated with choosing H_j is then

$$C_j = \frac{\sum_{i=1}^{n} C_{ji}\, p_i(\mathbf{y}) P(H_i)}{p(\mathbf{y})}$$

Since $p(\mathbf{y})$ does not depend on the hypothesis, the decision rule is to choose that hypothesis H_j for which

$$\lambda_j = \sum_{i=1}^{n} C_{ji}\, p_i(\mathbf{y}) P(H_i) \tag{5-20}$$

is a minimum. This is the Bayes test for multiple hypotheses.

Of particular interest is the case where $C_{ij} = 1$ for $i \neq j$, and $C_{ii} = 0$. Thus, there is no penalty for correct decisions, and all errors have equal weighting.

To use the Bayes rule compute

$$\lambda_j = \sum_{\substack{i=1 \\ i \neq j}}^{n} P(H_i) p_i(\mathbf{y}) \quad \text{for all} \quad j$$

and choose that hypothesis corresponding to the index j for which this is a minimum. Some typical terms are:

$$\begin{aligned}
\lambda_1 &= 0 + P(H_2)p_2(\mathbf{y}) + \cdots + P(H_n)p_n(\mathbf{y}) \\
\lambda_2 &= P(H_1)p_1(\mathbf{y}) + 0 + P(H_3)p_3(\mathbf{y}) + \cdots + P(H_n)p_n(\mathbf{y}) \\
\lambda_k &= P(H_1)p_1(\mathbf{y}) + \cdots + P(H_{k-1})p_{k-1}(\mathbf{y}) + 0 + P(H_{k+1})p_{k+1}(\mathbf{y}) + \cdots \\
&\quad + P(H_n)p_n(\mathbf{y}) \\
\lambda_n &= P(H_1)p_1(\mathbf{y}) + \cdots + P(H_{n-1})p_{n-1}(\mathbf{y}) + 0
\end{aligned}$$

Note that each λ with subscript k has only the term $P(H_k)p_k(\mathbf{y})$ missing from the right-hand side. Consequently

$$\lambda_k - \lambda_j = P(H_j)p_j(\mathbf{y}) - P(H_k)p_k(\mathbf{y}) \tag{5-21}$$

Without loss of generality, assume that λ_j is the minimum term. Then $\lambda_k - \lambda_j > 0$, or from Eq. (5-21)

$$P(H_j)p_j(\mathbf{y}) > P(H_k)p_k(\mathbf{y})$$

Therefore choosing the minimum λ_j corresponds to choosing the largest value of $P(H_j)p_j(\mathbf{y})$. The decision rule becomes: choose the hypothesis H_j for which $p_j(\mathbf{y})P(H_j)$ is a maximum.

Using the material of Sect. 5.2 we know that

$$\frac{p_j(\mathbf{y})P(H_j)}{p(\mathbf{y})} = P(H_j \mid \mathbf{y})$$

But $p(\mathbf{y})$ is a constant for any given $\mathbf{y}$. Therefore maximizing $p_j(\mathbf{y})P(H_j)$ is the same as maximizing $P(H_j \mid \mathbf{y})$. This is the a posteriori probability of hypothesis H_j given the sample values $\mathbf{y}$. Therefore using the costs $C_{ij} = 1$ for $i \neq j$, and

0 for $i = j$, the Bayes criterion reduces to a maximum a posteriori probability criterion.

An obvious extension occurs when all hypotheses are equally likely. Then the decision rule is to choose the hypotheses H_j for which $p_j(\mathbf{y})$ is a maximum. This case occurs in communication systems using M-ary alphabets where each signal is transmitted with equal probability. An example of such a system is given in Sect. 6.5.

EXAMPLE 5.8-1 Based on an n-dimension sample vector $\mathbf{y}$ design a test to choose between four hypotheses: H_1-mean is 1; H_2-mean is 2; H_3-mean is 3; and H_4-mean is 4. For each hypothesis the probability density function is Gaussian with variance σ^2. Assume all hypotheses are equally probable and $C_{ij} = 1$, $i \neq j$ and 0 for $i = j$.

The Bayes cost is minimized by choosing the hypothesis H_j corresponding to the maximum $p_j(\mathbf{y})$. From Example 5.7-1, the likelihood functions are

$$p_k(\mathbf{y}) = \left(\frac{1}{2\pi\sigma^2}\right)^{n/2} \exp\left[-\tfrac{1}{2}\sum_{i=1}^{n}\left(\frac{y_i - k}{\sigma}\right)^2\right], \qquad k = 1, 2, 3, 4$$

Therefore choose k such that the following is maximized:

$$\exp\left[-\tfrac{1}{2}\sum_{i=1}^{n}\left(\frac{y_i - k}{\sigma}\right)^2\right]$$

The above decision rule can be shown to be equivalent to the following: choose the hypothesis corresponding to the value of k which maximizes

$$\left(\frac{2}{n}\sum_{i=1}^{n} y_i k\right) - k^2$$

Writing out each term explicitly, choose the maximum of

$$\frac{2}{n}\sum_{i=1}^{n} y_i - 1, \qquad \frac{4}{n}\sum_{i=1}^{n} y_i - 4, \qquad \frac{6}{n}\sum_{i=1}^{n} y_i - 9, \qquad \frac{8}{n}\sum_{i=1}^{n} y_i - 16$$

In order, they correspond to the hypotheses H_1 through H_4. The decision rule becomes clear if, in turn, the kth likelihood function is assumed to be maximum. For example, if the likelihood function corresponding to H_1 is the largest, then it must be that

$$\frac{2}{n}\sum_{i=1}^{n} y_i - 1 > \frac{4}{n}\sum_{i=1}^{n} y_i - 4$$

or equivalently, the region in which H_1 is chosen satisfies

$$\frac{1}{n}\sum_{i=1}^{n} y_i < 1.5$$

The test statistic $(1/n) \sum_{i=1}^{n} y_i$ will be recognized as the sample mean and we again denote it by $\hat{m}_y$. The regions for each hypothesis may similarly be shown to be

$$\begin{aligned}
&\text{choose } H_1 \text{ if} \quad \hat{m}_y \leq 1.5 \\
&\text{choose } H_2 \text{ if} \quad 1.5 < \hat{m}_y \leq 2.5 \\
&\text{choose } H_3 \text{ if} \quad 2.5 < \hat{m}_y \leq 3.5 \\
&\text{choose } H_4 \text{ if} \quad \hat{m}_y > 3.5
\end{aligned}$$

The assignment of the equality is arbitrary since for continuous probability distributions there is a zero probability that the likelihood ratio will take on a value coincident with a boundary.

The regions in which each hypothesis is chosen are shown in Fig. 5-4

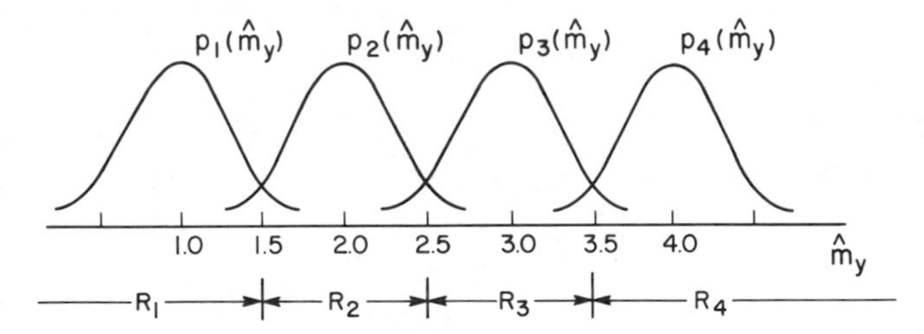

Fig. 5-4 Conditional density functions and acceptance regions for Example 5.8-1.

along with the conditional density functions for the test statistic $\hat{m}_y$. By noting the conditional density functions of the test statistic, the acceptance regions are obvious in the case of a maximum a posteriori probability criterion with equal a priori probabilities for the hypotheses.

5.9 Composite Hypothesis Testing (*1*)

Each hypothesis previously considered depended on only one "unknown" parameter, and that parameter could assume only one value. Treating the parameter in the context of a signal property, for each hypothesis the signal was known precisely and it was the addition of noise which caused uncertainty as to which hypothesis to choose. There are many cases, however, when the signal is not known precisely. For example, if the signal is a sine wave, its phase may be unknown. This is a common occurrence in signal processing and it is usually assumed that the phase is a random variable uniformly distributed over the interval $(0, 2\pi)$. Similar statements may be made about the amplitude or frequency. For these cases we could choose to generate a hypothesis for each possible value of the unknown parameter. Then, for example, the null hypothesis might be that no signal is present and the remaining hypotheses H_i, $i = 1, 2, \ldots$ might be that a signal is present and its phase is θ_i. The result would correspond to both detection and estimation of signal parameters. In many practical cases the estimation portion of the hypothesis may not be of interest and we wish only to decide whether a signal is present or not independently of the parameters which the signal may have if it is present. We shall discuss here hypotheses which include these random parameters. Such hypotheses are called *composite hypotheses*, the others are called *simple hypotheses*. To be more specific, the likelihood function associated with a hypothesis depends on one or more unknown parameters. If the sample space of the parameters is single valued, the hypothesis is simple. Otherwise, the hypothesis is composite.

We shall limit ourselves here to only two hypotheses although the approach may be generalized to multiple hypotheses. The approach will be to assign cost functions and to use the Bayes criterion. The results will then be applied to cases of special interest.

Let $\boldsymbol{\phi}(= \phi_1, \phi_2, \ldots, \phi_n)$ denote a vector of random parameters associated with hypothesis H_0. For example, ϕ_1 might denote phase, ϕ_2 amplitude, ϕ_3 frequency, etc. Let $\boldsymbol{\theta}$ represent the random parameters associated with hypothesis H_1. Here again, θ_1, θ_2, and θ_3 could represent phase, amplitude, and frequency. Let $w_0(\boldsymbol{\phi})$ and $w_1(\boldsymbol{\theta})$ represent the joint a priori probability density functions associated with $\boldsymbol{\phi}$ and $\boldsymbol{\theta}$ respectively. Cost functions will be applied to decisions but now these costs may also be functions of the variables $\boldsymbol{\phi}$ and $\boldsymbol{\theta}$. However C_{00} and C_{10} can be functions of only $\boldsymbol{\phi}$ and not $\boldsymbol{\theta}$, and C_{01} and C_{11} are functions of $\boldsymbol{\theta}$ and not $\boldsymbol{\phi}$.

Because of the uncertainty in the signal parameters, the density functions of the observations $\mathbf{y}$ are not only dependent upon the hypothesis but also upon the values of the unknown parameters. Therefore, denote these density functions as $p_1(\mathbf{y} \mid \boldsymbol{\theta})$ and $p_0(\mathbf{y} \mid \boldsymbol{\phi})$.

Denote the region over which H_0 is chosen as R_0, and the region over which H_1 is chosen as R_1. Then the average cost is†

$$\begin{aligned}
\bar{C} = P(H_0) &\int_{R_0} \int_{\{\boldsymbol{\phi}\}} p_0(\mathbf{y} \mid \boldsymbol{\phi}) w_0(\boldsymbol{\phi}) C_{00}(\boldsymbol{\phi}) \, d\boldsymbol{\phi} \, d\mathbf{y} \\
&+ P(H_0) \int_{R_1} \int_{\{\boldsymbol{\phi}\}} p_0(\mathbf{y} \mid \boldsymbol{\phi}) w_0(\boldsymbol{\phi}) C_{10}(\boldsymbol{\phi}) \, d\boldsymbol{\phi} \, d\mathbf{y} \\
&+ P(H_1) \int_{R_0} \int_{\{\boldsymbol{\theta}\}} p_1(\mathbf{y} \mid \boldsymbol{\theta}) w_1(\boldsymbol{\theta}) C_{01}(\boldsymbol{\theta}) \, d\boldsymbol{\theta} \, d\mathbf{y} \\
&+ P(H_1) \int_{R_1} \int_{\{\boldsymbol{\theta}\}} p_1(\mathbf{y} \mid \boldsymbol{\theta}) w_1(\boldsymbol{\theta}) C_{11}(\boldsymbol{\theta}) \, d\boldsymbol{\theta} \, d\mathbf{y}
\end{aligned} \tag{5-22}$$

† The integral $\int_{\{\boldsymbol{\theta}\}}$ indicates an n-fold integration over the possible range of $\boldsymbol{\theta}$.

We know that

$$\int_{R_1} p_0(\mathbf{y} \mid \boldsymbol{\phi})\, d\mathbf{y} = 1 - \int_{R_0} p_0(\mathbf{y} \mid \boldsymbol{\phi})\, d\mathbf{y}$$

and

$$\int_{R_1} p_1(\mathbf{y} \mid \boldsymbol{\theta})\, d\mathbf{y} = 1 - \int_{R_0} p_1(\mathbf{y} \mid \boldsymbol{\theta})\, d\mathbf{y} \tag{5-23}$$

Substituting these into Eq. (5-22) yields, after some manipulation,

$$\begin{aligned}\bar{C} = {} & P(H_0) \int_{\{\boldsymbol{\phi}\}} w_0(\boldsymbol{\phi}) C_{10}(\boldsymbol{\phi})\, d\boldsymbol{\phi} + P(H_1) \int_{\{\boldsymbol{\theta}\}} w_1(\boldsymbol{\theta}) C_{11}(\boldsymbol{\theta})\, d\boldsymbol{\theta} \\ & + \int_{R_0} \Big\{ P(H_1) \int_{\{\boldsymbol{\theta}\}} p_1(\mathbf{y} \mid \boldsymbol{\theta}) w_1(\boldsymbol{\theta}) [C_{01}(\boldsymbol{\theta}) - C_{11}(\boldsymbol{\theta})]\, d\boldsymbol{\theta} \\ & - P(H_0) \int_{\{\boldsymbol{\phi}\}} p_0(\mathbf{y} \mid \boldsymbol{\phi}) w_0(\boldsymbol{\phi}) [C_{10}(\boldsymbol{\phi}) - C_{00}(\boldsymbol{\phi})]\, d\boldsymbol{\phi} \Big\}\, d\mathbf{y}\end{aligned} \tag{5-24}$$

Using the Bayes criterion, the objective is to determine R_0 to minimize this average cost. Assume that

$$C_{10}(\boldsymbol{\phi}) - C_{00}(\boldsymbol{\phi}) > 0, \qquad C_{01}(\boldsymbol{\theta}) - C_{11}(\boldsymbol{\theta}) > 0$$

Then, since the first two terms are independent of R_0, $\bar{C}$ is a minimum when R_0 is chosen as the region in which the integrand of the last integral is negative. This corresponds to choosing the hypothesis H_0 whenever

$$\frac{P(H_0) \int_{\{\boldsymbol{\phi}\}} p_0(\mathbf{y} \mid \boldsymbol{\phi}) w_0(\boldsymbol{\phi}) [C_{10}(\boldsymbol{\phi}) - C_{00}(\boldsymbol{\phi})]\, d\boldsymbol{\phi}}{P(H_1) \int_{\{\boldsymbol{\theta}\}} p_1(\mathbf{y} \mid \boldsymbol{\theta}) w_1(\boldsymbol{\theta}) [C_{01}(\boldsymbol{\theta}) - C_{11}(\boldsymbol{\theta})]\, d\boldsymbol{\theta}} > 1$$

In terms of the hypothesis H_1, choose H_1 when

$$\frac{\int_{\{\boldsymbol{\theta}\}} p_1(\mathbf{y} \mid \boldsymbol{\theta}) w_1(\boldsymbol{\theta}) [C_{01}(\boldsymbol{\theta}) - C_{11}(\boldsymbol{\theta})]\, d\boldsymbol{\theta}}{\int_{\{\boldsymbol{\phi}\}} p_0(\mathbf{y} \mid \boldsymbol{\phi}) w_0(\boldsymbol{\phi}) [C_{10}(\boldsymbol{\phi}) - C_{00}(\boldsymbol{\phi})]\, d\boldsymbol{\phi}} \geq \frac{P(H_0)}{P(H_1)} \tag{5-25}$$

The expression on the left is the *cost likelihood ratio*, and the equation represents the general Bayes test for composite hypotheses.

If the cost functions are independent of the variables $\boldsymbol{\phi}$ and $\boldsymbol{\theta}$, the test becomes: choose H_1 when

$$\frac{\int_{\{\boldsymbol{\theta}\}} p_1(\mathbf{y} \mid \boldsymbol{\theta}) w_1(\boldsymbol{\theta})\, d\boldsymbol{\theta}}{\int_{\{\boldsymbol{\phi}\}} p_0(\mathbf{y} \mid \boldsymbol{\phi}) w_0(\boldsymbol{\phi})\, d\boldsymbol{\phi}} \geq \frac{P(H_0)(C_{10} - C_{00})}{P(H_1)(C_{01} - C_{11})} \tag{5-26}$$

Note that

$$p_1(\mathbf{y} \mid \boldsymbol{\theta}) w_1(\boldsymbol{\theta}) = p_1(\mathbf{y}, \boldsymbol{\theta}) \qquad \text{and} \qquad \int_{\{\boldsymbol{\theta}\}} p_1(\mathbf{y}, \boldsymbol{\theta})\, d\boldsymbol{\theta} = p_1(\mathbf{y})$$

Similar relations hold for hypothesis H_0. Using these relations, the decision rule of Eq. (5-26) can be stated: choose H_1 if

$$\frac{p_1(\mathbf{y})}{p_0(\mathbf{y})} \geq \frac{P(H_0)(C_{10} - C_{00})}{P(H_1)(C_{01} - C_{11})} \tag{5-27}$$

Therefore, for known a priori density functions $w_0(\boldsymbol{\phi})$ and $w_1(\boldsymbol{\theta})$, and for parameter independent cost functions, the composite hypothesis test reduces to a test of simple hypotheses and the previous results apply.

In many cases the null hypothesis, H_0, is simple and the alternative hypothesis, H_1, is composite. (An example is the radar detection problem when the alternative hypothesis H_1 assumes a randomly phased sine wave signal, and H_0 assumes no signal.) Then, if C_{01} and C_{11} are independent of the variables $\boldsymbol{\theta}$, the decision rule is choose H_1 if

$$\frac{\int_{\{\boldsymbol{\theta}\}} p_1(\mathbf{y} \mid \boldsymbol{\theta}) w_1(\boldsymbol{\theta})\, d\boldsymbol{\theta}}{p_0(\mathbf{y})} \geq \frac{P(H_0)(C_{10} - C_{00})}{P(H_1)(C_{01} - C_{11})} \tag{5-28}$$

For the latter case, the probability of the first kind is

$$P(D_1 \mid H_0) = \int_{R_1} p_0(\mathbf{y})\, d\mathbf{y}$$

and for a given value of $\boldsymbol{\theta}$ the probability of an error of the second kind is

$$P^{\theta}(D_0 \mid H_1) = \int_{R_0} p_1(\mathbf{y} \mid \boldsymbol{\theta})\, d\mathbf{y}$$

The superscript θ is a reminder that $P^{\theta}(D_0 \mid H_1)$ may be a function of $\boldsymbol{\theta}$.

5.10 Unknown A Priori Information

The preceding material on composite hypothesis testing assumed that the a priori density function for the random parameters associated with a hypothesis is known. In practice, this is not always the case, and alternatives must be considered.

Consider the case when the null hypothesis is simple, and the alternative hypothesis is composite and assume that the cost functions are independent of the parameters. For a *given* value of the parameter $\boldsymbol{\theta}$ (this makes the hypothesis "conditionally simple"), application of the Neyman–Pearson test will maximize $P^{\theta}(D_1 \mid H_1)$ subject to a constrained value of $P(D_1 \mid H_0)$. But, the probability $P^{\theta}(D_1 \mid H_1)$ may depend on $\boldsymbol{\theta}$. If the test maximizes $P^{\theta}(D_1 \mid H_1)$ independent of the value of $\boldsymbol{\theta}$, then by definition this test must be optimum for all values of $\boldsymbol{\theta}$ and, therefore, for all a priori density functions $w_1(\boldsymbol{\theta})$. Such a test is referred to as a *uniformly most powerful test*.

If a uniformly most powerful test does not exist, another strategy is to determine a *least favorable* (*1*) a priori distribution, if one exists, and proceed accordingly. In principle, this method is similar to the minimax criterion. To see conceptually how such a distribution might be determined, assume a distribution function, say $w(\boldsymbol{\theta})$, and apply the likelihood ratio test (the null hypothesis is assumed simple)

$$\frac{\int_{\{\boldsymbol{\theta}\}} p_1(\mathbf{y} \mid \boldsymbol{\theta}) w(\boldsymbol{\theta})\, d\boldsymbol{\theta}}{p_0(\mathbf{y})} \geq \lambda_o \tag{5-29}$$

Corresponding to the assumed a priori function, the average probability of correctly choosing H_1 is

$$E\{P^\theta(D_1 \mid H_0)\} = \overline{P(D_1 \mid H_1)} = \int_{\{\boldsymbol{\theta}\}} P^\theta(D_1 \mid H_1) w(\boldsymbol{\theta})\, d\boldsymbol{\theta} \tag{5-30}$$

or in terms of the density functions

$$\overline{P(D_1 \mid H_1)} = \int_{\{\boldsymbol{\theta}\}} \int_{R_1} p_1(\mathbf{y} \mid \boldsymbol{\theta}) w(\boldsymbol{\theta})\, d\mathbf{y}\, d\boldsymbol{\theta} \tag{5-31}$$

The probability density function which minimizes $\overline{P(D_1 \mid H_1)}$ for the test in Eq. (5-29) is the least favorable distribution. In the general case, such a solution, if it exists, may be difficult to determine. In specific cases, insight into the problem may produce quick results. For example, in certain detection problems discussed in later chapters the assumption that the phase of a signal is uniformly distributed over the interval $(0, 2\pi)$ corresponds to a least favorable distribution (*1*).

Still another alternative, in the event that the a priori distributions are unknown and a uniformly most powerful test does not exist, is to use the *maximum-likelihood* principle (*5*) for which the generalized likelihood ratio

$$\lambda(\mathbf{y}) = \frac{\max_{\boldsymbol{\theta}} p_1(\mathbf{y} \mid \boldsymbol{\theta})}{\max_{\boldsymbol{\phi}} p_0(\mathbf{y} \mid \boldsymbol{\phi})} \tag{5-32}$$

is used. That is, we find that value of $\boldsymbol{\theta}$, say $\hat{\boldsymbol{\theta}}$, which maximizes $p_1(\mathbf{y} \mid \boldsymbol{\theta})$ and similarly that value of $\boldsymbol{\phi}$, say $\hat{\boldsymbol{\phi}}$, which maximizes $p_0(\mathbf{y} \mid \boldsymbol{\phi})$. The generalized likelihood ratio is then

$$\lambda(\mathbf{y}) = \frac{p_1(\mathbf{y} \mid \hat{\boldsymbol{\theta}})}{p_0(\mathbf{y} \mid \hat{\boldsymbol{\phi}})} \tag{5-33}$$

It will be pointed out in Chap. 10 that $\hat{\boldsymbol{\theta}}$ and $\hat{\boldsymbol{\phi}}$ are maximum-likelihood estimates of $\boldsymbol{\theta}$ and $\boldsymbol{\phi}$ respectively. Thus, this method directs us to make maximum-likelihood estimates of the unknown parameters and to substitute these estimates in the likelihood functions. The estimates are therefore used in the same manner as the true parameter values had they been known. However, the result is not necessarily optimum.

EXAMPLE 5.10-1 We shall treat here a decision problem which does not result in a uniformly most powerful test. For the assumption of a Gaussian likelihood function with variance σ^2 for each hypothesis, we shall test the hypothesis that the mean is zero (H_0) against the hypothesis that the mean is not zero (H_1). For the present, the mean under H_1 may be positive or negative. For the sake of clarity, assume that only one sample is used. The results are easily extended to multiple samples.

The likelihood functions are

$$p_0(y) = \left(\frac{1}{2\pi\sigma^2}\right)^{1/2} \exp\left[-\frac{y^2}{2\sigma^2}\right]$$

and

$$p_1(y) = \left(\frac{1}{2\pi\sigma^2}\right)^{1/2} \exp\left[-\frac{(y-m)^2}{2\sigma^2}\right]$$

where m is the mean for H_1.

The likelihood ratio is

$$\lambda(y) = \exp\left(\frac{my - m^2/2}{\sigma^2}\right)$$

Using the log-likelihood ratio produces the decision rule to choose H_1 if

$$my - m^2/2 > \lambda_o$$

where λ_o is determined by the criterion. The rule may then be expressed as choose H_1 if

$$my > \lambda_o'$$

where $\lambda_o' = \lambda_o + m^2/2$. Therefore, if the mean were positive, the decision rule would be to choose H_1 if $y > \lambda_o'/m$. If the mean were negative, the decision rule would be to choose H_1 if $y < \lambda_o'/m$. Thus, not knowing the sign of the mean, the decision rule cannot be applied and no uniformly most powerful test exists.

To further demonstrate this, fix the error of the first kind $P(D_1 \mid H_0)$ at, say 0.1, and design the test as though the mean were positive. Then the one-sided threshold test as shown in Fig. 5-5A is appropriate. (If the assumption of positive mean is correct, this test is uniformly most powerful.) For such a test, the resulting power, $P(D_1 \mid H_1)$, is shown in Fig. 5-6 as a function of the true value of the mean. Note that if the mean is allowed to be negative, very poor performance results. We might similarly have designed the test

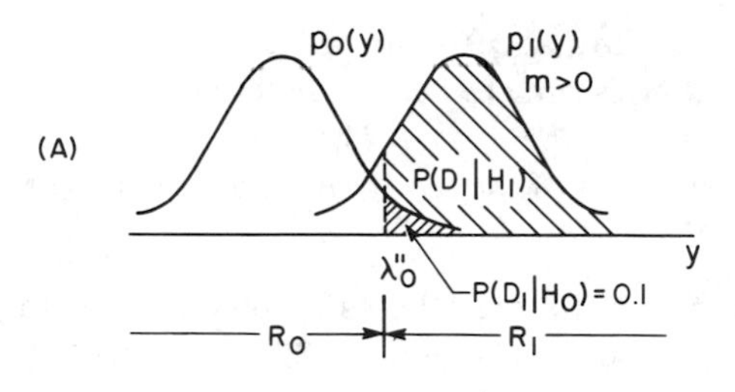

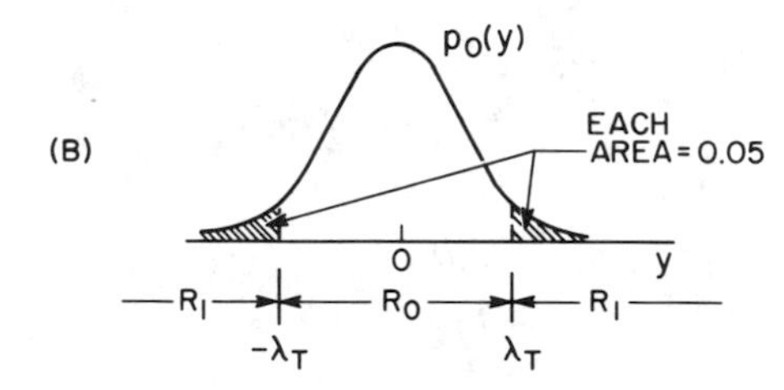

Fig. 5-5 Likelihood functions for Example 5.10-1. (A) One-sided test for $m > 0$. (B) Two-sided test for $m > 0$ or $m < 0$, $P(D_1 \mid H_0) = 0.1$.

as though the mean were negative. The resulting performance for this case is also shown in Fig. 5-6 and the same comments apply here as for the previous test. No other tests can be found to surpass the above tests when the assumptions are correct. However, these tests can obviously have very poor performance if our assumptions about the mean are incorrect.

A reasonable compromise in this example is to use a two-sided test as depicted in Fig. 5-5B. The decision regions are chosen so that the shaded portion of both tails add up to the constraint $P(D_1 \mid H_0) = 0.1$. The decision rule is to choose H_1 if $|y| > \lambda_T$. The resulting performance is also shown in

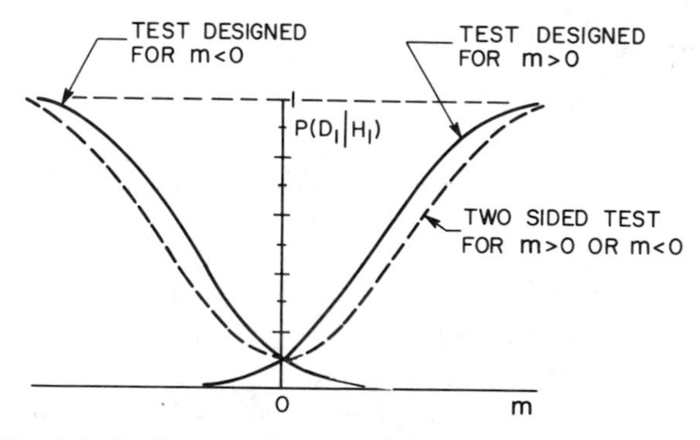

Fig. 5-6 Performance for tests of the mean. $P(D_1 \mid H_0)$ fixed.

Fig. 5-6. While the performance is not as good as the one-sided test with correct assumptions, it is much better than the one-sided test based on incorrect assumptions for the mean. Thus for a modest reduction in the maximum achievable power of the test we have greatly increased the minimum power.

EXAMPLE 5.10-2 For the hypotheses of the above example, determine a generalized-likelihood ratio test Eq. (5-32). Assume that n statistically independent samples are taken.

Since the null hypothesis is simple, the generalized likelihood ratio is

$$\lambda(\mathbf{y}) = \frac{\max_m p_1(\mathbf{y} \mid m)}{p_0(\mathbf{y})}$$

The likelihood function for hypothesis H_1 is

$$p(\mathbf{y} \mid m) = \left(\frac{1}{2\pi\sigma^2}\right)^{n/2} \exp\left[-\sum_{i=1}^{n} \frac{(y_i - m)^2}{2\sigma^2}\right]$$

It may be shown (see for example Chap. 10) that the maximum-likelihood estimate of m, denoted $\hat{m}$, is

$$\hat{m} = \frac{1}{n} \sum_{i=1}^{n} y_i$$

The generalized-likelihood ratio is then

$$\lambda(\mathbf{y}) = \exp\left[\sum_{i=1}^{n} \frac{(2y_i \hat{m} - \hat{m}^2)}{2\sigma^2}\right]$$

Omitting the mathematical details and not accounting for modifications to the threshold, the maximum-likelihood ratio test is to choose H_1 if

$$\hat{m}^2 > \lambda_o$$

or what is the same, choose H_1 if

$$\hat{m} > +\lambda_o^{1/2} \qquad \text{or} \qquad \hat{m} < -\lambda_o^{1/2}$$

which is just the two-sided test discussed in the previous example.

Exercises

5.1 Using just a single observation, what is the likelihood ratio receiver to choose between the hypotheses that for H_0 the sample is zero mean Gaussian with variance σ_0^2, and for H_1 the sample is zero mean Gaussian with variance σ_1^2. ($\sigma_1^2 > \sigma_0^2$.)

(a) In terms of the observation, what are the decision regions R_0 and R_1?

(b) What is the probability of choosing H_1 when H_0 is true?

5.2 Based on N independent samples design a likelihood ratio test to choose between

$$H_0: \quad r(t) = n(t)$$
$$H_1: \quad r(t) = 1 + n(t)$$
$$H_2: \quad r(t) = -1 + n(t)$$

where $n(t)$ is zero mean Gaussian with variance σ^2. Assume equal a priori probabilities for each hypothesis, no cost for correct decisions, and equal costs for any error. Show that the test statistic may be chosen to be $\bar{r} = (1/N)\sum_{i=1}^{N} r_i$. Find the decision regions for $\bar{r}$.

5.3 Design a likelihood ratio test to choose between

$$H_1: \quad p_1(y) = \frac{1}{(2\pi)^{1/2}} \exp\left(-\frac{y^2}{2\sigma^2}\right), \qquad -\infty < y < \infty$$

$$H_0: \quad p_0(y) = \begin{cases} \frac{1}{2}, & -1 \le y \le 1 \\ 0, & \text{otherwise} \end{cases}$$

(a) Assume that $\lambda_0 = 1$. In terms of y, and as a function of σ^2, what are the decision regions?

(b) Use a Neyman–Pearson test with $P(D_1 \mid H_0) = \alpha$. What are the decision regions?

5.4 (a) Design a likelihood-ratio test to choose between the hypotheses

H_1: The sample x is χ^2 distributed with n degrees of freedom.
H_0: The sample x is χ^2 distributed with two degrees of freedom.

Assume equal a priori probabilities and costs of error.

(b) Suppose the number of degrees of freedom in H_1 is a discrete random variable such that

$$P(n = N) = \tfrac{1}{2}, \qquad P(n = M) = \tfrac{1}{2}$$

What is the likelihood ratio test for this case?

5.5 Based on a single observation, use a minimax test to decide between

$$H_1: \quad r(t) = n(t)$$
$$H_0: \quad r(t) = 1 + n(t)$$

Assume $n(t)$ is Gaussian with zero mean and power σ^2. Assume $C_{00} = C_{11} = 0$, $C_{10} = C_{01} = 1$. What is the threshold in terms of the observation? What a priori probability for each hypothesis is implied by the solution?

5.6 For the preceding problem, assume $C_{10} = 3$ and $C_{01} = 6$.

(a) What a priori probability of each hypothesis would limit the maximum possible cost?

(b) What is the decision region in terms of the single observation?

5.7 Consider the receiver shown in Fig. 5-7. The hypotheses are

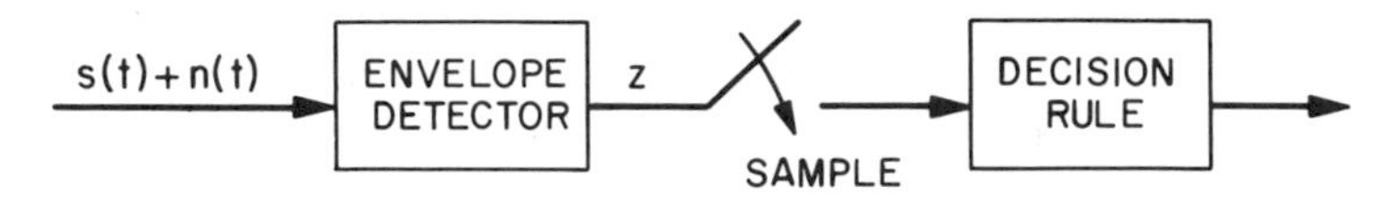

Fig. 5-7 Single sample receiver.

H_1: $s(t) = A\cos(\omega_c t + \theta)$
H_0: $s(t) = 0$

where A and ω_c are constants. The narrowband Gaussian noise has zero mean and variance σ_n^2.

(a) Assume an arbitrary value of λ_0. For a single sample, z, of the envelope, can the decision rule be implemented by comparing z to a threshold? Why?

(b) In closed form, what is the probability of choosing H_1 when H_0 is true? Using a Neyman–Pearson test with a fixed value of $P(D_1 \mid H_0)$, show that the threshold value against which z is compared is

$$[-2\sigma_n^2 \ln P(D_1 \mid H_0)]^{1/2}$$

(c) Determine the decision rule if M independent samples of z are used.

5.8 Consider the receiver shown in Fig. 5-8. The hypotheses are

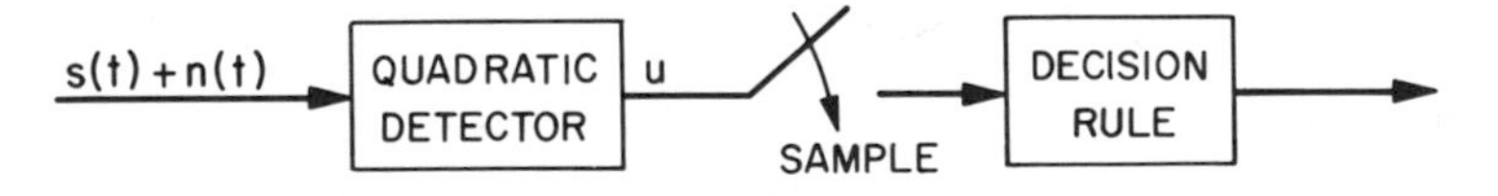

Fig. 5-8 Single sample receiver.

H_1: $s(t) = A\cos(\omega_c t + \theta)$
H_0: $s(t) = 0$

where A and ω_c are constants. The narrowband Gaussian noise has zero mean and variance σ_n^2.

(a) Assume an arbitrary value of λ_0. For a single sample of u, can the decision rule be implemented by comparing u to a threshold? Why? How would the performance of this system compare with that of part (a) of the preceding problem?

(b) Determine the decision rule if M independent samples of u are used.

5.9 For the receiver in Fig. 5-8, assume that the noise is zero mean Gaussian with variance σ_n^2 and, in addition, the signal is also zero mean Gaussian with variance σ_s^2. The null hypothesis is noise only.

(a) For a single sample, show that the decision rule can be implemented by comparing z to a threshold.

(b) Show that

$$P(D_1 \mid H_1) = [P(D_1 \mid H_0)]^{\sigma_n^2/(\sigma_n^2+\sigma_s^2)}$$

(c) Assume that M statistically independent samples of z are used. Show that the likelihood ratio receiver can be implemented by comparing $\lambda_T = \sum_{i=1}^{M} z_i^2$ to a threshold.

(d) Denote this threshold as V_T. Show that

$$P(D_1 \mid H_0) = \int_{V_T/\sigma_n^2}^{\infty} \frac{y^{M-1}e^{-y/2}}{2^M\Gamma(M)}\, dy$$

and

$$P(D_1 \mid H_1) = \int_{V_T/(\sigma_n^2+\sigma_s^2)}^{\infty} \frac{y^{M-1}e^{-y/2}}{2^M\Gamma(M)}\, dy$$

5.10 Assume that the four hypotheses to be tested are

H_0: χ^2 distribution with two degrees of freedom
H_1: χ^2 distribution with four degrees of freedom
H_2: χ^2 distribution with six degrees of freedom
H_3: χ^2 distribution with eight degrees of freedom.

Assume equal a priori probabilities, no cost for correct decisions, and equal cost for any error.

(a) Based on a single sample x show that the results of a likelihood ratio test is the following: choose

H_0 if $0 \le x < 2$
H_1 if $2 \le x < 4$
H_2 if $4 \le x < 6$
H_3 if $6 \le x$

(Neglect the problem of what to do if x is equal to 2, 4, or 6. It is not important in this example.)

(b) Assume M statistically independent samples x_i, $i = 1, \ldots, M$ are used. Show that the optimum test is the same as in part (a) except that x is replaced by $(\prod_{i=1}^{M} x_i)^{1/M}$.

5.11 Design a likelihood ratio test to choose between the hypotheses

H_0: The samples are χ^2 distributed with two degrees of freedom
H_1: The samples are χ^2 distributed with either four, six, or eight degrees of freedom.

Assume M statistically independent samples are used. Assume $P(H_0) = \frac{1}{4}$ and $P(H_1) = \frac{3}{4}$. Furthermore, the a priori probability of the degrees of freedom v is $P(v = 4) = P(v = 6) = P(v = 8) = \frac{1}{3}$.

References

1. Helstrom, C. W., "Statistical Theory of Signal Detection." Pergamon Press, Oxford, 1960.
2. Lawson, J. L., and Uhlenbeck, G. E., "Threshold Signals, Radiation Laboratory Series," Vol. 24. McGraw-Hill, New York, 1950.
3. Goldstein, H., "Classical Mechanics." Addison-Wesley, Reading, Massachusetts, 1959.
4. Middleton, D., and Van Meter, D., On optimum multiple alternative detection of signals in noise, *IRE Trans. Inform. Theory* (September 1955).
5. Davenport, W. B. Jr., and Root, W. L., "An Introduction to the Theory of Random Signals and Noise." McGraw-Hill, New York, 1958.

SUPPLEMENTARY BIBLIOGRAPHY

Kendal, M. G., and Stuart, A., "Inference and Relationship, The Advanced Theory of Statistics," Vol. 2. Hafner, New York, 1961.

Kullback, S., "Information Theory and Statistics." Wiley, New York, 1959.

Lehmann, E. L., "Testing Statistical Hypotheses." Wiley, New York, 1959.

Middleton, D., "An Introduction to Statistical Communication Theory." McGraw-Hill, New York, 1960.

Mood, A. M., and Graybill, F. A., "Introduction to the Theory of Statistics," 2nd Ed. McGraw-Hill, New York, 1963.

Wald, A., "Sequential Analysis." Wiley, New York, 1947; 5th Printing, 1959.

Chapter 6

Detection of Known Signals

6.1 Introduction†

The preceding chapters include most of the fundamentals required for statistical detection of signals in noise. By statistical detection we mean the application of probability and statistical tools to design "receivers" which discriminate noise corrupted signals from noise only, or which distinguish between different signals in the presence of noise. For our purposes a *receiver* is a mathematical description of the operation to be performed on the noise corrupted signal.

Our primary intent will be to design *optimum* receivers. It is usually necessary to add apologetic quotation marks around the word optimum. The reason is that an optimum receiver is strongly dependent on the assumptions and criteria of goodness. *By an optimum receiver we mean a receiver which best satisfies a given criteria under a given set of assumptions.* If either the criteria or assumptions change, there is generally no reason to believe that the form of the optimum receiver will remain unchanged. If the assumptions used in an analysis are not consistent with the conditions of a given environment, then the theoretical optimum receiver might perform poorly. In any event the optimum results serve a useful purpose as a standard against which other results may be compared.

There are a number of criteria of goodness which could be used. We shall

† For a status report on the many aspects of information theory consult Zadeh *et al.* (*1*) which also includes a substantial bibliography.

be interested primarily in the minimum error probability criterion for communication systems, and the Neyman–Pearson criterion for radar and sonar systems. Since these are special cases of the Bayes criterion, it should be obvious that hypothesis testing and likelihood ratios will be of importance (*2*). Their application here is in principle no different than in Chap. 5. In a sense we really need not go any further since we already have most of the fundamentals in hand. However the gap between knowledge of the fundamentals and its application is sometimes wide. A substantial portion of the remainder of this text is devoted to narrowing this gap.

Signals and Noise

In this chapter we shall discuss the detection of signals of known form to which noise is added.† That is, if a signal is present, its amplitude, frequency, phase, time of arrival, etc. are completely known. A hypothesis associated with such a signal will be simple. Although this is certainly an ideal it serves well as an introduction to the application of detection theory. Furthermore, some systems in practice do approach this ideal. In other cases the performance of the ideal systems may be used as standards against which nonideal systems are compared. Signals having unknown parameters will be discussed in remaining chapters.

The introductory material will assume that the noise is additive, white, and Gaussian. For white noise, the power spectral density is a constant $(N_o/2)$ over the entire frequency range. The corresponding autocorrelation function is a delta function $(N_o/2)\,\delta(\tau)$. The power of such a random process is theoretically infinite. In practice, however, the noise may have a constant spectral density over a wide but still finite range. As long as the noise is "flat" over a sufficiently broad band relative to the signal band of interest, a white noise assumption is reasonable.

6.2 A Binary Communication System

Consider a binary communication model in which one of two signals, $s_0(t)$ or $s_1(t)$, is received in the time interval $(0, T)$. At the receiver white Gaussian noise with zero mean and spectral density $N_o/2$ is unavoidably added to the signal. The observable signal is then one of the two forms

$$r(t) = \begin{Bmatrix} s_0(t) \\ \text{or} \\ s_1(t) \end{Bmatrix} + n(t) \tag{6-1}$$

† Consult the bibliography at the end of this chapter for references dealing with noise and its origins.

The objective is to design a receiver which operates on $r(t)$ and chooses one of the hypotheses:

$$H_0: \quad r(t) = s_0(t) + n(t)$$
$$H_1: \quad r(t) = s_1(t) + n(t)$$

For the criteria discussed in Chap. 5, the optimum decision rule is to compare the likelihood ratio to some threshold λ_o. For the moment, the particular criterion need not be specified since this changes only λ_o.

To get at a probabilistic description of the continuous signal, we first assume that m amplitude samples of the received signal are available, and then take the limit as $m \to \infty$. A different approach using a much different kind of sample is considered in Chap. 9.

The signal sampled at time t_k is

$$r(t_k) = s_i(t_k) + n(t_k) \tag{6-2}$$

The subscript $i\,(=0, 1)$ depends on which hypothesis is true. For convenience, denote these samples as r_k, s_{ik}, and n_k, so that

$$r_k = s_{ik} + n_k, \qquad 1 \le k \le m$$

For the finite dimension sample vector, the decision rule is to choose H_1 if

$$\lambda(\mathbf{r}) = \frac{p_1(r_1, r_2, \ldots, r_m)}{p_0(r_1, r_2, \ldots, r_m)} \ge \lambda_o$$

The receiver is illustrated in Fig. 6-1.

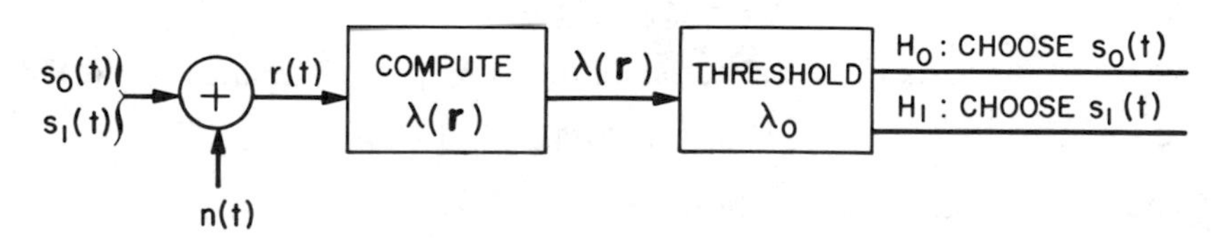

Fig. 6-1 Optimum receiver.

We shall momentarily assume that the noise is bandlimited white noise with power spectral density

$$S(\omega) = \begin{cases} N_o/2, & |\omega| < \Omega \\ 0, & \text{otherwise} \end{cases} \tag{6-3}$$

and autocorrelation function

$$R(\tau) = \frac{N_o \Omega}{2\pi} \frac{\sin \Omega\tau}{\Omega\tau} \tag{6-4}$$

These are shown in Fig. 6-2. The limit as $\Omega \to \infty$ is considered later.

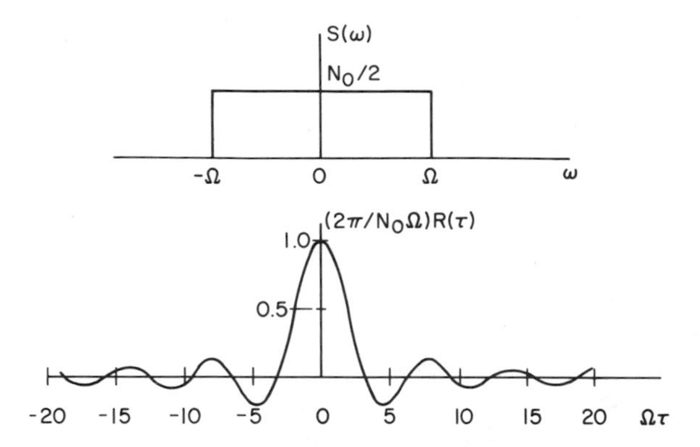

Fig. 6-2 Power spectral density and autocorrelation function of bandlimited white noise. The autocorrelation function has zeros at $\Omega\tau = k\pi$, $k = \pm 1, \pm 2, \ldots$.

The first zero of $R(\tau)$ occurs for $\tau = \pi/\Omega$. Therefore if the received signal is sampled at intervals $\Delta t = \pi/\Omega$, the samples are uncorrelated, and being Gaussian, they are statistically independent. In the interval $(0, T)$ we can take $m = T/\Delta t = \Omega T/\pi$ independent samples. To write explicitly the likelihood functions we must determine the mean and variance of the samples r_k. Since the noise has zero mean,

$$E\{r_k\} = E\{s_{ik} + n_k\} = s_{ik} \tag{6-5}$$

The variance of r_k is

$$E\{[r_k - E\{r_k\}]^2\} = E\{n_k^{\ 2}\} \tag{6-6}$$

This is just the variance of the noise, denote it $\sigma_n^{\ 2}$, which is equal to $R(0)$, or

$$\sigma_n^{\ 2} = N_o\Omega/2\pi \tag{6-7}$$

Therefore, the likelihood functions are

$$p_0(\mathbf{r}) = \left(\frac{1}{2\pi\sigma_n^{\ 2}}\right)^{m/2} \exp\left[-\sum_{k=1}^{m} \frac{(r_k - s_{0k})^2}{2\sigma_n^{\ 2}}\right] \tag{6-8}$$

and

$$p_1(\mathbf{r}) = \left(\frac{1}{2\pi\sigma_n^{\ 2}}\right)^{m/2} \exp\left[-\sum_{k=1}^{m} \frac{(r_k - s_{1k})^2}{2\sigma_n^{\ 2}}\right] \tag{6-9}$$

The likelihood ratio $\lambda(\mathbf{r})$ is

$$\lambda(\mathbf{r}) = \frac{p_1(\mathbf{r})}{p_0(\mathbf{r})} = \frac{\exp\left[-\sum_{k=1}^{m} \frac{(r_k - s_{1k})^2}{2\sigma_n^{\ 2}}\right]}{\exp\left[-\sum_{k=1}^{m} \frac{(r_k - s_{0k})^2}{2\sigma_n^{\ 2}}\right]} \tag{6-10}$$

or, upon combining terms,

$$\lambda(\mathbf{r}) = \exp\left\{-\tfrac{1}{2}\sum_{k=1}^{m}\left[\frac{2r_k s_{0k}}{\sigma_n^2} - \frac{2r_k s_{1k}}{\sigma_n^2} - \frac{(s_{0k}^2 - s_{1k}^2)}{\sigma_n^2}\right]\right\} \tag{6-11}$$

The decision rule is to choose H_1 if $\lambda(\mathbf{r}) \geq \lambda_o$, or in terms of the log-likelihood ratio, choose H_1 if

$$-\sum_{k=1}^{m}\frac{r_k s_{0k}}{\sigma_n^2} + \sum_{k=1}^{m}\frac{r_k s_{1k}}{\sigma_n^2} \geq \ln \lambda_o - \tfrac{1}{2}\sum_{k=1}^{m}\frac{(s_{0k}^2 - s_{1k}^2)}{\sigma_n^2} \tag{6-12}$$

To obtain the decision rule in terms of the continuous functions, we allow Δt to approach zero and m (and hence Ω) to approach infinity in such a way that $m\Delta t = T$, a constant. The variance of the noise σ_n^2 is equal to $N_o/(2\Delta t)$ since $\Omega = \pi/\Delta t$. Substitute this into Eq. (6-12) and consider the limit as $\Omega \to \infty$,

$$\lim_{\substack{\Delta t \to 0 \\ m \to \infty \\ m\Delta t = T}}\left[\sum_{k=1}^{m}\frac{2r_k s_{1k}\,\Delta t}{N_o} - \sum_{k=1}^{m}\frac{2r_k s_{0k}\,\Delta t}{N_o} \geq \ln \lambda_o - \sum_{k=1}^{m}\frac{(s_{0k}^2 - s_{1k}^2)\,\Delta t}{N_o}\right] \tag{6-13}$$

In the limit the summations become integrals and†

$$\frac{2}{N_o}\int_0^T r(t)s_1(t)\,dt - \frac{2}{N_o}\int_0^T r(t)s_0(t)\,dt \geq \ln \lambda_o - \frac{1}{N_o}\int_0^T [s_0^2(t) - s_1^2(t)]\,dt \tag{6-14}$$

The decision rule is then: choose H_1 if

$$\int_0^T r(t)s_1(t)\,dt - \int_0^T r(t)s_0(t)\,dt \geq V_T \tag{6-15}$$

where the threshold is

$$V_T = \tfrac{1}{2}N_o \ln \lambda_o - \tfrac{1}{2}\int_0^T [s_0^2(t) - s_1^2(t)]\,dt \tag{6-16}$$

Otherwise, choose H_0. The decision rule may be implemented as shown in Fig. 6-3. It is the well-known *correlation receiver*, so called because the input $r(t)$ is crosscorrelated with the signal $s_0(t)$ and $s_1(t)$.

For a Neyman–Pearson criterion, the threshold V_T is selected to satisfy the false alarm probability. For a minimum error probability criterion, λ_o is known beforehand so that V_T may be determined from Eq. (6-16).

† We have, of course, ignored any possibility of convergence difficulties.

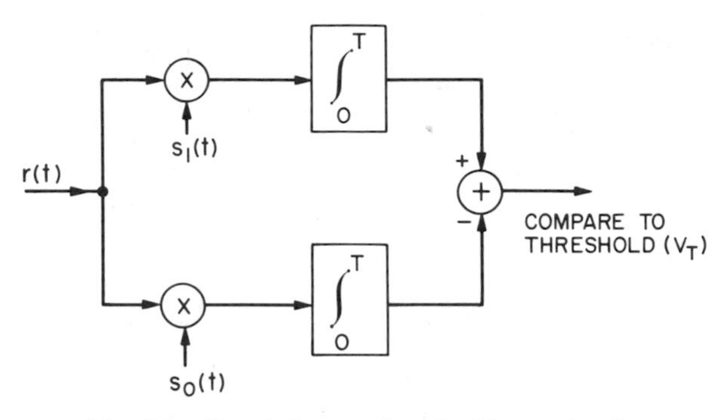

Fig. 6-3 Correlation receiver for binary signals.

Performance as a Communication Receiver

To determine the performance of the correlation receiver as applied to communications, assume that the a priori probabilities of H_0 and H_1 are each $\frac{1}{2}$, and that the cost of each kind of error is equal. This will result in a minimum error probability. With these assumptions $\lambda_o = 1$, so that the decision rule may be expressed: choose H_1 if

$$G = \int_0^T r(t)s_1(t)\,dt - \int_0^T r(t)s_0(t)\,dt + \tfrac{1}{2}\int_0^T [s_0{}^2(t) - s_1{}^2(t)]\,dt \geq 0 \tag{6-17}$$

For convenience, we have denoted the integral terms by G. To compute the error probability we require the density function for G conditioned on H_0 and H_1. Denote these density functions as $P_0(G)$ and $P_1(G)$ respectively. Then the error probability is

$$\begin{aligned} P_e &= P(D_1 \mid H_0)P(H_0) + P(D_0 \mid H_1)P(H_1) \\ &= \tfrac{1}{2}\int_0^\infty P_0(G)\,dG + \tfrac{1}{2}\int_{-\infty}^0 P_1(G)\,dG \end{aligned} \tag{6-18}$$

The first two terms of Eq. (6-17) are integrals of a Gaussian process. Thus G is a Gaussian random variable, and only its mean and variance are needed to specify its probability density function. Denote by $E_0\{G\}$ the mean of G given hypothesis H_0. Under H_0, $r(t) = s_0(t) + n(t)$, and

$$\begin{aligned} E_0\{G\} = E\bigg\{&\int_0^T [s_0(t) + n(t)]s_1(t)\,dt - \int_0^T [s_0(t) + n(t)]s_0(t)\,dt \\ &+ \tfrac{1}{2}\int_0^T [s_0{}^2(t) - s_1{}^2(t)]\,dt \end{aligned} \tag{6-19}$$

Since the mean of $n(t)$ is zero,

$$E_0\{G\} = \int_0^T s_0(t)s_1(t)\,dt - \int_0^T s_0^2(t)\,dt + \tfrac{1}{2}\int_0^T [s_0^2(t) - s_1^2(t)]\,dt$$

$$= -\tfrac{1}{2}\int_0^T [s_0(t) - s_1(t)]^2\,dt \qquad (6\text{-}20)$$

Now,

$$G - E_0\{G\} = \int_0^T n(t)[s_1(t) - s_0(t)]\,dt$$

and the variance, denoted $V_0\{G\} = E\{[G - E_0\{G\}]^2\}$, is

$$V_0\{G\} = E\left\{\int_0^T \int_0^T n(t)n(\tau)[s_1(t) - s_0(t)][s_1(\tau) - s_0(\tau)]\,dt\,d\tau\right\}$$

$$= \int_0^T \int_0^T E\{n(t)n(\tau)\}[s_1(t) - s_0(t)][s_1(\tau) - s_0(\tau)]\,dt\,d\tau$$

Since the noise is assumed to be stationary white Gaussian noise with spectral density $N_o/2$,

$$E\{n(t)n(\tau)\} = (N_o/2)\,\delta(t - \tau)$$

and, therefore, the variance of G given H_0 is

$$V_0\{G\} = \tfrac{1}{2}N_o \int_0^T [s_1(t) - s_0(t)]^2\,dt \qquad (6\text{-}21)$$

It may be shown similarly that if hypothesis H_1 is true, then

$$E_1\{G\} = \tfrac{1}{2}\int_0^T [s_0(t) - s_1(t)]^2\,dt \qquad (6\text{-}22)$$

and the variance denoted $V_1\{G\}$ is

$$V_1\{G\} = V_0\{G\} = \tfrac{1}{2}N_o \int_0^T [s_1(t) - s_0(t)]^2\,dt \qquad (6\text{-}23)$$

Now define

$$E = \tfrac{1}{2}\int_0^T [s_0^2(t) + s_1^2(t)]\,dt \qquad (6\text{-}24)$$

and

$$\rho = (1/E)\int_0^T s_0(t)s_1(t)\,dt \qquad (6\text{-}25)$$

Thus E is the average energy of the two signals, and ρ is a time crosscorrelation coefficient. We shall show that $|\rho| \leq 1$. Since the integrand in the following equation is everywhere greater than or equal to zero, it follows that

$$\int_0^T [s_0(t) \pm s_1(t)]^2 \, dt \geq 0$$

Expanding the integrand we get

$$\int_0^T [s_0{}^2(t) + s_1{}^2(t)] \, dt \pm 2 \int_0^T s_0(t)s_1(t) \, dt \geq 0$$

Using Eqs. (6-24) and (6-25) this becomes

$$2E \pm 2\rho E \geq 0 \qquad \text{or} \qquad 1 \pm \rho \geq 0$$

from which it follows that $|\rho| \leq 1$.

Using Eqs. (6-24) and (6-25) it may be shown that

$$E_0\{G\} = -E(1 - \rho) \tag{6-26}$$

and

$$E_1\{G\} = E(1 - \rho) \tag{6-27}$$

The variances under each hypothesis are equal and are given by

$$V\{G\} = N_o E(1 - \rho) \tag{6-28}$$

Knowing the means and variances and the fact that G is a Gaussian random variable enables us to write directly the probability density functions $P_0(G)$ and $P_1(G)$:

$$P_0(G) = \left[\frac{1}{2\pi N_o E(1 - \rho)}\right]^{1/2} \exp\left\{-\frac{[G + E(1 - \rho)]^2}{2N_o E(1 - \rho)}\right\} \tag{6-29}$$

and

$$P_1(G) = \left[\frac{1}{2\pi N_o E(1 - \rho)}\right]^{1/2} \exp\left\{-\frac{[G - E(1 - \rho)]^2}{2N_o E(1 - \rho)}\right\} \tag{6-30}$$

Using Eq. (6-29), it is not difficult to show that

$$P(D_1 \mid H_0) = \int_{[(1-\rho)E/N_o]^{1/2}}^{\infty} \frac{1}{(2\pi)^{1/2}} e^{-z^2/2} \, dz \tag{6-31}$$

It may easily be shown that $P(D_0 \mid H_1) = P(D_1 \mid H_0)$. Applying the above to Eq. (6-18), the error probability becomes

$$P_e = \int_{[(1-\rho)E/N_o]^{1/2}}^{\infty} \frac{1}{(2\pi)^{1/2}} e^{-z^2/2}\, dz \tag{6-32}$$

This important result shows the performance of the correlation receiver for detecting completely known signals in additive white Gaussian noise. The error performance depends on only three parameters:

1. the average signal energy
2. the level of the noise spectral density
3. the time crosscorrelation between signals.

The performance is otherwise independent of the particular signal waveforms used.

As $(1-\rho)E/N_o$ increases, the error probability decreases. For fixed E/N_o, the optimum system is that for which the correlation coefficient $\rho = -1$. This is attained only for $s_0(t) = -s_1(t)$. This is known as the optimum or *ideal binary communication system.*

The result, Eq. (6-32), is next applied to determine the error performance for three binary communication systems (*3*) of practical interest.

Coherent Phase Shift Keying (*CPSK*)†

In this system the binary signals are sine waves which are 180° out of phase with each other. That is, one signal is the negative of the other. For example, over an interval $(0, T)$

$$s_0(t) = A \sin \omega_c t \qquad \text{and} \qquad s_1(t) = A \sin(\omega_c t + \pi) = -A \sin \omega_c t$$

This is an example of an ideal binary system ($\rho = -1$). Since each signal has equal energy, $E = \int_0^T s_1{}^2(t)\, dt$. The receiver could be implemented as shown in Fig. 6-3. In this case the threshold V_T, defined in Eq. (6-16), is equal to zero. A little thought will show that the optimum receiver could be implemented with just one correlator and the output compared to zero.

The error probability is given by

$$P_e = \int_{(2E/N_o)^{1/2}}^{\infty} \frac{1}{(2\pi)^{1/2}} e^{-z^2/2}\, dz \tag{6-33}$$

which is shown in Fig. 6-4 as a function of E/N_o.

† Use of the word "coherent" for narrowband systems normally implies a knowledge of the signal phase.

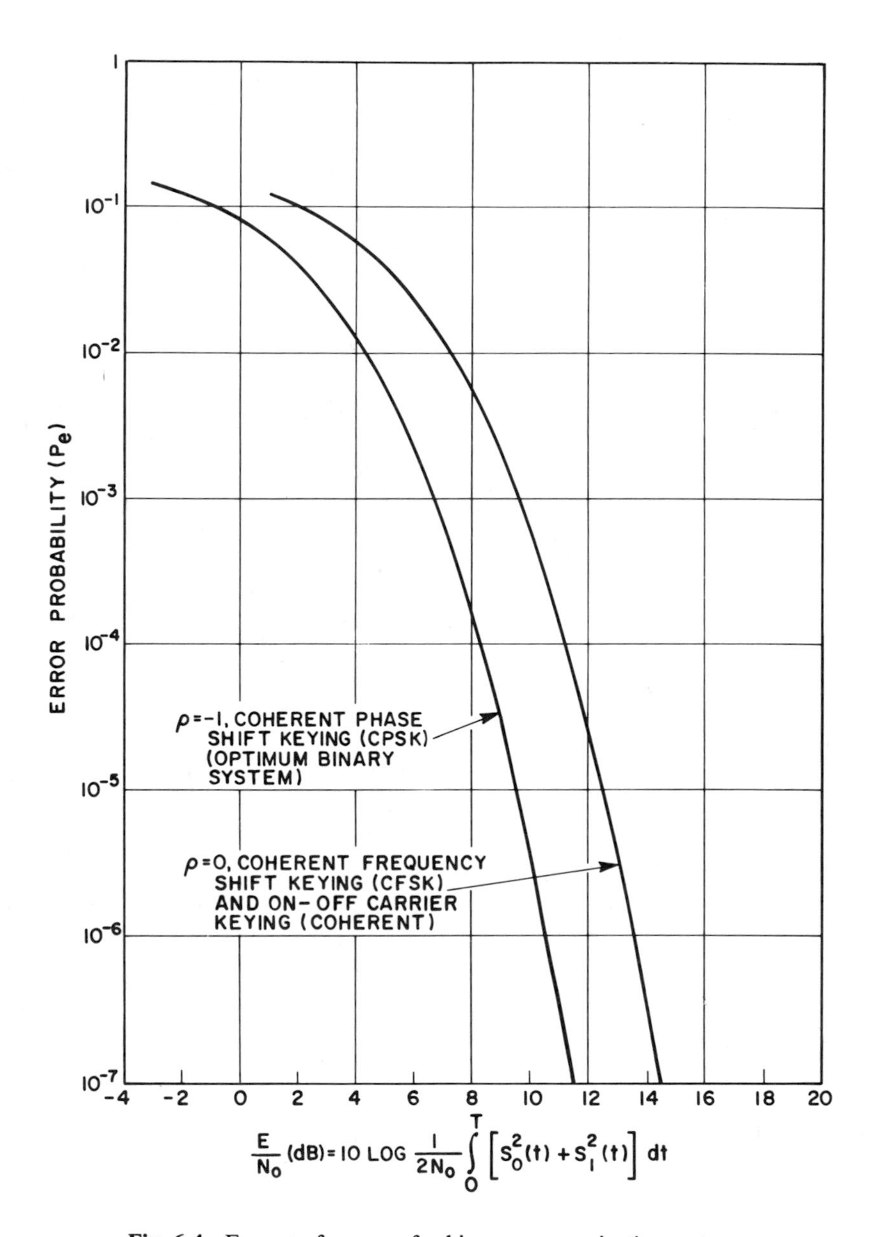

Fig. 6-4 Error performance for binary communication systems.

Coherent Frequency Shift Keying (CFSK)

The binary signals are given over an interval $(0, T)$ by

$$s_0(t) = A \sin \omega_0 t, \qquad 0 \le t \le T$$

and

$$s_1(t) = A \sin \omega_1 t, \qquad 0 \le t \le T$$

If the frequencies are chosen such that

$$\omega_1 - \omega_0 = n\pi/T \qquad \text{and} \qquad \omega_1 + \omega_0 = m\pi/T$$

where m and n are integers, then $\rho = 0$. (A choice of $\omega_1 - \omega_0 \approx (1.4)\pi/T$ would lead to a lower error probability; see Exercise 6.3). The signals are of equal energy and $E = \int_0^T s_1{}^2(t)\,dt$. The error probability is given by

$$P_e = \int_{(E/N_o)^{1/2}}^{\infty} \frac{1}{(2\pi)^{1/2}} e^{-z^2/2}\,dz \tag{6-34}$$

and is shown in Fig. 6-4. This system is 3 dB poorer than the optimum system because of the 2-to-1 difference in the value of $1 - \rho$.

On–Off Carrier Keyed System

The binary signals are

$$s_0(t) = 0, \qquad s_1(t) = B \cos \omega_0 t; \qquad 0 \le t \le T$$

and only one correlator is required for Fig. 6-3. It is obvious in this case that $\rho = 0$. However the signal energies are not equal and

$$E = \tfrac{1}{2} \int_0^T s_1{}^2(t)\,dt = E_1/2$$

where E_1 represents the energy of $s_1(t)$. From Eq. (6-16), the threshold, which in this case is signal dependent, is

$$V_T = \tfrac{1}{2} \int_0^T s_1{}^2(t)\,dt = E_1/2$$

The error probability is

$$P_e = \int_{(E_1/2N_o)^{1/2}}^{\infty} \frac{1}{(2\pi)^{1/2}} e^{-z^2/2}\,dz \tag{6-35}$$

Therefore, based on *average* signal energy E (to which the abscissa in Fig. 6-4 is related) the performance is the same as that of the CFSK system and is 3 dB worse than the optimum binary system (CPSK). If, however, the signal amplitude of $s_1(t)$ was constrained to be no higher than the signal amplitude for the CFSK case, the error performance of the on–off carrier keyed system would be less than the corresponding CFSK system.

Application to Radar

The signals just considered for the on–off carrier keyed system are typical of those used with a coherent radar (implying knowledge of the phase) and it is worthwhile discussing their application here. The hypothesis, H_0, is that no target is present so $r(t) = n(t)$. For hypothesis H_1, a target is present and $r(t) = s(t) + n(t)$. The analogy to the on–off carrier keyed system is therefore clear.

For this case, it is convenient to use as the test statistic

$$G' = \int_0^T r(t)s(t)\,dt$$

Then

$$E_0\{G'\} = 0 \qquad \text{and} \qquad E_1\{G'\} = E$$

Furthermore,

$$V\{G'\} = N_o E/2$$

It then follows that the false alarm probability is

$$P_{fa} = \int_{V_T}^{\infty} \frac{1}{(\pi N_o E)^{1/2}} \exp[-x^2/N_o E]\,dx$$

$$= \int_{\beta}^{\infty} \frac{1}{(2\pi)^{1/2}} e^{-z^2/2}\,dz \qquad (6\text{-}36)$$

where $\beta = V_T(2/N_o E)^{1/2}$ is chosen to give a specified false alarm probability. The probability of detection is

$$P_D = \int_{V_T}^{\infty} \frac{1}{(\pi N_o E)^{1/2}} \exp[-(x - E)^2/N_o E]\,dx$$

$$= \int_{\beta-(2E/N_o)^{1/2}}^{\infty} \frac{1}{(2\pi)^{1/2}} e^{-z^2/2}\,dz \qquad (6\text{-}37)$$

This is plotted in Fig. 7-5 of Chap. 7 where it is compared with the radar performance for signals having phase as a random parameter.

6.3 The Likelihood Functions

The limit of the log-likelihood ratio was used to derive the results of the preceding section. We next derive the limit of the likelihood functions for known signals in white Gaussian noise. This will provide a convenient starting point in many detection problems, including those for which the signal is not precisely known.

The likelihood functions for an n-dimensional vector $\mathbf{r}$ are given by Eqs. (6-8) and (6-9). With the substitutions $1/\sigma_n^2 = 2\Delta t/N_o$ the likelihood functions become

$$p_1(\mathbf{r}) = \left(\frac{\Delta t}{\pi N_o}\right)^{m/2} \exp\left[-\sum_{k=1}^{m} \frac{(r_k - s_{1k})^2 \Delta t}{N_o}\right]$$

and

$$p_0(\mathbf{r}) = \left(\frac{\Delta t}{\pi N_o}\right)^{m/2} \exp\left[-\sum_{k=1}^{m} \frac{(r_k - s_{0k})^2 \Delta t}{N_o}\right]$$

Using the arguments presented in Sect. 6.2, we take the limit as $\Omega \to \infty$. This implies $\Delta t \to 0$, $m \to \infty$, but $m\Delta t = T$. This produces the likelihood functions for "continuous sampling."

$$p_1(\mathbf{r}) = F \exp\left\{-(1/N_o) \int_0^T [r(t) - s_1(t)]^2 \, dt\right\} \tag{6-38}$$

and

$$p_0(\mathbf{r}) = F \exp\left\{-(1/N_o) \int_0^T [r(t) - s_0(t)]^2 \, dt\right\} \tag{6-39}$$

where F is some undetermined constant. The value of F is of no immediate interest since we generally consider only likelihood ratios. Since these functions vary inversely with the integral terms, a decision rule which chooses the largest $p_i(r)$ is the same as one which chooses the smallest of the integrals

$$\int_0^T [r(t) - s_i(t)]^2 \, dt \tag{6-40}$$

This integral is the integrated squared difference between the received signal $r(t)$ and the signal $s_i(t)$. Consequently, for known signals in white Gaussian noise, the signal which produces the least integrated squared difference also produces the largest likelihood function. An obvious extension of this result is seen when noise only is present. Then the likelihood function for the noise is

$$p(\mathbf{r}) = F \exp\left[-(1/N_o) \int_0^T n^2(t) \, dt\right] \tag{6-41}$$

and it is inversely proportional to $\int_0^T n^2(t) \, dt$.

6.4 Matched Filters

A particularly important topic in detection theory is the matched filter. Two cases will be considered, the matched filter for white noise, and the generalized matched filter for nonwhite or colored noise (*4–6*). A new criterion of goodness, signal-to-noise ratio, will be introduced. The equivalence to correlation type receivers will be pointed out.

Known Signals and White Noise (*7–9*)

Consider the correlator in the upper branch of the receiver shown in Fig. 6-3 and redrawn in Fig. 6-5. The output of the correlator at time $t = T$ is

$$e_1(T) = \int_0^T r(t)s_1(t)\, dt \tag{6-42}$$

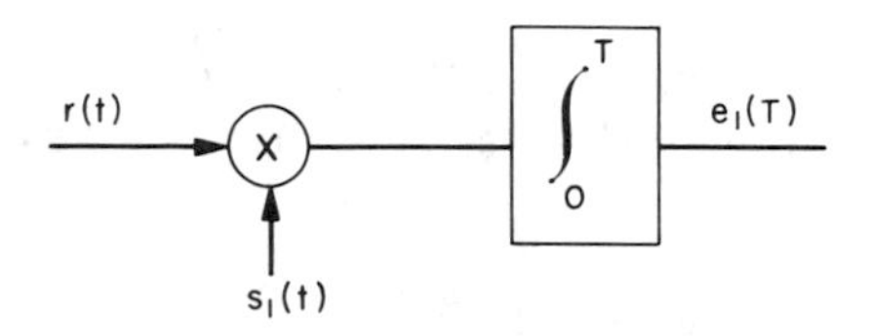

Fig. 6-5 Correlator.

We now inquire if $e_1(T)$ could be obtained by using a linear filter instead of the multiplier and integrator. Assume that such a filter does exist with a weighting function $h_1(t)$. Then the output of the filter in response to an input $r(t)$ is

$$e_1(t) = \int_0^t h_1(\tau)r(t - \tau)\, d\tau$$

In particular the output of the filter at time $t = T$ is given by

$$e_1(T) = \int_0^T h_1(\tau)r(T - \tau)\, d\tau \tag{6-43}$$

Suppose we make a fortuitous selection of $h_1(t)$ and assume that

$$h_1(t) = s_1(T - t), \qquad 0 \le t \le T \tag{6-44}$$

Substituting this into Eq. (6-43) we get

$$e_1(T) = \int_0^T s_1(T - \tau)r(T - \tau)\, d\tau = \int_0^T s_1(t)r(t)\, dt \tag{6-45}$$

Observe that Eq. (6-45) is identical to Eq. (6-42). Therefore the output of the correlator at time T is identical to the output of the filter also at time T. The filter defined by Eq. (6-44) is called a *matched filter*. It has the same form as the signal except for a reversal in time. An example of a signal and the corresponding matched filter is shown in Fig. 6-6.

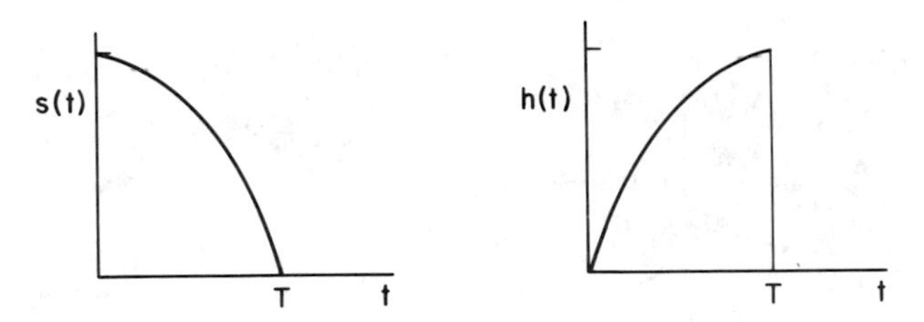

Fig. 6-6 A signal and the corresponding matched filter.

The above results applied to the signal $s_1(t)$. For $s_0(t)$, the matched filter would be

$$h_0(t) = s_0(T - t), \qquad 0 \le t \le T \tag{6-46}$$

These filters can be used to implement the operations of the correlation receiver of Fig. 6-3. The matched filter equivalent is shown in Fig. 6-7.

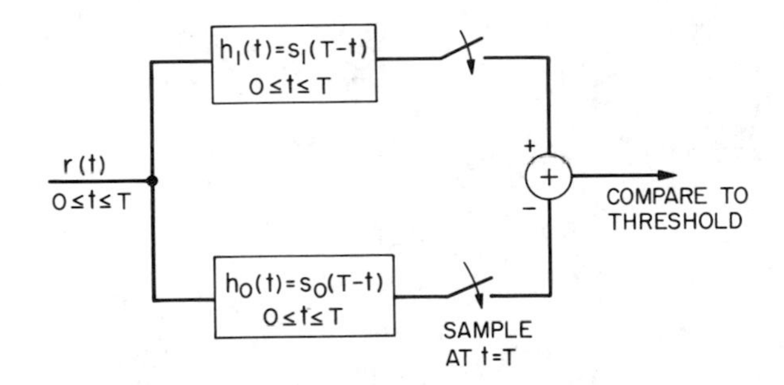

Fig. 6-7 Matched filter receiver.

We emphasize that the correlator output and the matched filter output are the same only at time T. For a sine wave input, the output of the correlator (with T replaced by t) is approximately a linear ramp for $0 \le t \le T$. On the other hand the matched filter output is approximately a sine wave amplitude modulated by a linear ramp for ($0 \le t \le T$) (see Sect. 3.5). These outputs are shown in Fig. 6-8.

It is of interest to compare the signal and the matched filter in the frequency domain. Suppose the signal has a Fourier transform

$$S(j\omega) = \int_0^T s(t)e^{-j\omega t}\, dt \tag{6-47}$$

The transfer function of the filter, denoted by $H(j\omega)$ is

$$H(j\omega) = \int_{-\infty}^{\infty} h(t)e^{-j\omega t}\, dt \tag{6-48}$$

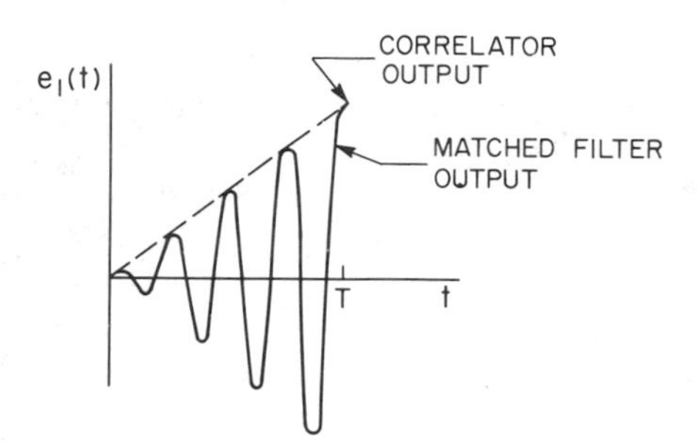

Fig. 6-8 Output of correlator and matched filter for sine wave input.

Substituting the matched filter condition, Eq. (6-44)

$$h(t) = \begin{cases} s(T-t), & 0 \le t \le T \\ 0, & \text{otherwise} \end{cases}$$

we get

$$H(j\omega) = \int_0^T s(T-t)e^{-j\omega t}\,dt$$

Changing the variable ($z = T - t$) we get

$$H(j\omega) = \int_0^T s(z)e^{-j\omega(T-z)}\,dz$$

or

$$H(j\omega) = e^{-j\omega T}\int_0^T s(z)e^{j\omega z}\,dz \tag{6-49}$$

If $s(z)$ is a real function, then the integral is $S^*(j\omega)$ so that the matched filter and signal are related in the frequency domain by the relation

$$H(j\omega) = e^{-j\omega T}S^*(j\omega) \tag{6-50}$$

Known Signals in Nonwhite Noise

We next discuss a standard engineering parameter called *signal-to-noise ratio* which is the ratio of the signal power to the noise power. In many instances, it may be defined as the ratio of the square of the average value of a random variable to the variance of the random variable. It will be shown that the matched filters defined above maximizes this ratio for the white noise case (*7*, *9*). This result is derived below as a special case of the generalized matched filter for nonwhite noise.

For the derivation of the generalized matched filter, start off by defining a new problem and a new criterion of goodness. This will lead to the general solution of the matched filter problem for nonwhite noise. The solution will then be used for the special case of white noise.

Consider a received waveform containing signal and noise, that is, $r(t) = s(t) + n(t)$. The signal is completely known and extends over the interval $(0, T)$. The noise, which is not necessarily white or Gaussian, has zero mean and autocorrelation function $R_n(\tau)$. It is therefore at least wide-sense stationary. We want to design a linear filter $h(t)$ such that the signal-to-noise ratio of the output at time T is a maximum.

The output of the filter at time T is

$$\begin{aligned} e(T) &= \int_0^T h(\tau)r(T-\tau)\,d\tau \\ &= \int_0^T h(\tau)s(T-\tau)\,d\tau + \int_0^T h(\tau)n(T-\tau)\,d\tau \end{aligned} \tag{6-51}$$

The signal and noise components are easily identified as

$$S(T) = \int_0^T h(\tau)s(T-\tau)\,d\tau \tag{6-52}$$

and

$$N(T) = \int_0^T h(\tau)n(T-\tau)\,d\tau \tag{6-53}$$

(Note that $S(T)$ is also the average value of $e(T)$. In statistics the ratio of standard deviation to average value of a random variable is called the coefficient of variation. Thus our criterion here might also be expressed as minimizing this coefficient of variation.) The noise component has zero mean, and its variance or power is

$$E\{N^2(T)\} = E\left\{\int_0^T\int_0^T h(\tau)h(z)n(T-\tau)n(T-z)\,d\tau\,dz\right\}$$

or in terms of its autocorrelation function

$$E\{N^2(T)\} = \int_0^T\int_0^T h(\tau)h(z)R_n(z-\tau)\,dz\,d\tau \tag{6-54}$$

The signal-to-rms noise ratio is therefore $S(T)/[E\{N^2(T)\}]^{1/2}$. For a meaningful solution, the ratio is maximized subject to the constraint that $S(T)$ is a constant. Then to maximize the ratio we must minimize $E\{N^2(T)\}$. The constraint is introduced by adding $\mu S(T)$ to $E\{N^2(T)\}$, where μ is a Lagrange multiplier (*10*). Thus

$$Q = E\{N^2(T)\} - \mu S(T) \tag{6-55}$$

is to be minimized with respect to the filter weighting function $h(t)$. The explicit expression for Q is

$$Q = \int_0^T \int_0^T h(\tau)h(z)R_n(z - \tau)\, dz\, d\tau - \mu \int_0^T h(\tau)s(T - \tau)\, d\tau \tag{6-56}$$

To accomplish the minimization, we use the calculus of variation (*10*). Assume that $h_0(x)$ is the optimum linear filter. Express

$$h(x) = h_0(x) + \alpha\varepsilon(x) \tag{6-57}$$

where $\varepsilon(x)$ is an arbitrary function defined for $0 \le x \le T$, and α is an arbitrary multiplier. If this expression is substituted into Eq. (6-56), then Q becomes a function of α for any given $\varepsilon(x)$. Further, it follows from Eq. (6-57) that $Q(\alpha)$ is a minimum at $\alpha = 0$. Thus

$$\left.\frac{\partial Q(\alpha)}{\partial \alpha}\right|_{\alpha=0} = 0 \tag{6-58}$$

The solution of this equation must be $h_0(x)$ since it was assumed to be the function which minimized Q. Substituting Eq. (6-57) into Eq. (6-56) yields

$$Q(\alpha) = \int_0^T \int_0^T [h_0(\tau) + \alpha\varepsilon(\tau)][h_0(z) + \alpha\varepsilon(z)]R_n(z - \tau)\, dz\, d\tau$$
$$- \mu \int_0^T [h_0(\tau) + \alpha\varepsilon(\tau)]s(T - \tau)\, d\tau$$

Then

$$\frac{\partial Q(\alpha)}{\partial \alpha} = \int_0^T \int_0^T [\varepsilon(\tau)h_0(z) + \varepsilon(z)h_0(\tau) + 2\alpha\varepsilon(\tau)\varepsilon(z)]R_n(z - \tau)\, dz\, d\tau$$
$$- \mu \int_0^T \varepsilon(\tau)s(T - \tau)\, d\tau \tag{6-59}$$

The second derivative is

$$\frac{\partial^2 Q(\alpha)}{\partial \alpha^2} = \int_0^T \int_0^T 2\varepsilon(\tau)\varepsilon(z)R_n(z - \tau)\, dz\, d\tau$$

For a positive definite function $R_n(z)$ (cf. Chap. 9) the integral is greater than zero for any arbitrary, nonzero function $\varepsilon(t)$. Therefore the solution we determine will indeed minimize $Q(\alpha)$.

Evaluating Eq. (6-59) at $\alpha = 0$,

$$\left.\frac{\partial Q(\alpha)}{\partial \alpha}\right|_{\alpha=0} = \int_0^T \int_0^T [\varepsilon(\tau)h_0(z) + \varepsilon(z)h_0(\tau)]R_n(z - \tau)\, dz\, d\tau$$
$$- \mu \int_0^T \varepsilon(\tau)s(T - \tau)\, d\tau = 0 \tag{6-60}$$

In the double integral, the first and second terms are equal, as the following equation shows:

$$\int_0^T \int_0^T \varepsilon(z)h_0(\tau)R_n(z-\tau)\,dz\,d\tau = \int_0^T \int_0^T \varepsilon(\tau)h_0(z)R_n(\tau - z)\,d\tau\,dz$$
$$= \int_0^T \int_0^T \varepsilon(\tau)h_0(z)R_n(z-\tau)\,d\tau\,dz$$

In this manipulation we have interchanged the variables z and τ and made use of the fact that $R_n(\tau - z) = R_n(z - \tau)$. Then Eq. (6-60) becomes

$$\int_0^T \int_0^T 2\varepsilon(\tau)h_0(z)R_n(\tau - z)\,dz\,d\tau - \mu \int_0^T \varepsilon(\tau)s(T-\tau)\,d\tau = 0$$

or, combining integrals,

$$\int_0^T d\tau\,\varepsilon(\tau)\left[\int_0^T 2h_0(z)R_n(\tau - z)\,dz - \mu s(T-\tau)\right] = 0 \qquad (6\text{-}61)$$

Since $\varepsilon(\tau)$ is arbitrary except for the range of its argument $(0, T)$ over which it is defined, it must be the bracketed term which causes this expression to be identically zero. That is,

$$\int_0^T h_0(z)R_n(\tau - z)\,dz = (\mu/2)s(T-\tau), \qquad 0 \le \tau \le T \qquad (6\text{-}62)$$

The constant multiplier $\mu/2$ changes only the gain of the filter and has the same influence on the signal and noise so that its value does not change the signal-to-noise ratio. We therefore arbitrarily set it to unity. Then Eq. (6-62) becomes the integral equation

$$\int_0^T h_0(z)R_n(\tau - z)\,dz = s(T-\tau), \qquad 0 \le \tau \le T \qquad (6\text{-}63)$$

This is the general expression for a matched filter. The filter $h_0(z)$ which satisfies this relation maximizes the signal-to-noise ratio for a known signal in noise having an autocorrelation function $R_n(\tau)$. Note that we have not used a Gaussian assumption. The solution of integral equations is discussed in Chap. 9.

As an aside, suppose that the filter memory (the time beyond which the impulse response is zero) were constrained to T_o where $T_o \le T$. Thus the filter has memory for a time less than the duration of the signal. We wish to maximize the signal-to-noise ratio at some time T_m where $T_o \le T_m \le T$. It is an easy matter to show that the optimum linear filter would be the solution of the equation

$$\int_0^{T_o} h_0(z)R_n(\tau - z)\,dz = s(T_m - \tau), \qquad 0 \le \tau \le T_o \qquad (6\text{-}64)$$

We now calculate the value of the maximum signal-to-noise ratio for the filter specified by Eq. (6-63). For this we use the power ratio $(S/N)_p \triangleq S^2(T)/E\{N^2(T)\}$. From Eqs. (6-52) and (6-54) and using the optimum filter $h_0(\tau)$ the maximum signal-to-noise ratio is

$$\left(\frac{S}{N}\right)_p = \frac{\int_0^T \int_0^T h_0(\tau)h_0(z)s(T-\tau)s(T-z)\,d\tau\,dz}{\int_0^T \int_0^T h_0(\tau)h_0(z)R_n(z-\tau)\,d\tau\,dz} \tag{6-65}$$

Rearranging terms gives

$$\left(\frac{S}{N}\right)_p = \frac{[\int_0^T h_0(\tau)s(T-\tau)\,d\tau][\int_0^T h_0(z)s(T-z)\,dz]}{\int_0^T d\tau\, h_0(\tau) \int_0^T dz\, h_0(z)R_n(z-\tau)}$$

The inner integral in the denominator is, by Eq. (6-63), equal to $s(T-\tau)$. Therefore

$$\left(\frac{S}{N}\right)_p = \frac{[\int_0^T h_0(\tau)s(T-\tau)\,d\tau][\int_0^T h_0(z)s(T-z)\,dz]}{\int_0^T h_0(\tau)s(T-\tau)\,d\tau} \tag{6-66}$$

The integral in the denominator cancels the first integral in the numerator so that the maximum signal-to-noise ratio is

$$(S/N)_p = \int_0^T h_0(z)s(T-z)\,dz \tag{6-67}$$

These solutions shall now be applied to the case where the noise is white. That is, the noise spectral density is a constant, $N_o/2$, and the autocorrelation function is $(N_o/2)\delta(\tau)$.

The optimum linear filter is given by Eq. (6-63). Using the autocorrelation function for the white noise we have

$$(N_o/2)\int_0^T h_0(z)\delta(\tau - z)\,dz = s(T-\tau)$$

so that

$$h_0(\tau) = (2/N_o)s(T-\tau) \tag{6-68}$$

This is the matched filter previously discussed for the white noise case. (The signal-to-noise ratio is not influenced by the multiplier $2/N_o$.)

For the white noise case, the maximum signal-to-noise ratio is found by substituting Eq. (6-68) into (6-67). Therefore

$$(S/N)_p = \frac{2}{N_o}\int_0^T s^2(T-z)\,dz = \frac{2}{N_o}\int_0^T s^2(t)\,dt \tag{6-69}$$

The signal energy, E, is given by

$$E = \int_0^T s^2(t)\,dt$$

so that the maximum signal-to-noise is

$$(S/N)_p = 2E/N_o \tag{6-70}$$

We see therefore that for the matched filter in the white noise case, the signal-to-noise ratio is dependent on only the signal energy and the noise spectral density, and is otherwise independent of the signal waveform, provided of course that the filter is matched to the signal. Such features of the signal as peak power, time duration, and bandwidth do not directly enter into the calculation of S/N.

Approximation to Matched Filter Solution

The equation for the matched filter, Eq. (6-63), is an integral equation (in particular a Fredholm integral equation of the first kind). A discussion of methods to solve such equations will be deferred to Chap. 9. But for now, we shall consider what may be approximations to the optimum solution. We do so by completely disregarding the constraint imposed in Eq. (6-63) that $0 \leq \tau \leq T$, and allowing the limits of integration to be $(-\infty, \infty)$. The equation can then be solved as a convolution. It is possible therefore that the resulting solution may be both suboptimum and nonphysically realizable. (Actually, this solution is one part of the total solution which is discussed in Chap. 9.)

The nonphysically realizable part need not be a serious detriment unless the signal processing must be done in real time. In practice, analysis is often performed on recorded data, and the analysis is carried out in nonreal time. We therefore have access not only to the "past history" of the signal but also to its "future." For such nonreal time analysis we can build filters which make use of all this data. This is obviously not possible in real time since physically realizable filters cannot respond to a signal before it occurs. The amount of "future" data available is limited to the delay between recording the data and analyzing it. If the delay is sufficiently great, then for mathematical convenience it can be assumed that the infinite future is available.

For our approximate solution we shall denote the matched filter by $w_o(z)$ and shall find that filter which maximizes the signal-to-noise ratio at some time, say T_o. The filter is then the solution of the equation

$$\int_{-\infty}^{\infty} w_o(z) R_n(\tau - z)\,dz = s(T_o - \tau), \qquad -\infty < \tau < \infty \tag{6-71}$$

In this form, the left side represents a convolution operation and its Fourier transform is a product of transforms. Denoting the transforms of $w_o(z)$, $R_n(z)$, and $s(z)$ by $W_o(j\omega)$, $S_n(\omega)$, and $S(j\omega)$ respectively, we then have

$$W_o(j\omega)S_n(\omega) = S^*(j\omega)e^{-j\omega T_o}$$

or

$$W_o(j\omega) = \frac{S^*(j\omega)e^{-j\omega T_o}}{S_n(\omega)} \tag{6-72}$$

This transfer function will be recognized as the matched filter transfer function for the white noise case divided by the actual power spectral density of the noise. Using a different approach we shall develop the same result below.

Whitening Filter

It is common practice in communication and radar systems where non-white noise is encountered to *prewhiten* the noise. Those regions of the spectrum in which the noise is large are attenuated, and conversely those regions where the noise is low are accentuated. In many circumstances prewhitening is an integral part of the optimum solution. In other circumstances it can produce results which are close enough to the optimum solution to be of value. For a large number of spectra encountered in practice, the prewhitening can be accomplished to a good approximation with physically realizable filters.

Consider the operations indicated in Fig. 6-9. The input consists of a

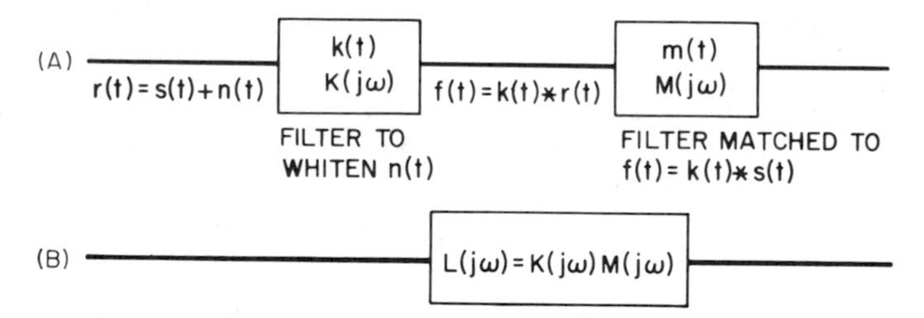

Fig. 6-9 (A) Combination of whitening and matched filters. (B) Single equivalent filter.

known signal $s(t)$ and additive noise having a nonwhite power spectral density $S_n(\omega)$. The filter $K(j\omega)$ is chosen to whiten the noise. The output of this filter is therefore noise with a white spectrum plus a known signal $f(t) = k(t)*s(t)$. Note that if the signal $s(t)$ is time limited, it does not necessarily follow that the transformed signal $f(t)$ is also time limited. In fact the duration of $f(t)$ will be the sum of the duration of $s(t)$ and the memory of $k(t)$. The latter is related to the zeros of $S_n(\omega)$.

The prewhitened signal is inserted into the filter $m(t)$ which is chosen to maximize the signal-to-noise ratio at some time T_o. Since the additive noise at the input to this filter is white, the best filter function to choose is the matched filter

$$M(j\omega) = F^*(j\omega)e^{-j\omega T_o} \tag{6-73}$$

where $F(j\omega)$ is the Fourier transform of $f(t) = s(t)*k(t)$. It is important to recognize that this filter might not be physically realizable. In fact, if the signal $f(t)$ is of infinite duration, the filter will not be physically realizable. This is easily seen from a discussion of Fig. 6-10 which shows a signal $f(t)$

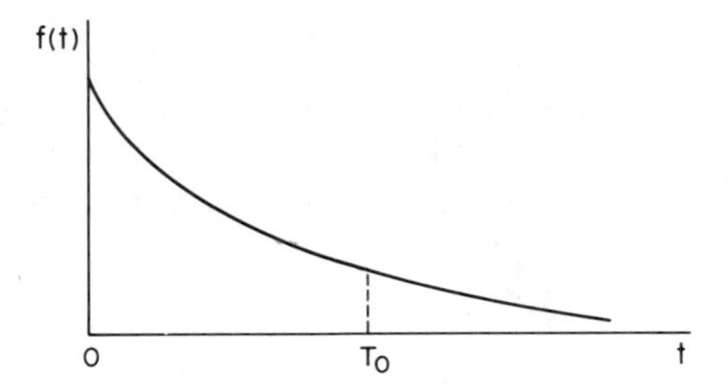

Fig. 6-10 Signal of infinite duration.

of infinite duration. If the signal-to-noise ratio is to be maximized at $t = T_o$, then that portion of the signal for which $t > T_o$ is in the "future" and is beyond the access of a physically realizable filter. To account for this we can modify the relation of Eq. (6-73) to be

$$M_{T_o}(j\omega) = F^*_{T_o}(j\omega)e^{-j\omega T_o} \tag{6-74}$$

where $F_{T_o}(j\omega)$ is the Fourier transform of

$$f_{T_o}(t) = \begin{cases} f(t), & 0 \le t \le T_o \\ 0, & \text{otherwise} \end{cases} \tag{6-75}$$

While such a filter is physically realizable, it must be recognized that it may be only an approximation to the optimum system.

If the processing is done in nonreal time and we allow a delay Δ in obtaining the signal-to-noise at the point T_o, the appropriate filter function becomes

$$M_{T_1}(j\omega) = F^*_{T_1}(j\omega)e^{-j\omega T_o} \tag{6-76}$$

where $T_1 = T_o + \Delta$, and $F_{T_1}(j\omega)$ is the Fourier transform of

$$f_{T_1}(t) = \begin{cases} f(t), & 0 \le t \le T_1 \\ 0, & \text{otherwise} \end{cases} \tag{6-77}$$

As the delay Δ increases, the filter eventually approaches that given in Eq. (6-73).

We now consider the whitening filter. If the spectral density $S_n(\omega)$ satisfies the Paley–Wiener criterion *(11)*

$$\int_{-\infty}^{\infty} \frac{|\ln S_n(\omega)|}{1+\omega^2}\, d\omega < \infty$$

then it may be factored into a product of two terms. The one term will contain poles and zeros in only the right half "p-plane," while the other will have poles and zeros in only the left half "p-plane." That is, the spectral density may be expressed as

$$S_n(\omega) = G_+(j\omega)G_-(j\omega) \tag{6-78}$$

where $G_+(j\omega)$ has no poles or zeros in the right half p-plane. The converse is true for $G_-(j\omega)$. It therefore follows that the inverse Fourier transform of $G_+(j\omega)$ will be zero for $t < 0$, and similarly, the transform of $G_-(j\omega)$ will be zero for $t > 0$. It is also true that $G_+{}^*(j\omega) = G_-(j\omega)$. Therefore, $S_n(\omega) = |G_+(j\omega)|^2$, and we choose

$$K(j\omega) = 1/G_+(j\omega) \tag{6-79}$$

as a physically realizable whitening filter.

It then follows that the combination of the whitening filter and the matched filter of Eq. (6-73) produces a combined filter

$$L(j\omega) = \frac{F^*(j\omega)e^{-j\omega T_0}}{G_+(j\omega)} = K(j\omega)M(j\omega) \tag{6-80}$$

However, $F(j\omega) = S(j\omega)K(j\omega) = S(j\omega)/G_+(j\omega)$ and

$$L(j\omega) = \frac{S^*(j\omega)e^{-j\omega T_0}}{G_+(j\omega)G_+{}^*(j\omega)} = \frac{S^*(j\omega)e^{-j\omega T_0}}{S_n(\omega)} \tag{6-81}$$

which is in agreement with the possibly nonphysically realizable filter of Eq. (6-72). As a reminder, the filters of Eqs. (6-74) or (6-76) may be better suited to Eq. (6-80).

EXAMPLE 6.4-1 For the RC spectral density $S_n(\omega) = 1/(\omega^2 + \omega_0{}^2)$, determine the whitening-matched filter combination impulse responses for a signal $s(t)$, $0 \le t < \infty$. Assume that the matched filter does not have to be physically realizable. Denote the Fourier transform of $s(t)$ by $S(j\omega)$. The transfer function of the combination filter is

$$L(j\omega) = S^*(j\omega)e^{-j\omega T_0}(\omega^2 + \omega_0{}^2)$$

We know that the Fourier transform of $S^*(j\omega)e^{-j\omega T_o}$ is $s(T_o - t)$, and the Fourier transform of $\omega^2 S^*(j\omega)e^{-j\omega T_o}$ is equal to $-d^2 s(T_o - t)/dt^2$. Therefore the impulse response is

$$l(t) = \omega_0{}^2 s(T_o - t) - \frac{d^2}{dt^2} s(T_o - t) \tag{6-82}$$

(It is pointed out in Chap. 9 that solutions such as this are part of the total optimum solution, and that in general delta functions at the end points must be included in order to identically satisfy the integral equation for the matched filter.)

6.5 An *M*-ary Communication System

As a further example of the application of hypothesis testing to the detection of known signals in white Gaussian noise, we shall consider a communication system which uses M possible waveforms. The previous cases considered were binary systems using only two waveforms. This example also demonstrates the use of multiple alternative hypothesis testing.

The problem may be stated as follows: A waveform $r(t)$, received in the time interval $(0, T)$, contains one of M signals, $s_i(t), i = 1, \ldots, M$, with equal probability. At the receiver zero mean, white Gaussian noise is added to the signal. The signals are assumed orthogonal with equal energy. That is,

$$\rho_{ij} = \int_0^T s_i(t)s_j(t)\,dt = \begin{cases} E, & i = j \\ 0, & i \neq j \end{cases} \tag{6-83}$$

Figure 6-11 shows an example of a set of orthogonal signals. We seek a decision rule to determine which signal is present using the criterion of minimum error probability.

Since we are to choose among M signals, we use multiple alternative hypothesis testing. The hypotheses are

H_1: $s_1(t)$ present

$\vdots$

H_M: $s_M(t)$ present

Since the signals are equiprobable and the costs of making errors are the same, the results of the Bayes test discussed in Sect. 5.8 may be used. That is, choose the hypothesis H_j for which $p_j(\mathbf{r})$ is a maximum, where $p_j(\mathbf{r})$ is the likelihood function assuming H_j is true. By an obvious extension of Eq. (6-38)

$$p_j(\mathbf{r}) \sim \exp\left\{-(1/N_o)\int_0^T [r(t) - s_j(t)]^2\,dt\right\}$$

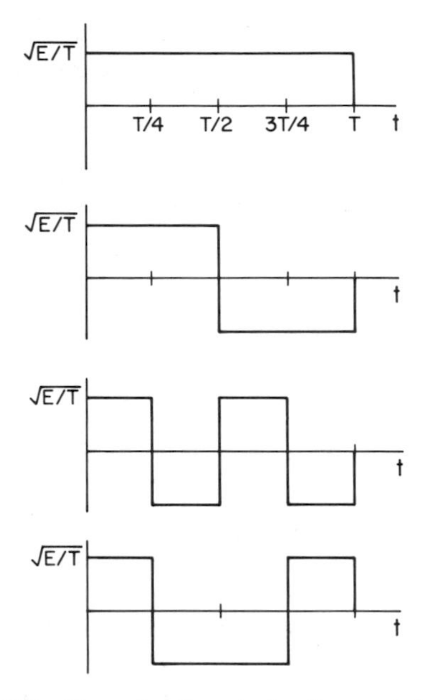

Fig. 6-11 Example of a set of orthogonal signals.

As before, $p_j(\mathbf{r})$ varies inversely with the integral, so that the decision rule is: choose that hypothesis H_j for which $\int_0^T [r(t) - s_j(t)]^2 \, dt$ is a minimum. The integrand may be written

$$[r(t) - s_j(t)]^2 = r^2(t) + s_j^2(t) - 2r(t)s_j(t)$$

The integral of the first and second terms is independent of which hypothesis is assumed to be true. The decision rule can therefore be expressed as: choose the hypothesis H_j for which $\int_0^T r(t)s_j(t) \, dt$ is a maximum. A correlation receiver which implements such a decision rule is shown in Fig. 6-12. As discussed previously, the receiver could also be implemented by a bank of matched filters with

$$h_i(\tau) = s_i(T - \tau), \qquad 0 \le \tau \le T$$

as shown in Fig. 6-13. The outputs of the matched filters are sampled at time T. The signal corresponding to the filter with the largest output is assumed to be present.

To determine the error probability, define

$$G_j = \int_0^T r(t)s_j(t) \, dt \tag{6-84}$$

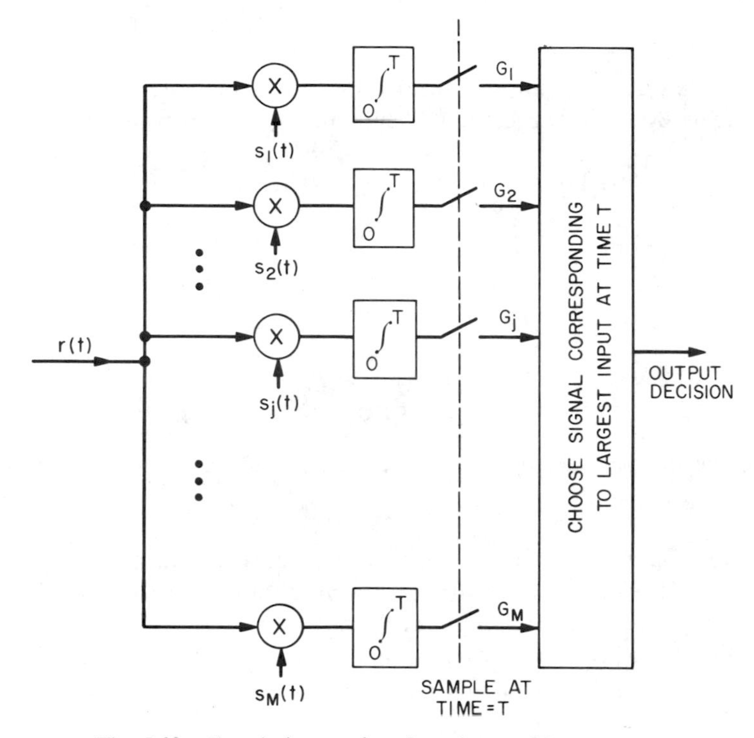

Fig. 6-12 Correlation receiver for coherent M-ary system.

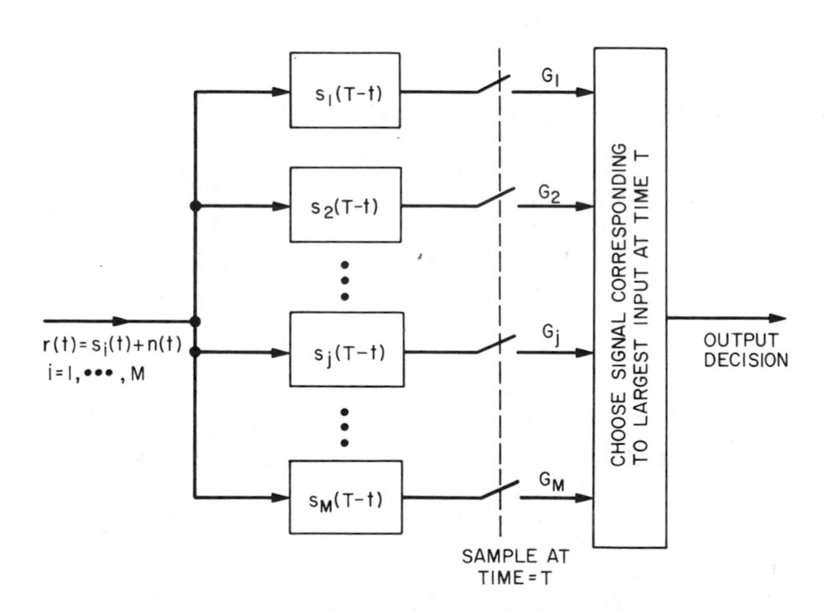

Fig. 6-13 Matched filter implementation of M-ary system.

This is the output of the jth correlator and is a Gaussian random variable. Denote as $E_i\{\cdot\}$ the expectation when hypothesis H_i is true. Then

$$E_i\{G_j\} = E\left\{\int_0^T [s_i(t) + n(t)]s_j(t)\,dt\right\} = \begin{cases} 0, & i \neq j \\ E, & i = j \end{cases} \tag{6-85}$$

and, furthermore,

$$E_k\{G_i G_j\} = 0 \qquad \text{for} \quad i \neq j$$

and

$$E_k\{G_j^2\} = \begin{cases} E^2 + N_o E/2, & j = k \\ N_o E/2, & j \neq k \end{cases}$$

It follows that the variance of the statistic G_j assuming hypothesis H_i is true is

$$V_i\{G_j\} = V\{G_j\} = N_o E/2 \qquad \text{all } i \text{ and } j$$

The variables $G_j, j = 1, \ldots, M$, are statistically independent Gaussian variables. The determination of the error probability is essentially a determination of the distribution of the maximum of a number of random variables. Without loss of generality assume that H_i is true. Then no error is made if $G_j < G_i$ for all $j \neq i$. Symbolically,

$$P\{\text{no error} \mid G_i = x\} = P\{G_1 < x, \ldots, G_{i-1} < x, G_{i+1} < x, \ldots, G_M < x \mid G_i = x\}$$

But, since the G_j are independent

$$P\{\text{no error} \mid G_i = x\} = [P\{G_j < x\}]^{M-1}, \qquad j \neq i$$

The probability of error is therefore

$$P_e = 1 - \int_{-\infty}^{\infty} P\{\text{no error} \mid G_i = x\} p_i(x)\,dx$$

$$= 1 - \int_{-\infty}^{\infty} p_i(x)[P\{G_j < x\}]^{M-1}\,dx$$

or

$$P_e = 1 - \int_{-\infty}^{\infty} \frac{1}{(2\pi)^{1/2}\sigma} \exp\left[-\frac{(x-E)^2}{2\sigma^2}\right] \left\{\int_{-\infty}^{x} dz\, \frac{1}{(2\pi)^{1/2}\sigma} \exp\left[-\frac{z^2}{2\sigma^2}\right]\right\}^{M-1} dx$$

where $\sigma^2 = N_o E/2$. This can be put in the equivalent form†

$$P_e = 1 - \int_{-\infty}^{\infty} \frac{1}{(2\pi)^{1/2}} e^{-z^2/2} \left[\int_{-\infty}^{z + (2E/N_o)^{1/2}} du\, \frac{1}{(2\pi)^{1/2}} e^{-u^2/2}\right]^{M-1} dz$$

† See Viterbi (*12*) for illustrations of the error performance.

6.6 Sampled Approach

The formulation at the beginning of this chapter included sampling a signal in a way which facilitated passing to the limiting case of "continuous sampling." The objective was to work with a signal having a continuous time parameter (*2*). There are occasions in practice, however, when signal processing is performed using time-sampled signals. For example, a sampled approach is necessary if the signal processing is accomplished by a digital computer. In other cases it may be more convenient analytically to work with the sampled version. Then, instead of integral equations, for example, we get matrix equations. Although this is generally not a trivial exercise if the orders of the matrices are large, the availability of large computing facilities and special purpose processing equipment can make the problems tractable.

The formal matrix expressions encountered for detecting signals in nonwhite noise are only slightly more complicated than those for which the noise samples are statistically independent. We therefore drop momentarily the assumption of white noise.

Consider the problem of detecting a single signal in noise. The hypotheses stated in sampled form are

$$\begin{aligned} H_1&: \quad r(t_k) = s(t_k) + n(t_k), \qquad k = 1, \ldots, M \\ H_0&: \quad r(t_k) = n(t_k), \qquad k = 1, \ldots, M \end{aligned}$$

The signal is assumed known, and the noise samples are zero mean Gaussian with covariance matrix $\mathbf{R}_n$. (The ijth element is $E\{n(t_i)n(t_j)\}$.) The covariance matrix is assumed to be positive definite so that its inverse exists. In vector notation the hypotheses may be expressed as

$$\begin{aligned} H_1&: \quad \mathbf{r} = \mathbf{s} + \mathbf{n} \\ H_0&: \quad \mathbf{r} = \mathbf{n} \end{aligned}$$

where

$$\mathbf{r} = \begin{bmatrix} r(t_1) \\ r(t_2) \\ \vdots \\ r(t_M) \end{bmatrix}, \qquad \mathbf{s} = \begin{bmatrix} s(t_1) \\ s(t_2) \\ \vdots \\ s(t_M) \end{bmatrix}, \qquad \mathbf{n} = \begin{bmatrix} n(t_1) \\ n(t_2) \\ \vdots \\ n(t_M) \end{bmatrix}$$

The sample vector $\mathbf{r}$ is Gaussian. From Eq. (4-22) the likelihood functions for the known signal case are

$$p_1(\mathbf{r}) = \frac{1}{(2\pi)^{n/2}|\mathbf{R}_n|^{1/2}} \exp[-\tfrac{1}{2}(\mathbf{r} - \mathbf{s})'\mathbf{R}_n^{-1}(\mathbf{r} - \mathbf{s})] \tag{6-86}$$

and

$$p_0(\mathbf{r}) = \frac{1}{(2\pi)^{n/2}|\mathbf{R}_n|^{1/2}} \exp(-\tfrac{1}{2}\mathbf{r}'\mathbf{R}_n^{-1}\mathbf{r}) \tag{6-87}$$

where the prime (′) denotes transpose, and $|\cdot|$ the determinant. From these, the likelihood ratio is

$$\lambda(\mathbf{r}) = \exp \tfrac{1}{2}(\mathbf{r}'\mathbf{R}_n^{-1}\mathbf{s} + \mathbf{s}'\mathbf{R}_n^{-1}\mathbf{r} - \mathbf{s}'\mathbf{R}_n^{-1}\mathbf{s})$$

Note that $\mathbf{s}'\mathbf{R}_n^{-1}\mathbf{r}$ is a scalar and is equal to its transpose so that $\mathbf{r}'\mathbf{R}_n^{-1}\mathbf{s} = \mathbf{s}'\mathbf{R}_n^{-1}\mathbf{r}$. The likelihood ratio becomes

$$\lambda(\mathbf{r}) = \exp(-\tfrac{1}{2}\mathbf{s}'\mathbf{R}_n^{-1}\mathbf{s}) \exp(\mathbf{r}'\mathbf{R}_n^{-1}\mathbf{s}) \tag{6-88}$$

The first term is not dependent on the observable $\mathbf{r}$, so the likelihood ratio is monotonically related to

$$\ln \lambda(\mathbf{r}) \sim \mathbf{r}'\mathbf{R}_n^{-1}\mathbf{s} = \mathbf{s}'\mathbf{R}_n^{-1}\mathbf{r} \tag{6-89}$$

If the noise samples are statistically independent and of equal variance, the test statistic is proportional to

$$\mathbf{r}'\mathbf{s} = \sum_{i=1}^{M} r(t_i)s(t_i) \tag{6-90}$$

and is a discrete version of a correlation receiver.

In the expression for the test statistic for correlated samples $\mathbf{s}'\mathbf{R}_n^{-1}\mathbf{r}$, the matrix $\mathbf{R}_n^{-1}$ can be considered to be an operator or filter which operates on the received sample vector $\mathbf{r}$. The result is further modified by the signal vector $\mathbf{s}$. It may be verified for this form of the test statistic that the computations cannot start before the entire vector $\mathbf{r}$ is received. In that sense, the filter is not "physically realizable." We shall develop below an equivalent form for the test statistic which does not have this feature.

To gain insight into the test statistic and to satisfy the preceding comments about "physical realizability," we shall derive an alternate representation of Eq. (6-89). To do so represent the covariance matrix as a product of two triangular matrices (*13*, *14*)

$$\mathbf{R}_n^{-1} = \mathbf{C}'\mathbf{C} \tag{6-91}$$

Here $\mathbf{C}$ is taken to be a lower triangular matrix of the same order as $\mathbf{R}_n^{-1}$, that is, a matrix whose elements above the main diagonal are identically equal to zero. It therefore may be written as

$$\mathbf{C} = \begin{bmatrix} c_{11} & 0 & \cdots & & 0 \\ c_{21} & c_{22} & 0 & & \\ c_{31} & c_{32} & c_{33} & \ddots & \vdots \\ \vdots & & & & 0 \\ c_{M1} & c_{M2} & c_{M3} & \cdots & c_{MM} \end{bmatrix} \tag{6-92}$$

The matrix $\mathbf{C}$ is unique if $\mathbf{R}_n^{-1}$ is symmetric and positive definite, which we have assumed. Furthermore, with $\mathbf{R}_n^{-1}$ positive definite, then $\mathbf{C}$ and $\mathbf{C}'$ are also positive definite. The fact that $\mathbf{C}$ is positive definite resolves the possible sign (± 1) ambiguity in the definition of $\mathbf{C}$. Thus the elements along the main diagonal will be positive.

We shall discuss a method for determining the triangular matrix $\mathbf{C}$ after investigating the consequence of such a matrix factorization. Substituting Eq. (6-91) into (6-89) produces the test statistic

$$\ln \lambda(\mathbf{r}) \sim \mathbf{s}'\mathbf{C}'\mathbf{C}\mathbf{r} = (\mathbf{C}\mathbf{s})'\mathbf{C}\mathbf{r} \tag{6-93}$$

This can be instrumented by operating with $\mathbf{C}$ on both the received data and the stored signal and correlating the results. Denoting the elements of $\mathbf{Cr}$ by α_i, and those of $\mathbf{Cs}$ by β_i, we get $\lambda(\mathbf{r}) \sim \sum \alpha_i \beta_i$. The operator $\mathbf{C}$ is a physically realizable but time varying filter. To see this, write out the terms of the column vector

$$\mathbf{Cr} = \begin{bmatrix} c_{11}r(t_1) \\ c_{21}r(t_1) + c_{22}r(t_2) \\ \vdots \\ c_{M1}r(t_1) + \cdots + c_{MM}r(t_M) \end{bmatrix}$$

The successive elements of $\mathbf{Cr}$ are generated in a manner which involves operating only on that data which has been received. The fact that the filter is time varying is seen from the fact that the coefficients along any given "diagonal" are not equal.

We note in passing

$$\mathbf{Cs} = \begin{bmatrix} c_{11}s(t_1) \\ c_{21}s(t_1) + c_{22}s(t_2) \\ \vdots \\ c_{M1}s(t_1) + \cdots + c_{MM}s(t_M) \end{bmatrix}$$

so that the test statistic is proportional to

$$\begin{aligned} \lambda(\mathbf{r}) \sim\ & c_{11}r(t_1)c_{11}s(t_1) \\ & + [c_{21}r(t_1) + c_{22}r(t_2)][c_{21}s(t_1) + c_{22}s(t_2)] \\ & \vdots \\ & + [c_{M1}r(t_1) + \cdots + c_{MM}r(t_M)][c_{M1}s(t_1) + \cdots + c_{MM}s(t_M)] \end{aligned}$$

An interesting feature of the operator $\mathbf{C}$ is that it decorrelates the received samples. To see this, consider what happens when noise only is present. The modified samples are given by $\mathbf{Cn}$. The covariance matrix of these samples is

$$E\{\mathbf{Cn}(\mathbf{Cn})'\} = E\{\mathbf{Cnn}'\mathbf{C}'\} = \mathbf{C}\mathbf{R}_n\mathbf{C}' \tag{6-94}$$

From Eq. (6-91)

$$\mathbf{R}_n = \mathbf{C}^{-1}(\mathbf{C}')^{-1} \tag{6-95}$$

It then follows that

$$E\{\mathbf{Cn}(\mathbf{Cn})'\} = \mathbf{I} \tag{6-96}$$

the identity matrix. Thus the operator **C** not only decorrelates the received samples, it also normalizes their variances. This too is called whitening. With the data prewhitened the problem becomes one of detecting a known signal (**Cs**) mixed with independent noise samples (at least for the Gaussian noise case) of equal variance. This is the discrete analog of detecting a known signal in white Gaussian noise.

Factorization into Triangular Matrices

To see how the elements of **C** may be determined,† denote its elements by c_{ij}, and those of $\mathbf{R}_n^{-1}$ by ϕ_{ij}. We first illustrate the method for a 3×3 matrix. For Eq. (6-91) to hold we must have

$$\begin{bmatrix} \phi_{11} & \phi_{12} & \phi_{13} \\ \phi_{12} & \phi_{22} & \phi_{23} \\ \phi_{13} & \phi_{23} & \phi_{33} \end{bmatrix} = \begin{bmatrix} c_{11} & c_{21} & c_{31} \\ 0 & c_{22} & c_{32} \\ 0 & 0 & c_{33} \end{bmatrix} \begin{bmatrix} c_{11} & 0 & 0 \\ c_{21} & c_{22} & 0 \\ c_{31} & c_{32} & c_{33} \end{bmatrix}$$

$$= \begin{bmatrix} c_{11}^2 + c_{21}^2 + c_{31}^2 & c_{21}c_{22} + c_{31}c_{32} & c_{31}c_{33} \\ c_{22}c_{21} + c_{32}c_{31} & c_{22}^2 + c_{32}^2 & c_{32}c_{33} \\ c_{33}c_{31} & c_{33}c_{32} & c_{33}^2 \end{bmatrix}$$

The unknown coefficients c_{ij} can be found by equating these matrices term by term. Solving for c_{33} first, we see that

$$c_{33} = (\phi_{33})^{1/2}, \quad c_{32} = \phi_{23}/c_{33}, \quad c_{31} = \phi_{13}/c_{33}$$

$$c_{22} = (\phi_{22} - c_{32}^2)^{1/2}, \quad c_{21} = (\phi_{12} - c_{31}c_{32})/c_{22}, \quad c_{11} = (\phi_{11} - c_{21}^2 - c_{31}^2)^{1/2}$$

If the c_{ij} are computed in the order shown, the expression for each involves only the c_{ij} values already computed. The c_{ij} for higher order matrices can be obtained in a similar manner. An alternative procedure is also given below (*13*).

Using the facts that $|\mathbf{R}_n^{-1}| = |\mathbf{C}'| \cdot |\mathbf{C}|$ and the determinant of a triangular matrix is the product of its diagonal elements, we get

$$|\mathbf{R}_n^{-1}| = (c_{11}c_{22} \cdots c_{MM})^2$$

Now, define the matrix $\boldsymbol{\phi}_i$ formed by omitting the first i rows and columns of $\mathbf{R}_n^{-1}$. For example, $\boldsymbol{\phi}_{M-1}$ is just the element ϕ_{MM}, $\boldsymbol{\phi}_{M-2}$ is the matrix

$$\begin{bmatrix} \phi_{M-1,M-1} & \phi_{M-1,M} \\ \phi_{M-1,M} & \phi_{M,M} \end{bmatrix}$$

† Triangular matrices can be used to find matrix inverses, see Westlake (*15*).

and $\boldsymbol{\phi}_0$ is the matrix $\mathbf{R}_n^{-1}$ itself. Then it can be shown that

$$|\boldsymbol{\phi}_i| = (c_{i+1,\,i+1}c_{i+2,\,i+2}\cdots c_{MM})^2 \tag{6-97}$$

By starting with $|\boldsymbol{\phi}_{M-1}|$ and progressing to $|\boldsymbol{\phi}_0|$, the diagonal elements of $\mathbf{C}$ can be computed successively. For example

$$|\boldsymbol{\phi}_{M-1}| = \phi_{MM} = c_{MM}^2$$

$$|\boldsymbol{\phi}_{M-2}| = c_{M-1,\,M-1}^2 c_{MM}^2, \qquad \text{etc.}$$

For the other elements, we have

$$\phi_{ij} = \sum_k c_{ki}\,c_{kj}$$

But c_{kj} is zero for $k < j$, c_{ki} is zero for $k < i$, and by our notation $i \leq j$. Therefore, for $i < j$

$$\phi_{ij} = \sum_{k=j}^{M} c_{ki}\,c_{kj} \tag{6-98}$$

These equations are solved sequentially by taking $j = M$, and progressing from $i = M - 1$ to 1. Then taking $j = M - 1$, and progressing from $i = M - 2$ to 1, etc.

Gram–Schmidt Orthogonalization

The above method of factoring the covariance matric $\mathbf{R}_n^{-1}$ into triangular matrices is similar to the Gram–Schmidt orthogonalization procedure for finding an orthonormal set of functions, or in the case at hand, finding a transformation to decorrelate and normalize random variables.

Suppose we have just two correlated random variables n_1 and n_2. We want a transformation such that the transformed variables are uncorrelated with unit variance. Consider the transformation

$$y_1 = a_{11}n_1, \qquad y_2 = a_{21}n_1 + a_{22}n_2$$

where we assume n_1 and n_2 are zero mean, having variance ${\sigma_1}^2$ and ${\sigma_2}^2$ respectively, and $E\{n_1 n_2\} = \sigma_{12}$. The coefficients are chosen so that $E\{{y_1}^2\} = E\{{y_2}^2\} = 1$, and $E\{y_1 y_2\} = 0$. Imposing these conditions produces

$$a_{11} = 1/\sigma_1$$

$$a_{21} = -\frac{\sigma_{12}}{\sigma_1}\left(\frac{1}{{\sigma_1}^2{\sigma_2}^2 - \sigma_{12}^2}\right)^{1/2}$$

$$a_{22} = \frac{\sigma_1}{({\sigma_1}^2{\sigma_2}^2 - \sigma_{12}^2)^{1/2}}$$

The procedure may be extended as more variables are considered. For example, we may also have

$$y_3 = a_{31}n_1 + a_{32}n_2 + a_{33}n_3$$
$$y_4 = a_{41}n_1 + a_{42}n_2 + a_{43}n_3 + a_{44}n_4$$
$$\vdots$$

As each new variable, say y_i, is added, i new coefficients are introduced. These are determined by solving the set of i equations

$$E\{y_i y_j\} = \delta_{ij}, \qquad j \le i$$

If there are M variable n_i, then M uncorrelated random variables y_i can be generated in this way provided that the covariance matrix of the n_i is positive definite.

To see the relationship between this approach and the determination of the triangular matrix $\mathbf{C}$, observe that the y_i variables can be expressed as

$$\mathbf{y} = \mathbf{An}$$

where

$$\mathbf{A} = \begin{bmatrix} a_{11} & 0 & \cdots & 0 \\ a_{21} & a_{22} & \cdots & \vdots \\ \vdots & & \ddots & 0 \\ a_{M1} & a_{M2} & \cdots & a_{MM} \end{bmatrix}$$

The conditions to be satisfied for the determination of $\mathbf{A}$ are given by the set of equations

$$E\{\mathbf{yy}'\} = \mathbf{I} = E\{\mathbf{Ann}'\mathbf{A}'\} = \mathbf{AR}_n\mathbf{A}'$$

But, from Eq. (6-95) this can be written as

$$\mathbf{AC}^{-1}(\mathbf{C}^{-1})'\mathbf{A}' = \mathbf{I} \qquad \text{or} \qquad \mathbf{AC}^{-1}(\mathbf{AC}^{-1})' = \mathbf{I}$$

Both $\mathbf{A}$ and $\mathbf{C}$ and hence $\mathbf{C}^{-1}$ are lower triangular, so that the product $\mathbf{AC}^{-1}$ is also lower triangular. Therefore the matrix $\mathbf{I}$ is factored into the product of a triangular matrix and its transpose. Since $\mathbf{I}$ is symmetric and positive definite, the triangular matrix is unique. Therefore the triangular matrix must be the unit matrix $\mathbf{I}$, so that

$$\mathbf{AC}^{-1} = \mathbf{I}$$

and therefore

$$\mathbf{A} = \mathbf{C}$$

Thus the determination of the triangular matrix $\mathbf{C}$ can be associated with the Gram–Schmidt orthogonalization procedure.

Performance

Denote the test statistic, Eq. (6-89), by $\lambda_T = \mathbf{s}'\mathbf{R}_n^{-1}\mathbf{r}$. Since the samples $\mathbf{r}$ are jointly Gaussian, and λ_T is generated by a linear operation on $\mathbf{r}$, it follows that λ_T is a Gaussian scalar. We need therefore only its mean and variance under each hypothesis to determine the receiver performance.

For hypothesis H_0,†

$$E_0\{\lambda_T\} = E\{\mathbf{s}'\mathbf{R}_n^{-1}\boldsymbol{n}\} = \text{tr } \mathbf{R}_n^{-1}E\{\boldsymbol{n}\mathbf{s}'\} = 0$$

The trace of a matrix, denoted by tr, is the sum of the elements along the main diagonal of the matrix. For hypothesis H_1,

$$E_1\{\lambda_T\} = E\{\mathbf{s}'\mathbf{R}_n^{-1}\mathbf{s} + \mathbf{s}'\mathbf{R}_n^{-1}\boldsymbol{n}\} = \mathbf{s}'\mathbf{R}_n^{-1}\mathbf{s}$$

The variance of λ_T is the same for each hypothesis and therefore $\sigma^2(\lambda_T) = E_0\{\lambda_T{}^2\}$. Now

$$E_0\{\lambda_T{}^2\} = E\{(\mathbf{s}'\mathbf{R}_n^{-1}\boldsymbol{n})^2\} = E\{\mathbf{s}'\mathbf{R}_n^{-1}\boldsymbol{n}\boldsymbol{n}'\mathbf{R}_n^{-1}\mathbf{s}\}$$

so that

$$\sigma^2(\lambda_T) = \mathbf{s}'\mathbf{R}_n^{-1}\mathbf{R}_n\mathbf{R}_n^{-1}\mathbf{s} = \mathbf{s}'\mathbf{R}_n^{-1}\mathbf{s}$$

Exercises

6.1 Design a receiver to choose between two hypotheses

H_1: $r(t) = s_1(t) + n(t)$
H_0: $r(t) = s_0(t) + n(t)$

using minimum error probability as the criterion. The signals $s_1(t)$ and $s_0(t)$ are shown in Fig. 6-14. The additive noise is white, Gaussian with power spectral density $N_o/2$. Assume equal a priori probabilities. Determine the probability of error for $E/N_o = 2$.

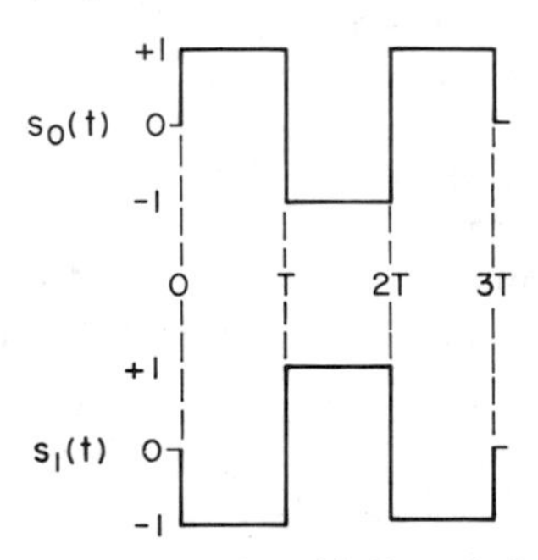

Fig. 6-14 Three-bit "words."

† These and other matrix relations involving expectations are developed in Chap. 11.

6.2 For the signals discussed in the preceding problem, each signal is a "word," and each word has three "bits." Suppose we detect each bit, one at a time. If at most one of the bits is in error, the word can still be decoded properly.

(a) What is the error probability of each bit?
(b) Given that we can "correct" a single bit error, what is the probability that the word is decoded in error?
(c) Compare the results with the preceding problem.

6.3 Consider a coherent frequency shift keying system with signals

$$\begin{aligned} s_1(t) &= \sin \omega_1 t, \qquad 0 \le t \le T \\ s_0(t) &= \sin \omega_0 t, \qquad 0 \le t \le T \end{aligned}$$

For white Gaussian noise, the optimum receiver was discussed in Sect. 6.2. Denote $\omega_d = \omega_1 - \omega_0$.

(a) Show that the error probability is minimized when the frequency difference is chosen such that

$$\omega_d/2\pi \simeq 0.7/T$$

Assume either $(\omega_1 + \omega_0)T = k\pi$, where k is an integer, or that $\omega_1 + \omega_0 \gg 0$.

(b) In terms of the required signal energy what is the improvement of this system over the system for which $\rho = 0$?

6.4 Consider a four-phase communication system with signals $(0 \le t \le T)$

$$\begin{aligned} s_0(t) &= A \sin \omega_0 t \\ s_1(t) &= A \sin(\omega_0 t + \pi/2) \\ s_2(t) &= A \sin(\omega_0 t + \pi) \\ s_3(t) &= A \sin(\omega_0 t + 3\pi/2) \end{aligned}$$

Assume the additive noise is white, Gaussian with spectral density $N_o/2$. Assume equal a priori probabilities and equal costs for errors.

(a) What is the optimum receiver? How many correlators are required?
(b) Show that the probability of a correct decision is

$$P_c = \left[\int_{-(E/N_o)^{1/2}}^{\infty} \frac{1}{(2\pi)^{1/2}} e^{-z^2/2}\, dz\right]^2$$

(c) How does this compare with the probability of correct decision for the binary coherent case?

6.5 Consider a ternary communication system with equally likely signals $(0 \le t \le T)$

H_0: $s_0(t) = 0$
H_1: $s_1(t) = A \sin \omega_0 t$
H_2: $s_2(t) = -A \sin \omega_0 t$

Assume white Gaussian noise with spectral density $N_o/2$.

(a) Design a likelihood ratio test to minimize the error probability.

(b) Show that

$$P(D_0 \mid H_0) = 2 \int_0^{(E/2N_o)^{1/2}} \frac{1}{(2\pi)^{1/2}} e^{-u^2/2} \, du$$

(c) Show that

$$P(D_1 \mid H_1) = P(D_2 \mid H_2) = \int_{-(E/2N_o)^{1/2}}^{\infty} \frac{1}{(2\pi)^{1/2}} e^{-u^2/2} \, du$$

and therefore that the probability of a correct decision is

$$P_c = \frac{2}{3} \left[2 \int_0^{(E/2N_o)^{1/2}} \frac{1}{(2\pi)^{1/2}} e^{-u^2/2} \, du + \frac{1}{2} \right]$$

(d) Compare this ternary system with the binary FSK case, and the four phase system of the previous exercise.

6.6 Determine the likelihood ratio receiver to choose between the hypotheses

H_1: $r(t) = A \cos \omega_1 t + B \cos(\omega_2 t + \phi) + n(t)$
H_0: $r(t) = B \cos(\omega_2 t + \phi) + n(t)$

where A, B, ω_1, ω_2, and ϕ are known constants. The noise is white, Gaussian, with power spectral density $N_o/2$. How is the performance of the receiver influenced by the signal $B \cos(\omega_2 t + \phi)$?

6.7 Consider the problem of detecting a noise-like signal in noise.

H_1: $r(t) = s(t) + n(t)$
H_0: $r(t) = n(t)$

Both $n(t)$ and $s(t)$ are zero mean Gaussian signals, bandlimited to $|\omega| < \Omega = 2\pi B$, and having power spectral densities $N_o/2$ and $S_o/2$ respectively. Assume $2BT$ samples spaced π/Ω apart are taken. Find the likelihood ratio receiver.

6.8 Consider the matched filter for a signal

$$s(t) = \begin{cases} A, & 0 \le t \le T \\ 0, & \text{otherwise} \end{cases}$$

within white Gaussian noise.

(a) What is the peak output signal-to-noise ratio?
(b) Suppose that instead of the matched filter, a filter

$$h(t) = \begin{cases} e^{-\alpha t}, & 0 \le t \le T \\ 0, & \text{otherwise} \end{cases}$$

is used. What is the peak output signal-to-noise ratio? What would you expect the optimum value of α to be?
(c) Suppose that a filter

$$h(t) = e^{-\alpha t}, \qquad t \ge 0$$

is used. What is the peak output signal-to-noise ratio? Show that the signal-to-noise ratio for this case is always less than or equal to that produced in part (b).

6.9 For the preceding problem, consider a Gaussian filter

$$h(\tau) = (1/\alpha) \exp\left(-\frac{(\tau - \tau_o)^2}{2\alpha^2}\right), \qquad -\infty < \tau < \infty, \quad \tau_o > 0$$

(Note, if $\tau_o \gg \alpha$ this can be approximated by a physically realizable filter.)
(a) At what value of time will the output signal-to-noise ratio be a maximum?
(b) Derive an expression for the signal-to-noise ratio.

6.10 Consider the signal $s(t) = 1 - \cos \omega_0 t$, $0 \le t \le 2\pi/\omega_0$ and the RC-filtered noise with power spectral density

$$S_n(\omega) = \frac{{\omega_1}^2}{\omega^2 + {\omega_1}^2}$$

(a) Find the generalized matched filter for this example by solving Eq. (6-81) with $T_0 = 2\pi/\omega_0$. The answer is

$$w(t) = 1 - \frac{{\omega_0}^2 + {\omega_1}^2}{{\omega_1}^2} \cos \omega_0 t, \qquad 0 \le t \le 2\pi/\omega_0$$

(b) What is the resulting maximum signal-to-noise ratio?

6.11 Consider the integral equation for the generalized matched filter for a signal $s(t) = \sin \omega_0 t$, $0 \le t \le 2\pi/\omega_0 = T$, and an autocorrelation function $(\omega_1/2)\, e^{-\omega_1|\tau|}$.
(a) Verify that the filter

$$h(t) = s(T - t) - \frac{1}{{\omega_1}^2} \frac{d^2}{dt^2} s(T - t)$$

does not identically solve the integral equation.

(b) Suppose that the delta functions

$$a\delta(t) + b\delta(t - T)$$

are added to the impulse response $h(t)$. Can the integral equation now be satisfied, and if so for what values of a and b?

6.12 For the preceding exercise,

(a) Find the maximum signal-to-noise ratio.

(b) For the same signal and noise conditions as in the previous exercise, assume that the filter

$$z(t) = -\left(\frac{{\omega_1}^2 + {\omega_0}^2}{{\omega_1}^2}\right) \sin \omega_0 t, \qquad 0 \leq t \leq 2\pi/\omega_0 = T$$

is used instead of the optimum filter. Determine the signal-to-noise ratio for this filter. Compare the optimum and suboptimum results for $\omega_0 = \omega_1$.

6.13 Determine the upper and lower triangular matrices whose product $\mathbf{C}'\mathbf{C}$ is the Markov covariance matrix

$$\begin{bmatrix} 1 & \rho & \rho^2 \\ \rho & 1 & \rho \\ \rho^2 & \rho & 1 \end{bmatrix}$$

References

1. Zadeh, L. A., Abramson, N., Balakrishnan, A. V., Braverman, D., Eden, M., Feigenbaum, E. A., Kailath, T., Lerner, R. M., Massey, J., Mueller, G. E., Peterson, W. W., Price, R., Sebestyen, G., Slepian, D., Thomasian, A. J., Turin, G. L., Report on progress in information theory in the U.S.A., *IEEE Trans. Inform. Theory* **IT-9,** No. 4, 221–264 (October 1963).
2. Peterson, W. W., Birdsall, T. G., and Fox, W. C., The theory of signal detectability, *1954 Symp. Inform. Theory, IRE Trans. Inform. Theory,* **PGIT-4,** 171–212 (September 1954). Also available as, W. W. Peterson and T. G. Birdsall, The Theory of Signal Detectability, Univ. of Michigan Tech. Rep. TR-13, Part 1 (June 1953). The General Theory, AD 16786; Part II, Applications With Gaussian Noise, AD 16787; W. C. Fox, Signal Detectability: A Unified Description of Statistical Methods Employing Fixed and Sequential Observation Processes, Univ. of Michigan Tech. Rep. TR-19 (December 1953).
3. Reiger, S., Error Probabilities of Binary Data Transmission Systems in the Presence of Random Noise, IRE Convention Record, Part 8—Information Theory (1958).
4. Bode, H. W., and Shannon, C. E., A simplified derivation of linear least square smoothing and prediction theory, *Proc. IRE* (April 1950).
5. Darlington, S., Linear least squares smoothing and prediction, with applications, *Bell Syst. Tech. J.* **37,** No. 5, 1221–1294 (1958).
6. Zadeh, L. A., and Ragazzini, J. R., Optimum filters for the detection of signals in noise, *Proc. IRE* (October 1952).

7. Turin, G. L., An introduction to matched filters, *IRE Trans. Inform. Theory*, 310–329 (June 1960).
8. North, D. O., An analysis of the factors which determine signal/noise discrimination in pulsed carrier systems, reproduced in *Proc. IEEE* (July 1963).
9. Dwork, B. M. Detection of a pulse superimposed on fluctuation noise, *Proc. IRE* 771–774 (July 1950).
10. Goldstein, H., "Classical Mechanics." Addison-Wesley, Reading, Massachusetts, 1959.
11. Wiener, N., "Extrapolation, Interpolation, and Smoothing of Stationary Time Series With Engineering Applications." The Technology Press of MIT and Wiley, New York, 1949.
12. Viterbi, A. J., "Principles of Coherent Communication." McGraw-Hill, New York, 1966.
13. Guillemin, E. A., "The Mathematics of Circuit Analysis." Wiley, New York, 1949.
14. Franklin, J. N., "Matrix Theory." Prentice-Hall, Englewood Cliffs, New Jersey, 1968.
15. Westlake, J. R., "A Handbook of Numerical Matrix Inversion and Solution of Linear Equations." Wiley, New York, 1968.

SUPPLEMENTARY BIBLIOGRAPHY

Texts:

Bennett, W. R., and Davey, J. R., "Data Transmission." McGraw-Hill, New York, 1965.

Golomb, S. W., ed., "Digital Communications With Space Applications." Prentice Hall, Englewood Cliffs, New Jersey, 1964.

Hancock, J. C., and Wintz, P. A., "Signal Detection Theory." McGraw-Hill, New York, 1966.

Kotel'nikov, V. A., "The Theory of Optimum Noise Immunity" (*Transl.* by R. A. Silverman). McGraw-Hill, New York, 1959.

Lee, Y. W., "Statistical Theory of Communication." Wiley, New York, 1960.

Middleton, D., "An Introduction to Statistical Communications Theory." McGraw-Hill, New York, 1960.

Middleton, D., "Topics in Communication Theory." McGraw-Hill, New York, 1965.

Pierce, J. R., "Symbols, Signals, and Noise: The Nature and Process of Communication." Harper, New York, 1961.

Schwartz, M., Bennett, W. R., and Stein, S., "Communication Systems and Techniques." McGraw-Hill, New York, 1966.

Van Trees, H. L., "Detection, Estimation, and Modulation Theory," Part I. Wiley, New York, 1968.

Wainstein, L. A., and Zubakov, V. D., "Extraction of Signals From Noise" (*Transl.* by R. A. Silverman). Prentice-Hall, Englewood Cliffs, New Jersey, 1962.

Woodward, P. M., "Probability and Information Theory, With Applications to Radar," 2nd Ed. Pergamon Press, Oxford, 1964.

Wozencraft, J. M., and Jacobs, I. M., "Principles of Communication Engineering." Wiley, New York, 1965.

Papers:

Arthurs, E., and Dym, H., On the optimum detection of digital signals in the presence of white gaussian noise—A geometric interpretation and a study of three basic data transmission systems, *IRE Trans. Commun. Syst.* **CS-10,** No. 4, 336–372 (December 1962).

Balakrishnan, A. V., A contribution to the sphere-packing problem of communication theory, *J. Math. Anal. Appl.* **3,** 485–506 (December 1961).

Cahn, C. R., Combined digital phase and amplitude modulation communication systems, *IRE Trans. Commun. Syst.* (September 1960).

Craig, J. W. Jr., Optimum approximation to a matched filter response, *Correspondence IRE Trans. Inform. Theory* (June 1960).

Faran, J. J. Jr., and Hills, R. Jr., Correlators for Signal Reception, Office of Naval Res., NR–384–903, Tech. Memo. 27, September 15 (1952).

Helstrom, C. W., The comparison of digital communication systems, *IRE Trans. Commun. Syst.* (September 1960).

Jacobs, I., Comparison of M-ary modulation systems, *Bell System Tech. J.* (May-June 1967).

Lawton, J. G., Comparison of binary data transmission systems, *Nat. Conf. Military Electron. 2nd* (1958).

Lindsey, W. C., Phase-shift-keyed signal detection with noisy reference signals, *IEEE Trans. Aerospace Electron. Systems* (July 1966).

Middleton, D., Statistical theory of signal detection, *IRE Trans. Inform. Theory* **PGIT-3,** 26–51 (March 1954).

Miller, K. S., and Bernstein, R. I., An analysis of coherent integration and its application to signal detection, *IRE Trans. Inform. Theory* **IT-3,** No. 4, 237–248 (December 1957).

Montgomery, G. F., A comparison of amplitude and angle modulation for narrowband communication of binary-coded messages in fluctuation noise, *Proc. IRE* (February 1954).

Nuttall, A. H., Error probabilities for equicorrelated M-Ary signals under phase-coherent and phase-incoherent reception, *IRE Trans. Inform. Theory* **IT-8,** No. 4, 305–314 (July 1962).

Slepian, D., Bounds on communication, *Bell System Tech. J.* (May 1963).

Stein, S., Unified analysis of certain coherent and noncoherent binary communications systems, *IEEE Trans. Inform. Theory* (January 1964).

Viterbi, A. J., On coded phase-coherent communications, *IRE Trans. Space Electron. Telemet.* **SET-7,** 3–14 (March 1961).

Texts and Papers on Sources of Noise:

Hogg, D. C., and Mumford, W. W., The effective noise temperature of the sky, *Microwave J.* (March 1960).

Kendig, P. M. Ambient noise in the sea and its measurement, Lecture 13, "Underwater Acoustics" (V. M. Albers, ed.). Plenum Press, New York, 1963.

Lawson, J. L., and Uhlenbeck, G. E., "Threshold Signals," Radiation Laboratory Series, Vol. 24. McGraw-Hill, New York, 1950.

Oliver, B. M., Thermal and quantum noise, *Proc. IRE* (May 1965).

Pierce, J. R., Physical sources of noise, *Proc. IRE* (May 1956).

Urick, R. J., "Principles of Underwater Sound for Engineers." McGraw-Hill, New York, 1967.

Van Der Ziel, A., "Noise." Prentice Hall, Englewood Cliffs, New Jersey, 1954.

Wenz, G. M., Acoustic ambient noise in the ocean: spectra and sources, *J. Acoust Soc. Amer.* **34,** No. 12, 1936–1956 (December 1962).

Chapter 7

Detection of Signals with Random Parameters

7.1 Introduction

In Chap. 6 the information bearing signal was assumed to be known exactly at the receiver. It is often found in practice however that in addition to the uncertainty created by the additive noise of a receiver there is an additional uncertainty created by the randomness of signal parameters. The usual cause of this randomness is distortion in the transmission medium, the channel. Thus it is possible that even if the additive noise could be removed, there will still remain an uncertainty about which signal was transmitted. In this chapter the signal parameters of praticular interest to us will be amplitude, phase, frequency, and time of arrival.

In this chapter the a priori density function for the random parameters is assumed known. When this is not the case, we have an estimation problem which is discussed in Chap. 10. Except in Sect. 7.10 only white noise is considered. This avoids premature complication and permits easier insight into the results. Besides, in practice this assumption is often satisfied or nearly so.

7.2 Signals With Random Phase

Perhaps the most common random signal parameter is the phase. This is especially true for narrowband signals. A typical example is a radar signal.

In many radar systems the received signal is of the form $A \sin(\omega_c t + \theta)$ and extends over a time interval $(0, T)$. Generally, the period of the signal

$2\pi/\omega_c$ is very much shorter than the time duration T of the signal. To predict the phase even to within a few degrees requires an extremely precise knowledge of the range to the reflecting target. In such cases, it is reasonable to assume that the phase is a random variable, and furthermore that it is uniformly distributed over the interval $(0, 2\pi)$. Such a distribution implies a complete lack of knowledge of the phase and is a least favorable distribution.

We set out to solve the following problem: a receiver is to be designed to choose between two signals $s_1(t)$ or $s_0(t)$ where

$$H_1: \quad s_1(t) = A \sin(\omega_c t + \theta), \qquad 0 \le t \le T$$

and

$$H_0: \quad s_0(t) = 0$$

The amplitude A, the frequency ω_c and the time of arrival are assumed to be known. The phase θ is a random variable having an a priori density function

$$w(\theta) = \begin{cases} 1/2\pi, & 0 \le \theta \le 2\pi \\ 0, & \text{elsewhere} \end{cases}$$

At the receiver, the signal is corrupted by additive white Gaussian noise with zero mean and spectral density $N_o/2$. The observable signal is

$$r(t) = \begin{Bmatrix} A \sin(\omega_c t + \theta) \\ \text{or} \\ 0 \end{Bmatrix} + n(t)$$

Assume that if costs are assigned, they are independent of θ. For now, we need not specify a particular criterion for the receiver knowing that a likelihood ratio test will be used. We proceed as outlined in Sect. 5.9 and in particular will use Eq. (5-28). That is, we choose H_1 if

$$\frac{\int_0^{2\pi} p_1(\mathbf{r} \mid \theta) w(\theta)\, d\theta}{p_0(\mathbf{r})} \ge \lambda_o$$

The likelihood functions were derived in Sect. 6.3. In particular, for the noise only case

$$p_0(\mathbf{r}) = F \exp\left[-(1/N_o) \int_0^T r^2(t)\, dt\right] \tag{7-1}$$

The *conditional* likelihood function for H_1 is easily seen to be an extension of Eq. (6-38) and is

$$p_1(\mathbf{r} \mid \theta) = F \exp\left\{-(1/N_o) \int_0^T [r(t) - A \sin(\omega_c t + \theta)]^2\, dt\right\} \tag{7-2}$$

The marginal likelihood function $p_1(\mathbf{r})$ is obtained by the average indicated in Eq. (5-28). That is

$$p_1(\mathbf{r}) = \int_0^{2\pi} p_1(\mathbf{r} \mid \theta) w(\theta)\, d\theta$$

$$= F \int_0^{2\pi} \exp\left\{ -\frac{1}{N_o} \int_0^T [r(t) - A \sin(\omega_c t + \theta)]^2 \, dt \right\} \frac{d\theta}{2\pi} \tag{7-3}$$

The integral in the exponential term when expanded becomes

$$\int_0^T [r^2(t) - 2r(t)A \sin(\omega_c t + \theta) + A^2 \sin^2(\omega_c t + \theta)]\, dt$$

or equivalently

$$\int_0^T [r^2(t) - 2r(t)A \sin(\omega_c t + \theta)]\, dt + A^2 T/2$$

For convenience, we assume $\omega_c T = k\pi$ where k is an integer. Then

$$\int_0^T A^2 \sin^2(\omega_c t + \theta)\, dt = A^2 T/2$$

If this assumption is not made, an additional term results. However, if $2\pi/\omega_c \ll T$, the additional term is quite small and may be neglected. The likelihood ratio is therefore

$$\lambda(\mathbf{r}) = \frac{\displaystyle\int_0^{2\pi} \exp\left\{ -(1/N_o) \int_0^T [r^2(t) - 2r(t)A \sin(\omega_c t + \theta)]\, dt \right\} \frac{d\theta}{2\pi}}{\displaystyle\exp(A^2 T/2N_o) \exp\left[-(1/N_o) \int_0^T r^2(t)\, dt \right]}$$

Canceling terms we get

$$\lambda(\mathbf{r}) = \exp(-A^2 T/2N_o) \int_0^{2\pi} \exp\left[\int_0^T \frac{2r(t)}{N_o} A \sin(\omega_c t + \theta)\, dt \right] \frac{d\theta}{2\pi} \tag{7-4}$$

Consider the exponential term. The sine function may be expanded so that

$$\int_0^T \frac{2r(t)A}{N_o} \sin(\omega_c t + \theta)\, dt = \frac{2A}{N_o} \cos\theta \int_0^T r(t) \sin \omega_c t\, dt$$

$$+ \frac{2A}{N_o} \sin\theta \int_0^T r(t) \cos \omega_c t\, dt \tag{7-5}$$

Define the terms q and θ_o by

$$q \cos \theta_o \triangleq \int_0^T r(t) \sin \omega_c t \, dt \tag{7-6}$$

and

$$q \sin \theta_o \triangleq \int_0^T r(t) \cos \omega_c t \, dt \tag{7-7}$$

where $q \geq 0$. Substituting these definitions into Eq. (7-5) produces

$$\int_0^T \frac{2A}{N_o} r(t) \sin(\omega_c t + \theta) \, dt = \frac{2Aq}{N_o} [\cos \theta \cos \theta_o + \sin \theta \sin \theta_o]$$

$$= \frac{2Aq}{N_o} \cos(\theta - \theta_o) \tag{7-8}$$

The likelihood ratio of Eq. (7-4) may then be written

$$\lambda(\mathbf{r}) = \exp\left[-\frac{A^2 T}{2N_o}\right] \int_0^{2\pi} \exp\left[\frac{2Aq}{N_o} \cos(\theta - \theta_o)\right] \frac{d\theta}{2\pi} \tag{7-9}$$

Recalling that

$$I_o(x) = \int_0^{2\pi} e^{x \cos(\theta - \alpha)} \frac{d\theta}{2\pi}$$

the likelihood ratio becomes

$$\lambda(\mathbf{r}) = \exp\left[-\frac{A^2 T}{2N_o}\right] I_o\left(\frac{2Aq}{N_o}\right) \tag{7-10}$$

The decision rule is therefore to choose H_1 if

$$\exp\left[-\frac{A^2 T}{2N_o}\right] I_o\left(\frac{2Aq}{N_o}\right) \geq \lambda_o \tag{7-11}$$

or the equivalent

$$I_o\left(\frac{2Aq}{N_o}\right) \geq \lambda_o \exp\left[\frac{A^2 T}{2N_o}\right] \tag{7-12}$$

The modified Bessel function $I_o(2Aq/N_o)$, is a monotonically increasing function of q. Therefore, the decision rule may be based on q or q^2 rather than $I_o(2Aq/N_o)$. Knowing A, T, and N_o we may determine a threshold for q. Denoting it by η, the decision rule becomes: choose H_1 if

$$q \geq \eta \qquad \text{or} \qquad q^2 \geq \eta^2 \tag{7-13}$$

If λ_o is specified, then η is the solution to the equation

$$\exp\left[-\frac{A^2T}{2N_o}\right]I_o\left(\frac{2A\eta}{N_o}\right) = \lambda_o \tag{7-14}$$

If a Neyman–Pearson test is used, the threshold η is selected from a consideration of the false alarm probability. The random variable q (or q^2) may be determined from Eqs. (7-6) and (7-7). Thus,

$$\begin{aligned} q^2 &= q^2 \cos^2 \theta_o + q^2 \sin^2 \theta_o \\ &= \left[\int_0^T r(t) \sin \omega_c t \, dt\right]^2 + \left[\int_0^T r(t) \cos \omega_c t \, dt\right]^2 \end{aligned} \tag{7-15}$$

7.3 The Quadrature Receiver and Equivalent Forms

Equation (7-15) indicates the operation to be performed by the receiver to obtain the test statistic q^2. This operation is illustrated in Fig. 7-1. This is the well-known *quadrature receiver*.

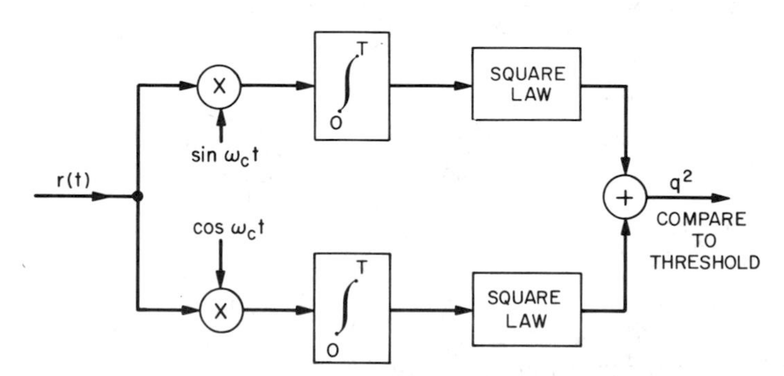

Fig. 7-1 Quadrature receiver.

We shall derive two equivalent forms of the quadrature receiver. The first form is easily determined by considering the material in Sect. 6.4 on the matched filter. In particular, we found that a correlator as shown in Fig. 6-5 may be replaced by a matched filter having an impulse response given by $h(t) = s(T - t)$, $0 \leq t \leq T$, and sampling its output at $t = T$. In the case at hand, the signals are of the form $\sin \omega_c t$ and $\cos \omega_c t$, $0 \leq t \leq T$. It is then readily seen that the quadrature receiver of Fig. 7-1 can be replaced by the receiver shown in Fig. 7-2. In one branch there is a filter matched to a signal, $\sin \omega_c t$, and in the other branch there is a filter matched to a 90° phase shifted version of the same signal. The filter outputs at $t = T$ are squared and added together.

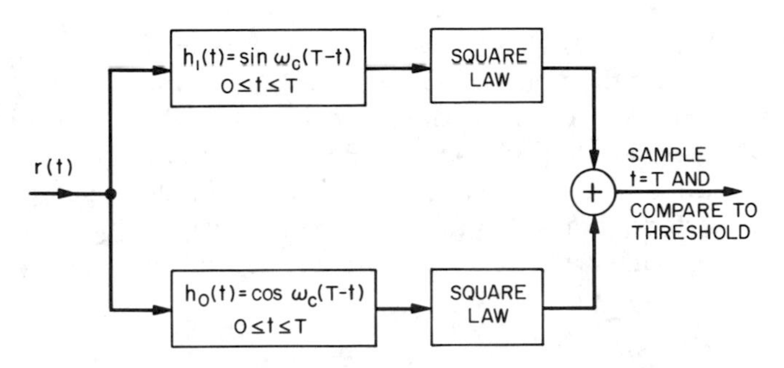

Fig. 7-2 Alternate form of quadrature receiver.

There is still another more important form of the quadrature receiver. Suppose we have a filter which is matched except for phase, ϕ, to the signal. It is left as an exercise to show that the choice of phase will make no difference (*1*). Therefore we shall choose $\phi = 0$. Then the weighting function is $\sin \omega_c(T - t)$ for $0 \le t \le T$. The output of such a filter as a function of time in response to the signal $r(t)$ is

$$e(t) = \int_0^t r(\tau) \sin \omega_c(T - t + \tau)\, d\tau$$

$$= \sin \omega_c(T - t) \int_0^t r(\tau) \cos \omega_c \tau\, d\tau + \cos \omega_c(T - t) \int_0^t r(\tau) \sin \omega_c \tau\, d\tau \tag{7-16}$$

The envelope of $e(t)$ evaluated at time T is

$$\left\{\left[\int_0^T r(\tau) \cos \omega_c \tau\, d\tau\right]^2 + \left[\int_0^T r(\tau) \sin \omega_c \tau\, d\tau\right]^2\right\}^{1/2}$$

But this is just q. Therefore, the output at time T of a filter matched to any arbitrary phase of the signal followed by an envelope detector is the same as the output of the quadrature receiver. This form of the receiver is shown in Fig. 7-3. The combination of the matched filter and the envelope detector is often called the *incoherent matched filter*. Since the phase matching is arbitrary we could also have used the matched filters $\cos \omega_c t$, or $\sin \omega_c t$, $0 \le t \le T$, in Fig. 7-3.

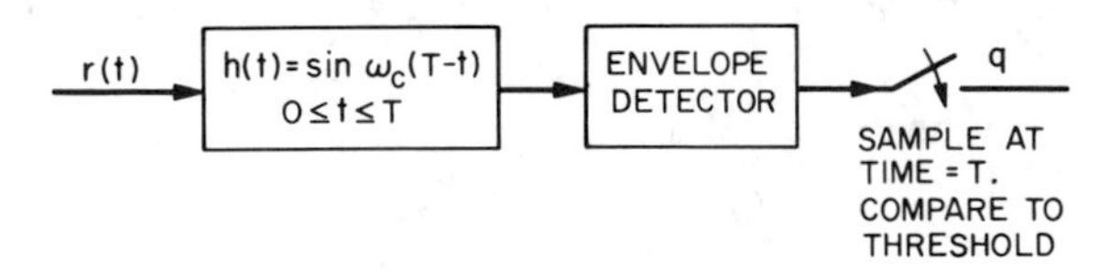

Fig. 7-3 Matched filter equivalent to quadrature receiver (incoherent matched filter).

Suppose that the signal has a slowly varying envelope instead of a constant envelope. That is, the signal is $s(t) = A(t)\sin(\omega_c t + \theta)$, $0 \le t \le T$, and the variations of $A(t)$ are very slow compared to $2\pi/\omega_c$. Then it may be shown (*2*) that the equivalent quadrature receiver is as shown in Fig. 7-4. The envelope of the filter is matched to the envelope of the signal. The phase matching is arbitrary provided the envelope is slowly changing. (For more complicated signals, the results can be expressed in terms of the complex envelope; see Chap. 9.)

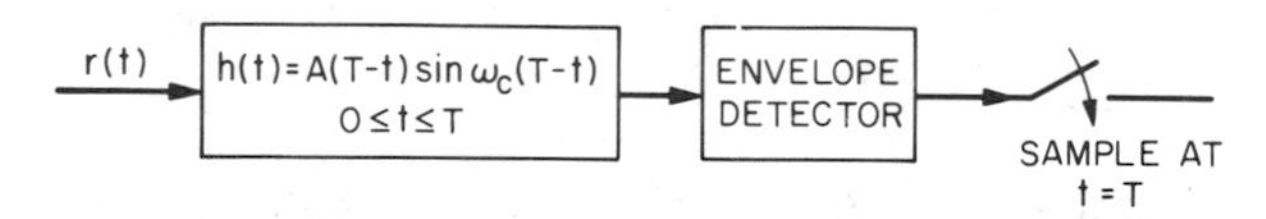

Fig. 7-4 Equivalent quadrature receiver for slowly changing signal envelope.

An intuitive discussion of the necessity for the envelope detector may shed some light on the incoherent matched filter. The output of a filter matched to a sine wave was shown in Fig. 6-8. It reaches a positive peak at the sampling instant T. If however the phase of the filter is not matched to that of the signal, the peak will occur at a time different than T. In fact, if the phases differ by 180°, a negative peak appears at the sampling instant.† To avoid such poor sampling in the absence of a priori phase knowledge, it is reasonable to extract the envelope of the output since this is independent of the phase mismatch. Indeed, this is precisely the operation indicated by the optimum receiver. The lack of phase knowledge does however incur a slight performance penalty.

7.4 Receiver Operating Characteristics

The objective here is to determine the performance of the quadrature receiver. A figure showing the probability of detection versus the probability of false alarm with signal-to-noise ratio as a parameter is usually referred to as the *receiver operating characteristics* (ROC). To determine the performance we use the test statistic q. Defining

$$x \triangleq \int_0^T r(t) \sin \omega_c t \, dt \qquad \text{and} \qquad y \triangleq \int_0^T r(t) \cos \omega_c t \, dt \tag{7-17}$$

the statistic q may be expressed as

$$q = [x^2 + y^2]^{1/2} \tag{7-18}$$

† See Exercise 7.1.

To determine the probability density function for q, we require the probability density functions for x and y. Since the noise is Gaussian, x and y are also Gaussian. Based on the material in Chap. 4 and in particular Sects. 4.3 and 4.4 we can certainly anticipate that q will have a Rician density function.

For hypotheses H_1, the expected values of x and y for a *given* value of the phase are

$$E_1\{x \mid \theta\} = (AT/2)\cos\theta \qquad \text{and} \qquad E_1\{y \mid \theta\} = (AT/2)\sin\theta \tag{7-19}$$

The quantity $x - E_1\{x \mid \theta\}$ is given by

$$\int_0^T n(t)\sin\omega_c t\,dt$$

where $n(t)$ is the sample function of the noise. Therefore the variance of x, denoted $V\{x \mid \theta\}$, is

$$\begin{aligned} V\{x \mid \theta\} &= E\left\{\int_0^T \int_0^T n(t)n(\tau)\sin\omega_c t\sin\omega_c \tau\,dt\,d\tau\right\} \\ &= \int_0^T \int_0^T R_n(t-\tau)\sin\omega_c t\sin\omega_c\tau\,dt\,d\tau \end{aligned}$$

Since white Gaussian noise with autocorrelation function $(N_o/2)\delta(\tau)$ is assumed,

$$V\{x \mid \theta\} = (N_o/2)\int_0^T \sin^2\omega_c t\,dt \simeq N_o T/4 \tag{7-20}$$

It may be similarly shown that the variance of y is equal to that of x. We now denote the variance by $\sigma_T{}^2$, that is,

$$\sigma_T{}^2 = N_o T/4 \tag{7-21}$$

The quantities x and y are taken from channels which are in phase quandrature to each other. It is not surprising therefore that they may be shown to be uncorrelated, and since they are Gaussian, they are statistically independent. Therefore the density function for x and y is

$$p_1(x, y \mid \theta) = \frac{1}{2\pi\sigma_T{}^2}\exp\left\{-\frac{1}{2\sigma_T{}^2}\left[\left(x - \frac{AT}{2}\cos\theta\right)^2 + \left(y - \frac{AT}{2}\sin\theta\right)^2\right]\right\}$$

Now, using the transformation

$$x = q\cos\theta_o \qquad \text{and} \qquad y = q\sin\theta_o$$

where $\theta_o = \tan^{-1} y/x \ (0 \le \theta_o \le 2\pi)$, the density function for q and θ_o becomes

$$p_1(q, \theta_o \mid \theta) = \frac{q}{2\pi\sigma_T{}^2}\exp\left\{-\frac{1}{2\sigma_T{}^2}\left[q^2 + \left(\frac{AT}{2}\right)^2 - ATq\cos(\theta - \theta_o)\right]\right\}$$

Integrating this over θ_o gives (see Sect. 4.4)

$$p_1(q \mid \theta) = \frac{q}{\sigma_T^2} \exp\left[\frac{q^2 + (AT/2)^2}{-2\sigma_T^2}\right] I_o\left(\frac{qAT}{2\sigma_T^2}\right)$$

It then follows that

$$p_1(q) = \int_0^{2\pi} p_1(q \mid \theta) \frac{d\theta}{2\pi} = \frac{q}{\sigma_T^2} \exp\left[\frac{q^2 + (AT/2)^2}{-2\sigma_T^2}\right] I_o\left(\frac{qAT}{2\sigma_T^2}\right) \tag{7-22}$$

This is the required density function. For hypothesis H_0, $A = 0$ and $I_o(0) = 1$, so that

$$p_0(q) = \frac{q}{\sigma_T^2} \exp\left[-\frac{q^2}{2\sigma_T^2}\right] \tag{7-23}$$

Equations (7-22) and (7-23) are the Rician and Rayleigh distributions respectively. In particular, Eqs. (7-22) and (7-23) are identical to Eqs. (4-47) and (4-52) with σ^2 replaced by σ_T^2, and the equivalent sine wave amplitude replaced by $AT/2$. Thus the equivalent signal-to-noise ratio is $(AT/2)^2/2\sigma_T^2 = E/N_o$ where E is the signal energy $A^2T/2$.

The probability of false alarm, now denoted by P_{fa} instead of $P(D_1 \mid H_0)$, is

$$P_{fa} = \int_\eta^\infty p_0(q)\, dq = \int_\eta^\infty \frac{q}{\sigma_T^2} \exp\left[-\frac{q^2}{2\sigma_T^2}\right] dq = e^{-\beta^2/2} \tag{7-24}$$

where

$$\beta = \eta/\sigma_T \tag{7-25}$$

is a normalized threshold. The probability of detection, now denoted by P_D instead of $P(D_1 \mid H_1)$, is

$$P_D = \int_\eta^\infty \frac{q}{\sigma_T^2} \exp\left[\frac{q^2 + (AT/2)^2}{-2\sigma_T^2}\right] I_o\left(\frac{qAT}{2\sigma_T^2}\right) dq \tag{7-26}$$

By a change of variable ($z = q/\sigma_T$), and denoting $\alpha^2 = 2E/N_o$ it can be shown that

$$P_D = \int_\beta^\infty z \exp\left[\frac{z^2 + \alpha^2}{-2}\right] I_o(\alpha z)\, dz \tag{7-27}$$

This is the Marcum Q-function (*3, 4*) discussed in Sect. 4.4. It is one minus the cumulative probability distribution of the Rician variable and is shown in Figs. 4-7 and 4-8.

Equations (7-24) and (7-27) completely specify the receiver operating characteristics. These have been evaluated and are shown by the dashed curves in Fig. 7-5 for a wide range of P_{fa} and E/N_o.† The latter is equal to

† Computer programs written by G. H. Robertson were used to compute the Q function. See also Peterson and Birdsall (*1*), Skolnik (*5*), and Helstrom (*6*).

$\alpha^2/2$ and is the signal-to-noise ratio at the sampling instant. For comparison, the performance for a coherent system (phase known) is also shown. These results were derived in Sect. 6.2, Eqs. (6-36) and (6-37). The difference in E/N_o over a substantial portion of these curves is less than 1 dB. This is the effect of not knowing or not using phase information. While this difference for a single signal comparison is not very much, the difference between the coherent and incoherent receivers becomes greater when many sequentially detected pulses are considered. This is discussed further in Chap. 8.

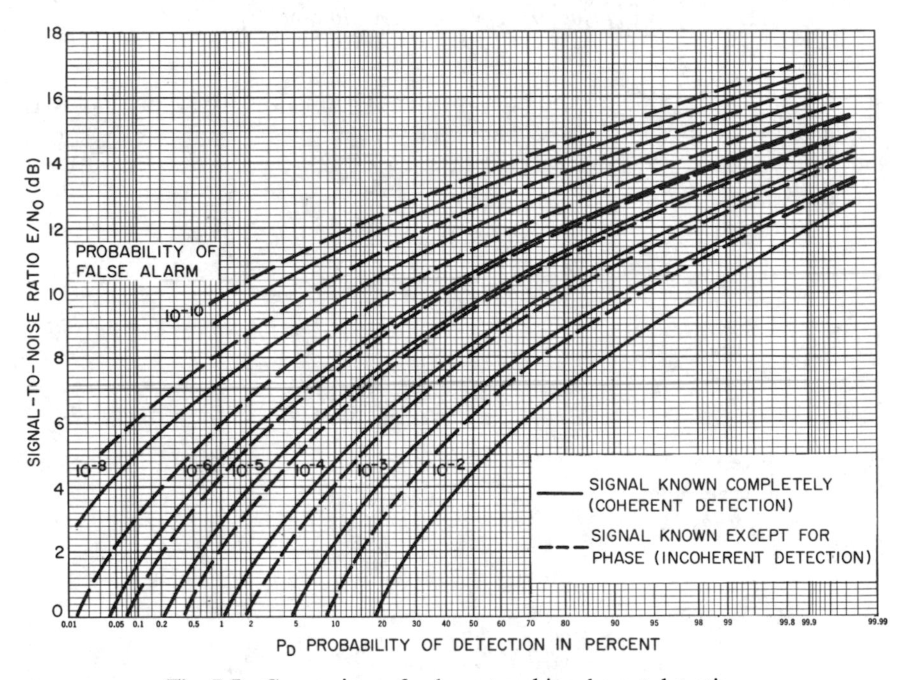

Fig. 7-5 Comparison of coherent and incoherent detection.

7.5 Signals With Random Phase and Amplitude

We now treat the case where both the signal amplitude and phase are random. The problem is conceptually the same as that for random phase.

As in the previous section, the hypotheses are

$$H_1: \quad r(t) = s(t) + n(t), \qquad 0 \le t \le T$$
$$H_0: \quad r(t) = n(t), \qquad 0 \le t \le T$$

where $s(t) = A \sin(\omega_c t + \theta)$. The amplitude and phase are random variables although they are assumed to be constant over any particular observation interval $(0, T)$. The frequency ω_c and the time of arrival are known precisely.

The noise is white and Gaussian with zero mean and spectral density $N_o/2$. We further assume that the a priori distributions of the amplitude and phase are known, so that the decision rule is given by Eq. (5-28). That is, choose H_1 if

$$\frac{\int_{\{\boldsymbol{\theta}\}} p_1(\mathbf{r} \mid \boldsymbol{\theta}) w(\boldsymbol{\theta})\, d\boldsymbol{\theta}}{p_0(\mathbf{r})} \geq \lambda_o$$

where $\boldsymbol{\theta} = (\theta, A)$. To be more specific, choose H_1 if

$$\frac{\int_{\{A\}} \int_{\{\theta\}} p_1(\mathbf{r} \mid \theta, A) w_1(\theta) w_2(A)\, d\theta\, dA}{p_0(\mathbf{r})} \geq \lambda_o \tag{7-28}$$

where $w_1(\theta)$ and $w_2(A)$ are the a priori density functions for θ and A. It is assumed that θ and A are independent random variables. This is often justified and represents a least favorable situation (*6*).

The *conditional likelihood ratio* is

$$\lambda(\mathbf{r} \mid A) = \frac{\int_{\{\theta\}} p_1(\mathbf{r} \mid \theta, A) w_1(\theta)\, d\theta}{p_0(\mathbf{r})}$$

From Eq. (7-10) this is

$$\lambda(\mathbf{r} \mid A) = \exp\left[-\frac{A^2 T}{2N_o}\right] I_o\left(\frac{2Aq}{N_o}\right) \tag{7-29}$$

where q is given by Eq. (7-15). We must still average over all A to determine the unconditioned likelihood ratio

$$\lambda(\mathbf{r}) = \int_{\{A\}} \lambda(\mathbf{r} \mid A) w_2(A)\, dA$$

Explicitly, this is

$$\lambda(\mathbf{r}) = \int_{\{A\}} \exp\left[-\frac{A^2 T}{2N_o}\right] I_o\left(\frac{2Aq}{N_o}\right) w_2(A)\, dA \tag{7-30}$$

At this point, assume that the amplitude A has a Rayleigh density function with

$$w_2(A) = \frac{A}{A_o^2} \exp\left[-\frac{A^2}{2A_o^2}\right], \qquad A \geq 0$$

This case is called slow Rayleigh fading. It is similar to the problem of detecting a Gaussian signal in Gaussian noise. See Exercises 6.7 and 7.9 for versions of such a problem.

The likelihood ratio is then

$$\lambda(\mathbf{r}) = \int_0^\infty \frac{A}{A_o^2} \exp\left[-\frac{A^2}{2}\left(\frac{1}{A_o^2} + \frac{T}{N_o}\right)\right] I_o\left(\frac{2Aq}{N_o}\right) dA \tag{7-31}$$

Carrying out the integration produces (7)

$$\lambda(\mathbf{r}) = \frac{N_o}{N_o + TA_o^2} \exp\left[\frac{2A_o^2 q^2}{N_o(N_o + TA_o^2)}\right] \tag{7-32}$$

Using the log-likelihood ratio, we get the decision rule: choose H_1 if

$$\ln \frac{N_o}{N_o + TA_o^2} + \frac{2A_o^2 q^2}{N_o(N_o + TA_o^2)} \geq \ln \lambda_o$$

or the equivalent: choose H_1 if

$$q \geq \left\{\frac{N_o(N_o + TA_o^2)}{2A_o^2} \ln\left[\frac{\lambda_o(N_o + TA_o^2)}{N_o}\right]\right\}^{1/2} \triangleq \eta \tag{7-33}$$

Therefore, the optimum receiver for this case is the matched filter followed by an envelope detector, Fig. 7-3. Intuitively this result is not surprising. Furthermore the result is independent of the a priori distribution of the amplitude. To show this consider the likelihood ratio of Eq. (7-29). For any value of A, the likelihood ratio is maximized if q is maximized so the decision rule is to choose H_1 if q is greater than some threshold. Thus, the statistic q provides a uniformly most powerful test with respect to amplitude. Even if the a priori amplitude distribution is unknown, we are assured that the statistic q is optimum. (See Exercises 7.6, 7.7, and 7.8.)

Error Performance

We shall find the error performance for a Neyman–Pearson test. The decision is based on the test statistic q. The threshold is η and is chosen to satisfy a given false alarm probability. If no signal is present, then from Eq. (7-23)

$$p_0(q) = \frac{q}{\sigma_T^2} \exp\left[-\frac{q^2}{2\sigma_T^2}\right]$$

The false alarm probability is, by Eq. (7-24),

$$P_{fa} = e^{-\beta^2/2}$$

where $\beta = \eta/\sigma_T$ and $\sigma_T^2 = N_o T/4$. For a given false alarm probability, this equation can be solved for the threshold η.

For a given value of A, the probability of detection is a function of A. Denote this by $P_D(A)$. The average probability of detection is obtained by averaging this over all A. Thus,

$$P_D = \int_{\{A\}} P_D(A) w_2(A)\, dA \tag{7-34}$$

Now, $P_D(A)$ is given by Eq. (7-26) for a given value of A and therefore

$$P_D(A) = \int_\eta^\infty \frac{q}{\sigma_T} \exp\left[\frac{q^2 + (AT/2)^2}{-2\sigma_T^2}\right] I_o\left(\frac{qAT}{2\sigma_T^2}\right) dq$$

Assuming as above that $w_2(A)$ is Rayleigh distributed, the probability of detection becomes (7) (see also Exercise 7.9)

$$P_D = \exp\left[-\frac{2\eta^2}{T(N_o + TA_o^2)}\right] \tag{7-35}$$

For a given signal level A, the signal energy over an interval T is equal to $A^2T/2$. The average energy over all such intervals is then

$$E_{av} = \int_0^\infty \frac{A^2T}{2} \frac{A}{A_o^2} \exp\left(-\frac{A^2}{2A_o^2}\right) dA = A_o^2 T \tag{7-36}$$

Substituting this into Eq. (7-35) and using $\eta = \beta\sigma_T$ and $\sigma_T^2 = N_o T/4$, the probability of detection may be expressed

$$P_D = [e^{-\beta^2/2}]^{1/(1+\varepsilon)} = (P_{fa})^{1/(1+\varepsilon)} \tag{7-37}$$

where $\varepsilon = A_o^2 T/N_o = E_{av}/N_o$.

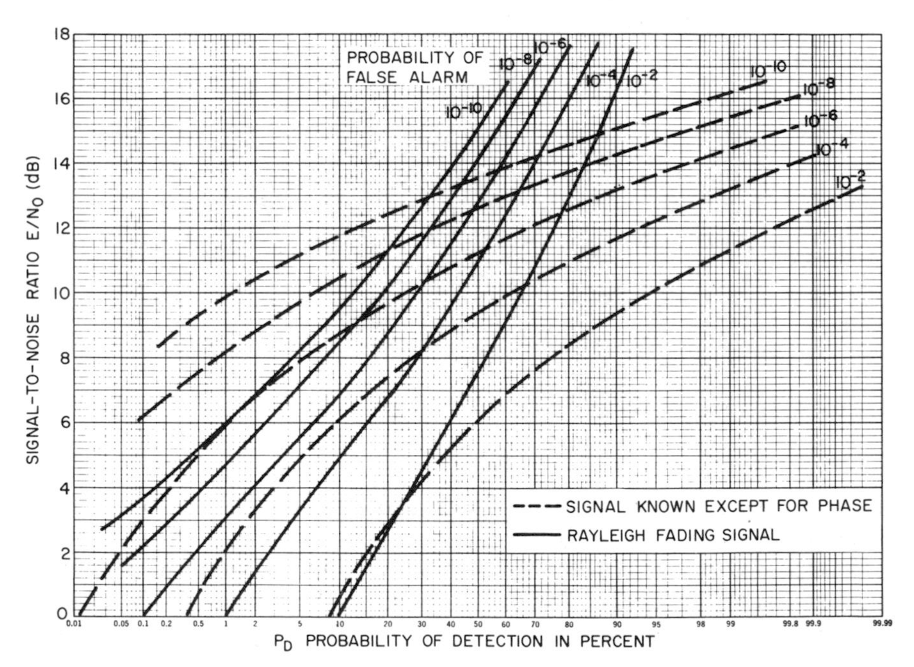

Fig. 7-6 Detection performance for a Rayleigh fading signal, and a signal known except for phase.

This result is plotted in Fig. 7-6 along with the results for constant amplitude signals with random phase. At high values of signal-to-noise ratio the amplitude fading causes a degradation; at low values of signal-to-noise ratio the converse is true. This is discussed further in the next section.

7.6 Noncoherent Frequency Shift Keying

We now apply a likelihood ratio test to a communications problem involving binary frequency shift keying (FSK). One of two frequencies is transmitted with equal probability to a receiver. That is, over an interval $(0, T)$,

$$s_0(t) = A \sin(\omega_0 t + \phi) \qquad \text{or} \qquad s_1(t) = A \sin(\omega_1 t + \theta) \tag{7-38}$$

The phases ϕ and θ are statistically independent random variables, uniformly distributed over the interval $(0, 2\pi)$. Two cases are considered for the amplitude. In the first case the amplitudes are known and equal. In the second case they are assumed to be random variables. This latter case will be related to a particular fading model. In either case white Gaussian noise with spectral density $N_o/2$ is added to the signal.

Known Amplitude and Random Phase

The hypotheses are

$$H_0: \quad r(t) = s_0(t) + n(t)$$
$$H_1: \quad r(t) = s_1(t) + n(t)$$

As before, the conditional likelihood functions may be found from Eq. (6-38) or more directly from Eq. (7-3). They are

$$p_1(\mathbf{r}) = \int_0^{2\pi} F \exp\left\{-\frac{1}{N_o}\int_0^T [r(t) - A\sin(\omega_1 t + \theta)]^2\, dt\right\} \frac{d\theta}{2\pi}$$

and

$$p_0(\mathbf{r}) = \int_0^{2\pi} F \exp\left\{-\frac{1}{N_o}\int_0^T [r(t) - A\sin(\omega_0 t + \phi)]^2\, dt\right\} \frac{d\phi}{2\pi}$$

It is not necessary to evaluate these likelihood functions again. Observe that dividing these likelihood functions by a constant will have no influence on the likelihood ratio. In particular, divide by

$$F \exp\left[-(1/N_o)\int_0^T r^2(t)\, dt\right]$$

The result of this operation is to produce functions like that in Eq. (7-4) which when evaluated become, as shown in Eq. (7-10),

$$p_1(\mathbf{r}) \sim \exp\left(-\frac{A^2T}{2N_o}\right)I_o\left(\frac{2Aq_1}{N_o}\right) \tag{7-39}$$

and

$$p_0(\mathbf{r}) \sim \exp\left(-\frac{A^2T}{2N_o}\right)I_o\left(\frac{2Aq_0}{N_o}\right) \tag{7-40}$$

where q_1 and q_0 are given by Eq. (7-15) except with ω_c replaced by ω_1 and ω_0 respectively. The likelihood ratio is then

$$\lambda(\mathbf{r}) = \frac{I_o(2Aq_1/N_o)}{I_o(2Aq_0/N_o)} \tag{7-41}$$

Since we are interested in minimum error probability, the decision rule is to choose H_1 if

$$I_o(2Aq_1/N_o) \geq I_o(2Aq_0/N_o)$$

or equivalently, choose H_1 if

$$q_1 \geq q_0 \tag{7-42}$$

By analogy with Fig. 7-3 the receiver for this case is easily seen to be as shown in Fig. 7-7.

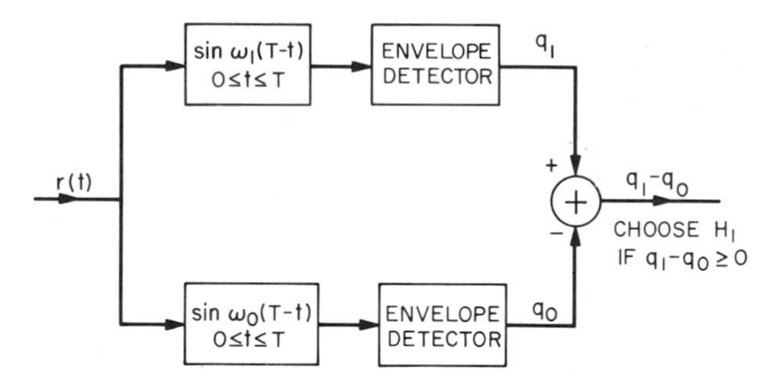

Fig. 7-7 Noncoherent frequency shift keying (FSK) receiver.

From the symmetry of the receiver it should be apparent that the probabilities of each type of error are equal. Since the a priori probabilities are also equal, the overall error probability is equal to

$$P_e = P(D_0 \mid H_1) = P(D_1 \mid H_0)$$

Therefore to calculate the error probability, we may assume without loss of generality that $s_1(t)$ is sent. For a *given value* of q_1, an error is made if $q_0 > q_1$. Thus, the conditional error probability is

$$P_e(q_1) = \int_{q_1}^{\infty} p(q_0 \mid q_1)\, dq_0$$

where $p(q_0 \mid q_1)$ is the probability density function for q_0 given that H_1 is true and given a value of q_1. The average error probability is found by averaging over all q_1. That is

$$P_e = \int_0^{\infty} P_e(q_1) p_1(q_1)\, dq_1 = \int_0^{\infty} dq_1\, p_1(q_1) \int_{q_1}^{\infty} dq_0\, p(q_0 \mid q_1)$$

Since both $s_0(t)$ and $s_1(t)$ are time limited, their energy extends over the frequency range $(-\infty, \infty)$. Consequently, when one of the signals is sent, some of its energy spills over into the matched filter corresponding to the other signal [see Turin (*8*)]. This tends to make q_0 and q_1 statistically dependent. However, if the difference $|\omega_0 - \omega_1|$ is large, the spillover is small and may be neglected. Assume that this condition is satisfied so that $p(q_0 \mid q_1) = p(q_0)$. With this assumption the error probability may be written

$$P_e = \int_0^{\infty} dq_1\, p_1(q_1) \int_{q_1}^{\infty} dq_0\, p(q_0) \tag{7-43}$$

The density function for q_1 is given by Eq. (7-22). Because of the assumption that q_0 is independent of q_1, it follows that q_0 is due to noise alone. The density function $p(q_0)$ is therefore identical to that given by Eq. (7-23). Substituting these density functions into Eq. (7-43) yields

$$P_e = \int_0^{\infty} dq_1 \int_{q_1}^{\infty} dq_0\, \frac{q_0 q_1}{\sigma_T^{\,4}} \exp\left[\frac{q_0^{\,2} + q_1^{\,2} + (AT/2)^2}{-2\sigma_T^{\,2}}\right] I_0\left(\frac{q_1 AT}{2\sigma_T^{\,2}}\right) \tag{7-44}$$

As before, $\sigma_T^{\,2} = N_0 T/4$. Integrating first with respect to q_0 produces

$$P_e = \int_0^{\infty} \frac{q_1}{\sigma_T^{\,2}} \exp\left[\frac{2q_1^{\,2} + (AT/2)^2}{-2\sigma_T^{\,2}}\right] I_0\left(\frac{q_1 AT}{2\sigma^2}\right) dq_1$$

The integral may be evaluated to give (*9*)

$$P_e = \tfrac{1}{2} e^{-E/2N_0} \tag{7-45}$$

where $E = A^2T/2$ is the signal energy. For an error probability of 10^{-4}, this system requires about 4 dB more power than the optimum binary system (coherent phase shift keying). The performance for this system is compared with others in Fig. 7-8 and will be discussed presently.

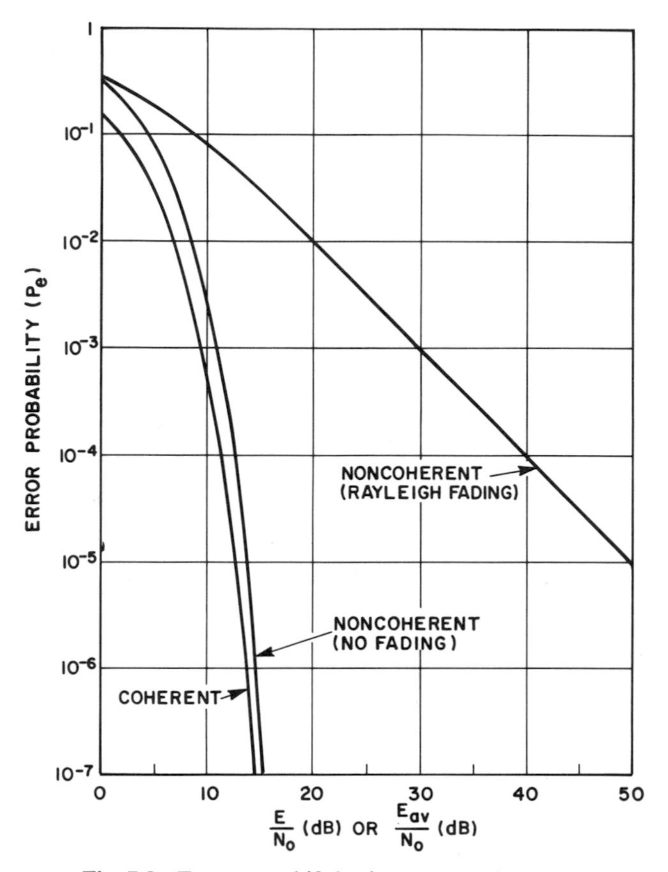

Fig. 7-8 Frequency shift keying error performance.

Random Amplitude–Rayleigh Fading

In many transmission media, the received signal is composed of the vector sum of many sine waves whose phases change slowly with time. Examples of "multipath" channels which experience such behavior include propagation† via the troposphere, ionsphere, and radar backscatter from complex targets (*3*). The vector sum signal will have an envelope which changes with time and is called a fading signal. Since the signal is the sum of many sine waves it is reasonable to model it as a Gaussian random process. If this is the case, then the envelope is Rayleigh distributed—hence Rayleigh fading. While this fading model is often assumed in practice, its application to any particular case must be carefully assessed.

† See, for example, Turin (*10*), Crawford (*11*), Rice (*12*), Sunde (*13*), Bullington (*14*), and Kaylor (*15*).

We shall apply the Rayleigh fading model to an FSK communication system. The signals are

$$s_0(t) = B \sin(\omega_0 t + \phi), \qquad 0 \le t \le T$$

or

$$s_1(t) = A \sin(\omega_1 t + \theta), \qquad 0 \le t \le T$$

The a priori probabilities are equal and the additive noise is white, Gaussian with spectral density $N_o/2$. The amplitudes are assumed to remain constant over the time interval $(0, T)$. This is called *slow fading*. That is, if the signal duration is extended it would look like a sample function of a narrowband Gaussian process. However, a short segment of the sample would look like a pure sine wave. Other short segments would also look like sine waves but of different amplitudes and phases.

Assume that the signal parameters are identically distributed, then for the amplitudes

$$w_2(B) = \frac{B}{A_o^2} \exp\left(-\frac{B^2}{2A_o^2}\right) \qquad \text{and} \qquad w_2(A) = \frac{A}{A_o^2} \exp\left(-\frac{A^2}{2A_o^2}\right)$$

and for the phases

$$w_1(\theta) = w_1(\phi) = 1/2\pi, \qquad 0 \le \theta, \phi \le 2\pi$$

The likelihood ratio is

$$\lambda(\mathbf{r}) = \frac{\int_{\{A\}} \int_{\{\theta\}} p_1(\mathbf{r} \mid A, \theta) w_1(\theta) w_2(A)\, d\theta\, dA}{\int_{\{B\}} \int_{\{\phi\}} p_0(\mathbf{r} \mid B, \phi) w_1(\phi) w_2(B)\, d\phi\, dB} \tag{7-46}$$

We can use the results of the preceding case, and in particular Eq. (7-32), to write down the averaged likelihood ratio

$$\lambda(\mathbf{r}) = \frac{\dfrac{N_o}{N_o + TA_o^2} \exp\left[\dfrac{2A_o^2 q_1^2}{N_o(N_o + TA_o^2)}\right]}{\dfrac{N_o}{N_o + TA_o^2} \exp\left[\dfrac{2A_o^2 q_0^2}{N_o(N_o + TA_o^2)}\right]} \tag{7-47}$$

where q_1 and q_0 are given by Eq. (7-15) with ω_c replaced by ω_1 and ω_0 respectively. Canceling the multipliers, and using the log-likelihood ratio, the decision rule becomes: choose H_1 if

$$q_1^2 - q_0^2 \ge \frac{N_o(N_o + TA_o^2)}{2A_o^2} \ln \lambda_o \tag{7-48}$$

Since the a priori probabilities are equal, $\ln \lambda_o = 0$, and the decision rule becomes: choose H_1 if $q_1 > q_0$. The optimum receiver for minimum error probability is the same as for the previous case which is given in Fig. 7-7.

In determining the error performance, we shall again assume that q_1 and q_0 are statistically independent. Without loss of generality, assume that the signal $s_1(t)$ is present. Then, for a *given* value of A the error probability is given by Eq. (7-45), that is,

$$P_e(A) = \tfrac{1}{2}\exp[-E/2N_o] = \tfrac{1}{2}[\exp - A^2T/4N_o]$$

The average error probability is obtained by averaging over all A. That is

$$P_e = \int_0^\infty w_2(A)P_e(A)\,dA$$

Carrying out the integration we get

$$P_e = 1/(2+\varepsilon) \tag{7-49}$$

where $\varepsilon = A_o{}^2T/N_o = E_{av}/N_o$.

It is of some interest to compare the various cases of frequency shift keying which have been considered; namely, coherent FSK, noncoherent FSK, and noncoherent FSK with Rayleigh fading. The error probability for coherent FSK is given by Eq. (6-34):

$$P_e = \int_{(E/N_o)^{1/2}}^\infty \frac{1}{(2\pi)^{1/2}}\, e^{-z^2/2}\,dz$$

The error probability for noncoherent FSK is given by Eq. (7-45) and is

$$P_e = \tfrac{1}{2}e^{-E/2N_o}$$

The error probability for noncoherent FSK with Rayleigh fading is given in Eq. (7-49). The performance for these three systems is plotted in Fig. 7-8. The harmful effects of fading is readily apparent. For the nonfading cases, the error probability decreases exponentially with increasing E/N_o; whereas in the fading case the error probability decreases inversely with E_{av}/N_o. For the latter case, little advantage is gained by increasing the signal energy to improve performance and other techniques must be used. One such technique is diversity combining which is discussed in Chap. 8.

It is instructive to look at the case of fading signals from a simplified point of view. This allows one to gain insight into the behavior of fading signals and why the performance is degraded for high signal-to-noise ratios and improved for very low signal-to-noise ratios. Consider the idealized error performance curve of Fig. 7-9A where it is assumed that the error probability is 1 for $E/N_o < \xi_L$ and zero for $E/N_o > \xi_H$. The conditional

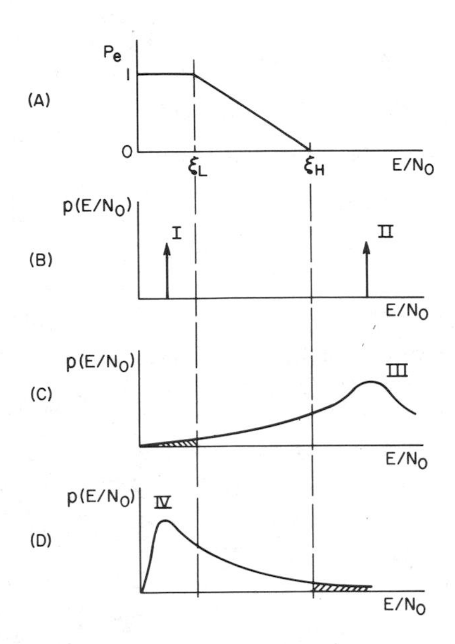

Fig. 7-9 An idealized error performance curve (A), and probability density functions for (B) constant amplitude signals, (C) a fading signal of high average E/N_0, and (D) a fading signal of low average E/N_0.

error probability is a function of E/N_0 and must be averaged over the appropriate probability distribution for E/N_0. In Fig. 7-9B for Cases I and II, the probability density is a delta function (E/N_0 a constant). For these cases, the error probabilities are 1 and 0 respectively. In Fig. 7-9C we depict a fading case which has the same average E/N_0 as in Case II. Note however that E/N_0 will occasionally be less than ξ_L as indicated by the shaded region. During the time that E/N_0 is within this region, the error probability is 1. If this area represented say 0.001, the average error probability would be greater than 0.001. Therefore, although Cases II and III had equal average E/N_0 the error probability for III is considerably higher. Another example is shown in Fig. 7-9D except that now the average E/N_0 is very low, representing a system with poor performance. For Case I, which has the same average E/N_0 as does Case IV, the probability of error is 1. But in Case IV, E/N_0 is occasionally in the region where $P_e < 1$ and the error performance will be improved. The improvement is not as dramatic as the degradation for Case III.

This idealized example should demonstrate that it is not only the average E/N_0 which is important to system performance but also the probability distribution function for E/N_0. This will be further demonstrated in Chap. 8 when diversity systems are discussed.

7.7 Signals With Random Frequency

The preceding material can be easily extended to signals having a random frequency and random time of arrival. In this section consider the random frequency case. Such signals are commonly encountered in practice. For example, in radar a signal, quite often a segment of a sine wave of known frequency, is radiated. The signal may be reflected by a moving target and reradiated back to the radar receiver. The frequency of the signal reradiated by the moving target will differ from that of the incident frequency by an amount called the doppler frequency. If the incident signal frequency is ω, the doppler frequency is, to a good approximation,

$$\omega_{\mathrm{d}} = 2v\omega/c$$

where v is the radial velocity of the target or reflecting surface, and c is the velocity of propagation. If the radial velocity component of the target is unknown, it then follows that the frequency of the reradiated signal is also unknown. We are therefore lead to ask what the optimum receiver is for such cases.

In analogy with previous results, we consider the hypotheses

$$\begin{aligned} H_1&: \quad r(t) = A\sin(\omega t + \theta) + n(t), \qquad 0 \le t \le T \\ H_0&: \quad r(t) = n(t), \qquad 0 \le t \le T \end{aligned}$$

The phase is assumed uniformly distributed, and the noise white and Gaussian. The signal amplitude and time of arrival are assumed known. The frequency is assumed to be a random variable with an a priori density function $p(\omega)$, $\omega_l \le \omega \le \omega_h$.

We can immediately use Eq. (7-10) to get the likelihood function conditioned on ω. Then

$$\lambda(\mathbf{r} \mid \omega) = \exp(-E/N_o)I_o(2Aq/N_o) \tag{7-50}$$

where $E = A^2T/2$ and

$$q^2 = \left[\int_0^T r(t)\sin\omega t\,dt\right]^2 + \left[\int_0^T r(t)\cos\omega t\,dt\right]^2$$

The average likelihood function is

$$\lambda(\mathbf{r}) = \int_{\omega_l}^{\omega_h} \lambda(\mathbf{r} \mid \omega)p(\omega)\,d\omega \tag{7-51}$$

It does not appear that the integral can be evaluated in a compact form. This need not discourage us. If $\lambda(\mathbf{r} \mid \omega)$ could be generated as a continuous function of frequency then $\lambda(\mathbf{r})$ could be found by instrumenting the indicated integration. To gain insight we shall proceed in a slightly different manner, making an approximation which is often justified. Replace the probability density function $p(\omega)$ by a discrete density function with an arbitrarily fine granularity. We therefore approximate $p(\omega)$ by

$$\sum_{i=1}^{M} P(\omega_i)\,\delta(\omega - \omega_i)$$

where, for a suitably chosen frequency increment $\Delta\omega$,

$$P(\omega_i) = p(\omega_i)\,\Delta\omega$$

and

$$\omega_i = \omega_l + i(\Delta\omega), \qquad i = 1, \ldots, (\omega_h - \omega_l)/\Delta\omega \triangleq M$$

Then

$$\lambda(\mathbf{r}) = \sum_{i=1}^{M} \lambda(\mathbf{r} \mid \omega_i) P(\omega_i)$$

The implementation of this receiver is shown in Fig. 7-10. The front end of this receiver is much like a spectrum analyzer† used in practice. For the assumptions at hand, the matched filters have the impulse functions $h(t) = \sin \omega_i t$, $i = 1, \ldots, M$, and $0 \leq t \leq T$. For other signal assumptions, an appropriate form for the matched filter can be used.

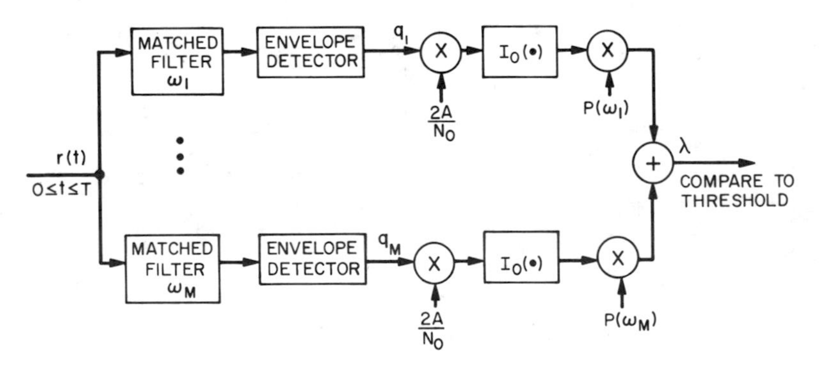

Fig. 7-10 Optimum receiver for constant amplitude signal of unknown frequency.

For small signal-to-noise ratios, the Bessel function may be approximated by

$$I_o(2Aq_i/N_o) \approx 1 + (Aq_i/N_o)^2$$

† Consult references and bibliography in Chap. 3.

With this approximation and assuming that the a priori frequency distribution is uniform, $P(\omega_i) = 1/M$, the likelihood function is

$$\lambda(\mathbf{r}) = (1/M)\sum_{i=1}^{M} \exp(-A^2T/2N_o)[1 + (Aq_i/N_o)^2]$$

The decision rule may therefore be expressed as: choose H_1 if

$$\sum q_i^2 \geq V_T$$

where V_T is the threshold. For this case, the receiver is shown in Fig. 7-11. This is analogous, except for summing the outputs, to a spectrum analyzer with a quadratic detector.

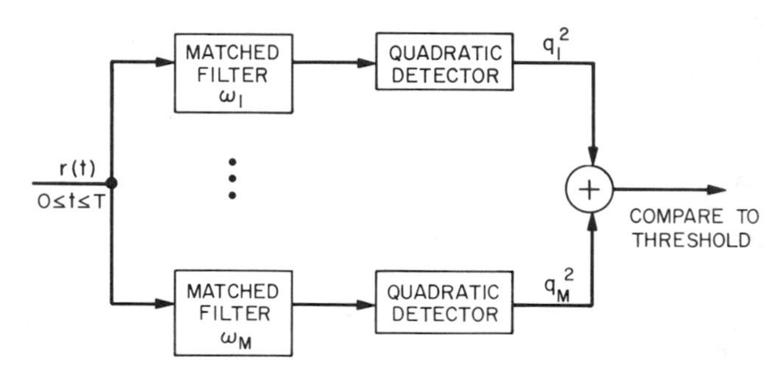

Fig. 7-11 Approximation to optimum receiver for low signal-to-noise ratio and uniform a priori discrete density function for frequency.

Rayleigh Fading Amplitude

For this case we make the same assumptions as above except that the amplitude A, while fixed for any given interval $(0, T)$, changes from interval to interval. Its distribution is given by

$$w(A) = \frac{A}{A_o^2}\exp\left[-\frac{A^2}{2A_o^2}\right]$$

Then, from Eq. (7-32) and denoting $\varepsilon = A_o^2T/N_o = E_{av}/N_o$

$$\lambda(\mathbf{r}\mid\omega) = \frac{1}{1+\varepsilon}\exp\left[\frac{(2A_o^2/N_o^2)q^2}{1+\varepsilon}\right]$$

and as before

$$\lambda(\mathbf{r}) = \int \lambda(\mathbf{r}\mid\omega)p(\omega)\,d\omega$$

Using the discrete approximation to the continuous frequency density function we get

$$\lambda(\mathbf{r}) = \frac{1}{1+\varepsilon} \sum_{i=1}^{M} P(\omega_i) \exp\left[\frac{(2A_o^2/N_o^2)q_i^2}{1+\varepsilon}\right] \tag{7-52}$$

This may be implemented as shown in Fig. 7-12. The similarity of this receiver

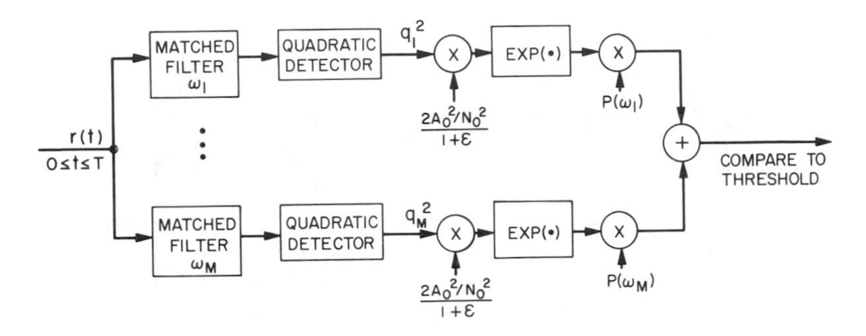

Fig. 7-12 Optimum receiver for Rayleigh fading signal of unknown frequency.

to that of Fig. 7-10 should be noted.

It is interesting to consider an approximation to this optimum receiver (*16, 17*). In particular, consider the detection of just two equally likely signals. In such a case the likelihood ratio can be manipulated to yield the decision rule: choose H_1 if

$$e^{Q_1^2} + e^{Q_2^2} \geq V \tag{7-53}$$

where

$$Q_i = \left[\frac{2A_o^2/N_o^2}{1+\varepsilon}\right]^{1/2} q_i$$

and V is the threshold. The decision region in terms of Q_1 and Q_2 is shown in Fig. 7-13 for several threshold values. Note that for the higher threshold values the decision region is nearly square. Hence, we may consider approximating the decision boundary by the dashed lines which are parallel to the axes. With the approximate boundaries, the decision rule is: choose H_1 if

$$\max\{q_1, q_2\} \geq V_T \tag{7-54}$$

where V_T is the threshold. This approximate form of the optimum receiver may be implemented as shown in Fig. 7-14.

The preceding result may be generalized to M frequencies. Then, an approximate form of the optimum receiver (such as shown in Fig. 7-12)

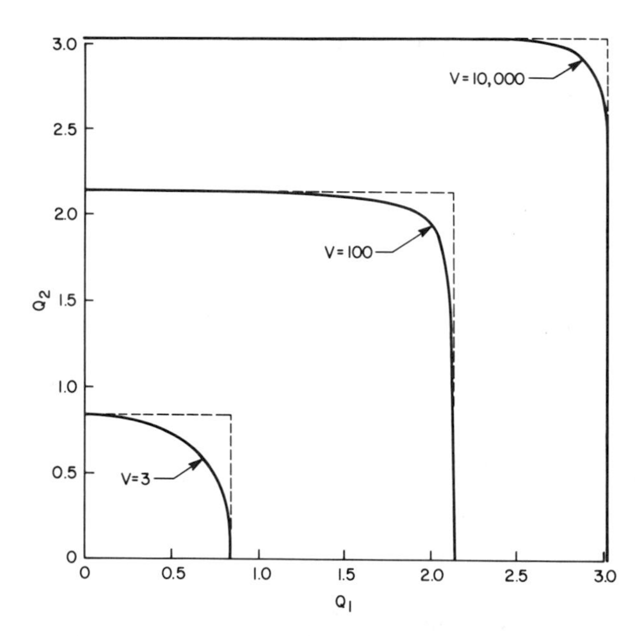

Fig. 7-13 Decision regions for detecting signals with random frequency; optimum decision regions separated by solid curves, approximate decision regions separated by dashed curves.

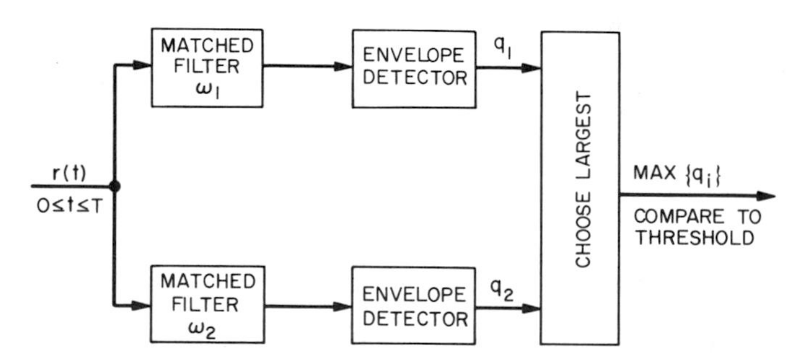

Fig. 7-14 Approximate form of optimum receiver (Fig. 7-12) for two frequencies.

consists of a bank of matched filters and detectors. The largest output is selected, and compared to a threshold (see Fig. 7-15). This form of the receiver is commonly used. It does have an advantage in the event that the noise level is unknown. In such a case it is difficult to set the threshold to control the false alarm probability. Having the output of each detector available allows a comparison of each. If one is substantially greater than the others, this is a relatively clear indication that a signal is present. Such a decision can be made with less than perfect knowledge of the noise level

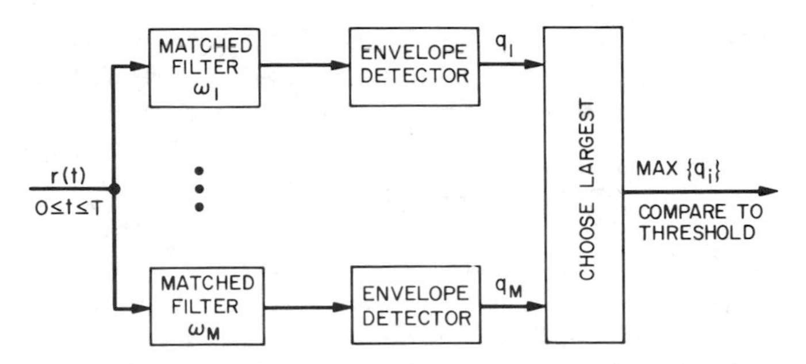

Fig. 7-15 Receiver for detecting signal with one of M frequencies or noise only (see also maximum likelihood principle, Chap. 10).

while maintaining control of the false alarms. The problem can be viewed in still another way. Since signal is present in at most one channel, the collection of the other channels are used to estimate the noise level and from that the threshold.

The above approximation was made for a receiver with an exponential nonlinearity. It may be verified that a similar approximation is valid for the Bessel function nonlinearity used in Fig. 7-10.

Alternate Approach for Random Frequency

In the previous approach to the signal with a random frequency, there was only one composite alternative hypothesis and we averaged over the frequency variable. In so doing we chose not to recognize what frequency the signal actually had. However, for the approximation shown in Fig. 7-14 for the two-frequency case or its generalization in the M-frequency case, Fig. 7-15, we have an opportunity to determine the frequency of the signal since we look at the detected output of each filter.

The alternate approach which we take here will require that we both detect a signal and identify its frequency. This problem can be related to estimating the signal frequency. However, we shall delay such a discussion until Chap. 10, and deal with the present problem in the context of a multiple alternative hypothesis test.

Assume that the signal frequency is one of M possible values. (If the frequency is a continuous random variable, approximate its density function by a set of discrete probabilities as discussed above.) To each discrete frequency ω_i assign a hypothesis, H . That is

$$\begin{aligned} H_0&: \ r(t) = n(t) \\ H_1&: \ r(t) = A \sin(\omega_1 t + \theta_1) + n(t) \\ &\ \ \vdots \\ H_M&: \ r(t) = A \sin(\omega_M t + \theta_M) + n(t) \end{aligned}$$

Assume the phase is uniformly distributed. For simplicity assume also that the frequencies are equally likely to occur so that $P(\omega_i) = 1/M$. The likelihood ratio comparing the ith hypothesis to the null hypothesis is

$$\lambda_i = \frac{p_i(\mathbf{r})}{p_0(\mathbf{r})} = e^{-E/N_o} I_o\left(\frac{2Aq_i}{N_o}\right), \qquad i = 1, \ldots, M$$

where $E = A^2T/2$. If none of these is greater than the threshold, we select H_0. Otherwise, we select the hypothesis corresponding to the maximum λ_i. For equal amplitudes, and equal probabilities for each possible frequency, the decision rule can be stated in terms of the q_i as follows: choose H_i if the largest q_i is greater than the threshold and choose H_0 otherwise. This is shown in Fig. 7-15. Thus, for all frequencies equally likely, the optimum receiver is the same as the approximate form of the receiver when all the signals are grouped in one composite hypothesis. A similar result can be obtained for fading signals.

Another approach to this problem is taken in Sect. 10.9 where the maximum likelihood principle is considered.

7.8 Signals With Random Time of Arrival

The case for random time of arrival is quite similar to that above. The hypotheses are

$$H_1: \quad r(t) = s(t - \tau) + n(t)$$
$$H_0: \quad r(t) = n(t)$$

where $s(t) = A \sin(\omega t + \theta), 0 \le t \le T$. The probability density for time of arrival, $p(\tau)$, is defined for $0 \le \tau \le \tau_m$. The likelihood function conditioned on τ is

$$\lambda(\mathbf{r} \mid \tau) = e^{-E/N_o} I_o\left(\frac{2Aq(\tau + T)}{N_o}\right) \tag{7-55}$$

where $E = A^2T/2$, and

$$q^2(\tau + T) = \left[\int_\tau^{\tau+T} r(t) \sin \omega(t - \tau)\, dt\right]^2 + \left[\int_\tau^{\tau+T} r(t) \cos \omega(t - \tau)\, dt\right]^2 \tag{7-56}$$

Then

$$\lambda(\mathbf{r}) = \int_0^{\tau_m} \lambda(\mathbf{r} \mid \tau) p(\tau)\, d\tau$$
$$= \int_T^{T+\tau_m} \exp\left(-\frac{E}{N_o}\right) I_o\left[\frac{2A}{N_o} q(u)\right] p(u - T)\, du \tag{7-57}$$

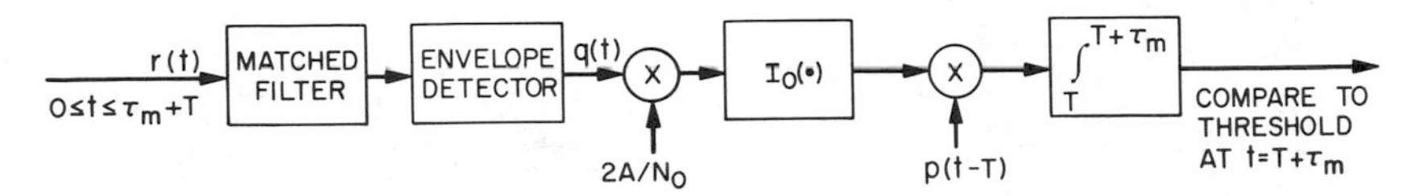

Fig. 7-16 Optimum receiver for detecting signal with random time of arrival.

This is shown in Fig. 7-16.

To draw a parallel to the random frequency case, assume that the random time of arrival is quantized into a discrete set of equally likely delays τ_i, $i = 1, \ldots, m$. Then the receiver can be implemented as shown in Fig. 7-17. In analogy to the argument used for the random frequency case, an approximation to this receiver could be realized by comparing the largest value of $q(\tau_i + T)$ to a threshold. If it does not exceed the threshold, then H_0 is chosen.

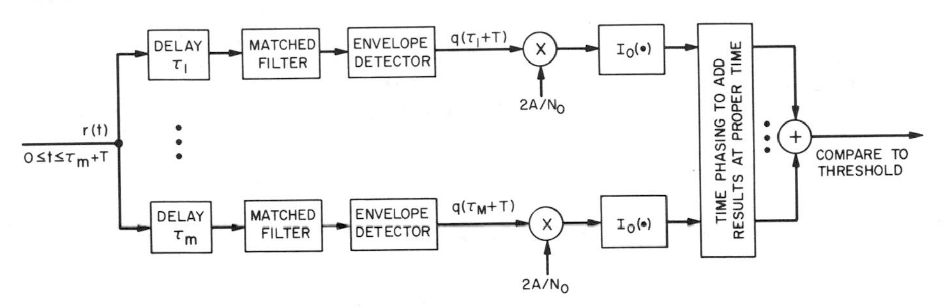

Fig. 7-17 One method to implement optimum receiver for detecting signal with discrete arrival times. Results could also be realized with a form such as is illustrated in Fig. 7-16.

We can also take a multiple alternative hypothesis approach to determine if a signal is present, and if so, to determine its time of arrival. The hypotheses are

$$\begin{aligned} H_0&: \; r(t) = n(t) \\ H_1&: \; r(t) = s(t - \tau_1) + n(t) \\ &\;\;\vdots \\ H_m&: \; r(t) = s(t - \tau_m) + n(t) \end{aligned}$$

For a signal $s(t) = A \sin(\omega t + \theta)$, $0 \le t \le T$, and with equally likely arrival times the receiver may be implemented by first choosing the largest value of $q(\tau_i + T)$, the output of the envelope detector shown in Fig. 7-17. This value is compared to a threshold, and if it is less, H_0 is chosen. Otherwise the hypothesis H_i corresponding to $q(\tau_i + T)$ is chosen.

If the time of arrival granularity is fine (say $\ll T$) and still maintains its uniform probability distribution, then the receiver can be implemented as shown in Fig. 7-18. Similar comments apply for a Rayleigh fading signal.

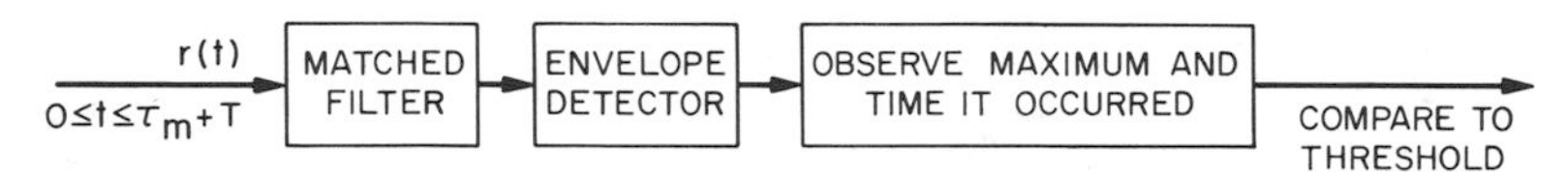

Fig. 7-18 Optimum receiver for both detecting signal and estimating its time of arrival. (Arbitrarily fine granularity on time of arrival.)

7.9 Random Frequency and Time of Arrival

This case is similar to the preceding cases so we shall just outline the procedure. Using the multiple alternative hypothesis approach for discrete frequency and time of arrival, the hypotheses are

$$H_0:\ r(t) = n(t)$$
$$H_{ij}:\ r(t) = s(t - \tau_i;\ \omega_j) + n(t), \qquad 0 \le \tau_i \le \tau_m, \qquad \omega_1 \le \omega_j \le \omega_M$$

where $s(t, \omega) = A \sin(\omega t + \theta)$, $0 \le t \le T$. Assume that frequency and time of arrival are quantized and are statistically independent with uniform distributions. Then

$$\lambda(\mathbf{r} \mid \tau_i, \omega_j) = e^{-E/N_o} I_0\left(\frac{2Aq_j(\tau_i + T)}{N_o}\right)$$

where

$$q_j^2(\tau_i + T) = \left[\int_{\tau_i}^{\tau_i + T} r(t) \sin \omega_j(t - \tau_i)\, dt\right]^2 + \left[\int_{\tau_i}^{\tau_i + T} r(t) \cos \omega_j(t - \tau_i)\, dt\right]^2$$

Assuming arbitrarily fine quantization for the time of arrival, the receiver may be implemented as shown in Fig. 7-19.

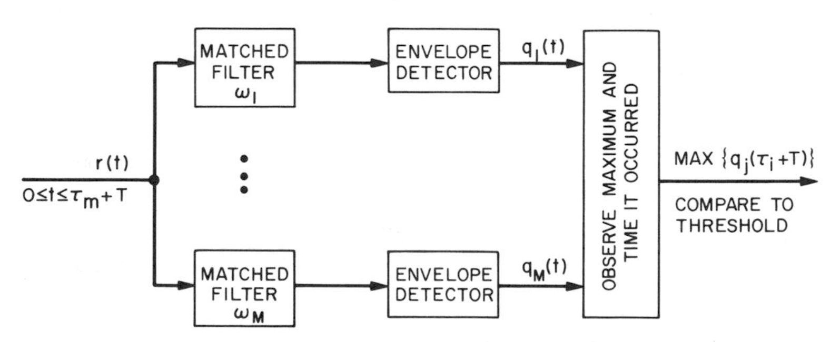

Fig. 7-19 Optimum receiver for both detecting signal and estimating its frequency and time of arrival. (See also maximum likelihood principle, Chap. 10).

7.10 Sampled Approach

As in Sect. 6.6, we shall again consider the problem of detecting signals when the data consists of samples taken at discrete time instants. We limit ourselves to the problem of detecting a sinusoidal signal with random phase. Thus the hypotheses are

$$\begin{aligned} H_1&: r(t_k) = s(t_k) + n(t_k), \qquad k = 1, \ldots, M \\ H_0&: r(t_k) = n(t_k) \end{aligned}$$

or in vector form

$$\begin{aligned} H_1&: \quad \mathbf{r} = \mathbf{s} + \mathbf{n} \\ H_0&: \quad \mathbf{r} = \mathbf{n} \end{aligned}$$

The signal samples are

$$s(t_k) = A(t_k) \sin(\omega t_k + \theta)$$

where θ is uniform $(0, 2\pi)$. The noise covariance matrix is

$$\mathbf{R}_n = E\{\mathbf{nn}'\}$$

The elements of the inverse matrix $\mathbf{R}_n^{-1}$ are denoted ϕ_{ij}. From Eq. (6-88) the conditional likelihood function is

$$\lambda(\mathbf{r} \mid \theta) = \exp\left(-\tfrac{1}{2}\mathbf{s}'(\theta)\mathbf{R}_n^{-1}\mathbf{s}(\theta)\right) \exp\left(\mathbf{r}'\mathbf{R}_n^{-1}\mathbf{s}(\theta)\right) \tag{7-58}$$

Before determining the average likelihood ratio we wish to argue that the first term is substantially independent of phase and to determine its approximate equivalent. Denote

$$\psi = \mathbf{s}'(\theta)\mathbf{R}_n^{-1}\mathbf{s}(\theta)$$

and replace $\mathbf{R}_n^{-1}$ by the product of $\mathbf{C}'\mathbf{C}$, where $\mathbf{C}$ is a lower triangular matrix as discussed in Sect. 6.6. Then

$$\psi = [\mathbf{Cs}(\theta)]'\mathbf{Cs}(\theta)$$

Since $\mathbf{Cs}(\theta)$ is a column vector, ψ is the sum of the squares of the individual elements. Therefore

$$\begin{aligned} \psi = {} & [c_{11}A_1 \sin(\omega t_1 + \theta)]^2 \\ & + [c_{21}A_1 \sin(\omega t_1 + \theta) + c_{22}A_2 \sin(\omega t_2 + \theta)]^2 \\ & \vdots \\ & + [c_{M1}A_1 \sin(\omega t_1 + \theta) + \cdots + c_{MM}A_M \sin(\omega t_M + \theta)]^2 \end{aligned}$$

where A_i denotes $A(t_i)$. The square of each term such as $\sin(\omega t_i + \theta)$ is $\frac{1}{2} + \frac{1}{2}\sin 2(\omega t_i + \theta)$. The cross terms will produce terms like $\cos \omega(t_i - t_j)$ and $\cos[\omega(t_i + t_j) + 2\theta]$. We argue, with more intuition than rigor, that the sum of terms involving the constant $\frac{1}{2}$ and terms like $\cos \omega(t_i - t_j)$ will be greater than the sum of terms involving the phase θ. (We are of course assuming that the interval of time over which the samples are taken is much longer than the period of the sine wave signal.) We assume therefore that ψ is substantially independent of θ. To approximate ψ, replace it by its average value,

$$E_\theta\{\psi\} = E_\theta\{\mathbf{s}'(\theta)\mathbf{R}_n^{-1}\mathbf{s}(\theta)\}$$

$$= \operatorname{tr}[\mathbf{R}_n^{-1}E_\theta\{\mathbf{s}(\theta)\mathbf{s}'(\theta)\}] \tag{7-59}$$

where $E_\theta\{\mathbf{s}(\theta)\mathbf{s}'(\theta)\}$ is the matrix:

$$\frac{1}{2}\begin{bmatrix} {A_1}^2 & A_1A_2\cos\omega(t_1-t_2) & \cdots & A_1A_M\cos\omega(t_1-t_M) \\ A_1A_2\cos\omega(t_1-t_2) & {A_2}^2 & \cdots & A_2A_M\cos\omega(t_2-t_M) \\ \vdots & \vdots & & \vdots \\ A_1A_M\cos\omega(t_1-t_M) & & \cdots & {A_M}^2 \end{bmatrix}$$

Since the first exponential in Eq. (7-58) is independent of θ, it may be used to modify the threshold rather than the test statistic. Then,

$$\lambda(\mathbf{r}\mid\theta) \sim \exp\bigl(\mathbf{r}'\mathbf{R}_n^{-1}\mathbf{s}(\theta)\bigr)$$

Now,

$$\exp\bigl(\mathbf{r}'\mathbf{R}_n^{-1}\mathbf{s}(\theta)\bigr) = \exp\left[\sum_{k,l}^{M} r_k\,\phi_{kl}\,A_l \sin(\omega t_l + \theta)\right]$$

$$= \exp\left[(\cos\theta)\sum_{k,l}^{M} r_k\,\phi_{kl}\,A_l \sin\omega t_l + (\sin\theta)\sum_{k,l}^{M} r_k\,\phi_{kl}\,A_l\cos\omega t_l\right]$$

In analogy with the time continuous case, this may be expressed as

$$\lambda(\mathbf{r}\mid\theta) \sim \exp[q\cos(\theta - \theta_0)]$$

where

$$q^2 = \left(\sum_{k,l}^{M} r_k\,\phi_{kl}\,A_l\sin\omega t_l\right)^2 + \left(\sum_{k,l}^{M} r_k\,\phi_{kl}\,A_l\cos\omega t_l\right)^2$$

This may be written in a compact form. Define the vectors

$$\mathscr{S} = \begin{bmatrix} A_1\sin\omega t_1 \\ \vdots \\ A_M\sin\omega t_M \end{bmatrix} \quad\text{and}\quad \mathscr{C} = \begin{bmatrix} A_1\cos\omega t_1 \\ \vdots \\ A_M\cos\omega t_M \end{bmatrix}$$

Then

$$q^2 = [\mathbf{r}'\mathbf{R}_n^{-1}\mathscr{S}]^2 + [\mathbf{r}'\mathbf{R}_n^{-1}\mathscr{C}]^2 \tag{7-60}$$

Proceeding as in Sect. 7.2, we find that the average likelihood ratio is

$$\lambda(\mathbf{r}) = \int_0^{2\pi} \lambda(\mathbf{r} \mid \theta) p(\theta)\, d\theta \sim I_0(q)$$

We may therefore use q or q^2 as the test statistic.

Factoring $\mathbf{R}_n^{-1}$ into its triangular matrices as in Sect. 6.6, we get

$$q^2 = [(\mathbf{Cr})'\mathbf{C}\mathscr{S}]^2 + [(\mathbf{Cr})'\mathbf{C}\mathscr{C}]^2 \tag{7-61}$$

The receiver is therefore analogous to the quadrature receiver already discussed except for the prewhitening filter $\mathbf{C}$.

For independent noise samples with equal variance σ^2, $\mathbf{R}_n^{-1}$ is diagonal with elements $1/\sigma^2$. Then the test statistic from Eq. (7-60) is

$$q^2 \sim [\mathbf{r}'\mathscr{S}]^2 + [\mathbf{r}'\mathscr{C}]^2$$

Exercises

7.1 For the detection of a known signal such as $A \sin(\omega_c t + \theta)$ in white Gaussian noise, where θ is known, a correlation receiver was optimum. Suppose we believed that θ was zero, when in fact it was not.

(a) For the correlation receiver, find the probability of detection as a function of θ and compare it to case where θ is correctly known. (Assume a radar type problem where the null hypothesis is noise only.)

(b) Also show that depending on the value of θ, the probability of detection may be less than the probability of false alarm.

7.2 For the incoherent matched filter of a sine wave, (a matched filter followed by an envelope detector), show that the choice of phase for the filter is arbitrary.

7.3 Consider the detection of a narrowband signal of the form

$$s(t) = Af(t)\sin(\omega_c t + \theta)$$

$0 \le t \le T$, $2\pi/\omega_c \ll T$, where the envelope $f(t)$ is slowly varying. The additive noise is white Gaussian. Show that the corresponding incoherent matched filter consists of a linear filter of impulse response $h(\tau) = f(T - \tau)\sin \omega_c \tau$ followed by an envelope detector. (Make the usual narrowband assumptions.)

7.4 For the signal and noise of the previous problem, determine the likelihood function $p_1(q)$ where

$$q^2 = \left[\int_0^T r(t)f(t) \sin \omega_c t\, dt\right]^2 + \left[\int_0^T r(t)f(t) \cos \omega_c t\, dt\right]^2$$

Make the usual narrowband assumptions which follow from the fact that $2\pi/\omega_c \ll T$.

7.5 For the hypotheses

$$H_1: \quad r(t) = A \sin(\omega_c t + \theta) + n(t), \qquad 0 \leq t \leq T$$
$$H_0: \quad r(t) = n(t), \qquad 0 \leq t \leq T$$

assume that A and ω_c are constants, $n(t)$ is white Gaussian noise of spectral density $N_o/2$. Assume that the phase is a random variable with density function (*18*) $p(\theta) = e^{v \cos \theta}/2\pi I_o(v)$. Using the Neyman–Pearson criterion, show that the receiver can be implemented as shown in Fig. 7-20 (*19*). Use the fact that

$$\int_0^{2\pi} \exp[(v + q_c) \cos \theta - q_s \sin \theta] \frac{d\theta}{2\pi} = I_o\{[(v + q_c)^2 + q_s^2]^{1/2}\}$$

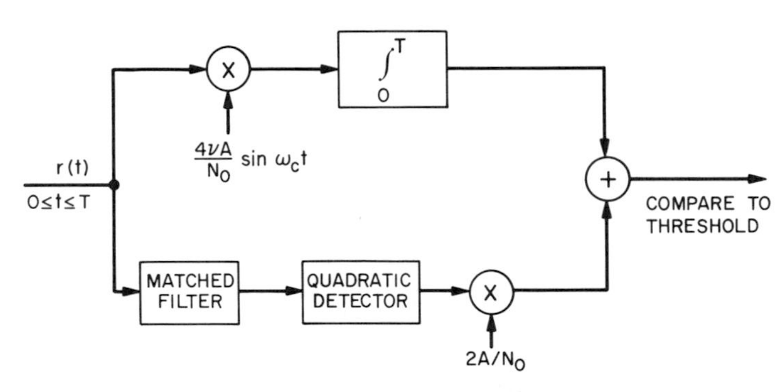

Fig. 7-20 Neyman–Pearson receiver for a particular distribution of phase (Exercise 7.5).

7.6 Consider the hypotheses

$$H_1: \quad r(t) = A \sin(\omega_c t + \theta) + n(t)$$
$$H_0: \quad r(t) = n(t)$$

where ω_c is a constant, θ is a random variable uniformly distributed $(0, 2\pi)$, $n(t)$ is white Gaussian noise with spectral density $N_o/2$.

(a) Let A be a discrete random variable such that $P(A = 0) = 1 - p$ and $P(A = A_o) = p$. Determine the likelihood ratio using the Neyman–Pearson criterion. Can q be used as the test statistic?

(b) Show that the probability of detection is $P_D = (1 - p)P_{fa} + pP_D(A_o)$ where P_{fa} is the probability of false alarm and $P_D(A_o)$ is the probability of detection for a constant amplitude signal of level A_o.

7.7 For the detection problem of the preceding exercise, assume that the amplitude has a distribution

$$p(A) = (1 - p)\,\delta(A) + p \frac{A}{A_o^2} \exp\left(-\frac{A^2}{2A_o^2}\right), \qquad A \geq 0, \quad p \neq 0$$

(a) Determine the likelihood ratio using the Neyman–Pearson criterion. Can q be used as the test statistic?

(b) Show that the probability of detection is

$$P_D = (1 - p)P_{fa} + pP_{fa}^{u}$$

where

$$u = \frac{1}{1 + (TA_o^2/N_o)}$$

Note that as A_o becomes arbitrary large, P_D approaches p (assuming low P_{fa}).

7.8 For the detection problem of Exercise 7.6, assume that

$$p(A) = \sum_{i=1}^{M} p_i \, \delta(A - A_i)$$

Can q be used as the test statistic?

7.9 The probability of detection of a Rayleigh fading signal is given by Eq. (7-35). Rederive this result by considering the following problem. A signal

$$r(t) = a(t)\cos(\omega_c t + \phi(t)) + n(t), \qquad -\infty < t < \infty$$

is passed through a linear filter, $h(\tau) = \cos \omega_c \tau$, $0 \le \tau \le T$, followed by an envelope detector. The signal $a(t)\cos(\omega_c t + \phi(t))$ is a Gaussian process, and $R_a(\tau)$ and $\phi(t)$ are substantially constant over any interval T. Make any reasonable narrowband assumption.

(a) Find the probability that a sample of the detector output is greater than a threshold η.

(b) This result is the same as Eq. (7-35). Explain why this is so.

7.10 Consider the M-ary noncoherent frequency shift keying problem:

$$H_1\colon \; r(t) = A_1 \sin(\omega_1 t + \theta_1) + n(t)$$
$$H_2\colon \; r(t) = A_2 \sin(\omega_2 t + \theta_2) + n(t)$$
$$\vdots$$
$$H_m\colon \; r(t) = A_m \sin(\omega_m t + \theta_m) + n(t)$$

Assume the hypotheses are equally likely, cost functions equal, phases uniformly distributed $(0, 2\pi)$, and $n(t)$ zero mean white Gaussian noise.

(a) Assume that the amplitudes are equal, $A_i = A$. Design a receiver to minimize the probability of error.

(b) Assume that the output of the parallel bank of filters in the receiver are statistically independent. The probability of error is independent of which hypothesis is true. Show that the error probability is equal to

$$P_e = 1 - \sum_{k=0}^{M-1} \frac{(-1)^k \binom{M-1}{k}}{1+k} \exp\left(-\frac{k}{k+1}\frac{E}{N_o}\right)$$

7.11 Repeat the preceding problem but assume the A_i are Rayleigh distributed

$$p(A_i) = \frac{A_i}{A_o^2} \exp\left(-\frac{A_i^2}{2A_o^2}\right)$$

Show that the error probability is given by

$$P_e = 1 - \sum_{k=0}^{M-1} \frac{(-1)^k \binom{M-1}{k}}{1+k} \frac{1}{(1+[k/(k+1)]\varepsilon)}$$

where $\varepsilon = A_o^2 T/N_o$.

7.12 Consider the detection problem

$$H_1: \quad r(t) = A \cos \omega_1 t + B \cos(\omega_2 t + \phi) + n(t), \qquad 0 \le t \le T$$
$$H_0: \quad r(t) = B \cos(\omega_2 t + \phi) + n(t), \qquad 0 \le t \le T$$

where A, B, ω_1, ω_2 are known constants, $n(t)$ is white Gaussian, and ϕ is uniformly distributed over $(0, 2\pi)$. If

$$\int_0^T \cos \omega_1 t \cos \omega_2 t \, dt = \int_0^T \cos \omega_1 t \sin \omega_2 t \, dt = 0$$

show that the optimum receiver can use $\int_0^T r(t) \cos \omega_1 t \, dt$ as the test statistic. Discuss.

7.13 Redo the preceding problem with the change

$$H_1: \quad r(t) = A \cos(\omega_1 t + \theta) + B \cos(\omega_2 t + \phi) + n(t)$$

where θ is uniformly distributed $(0, 2\pi)$ and is statistically independent of ϕ. Assume that

$$\int_0^T \cos(\omega_1 t + \theta) \cos(\omega_2 t + \phi) \, dt = 0$$

for all θ and ϕ. Show that the optimum receiver can use as the test statistic q, where

$$q^2 = \left[\int_0^T r(t) \cos \omega_1 t \, dt\right]^2 + \left[\int_0^T r(t) \sin \omega_1 t \, dt\right]^2$$

Discuss.

7.14 (a) For the following series of problems, the hypotheses are of the form

$$H_1: \ r(t) = A\cos(\omega_0 t + \theta) + [1 + m\cos(\omega_0 t + \phi)]s(t) + n(t)$$
$$H_0: \ r(t) = n(t)$$

where $s(t)$ and $n(t)$ are bandlimited Gaussian processes, $|\omega| \leq 2\pi B = \Omega$, and spectral density $S_o/2$ and $N_o/2$ respectively. Also, $\omega_0 \ll \Omega$ and $m \ll 1$. (The signal consists of an amplitude modulated noiselike signal and the modulating signal is also present as an additive term.) Assume the received signal is sampled at intervals $1/2B$ and the total number of samples is $2BT$. Assume these samples are statistically independent! Define V_n and $V_{sn}(i)$ as the variance of $r_i(= r(t_i))$ under hypotheses H_0 and H_1 (for given θ and ϕ) respectively and assume that $\prod_{i=1}^{2BT} [V_n/V_{sn}(i)]^{1/2}$ is sensibly independent of ϕ.

(i) Show that the conditional likelihood ratio is monotonically related to

$$\lambda(\mathbf{r} \mid \theta, \phi) \sim \exp - \tfrac{1}{2} \sum \left\{ \frac{[r_i - A\cos(\omega_0 i/2B + \theta)]^2}{V_{sn}(i)} - \frac{r_i^2}{V_n} \right\}$$

7.14 (b) For $A = 0$ and ϕ known, show that the likelihood ratio receiver can be implemented as shown in Fig. 7-21. Because $m \ll 1$, assume that

$$\frac{1}{V_n} - \frac{1}{V_{sn}(i)} \approx \alpha\left[\beta + \cos\left(\frac{\omega_0 i}{2B} + \phi\right)\right]$$

where

$$\alpha = \frac{2mS_o}{(S_o + N_o)^2 B} \qquad \text{and} \qquad \beta = \left(\frac{S_o}{N_o}\right)\left(\frac{S_o + N_o}{2mS_o}\right)$$

Discuss the receiver qualitatively.

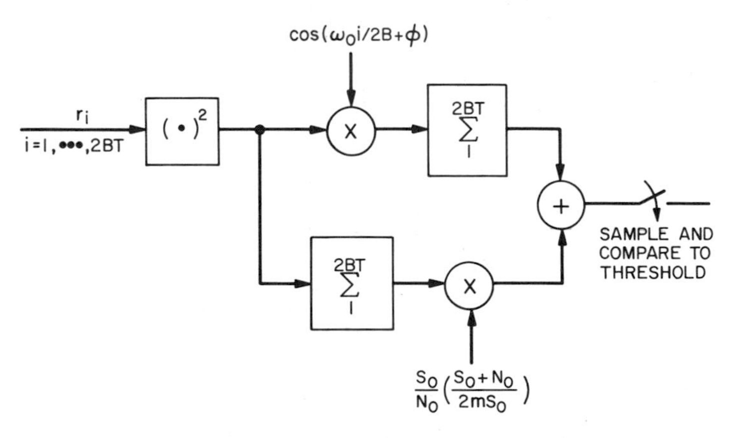

Fig. 7-21 Maximum likelihood receiver ($A = 0$, ϕ known).

7.14 (c) Assume that $A = 0$, and that ϕ is uniformly distributed $(0, 2\pi)$. Show that the receiver may be implemented as shown in Fig. 7-22. Use the small signal assumption that $\ln I_o(v) \simeq v^2/4$. Discuss.

7.14 (d) Assume that A is a constant not equal to zero, and that $\theta = \phi$, a constant. Show that receiver may be implemented as shown in Fig. 7-23. Hint: Use previous assumption for $1/V_n - 1/V_{sn}(i)$ and assume

$$\sum \frac{r_i \cos(\omega_0 i/2B + \phi)}{V_{sn}(i)} \approx \frac{1}{(S_o + N_o)B} \sum r_i \cos(\omega_0 i/2B + \phi)$$

and assume $\sum [\cos^2(\omega_0 i/2B + \phi)]/V_{sn}(i)$ is a constant. Discuss.

7.14 (e) Assume A is a constant not equal to zero, and $\theta = \phi$ is uniformly distributed $(0, 2\pi)$. Using the usual small signal approximation for $I_o(\cdot)$, show that the receiver may be implemented as shown in Fig. 7-24. Discuss.

7.14 (f) Assume A is a constant not equal to zero, and that θ and ϕ are known but unequal. Show that the receiver may be implemented as shown in Fig. 7-25. Discuss.

7.14 (g) Assume A is a constant not equal to zero and that θ and ϕ are statistically independent and uniformly distributed $(0, 2\pi)$. Using the low signal approximation for $I_o(\cdot)$, show that the receiver may be implemented as shown in Fig. 7-26.

7.14 (h) Assume A is a constant not equal to zero, and that $\phi = \theta + \delta$, where δ is known, and θ is uniformly distributed $(0, 2\pi)$. Using the low signal approximation for $I_o(\cdot)$, show that the receiver may be implemented as shown in Fig. 7-27. Discuss.

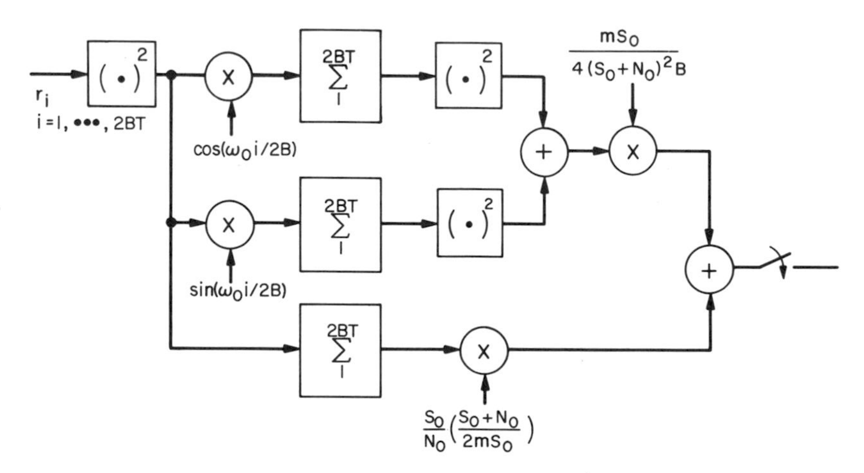

Fig. 7-22 Maximum likelihood receiver [$A = 0$, ϕ uniform $(0, 2\pi)$].

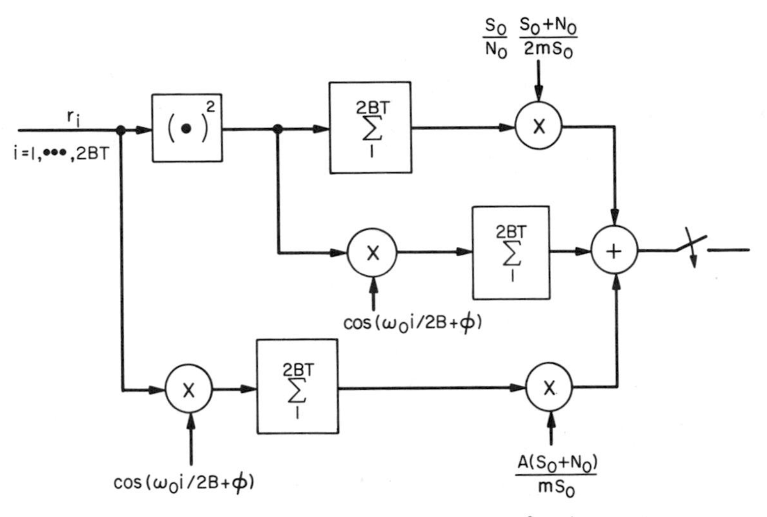

Fig. 7-23 Maximum-likelihood receiver ($\theta = \phi$ known).

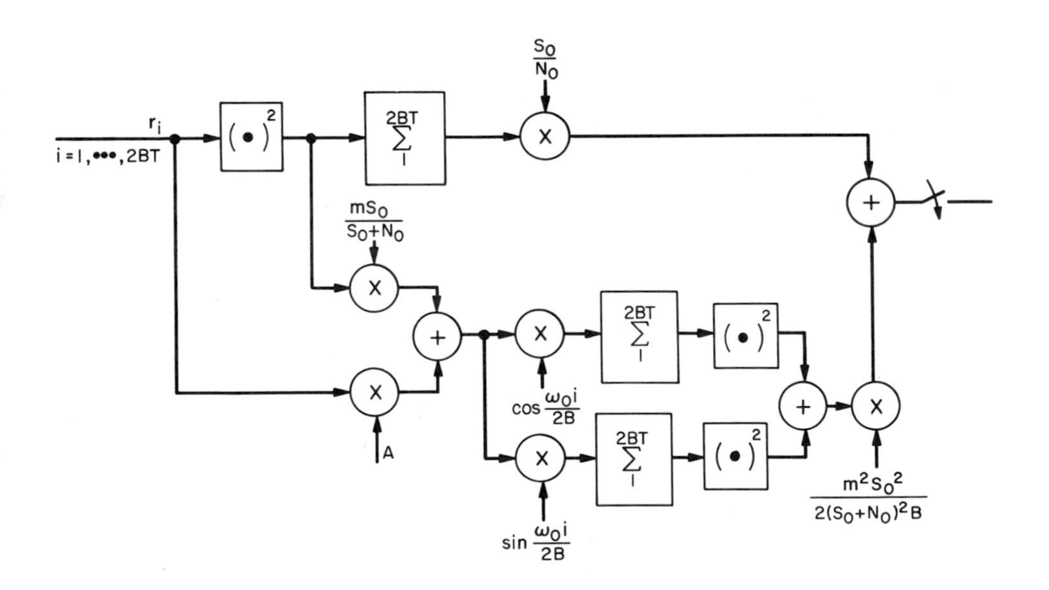

Fig. 7-24 Maximum-likelihood receiver [$\theta = \phi$ uniform $(0, 2\pi)$].

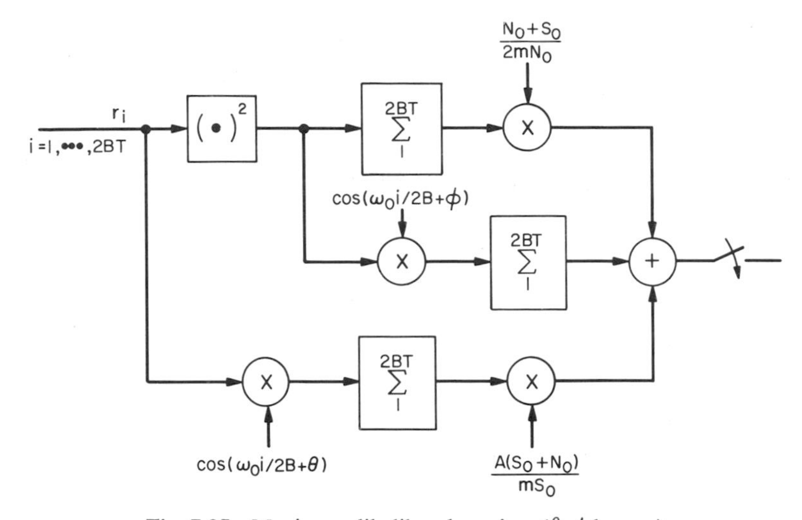

Fig. 7-25 Maximum-likelihood receiver (θ, ϕ known).

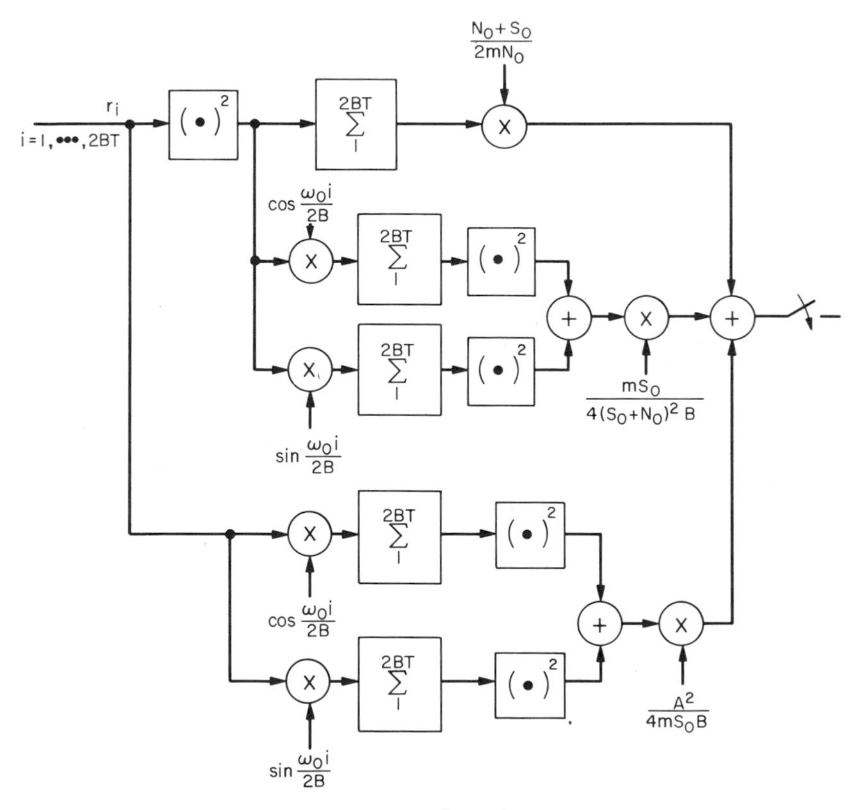

Fig. 7-26 Maximum likelihood receiver [θ and ϕ statistically independent and uniform $(0, 2\pi)$].

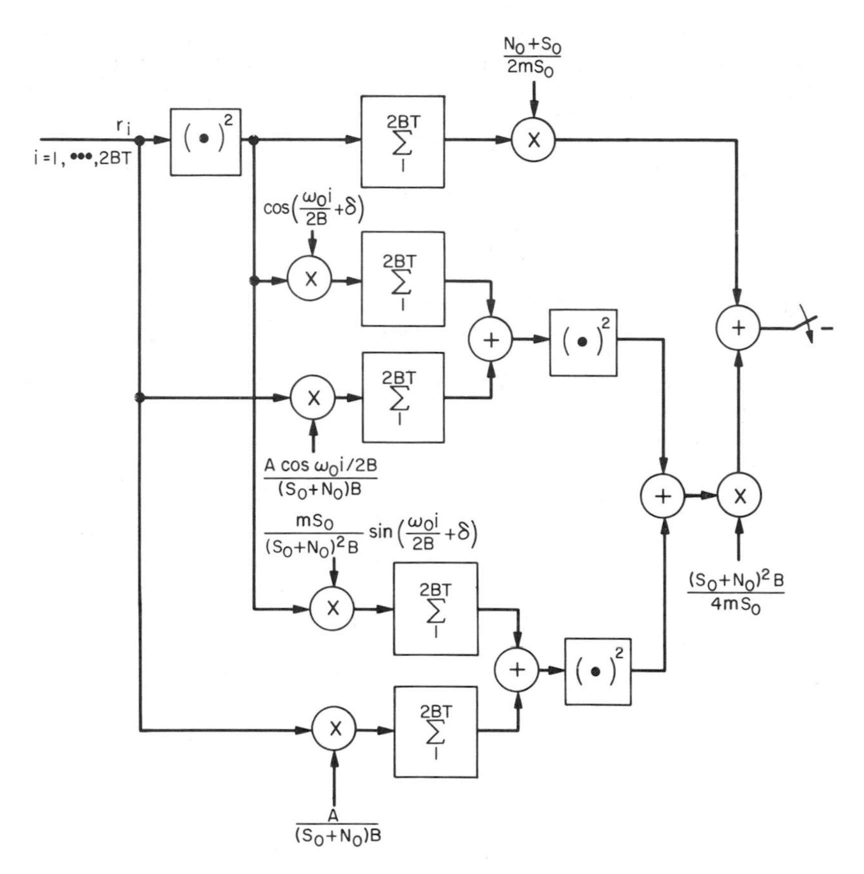

Fig. 7-27 Maximum likelihood receiver [$\phi = \theta + \delta$, δ known, θ uniform $(0, 2\pi)$].

References

1. Peterson, W. W., Birdsall, T. G., and Fox, W. C., The theory of signal detectability, 1954 Symposium on Information Theory, *IRE Trans. Inform. Theory* **PGIT-4,** 171–212 (September, 1954).
2. Abramson, N., and Farison, J., On Applied Decision Theory, Stanford Electronics Laboratory, Tech. Rep. 2005–2, AD 441210 (September 1962).
3. Marcum, J. I., and Swerling, P., A statistical theory of target detection by pulsed radar, *IRE Trans. Inform. Theory* (April 1960).
4. Marcum, J. I., Table of *Q*-Functions, Rand Corp. Rep. RM-339 (January 1950).
5. Skolnik, M. I., "Introduction to Radar Systems." McGraw-Hill, New York, 1962.
6. Helstrom, C. W., "Statistical Theory of Signal Detection." Pergamon Press, Oxford, 1960.
7. Rubin, W. L., and DiFranco, J. V., Radar Detection, Electro-Technology (April 1964).

8. Turin, G. L., The Effects of Pulse Shape and Frequency Separation on FSK Transmission Through Fading, 1958 IRE Nat. Convention Record, Part 8, pp. 217–224 (1958).
9. Reiger, S., Error Probabilities of Binary Data Transmission Systems in the Presence of Random Noise, IRE Convention Record, Part 8—Information Theory (1953).
10. Turin, G. L., Communication Through Noisy, Random Multipath Channels, IRE Convention Record, Part 4, pp. 154–166 (1956).
11. Crawford, A. B., Hogg, D. C., and Kummer, W. H., Studies in troposphere propagation beyond the horizon, *Bell System Tech. J.* (September 1959).
12. Rice, S. O., Distribution of the duration of fades in radio transmission: Gaussian noise model, *Bell System Tech. J.* (May 1958).
13. Sunde, E. D., Digital troposcatter transmission and modulation theory, *Bell System Tech. J.* (January 1964).
14. Bullington, K., Radio propagation fundamentals, *Bell System Tech. J.* **36,** No. 3, 593–626 (May 1957).
15. Kaylor, R. L., A statistical study of selective fading of super-high frequency radio signals, *Bell System Tech. J.* 1187–1202 (September 1953).
16. Brennan, L. E., Reed, I. S., and Sollfrey, W., A comparison of average likelihood and maximum likelihood ratio tests for detecting radar targets of unknown doppler frequency, *IEEE Trans. Inform. Theory* **IT-4,** No. 1, 104–110 (January 1968). Also, Rand Corp. Rep., RM-5198-PR, AD 646366 (December 1966).
17. Hill, F. S. Jr., Detection of Sonar Sinusoids of Unknown Frequency and Known or Unknown Phase, Rand Corp. Rep. RM-4809-ARPA, AD 627645 (December 1965).
18. Viterbi, A. J., Phase locked loop dynamics in the presence of noise by fokker-planck techniques, *Proc. IEEE* **51,** No. 12, 1737–1753 (December 1963).
19. Viterbi, A. J., "Principles of Coherent Communication." McGraw-Hill, New York, 1966.

SUPPLEMENTARY BIBLIOGRAPHY

Texts:

Baghdady, E. J., ed., "Lectures on Communication System Theory." McGraw-Hill, New York, 1961.

Berkowitz, R. S., ed. "Modern Radar, Analysis, Evaluation, and System Design." Wiley, New York, 1965.

Middleton, D., "Topics in Communication Theory." McGraw-Hill, New York, 1965.

Selin, I., "Detection Theory." Princeton Univ. Press, Princeton, New Jersey, 1965.

Schwartz, M., Bennett, W. R., and Stein, S., "Communication Systems and Techniques." McGraw-Hill, New York, 1966.

Van Trees, H. L., "Detection, Estimation, and Modulation Theory," Part I. Wiley, New York, 1968.

Wainstein, L. A., and Zubakov, V. D., "Extraction of Signals From Noise." Prentice Hall, Englewood Cliffs, New Jersey, 1962.

Woodward, P. M., "Probability and Information Theory With Applications to Radar," 2nd Ed., Pergamon Press, Oxford, 1964.

Wozencraft, J. M., and Jacobs, I. M., "Principles of Communication Engineering." Wiley, New York, 1965.

Papers:

Akima, H., The Error Rates in Multiple FSK Systems and the Signal-to-Noise Characteristics of FM and PCM-FS Systems, Nat. Bur. of Std. Tech. Note 167, March 25 (1963).

Anderson, R. R., Bennett, W. R., Davey, J. R., and Salz, J., Differential detection of binary FM, *Bell System Tech. J.* (January 1965).

Arthurs, E., and Dym, H., On the optimum detection of digital signals in the presence of white gaussian noise—A geometric interpretation and a study of three basic data transmission systems, *IRE Trans. Commun. Systems* **CS-10**, No. 4, 336–372 (December 1962).

Fralick, S. C., Slenkovich, G. L., and Wilson, D. L., An adaptive receiver for signals of unknown frequency, *IEEE Trans. Commun. Techn.*, October 1968.

Grettenberg, T. L., Signal selection in communication and radar systems, *IEEE Trans. Inform. Theory*, **IT-9**, No. 4, 265–275 (October 1963).

Levesque, A. H., Detection of Signals of Unknown Frequency, Appendix C of: Processing of Data From Sonar Systems, Vol. III by M. Kanefsky, A. H. Levesque, P. M. Schultheiss, F. B. Tuteur, Yale Univ. Rep. U417–65–033, AD 476591 (August 21, 1965).

Mazo, J. E., and Salz, J., Probability of error for quadratic detectors, *Bell System Tech. J.* (November 1965).

Mazo, J. E., and Salz, J., Theory of error rates for digital FM, *Bell System Tech. J.* (November 1966).

Morakis, J. C., Bandwidth Optimization for Frequency Shift Keying, Goddard Space Flight Center, NASA TN D-3000 (September 1965).

Nuttall, A. H., Error probabilities for equicorrelated M-Ary signals under phase-coherent and phase-incoherent reception, *IRE Trans. Inform. Theory* **IT-8,** No. 4, 305–314 (July 1962).

Price, R., Optimum detection of random signals in noise with application to scatter-multipath communications, I, *IRE Trans. Inform. Theory*, **IT-2,** No. 4, 125–135 (December 1956).

Scholtz, R. A., and Weber, C. L., Signal design for phase incoherent communication, *IEEE Trans. Inform. Theory* **IT-12,** No. 4, 456–463 (October 1966).

Shein, N., Error probability for transmission of M-Ary equicorrelated signals over a phase-incoherent Rayleigh-Fading Channel, correspondence, *IEEE Trans. Inform. Theory* **IT-11,** No. 3, 449, 450 (July 1965).

Swerling, P., Detection of radar echoes in noise revisited, *IEEE Trans. Inform. Theory* **IT-12,** No. 3, 348–361 (July 1966). Also Rand Corp., RM–4586–PR, AD 616700 (May 1965).

Turin, G. L., Error probabilities for binary symmetric ideal reception through nonselective slow fading and noise, *Proc. IRE* (September 1958).

Turin, G. L., The asymptotic behavior of ideal M-Ary systems, *Proc. IRE* **27,** 93–94 (January 1959).

Tuteur, F. B., Detection of Wide-Band Signals Modulated by a Low Frequency Sinusoid, R. A. McDonald, Evaluation of N/S Ratio for a Correlation Detector With Assumed Special Signal Properties, Appendices 5 and 7 of: Processing of Data From Sonar Systems, R. A. McDonald, P. M. Schultheiss, F. B. Tuteur, and T. Usher, Jr., Yale Univ. Rep. No. U417–63–045, AD 420575 (September 1963).

Chapter 8

Multiple Pulse Detection of Signals

8.1 Introduction

In the preceding chapter it was assumed that only a single pulse or received signal was available at the receiver. Based on the signal received in a single time interval, a decision was made that a target was present or not, or that one of several possible signals was present. In many applications, more than one such signal is available to the receiver. For example, in a radar system a number of pulses are transmitted sequentially. The receiver observes the returns of all these pulses before making a decision. In this case the pulses are usually received in different time intervals. A similar circumstance also occurs in communication systems. Take for example the fading FSK system of the last chapter. The error performance is very poor. To get an acceptable error rate, the signal energy would have to be increased prohibitively. In such a case, parallel channels may be used to transmit the same information. For example, when a signal is sent, it may be simultaneously transmitted on four different carrier frequencies sufficiently separated to make the four received signals independent. At the receiver the four signals are "combined" before a decision is made. Such a system employs what is called frequency diversity. Often a substantial improvement in the error performance is realized. Some other forms of diversity are time, space, and polarization diversity.

In this chapter we shall consider the detection problem where many signals are available. Generally, the signals carry the same information and in that sense they are redundant. For example, in the radar case the signals received in the different time intervals are either all assumed to contain a signal reflected by a target or they are all considered to be due to noise. Therefore each signal contains the information that a target is present or not. The benefit to be gained from many pulses derives from the usual assumption that the noise accompanying each is generally statistically independent from that of all others. Furthermore, when considering fading signals, the greatest advantage is obtained when the signals fade independently of each other.

Conceptually, dealing with such signals is exactly the same as for single pulses and we form a likelihood ratio based on a set of M received signal vectors. We shall apply this to several cases of practical interest and generate the appropriate receivers and determine their performance.

8.2 Known Signals

To demonstrate the techniques to be applied for multiple observations, the case of completely known signals will be considered first. Later, some uncertainty in the signal parameters will be allowed.

The problem to be considered is as follows: A radar transmits M successive pulses. The return pulses at the receiver are denoted $s_1(t), s_2(t), \ldots, s_M(t)$. The additive noise sample functions are denoted $n_1(t), \ldots, n_M(t)$. It is assumed that the signals $s_i(t)$ are completely known and that each exists over an interval T seconds long. The noise waveforms are sample functions of white Gaussian noise with spectral density $N_0/2$. The received signals, $r_i(t) = s_i(t) + n_i(t)$, are assumed to be statistically independent. A decision based on these M received signals is to be made to determine if a target is present. The hypotheses are

$$\begin{aligned} &H_1: \quad r_i(t) \text{ contains a signal and noise } (s_i(t) \neq 0) \\ & \qquad\qquad\qquad\qquad\qquad\qquad\qquad\qquad (i = 1, \ldots, M) \\ &H_0: \quad r_i(t) \text{ consists of noise alone, i.e., } (s_i(t) = 0) \end{aligned}$$

The likelihood ratio is

$$\lambda(\mathbf{r}) = \frac{p_1(\mathbf{r}_1, \mathbf{r}_2, \ldots, \mathbf{r}_M)}{p_0(\mathbf{r}_1, \mathbf{r}_2, \ldots, \mathbf{r}_M)} \tag{8-1}$$

Since the received signals were assumed independent, the likelihood function may be written as

$$\lambda(\mathbf{r}) = \frac{p_1(\mathbf{r}_1) \cdot p_1(\mathbf{r}_2) \cdots p_1(\mathbf{r}_M)}{p_0(\mathbf{r}_1) \cdot p_0(\mathbf{r}_2) \cdots p_0(\mathbf{r}_M)} \tag{8-2}$$

or

$$\lambda(\mathbf{r}) = \prod_{i=1}^{M} \frac{p_1(\mathbf{r}_i)}{p_0(\mathbf{r}_i)} = \prod_{i=1}^{M} \lambda_i(\mathbf{r}_i) \tag{8-3}$$

where the symbol $\prod_{i=1}^{M}$ denotes an M-fold product, and $\lambda_i(\mathbf{r}_i)$ represents the likelihood ratio for the ith pulse. Consequently, the likelihood ratio for M statistically independent pulses is the product of the individual likelihood ratios.

Using Eqs. (6-38) and (6-39), the likelihood function for the ith pulse is easily seen to be

$$p_1(\mathbf{r}_i) = F \exp\left\{-(1/N_o) \int_0^T [r_i(t) - s_i(t)]^2 \, dt\right\} \tag{8-4}$$

and

$$p_0(\mathbf{r}_i) = F \exp\left[-(1/N_o) \int_0^T r_i^2(t) \, dt\right] \tag{8-5}$$

The likelihood ratio for the ith pulse is therefore

$$\lambda_i(\mathbf{r}) = \exp\left[-(1/N_o) \int_0^T s_i^2(t) \, dt\right] \exp\left[(2/N_o) \int_0^T r_i(t) s_i(t) \, dt\right]$$

The overall likelihood ratio becomes

$$\lambda(\mathbf{r}) = \prod_{i=1}^{M} \exp(-E_i/N_o) \prod_{i=1}^{M} \exp\left[(2/N_o) \int_0^T r_i(t) s_i(t) \, dt\right] \tag{8-6}$$

where E_i is the signal energy of the ith pulse. Using the log-likelihood ratio, the decision rule becomes: choose H_1 if

$$-(1/N_o) \sum_{i=1}^{M} E_i + (2/N_o) \sum_{i=1}^{M} \int_0^T r_i(t) s_i(t) \, dt \geq \ln \lambda_o \tag{8-7}$$

where λ_o is the threshold which is determined by the criterion we are attempting to satisfy. Rearranging this inequality produces the rule: choose H_1 if

$$\sum_{i=1}^{M} \int_0^T r_i(t) s_i(t) \, dt \geq \tfrac{1}{2} N_o \ln \lambda_o + \tfrac{1}{2} \sum_{i=1}^{M} E_i \tag{8-8}$$

The receiver design is seen to be similar to that for the single pulse case except that there are M such receivers. The outputs of these receivers are added and compared to a threshold. One method of implementing this receiver is shown in Fig. 8-1. Note however that the time of occurrence of the M received signals has not been explicitly accounted for in either Eq. (8-8) or Fig. 8-1. If the pulses occur sequentially, appropriate time delays must be added to the signals before they are summed. Depending on the application this might be accomplished by a beam-forming phased array antenna or simply a tapped delay line with successive taps separated by the pulse spacing. It should be mentioned that the time integration in Fig. 8-1 could also be done in each input branch immediately after the multipliers and before the summing

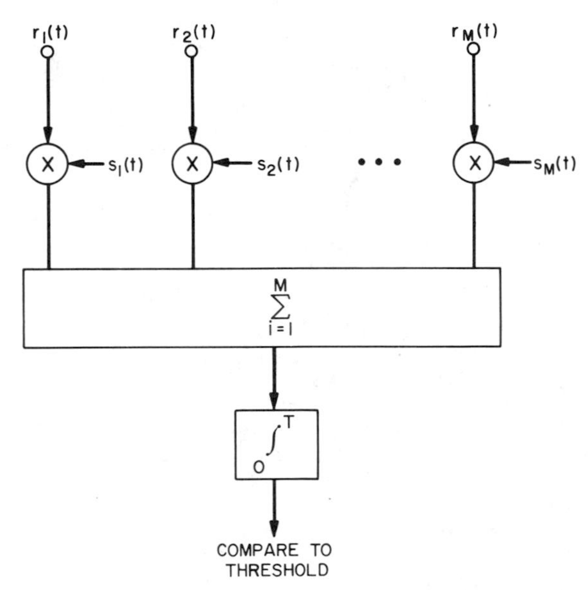

Fig. 8-1 Correlation receiver for known signals in white Gaussian noise.

device. It is then clear that the correlators may be replaced by matched filters and the outputs sampled at time T. Note the contrast of this receiver and that shown in Fig. 6-12 for the M-ary communication system.

It is left as an exercise to show that the performance of the receiver shown in Fig. 8-1 is given by the set of equations

$$P_{\mathrm{fa}} = \int_{\beta}^{\infty} \frac{1}{(2\pi)^{1/2}} e^{-z^2/2}\, dz \qquad \text{and} \qquad P_{\mathrm{D}} = \int_{\beta - (2E_T/No)^{1/2}}^{\infty} \frac{1}{(2\pi)^{1/2}} e^{-z^2/2}\, dz \tag{8-9}$$

where $E_T = \sum_{i=1}^{M} E_i$ and β is adjusted to produce an acceptable probability of false alarm. These results are therefore given by the coherent curves of Fig. 7-5 if E/N_o is replaced by E_T/N_o. For the special case of all signal energies equal we have $E_T/N_o = ME/N_o$ and there is a 3 dB increase in the equivalent performance for each doubling of the number of signals.

8.3 Signals With Random Parameters

We shall again investigate two cases. In each the phase is assumed to be a random variable uniformly distributed over the interval $(0, 2\pi)$. The amplitude in the first case is constant whereas the amplitude in the second is assumed to be a random variable. In both cases, the receiver noise is taken to be white and Gaussian. We shall discuss the radar type problem so that a choice is to be made between a target being present (H_1) and no target (H_0). It is assumed that M radar returns (pulses) are available and based on these pulses, the decision is made. Complete knowledge of the arrival time and frequency of each pulse is assumed.

Known and Equal Amplitudes, Unknown Phase

The signal, if present, is $r_i(t) = A \sin(\omega_c t + \theta_i)$, $(i = 1, \ldots, M)$ where A is constant and θ_i is uniformly distributed $(0, 2\pi)$. Each of the M pulses is assumed statistically independent. The likelihood ratio for the ith pulse is given by Eq. (7-10) and is

$$\lambda_i(\mathbf{r}) = \exp\left(-\frac{A^2T}{2N_o}\right) I_o\left(\frac{2Aq_i}{N_o}\right)$$

where q_i in analogy to Eq. (7-15) is

$$q_i = \left\{\left[\int_0^T r_i(t) \sin \omega_c t \, dt\right]^2 + \left[\int_0^T r_i(t) \cos \omega_c t \, dt\right]^2\right\}^{1/2} \tag{8-10}$$

Due to the pulse-to-pulse independence, the overall likelihood ratio is

$$\lambda(\mathbf{r}) = \prod_{i=1}^{M} \lambda_i(\mathbf{r}) = \exp\left[-\frac{MA^2T}{2N_o}\right] \prod_{i=1}^{M} I_o\left(\frac{2Aq_i}{N_o}\right) \tag{8-11}$$

Using the log-likelihood ratio, the decision rule is to choose H_1 if

$$-\frac{MA^2T}{2N_o} + \sum_{i=1}^{M} \ln I_o\left(\frac{2Aq_i}{N_o}\right) \geq \ln \lambda_o$$

or

$$\sum_{i=1}^{M} \ln I_o\left(\frac{2Aq_i}{N_o}\right) \geq \ln \lambda_o + \frac{MA^2T}{2N_o} \triangleq V_T \tag{8-12}$$

This equation describes the receiver which implements the likelihood ratio. For *each* pulse the receiver determines $\ln I_o(2Aq_i/N_o)$. We have previously seen that q_i may be implemented by a matched filter having an impulse response $h(t) = \sin \omega_c(T - t)$ followed by an envelope detector. Therefore the receiver block diagram is as shown in Fig. 8-2. This then is the "optimum"

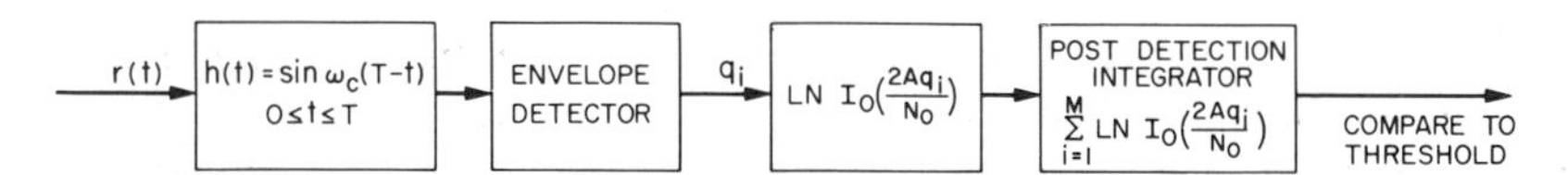

Fig. 8-2 Receiver for incoherent detection of a train of M pulses.

receiver. However, this receiver is moderately complex, particularly in the implementation of the logarithm and the Bessel function. Fortunately, approximations (*1*) may be made which do not seriously affect the overall performance. For small signal-to-noise ratios,

$$I_o\left(\frac{2Aq_i}{N_o}\right) \approx 1 + \left(\frac{Aq_i}{N_o}\right)^2 \tag{8-13}$$

and therefore†

$$\ln I_o\left(\frac{2Aq_i}{N_o}\right) \approx \ln\left[1 + \left(\frac{Aq_i}{N_o}\right)^2\right] \approx \left(\frac{Aq_i}{N_o}\right)^2 \tag{8-14}$$

Substituting this into Eq. (8-12) produces the decision rule: choose H_1 if

$$\sum_{i=1}^{M} q_i^2 \geq (N_o^2/A^2)V_T \tag{8-15}$$

This receiver is therefore implemented with a square-law or quadratic detector whose output is proportional to the square of the envelope of its input. The receiver for small signal-to-noise ratios is shown in Fig. 8-3 A.

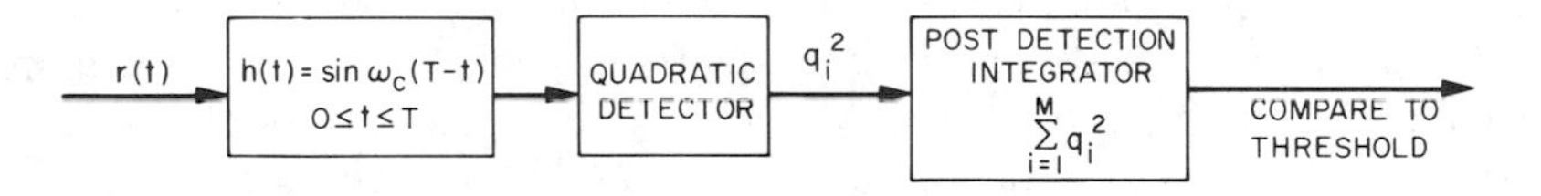

(A) APPROXIMATION TO OPTIMUM RECEIVER FOR LOW SIGNAL-TO-NOISE RATIOS

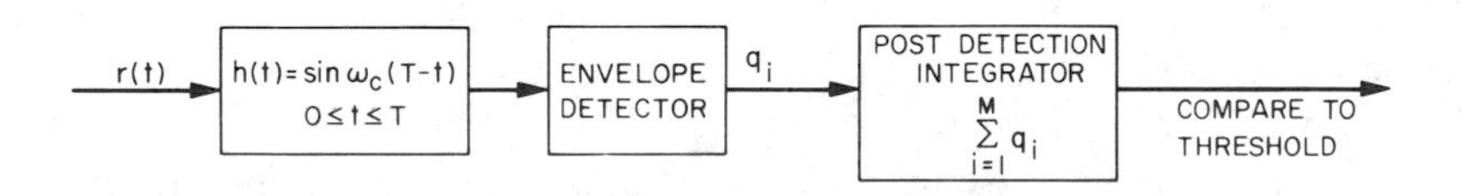

(B) APPROXIMATION TO OPTIMUM RECEIVER FOR HIGH SIGNAL-TO-NOISE RATIOS

Fig. 8-3 Approximate forms of the optimum receiver for noncoherent detection of a train of M pulses.

For large signal-to-noise ratios we may make the approximation

$$I_o\left(\frac{2Aq_i}{N_o}\right) \approx \frac{\exp(2Aq_i/N_o)}{(4\pi Aq_i/N_o)^{1/2}} \tag{8-16}$$

† These approximations are valid for small values of the argument. However, the random variable q_i can take on any positive value including values so large that the approximation may become poor. However, for sufficiently low signal-to-noise ratios, the probability of this occurring is negligible and the approximation is valid "almost always."

so that

$$\ln I_0\left(\frac{2Aq_i}{N_0}\right) \approx \frac{2Aq_i}{N_0} - \frac{1}{2}\ln\left(\frac{4\pi Aq_i}{N_0}\right) \approx \frac{2Aq_i}{N_0} \tag{8-17}$$

The decision rule for large signal-to-noise ratios is then: choose H_1 if

$$\sum_{i=1}^{M} q_i \geq \frac{N_0}{2A} V_T \tag{8-18}$$

This receiver is shown in Fig. 8-3B, and uses a linear detector whose output is proportional to the envelope of the input. The difference in performance between the linear and quadratic detectors will be discussed shortly. For the present it is sufficient to say that there is little difference in their relative performance. In radar terminology, the output of the detector is referred to as "video." The summation of the signals at the detector output is often referred to as video integration or postdetection integration or summation.

Performance for Quadratic Detector

In practice, it is often easier to implement an envelope detector than a quadratic detector. However the analysis for an envelope detector is quite difficult. This arises in part because the probability density of the sum of Rayleigh or Rician variables is not known in closed form. On the other hand, the quadratic detector is analytically tractable. For this reason, analysis is most often based on a quadratic detector assumption. This is reasonable since the difference in performance is negligible for most cases of interest.

We shall discuss here the analysis for a quadratic detector but shall display below Robertson's (*2*) results for the linear detector and his comparison of the linear and quadratic detectors.

For the quadratic detector, the statistic is $Q = \sum_{i=1}^{M} q_i^2$. For convenience we shall use the normalized statistic

$$G = \sum_{i=1}^{M} q_i^2/\sigma_T^2 \tag{8-19}$$

where $\sigma_T^2 = N_0 T/4$. Now, $q_i^2 = x_i^2 + y_i^2$ where x_i and y_i follow from Eq. (7-17). That is

$$x_i = \int_0^T r_i(t) \sin \omega_c t \, dt \qquad \text{and} \qquad y_i = \int_0^T r_i(t) \cos \omega_c t \, dt$$

For a given value of the phase θ, or when noise only is present these are Gaussian variables with variances equal to $\sigma_T^2 = N_0 T/4$ (Eq. 7-20). With signal present, the expected values are $(AT/2)\cos\theta$ and $(AT/2)\sin\theta$ respectively (Eq. 7-19). From the results of Chap. 4, it follows that G is noncentrally χ^2

distributed with $2M$ degrees of freedom and has a noncentral parameter equal to

$$\nu = MA^2T^2/4\sigma_T{}^2 = 2ME/N_o \tag{8-20}$$

With no signal present G has a central χ^2 distribution with $2M$ degrees of freedom. Therefore, the distributions for G are from Eq. (4-77)

$$p_1(G) = \frac{1}{2}\left(\frac{G}{\nu}\right)^{(M-1)/2} \exp\left(-\frac{G}{2} - \frac{\nu}{2}\right) I_{M-1}((G\nu)^{1/2}) \tag{8-21}$$

and from Eq. (4-65)

$$p_0(G) = \frac{1}{2^M \Gamma(M)} G^{M-1} e^{-G/2} \tag{8-22}$$

Denoting the decision threshold as G_T, the false alarm probability is

$$\begin{aligned} P_{fa} &= \int_{G_T}^{\infty} \frac{G^{M-1} e^{-G/2}}{2^M \Gamma(M)}\, dG = 1 - \int_0^{G_T} \frac{G^{M-1} e^{-G/2}}{2^M \Gamma(M)}\, dG \\ &= 1 - I\left(\frac{G_T}{2M^{1/2}}, M-1\right) \end{aligned} \tag{8-23}$$

where $I(u, p)$ is Pearson's form of the incomplete Gamma function (*3*) and is tabulated. Pachares (*4*) has also tabulated the false alarm probability as a function of the number of pulses (M) and the threshold setting against which $\sum_{i=1}^{M} q_i{}^2$ is compared. (See also Fig. 4-11 for the cumulative χ^2 distribution.)

The probability of detection is

$$P_D = \int_{G_T}^{\infty} \frac{1}{2}\left(\frac{G}{\nu}\right)^{(M-1)/2} \exp\left(-\frac{G}{2} - \frac{\nu}{2}\right) I_{M-1}((G\nu)^{1/2})\, dG \tag{8-24}$$

After some work this can be put in the form

$$P_D = Q_M\left(\left(\frac{2ME}{N_o}\right)^{1/2}, (G_T)^{1/2}\right) \tag{8-25}$$

where

$$Q_M(\alpha, \beta) = \int_{\beta}^{\infty} z\left(\frac{z}{\alpha}\right)^{M-1} \exp\left(\frac{z^2 + \alpha^2}{-2}\right) I_{M-1}(\alpha z)\, dz \tag{8-26}$$

is the generalized Marcum Q-function (*5*), $I_N(x)$ is the modified Bessel function of order N, and $\alpha = \nu^{1/2}$. The function is, of course, one minus the cumulative probability distribution of the noncentral χ^2 distribution. Selected plots of this can be found in the literature (*5–8*). A particular series expansion used by Robertson (*9*) makes this function amenable to computer evaluation.

Performance for Linear Detector

As already mentioned, an analytical determination of the performance of the receiver which uses a linear detector is difficult. Nevertheless, the performance can be determined by representing the probability density function by a Gram–Charlier series (*5*, *6*, *10*) which requires a knowledge of the moments of the distribution. The greater the number of moments used, the greater is the accuracy inherent in this method. Thus, before presenting Robertson's results for the linear detector, we shall digress for a discussion of the Gram–Charlier series.

Gram–Charlier Series (*5*, *6*, *10*)

We shall eventually show that this is a series expansion of a probability density function in terms of the Gaussian function, its derivatives, and the moments of the original density function. (This provides one method of reconstructing a density function from knowledge of its moments only.) In the derivation below we confine ourselves to only those distributions for which the moments of all orders exist.

We wish to derive an expansion for a probability density function $w(x)$. Consider the definition of the characteristic function

$$C(j\omega) = \int w(x) e^{j\omega x}\, dx$$

and expand the exponential in a power series in $j\omega$. For notational convenience substitute p for $j\omega$. Then the natural logarithm of the characteristic function is

$$\begin{aligned}\ln C(p) &= \ln \int w(x)\left[1 + px + \frac{p^2x^2}{2!} + \cdots + \frac{p^nx^n}{n!} + \cdots\right] dx \\ &= \ln\left(1 + p\mu_1 + \frac{p^2\mu_2}{2!} + \cdots + \frac{p^n\mu_n}{n!} + \cdots\right)\end{aligned}$$

where the moments $\mu_n = E\{x^n\}$ are assumed finite for all n. Now, expanding $\ln C(p)$ in a Taylor series in p about the point $p = 0$ we get

$$\ln C(p) = \mu_1 p + \frac{\sigma^2}{2!} p^2 + \sum_{k=3}^{\infty} \frac{\gamma_k p^k}{k!} \tag{8-27}$$

where σ^2 is the variance of x, and γ_k is the kth cumulant or semiinvariant (*6*). We shall not dwell on these cumulants since the final series expansion will not contain them explicitly. Denoting

$$r(p) = \sum_{k=3}^{\infty} \frac{\gamma_k p^k}{k!} \tag{8-28}$$

the characteristic function becomes

$$C(p) = \exp\left(\mu_1 p + \frac{\sigma^2}{2} p^2\right) e^{r(p)}$$

Expand the second exponential

$$e^{r(p)} = 1 + r(p) + \frac{r^2(p)}{2!} + \cdots + \frac{r^n(p)}{n!} + \cdots$$

$$= 1 + \sum_{i=1}^{\infty} \frac{r^i(p)}{i!} = 1 + \sum_{i=1}^{\infty} c_i p^i \tag{8-29}$$

To determine the c_i we write out a few typical terms

$$r(p) = \sum_{k=3}^{\infty} \frac{\gamma_k p^k}{k!} = \frac{\gamma_3 p^3}{3!} + \frac{\gamma_4 p^4}{4!} + \frac{\gamma_5 p^5}{5!} + \cdots + \frac{\gamma_n p^n}{n!} + \cdots$$

$$\frac{r^2(p)}{2!} = \frac{1}{2!} \sum_{i=3}^{\infty} \sum_{k=3}^{\infty} \frac{\gamma_i \gamma_k p^{k+i}}{i!\,k!} = \frac{\gamma_3{}^2 p^6}{2!(3!)^2} + \frac{\gamma_3 \gamma_4 p^7}{3!\,4!} + \frac{\gamma_4{}^2 p^8}{2!(4!)^2} + \cdots$$

$$\frac{r^3(p)}{3!} = \frac{1}{3!} \sum_{i=3}^{\infty} \sum_{j=3}^{\infty} \sum_{k=3}^{\infty} \frac{\gamma_i \gamma_j \gamma_k}{i!\,j!\,k!} p^{i+j+k} = \frac{\gamma_3{}^3}{(3!)^4} p^9 + \frac{\gamma_4 \gamma_3{}^2 p^{10}}{4!(3!)^2 2!} + \cdots$$

Grouping terms in equal powers of p produces coefficients such as

$$c_1 = 0, \qquad c_6 = \frac{\gamma_6}{6!} + \frac{\gamma_3{}^2}{2!(3!)^2}$$

$$c_2 = 0, \qquad c_7 = \frac{\gamma_7}{7!} + \frac{\gamma_3 \gamma_4}{3!\,4!}$$

$$c_3 = \frac{\gamma_3}{3!}, \qquad c_8 = \frac{\gamma_8}{8!} + \frac{\gamma_4{}^2}{2!(4!)^2}$$

$$c_4 = \frac{\gamma_4}{4!}, \qquad c_9 = \frac{\gamma_9}{9!} + \frac{\gamma_3{}^3}{(3!)^2} + \frac{\gamma_4 \gamma_5}{4!\,5!}$$

$$c_5 = \frac{\gamma_5}{5!},$$

In terms of the original notation, $j\omega$, the characteristic function may now be expressed as

$$C(j\omega) = \exp(j\omega\mu_1 - \tfrac{1}{2}\omega^2\sigma^2)\left[1 + \sum_{i=3}^{\infty} c_i (j\omega)^i\right] \tag{8-30}$$

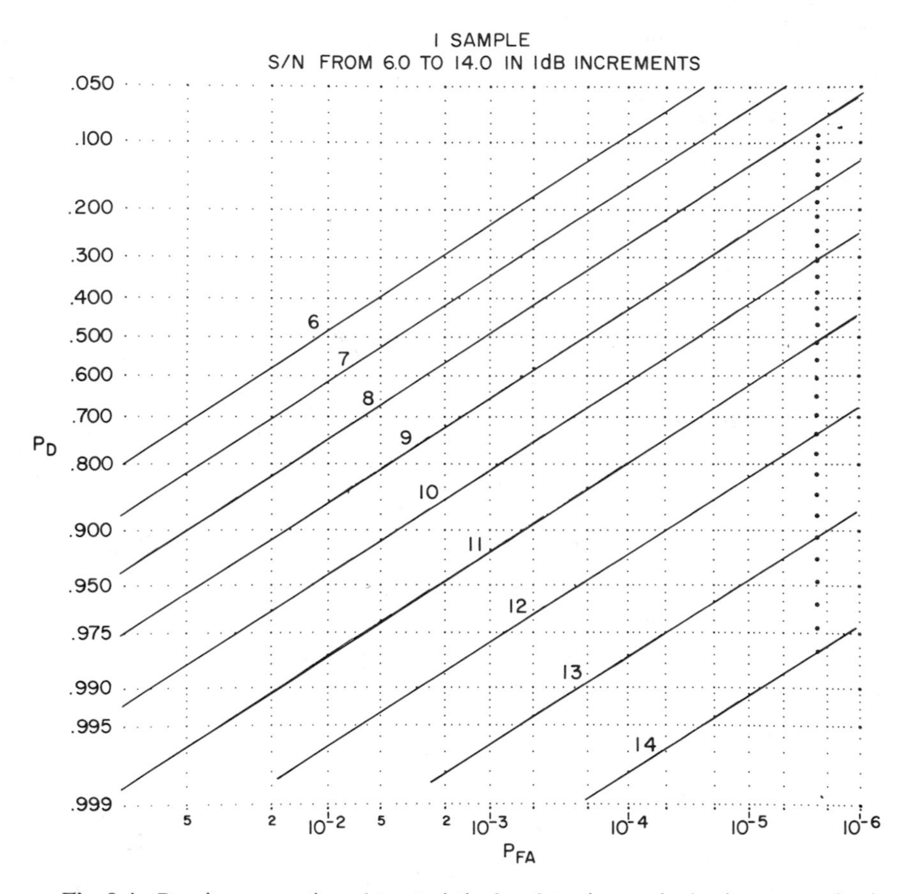

Fig. 8-4 Receiver operating characteristic for detecting a single sinewave pulse in white Gaussian noise. [From Robertson (*2*); copyright, 1967, The American Telephone and Telegraph Co., reprinted by permission.]

We are now in a position to take the inverse Fourier transform of $C(j\omega)$ to produce the density function $w(x)$. Note that the exponential multiplier is the characteristic function of a Gaussian process. Denoting

$$\phi(x) = \frac{1}{(2\pi)^{1/2}} e^{-x^2/2}$$

the multiplier has a transform $(1/\sigma)\ \phi((x - \mu_1)/\sigma)$. Furthermore, the Fourier transform of the product of the multiplier and a term like $(j\omega)^n$ is equal to

$$\frac{(-1)^n}{\sigma} \frac{d^n}{dx^n} \phi\left(\frac{x - \mu_1}{\sigma}\right) = \frac{(-1)^n}{\sigma^{n+1}} \phi^{(n)}\left(\frac{x - \mu_1}{\sigma}\right) \tag{8-31}$$

where $\phi^{(n)}((x - \mu_1)/\sigma)$ is the nth derivative of $\phi(t)$ evaluated at $t = (x - \mu_1)/\sigma$.

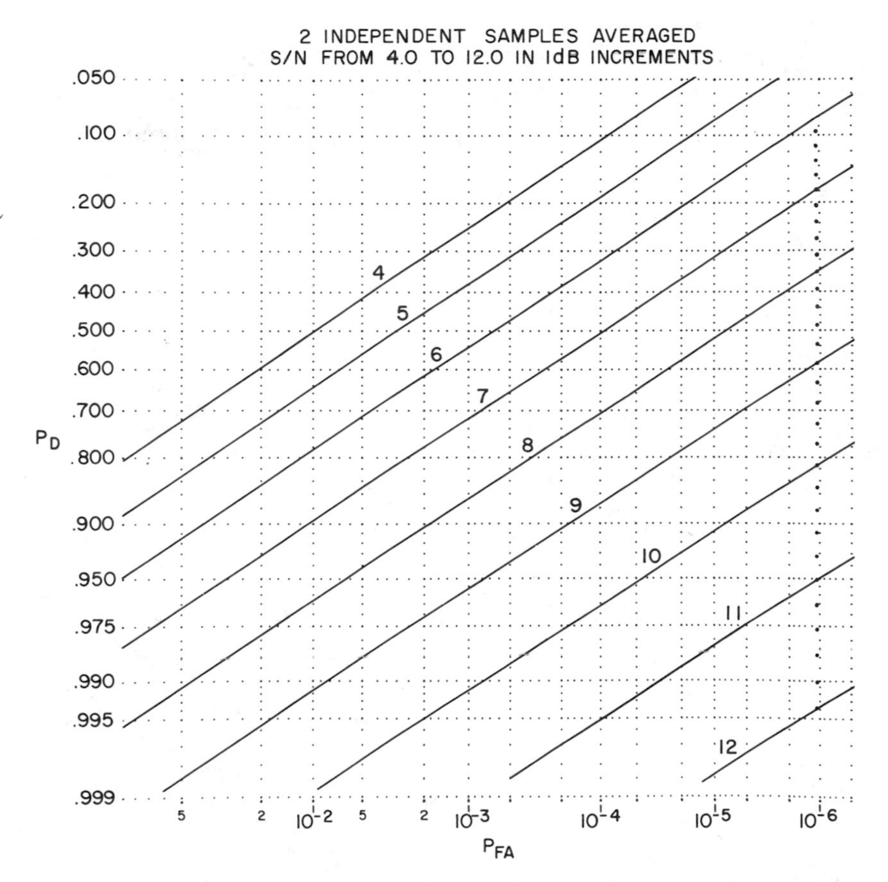

Fig. 8-5 Receiver operating characteristics for detecting sinewaves in white Gaussian noise (2 samples averaged). [From Robertson (2); copyright, 1967, The American Telephone and Telegraph Co., reprinted by permission.]

The density function may now be expressed as

$$w(x) = \frac{1}{\sigma}\left[\phi\left(\frac{x-\mu_1}{\sigma}\right) + \sum_{n=3}^{\infty} \frac{(-1)^n c_n}{\sigma^n}\,\phi^{(n)}\left(\frac{x-\mu_1}{\sigma}\right)\right] \tag{8-32}$$

This is the *Gram–Charlier* series and it is seen to be an expansion of a density function in terms of the Gaussian function and its derivatives. An expansion of this type has much intuitive appeal particularly when the random variable x is the sum of a number of random variables. As the number of variables in such a sum increases, the central limit theorem tells us that subject to conditions, cf Chap. 4, the sum is asymptotically Gaussian. The first term of the Gram–Charlier series exhibits the Gaussian form.

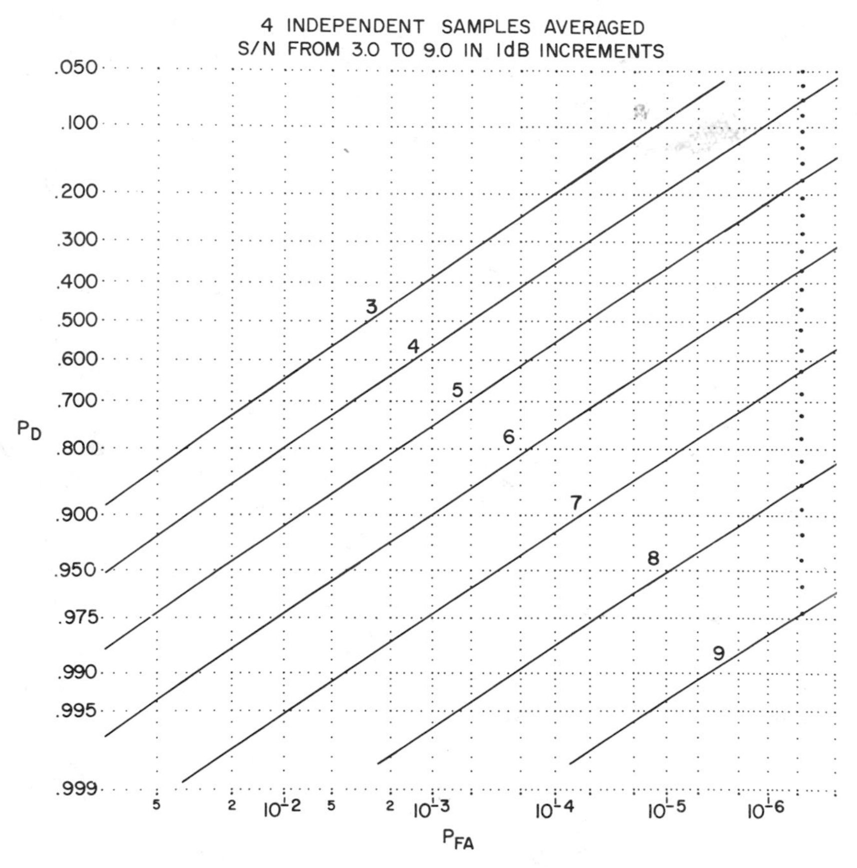

Fig. 8-6 Receiver operating characteristics for detecting sinewaves in white Gaussian noise (4 samples averaged). [From Robertson (*2*); copyright, 1967, The American Telephone and Telegraph Co., reprinted by permission.]

We now normalize the representation for the density function and derive a recursive way to generate the series coefficients. Express the series as

$$w(x) = \frac{1}{\sigma}\left[\phi\left(\frac{x-\mu_1}{\sigma}\right) + a_3\,\phi^{(3)}\left(\frac{x-\mu_1}{\sigma}\right) + \cdots + a_n\,\phi^{(n)}\left(\frac{x-\mu_1}{\sigma}\right) + \cdots\right] \tag{8-33}$$

We wish to derive explicit expressions for the coefficients a_n. To simplify the discussion, define the normalized variable

$$y = \frac{x-\mu_1}{\sigma}$$

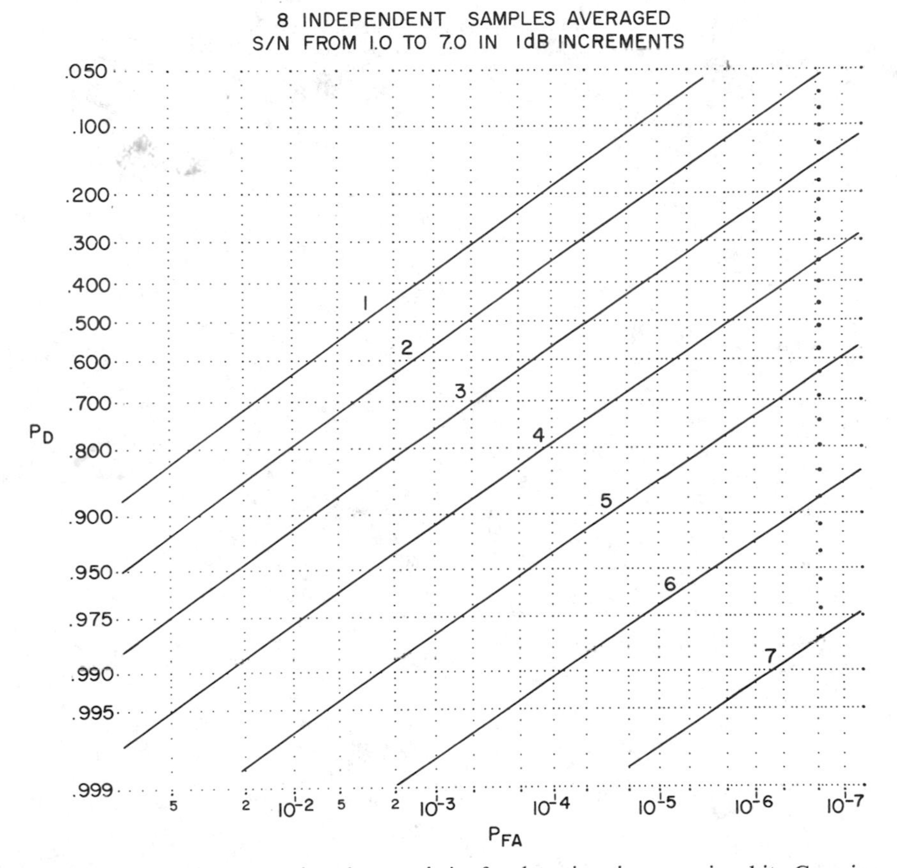

Fig. 8-7 Receiver operating characteristics for detecting sinewaves in white Gaussian noise (8 samples averaged). [From Robertson (*2*); copyright, 1967, The American Telephone and Telegraph Co., reprinted by permission.]

and denote its moments as

$$\nu_n = E\{y^n\} = E\left\{\left(\frac{x - \mu_1}{\sigma}\right)^n\right\}$$

These are the normalized central moments of x, and in particular, $\nu_1 = 0$ and $\nu_2 = 1$. With this change of variables we get the series

$$w(y) = \phi(y) + a_3\phi^{(3)}(y) + a_4\phi^{(4)}(y) + \cdots + a_n\phi^{(n)}(y) + \cdots \tag{8-34}$$

By direct evaluation of the derivatives it may be shown that

$$\phi^{(n)}(y) = (-1)^n\phi(y)H_n(y)$$

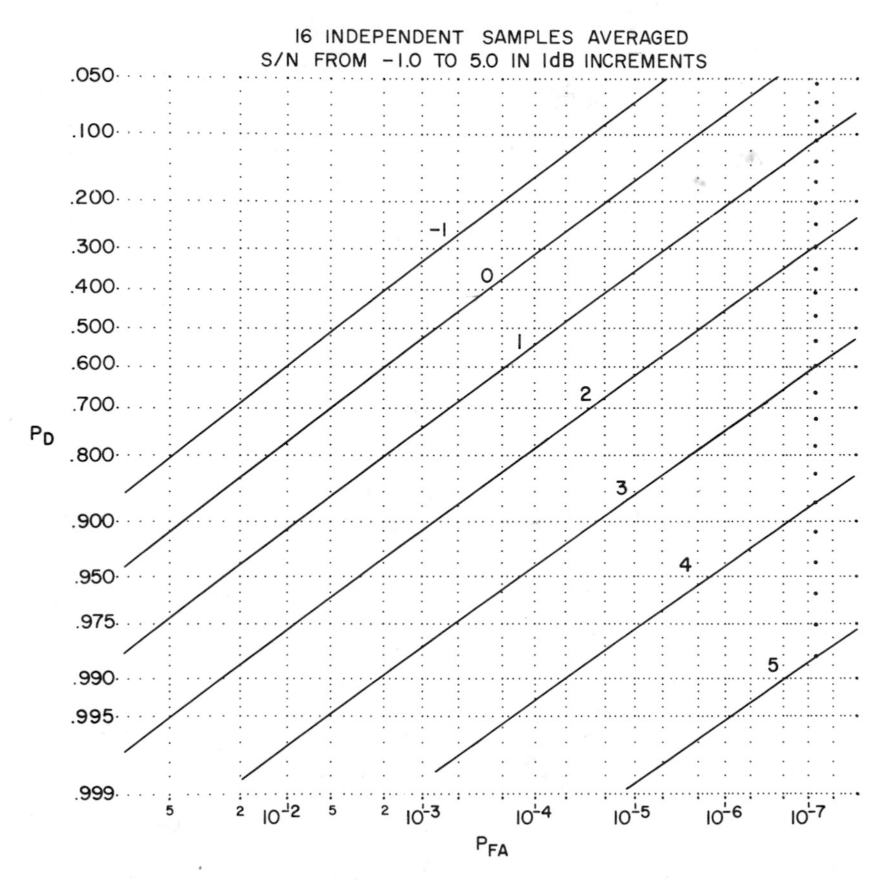

Fig. 8-8 Receiver operating characteristics for detecting sinewaves in white Gaussian noise (16 samples averaged). [From Robertson (*2*); copyright, 1967, The American Telephone and Telegraph Co., reprinted by permission.]

where the $H_n(y)$ are the Hermite Polynomials (*1*, *10*). Some typical polynomials are (*10*)

$$H_0(y) = 1, \qquad H_4(y) = y^4 - 6y^2 + 3$$

$$H_1(y) = y, \qquad H_5(y) = y^5 - 10y^3 + 15y$$

$$H_2(y) = y^2 - 1, \qquad H_6(y) = y^6 - 15y^4 + 45y^2 - 15$$

$$H_3(y) = y^3 - y, \qquad H_7(y) = y^7 - 21y^5 + 105y^3 - 105y$$

A recursive relation for these polynomials is

$$H_{n+1}(y) = yH_n(y) - nH_{n-1}(y)$$

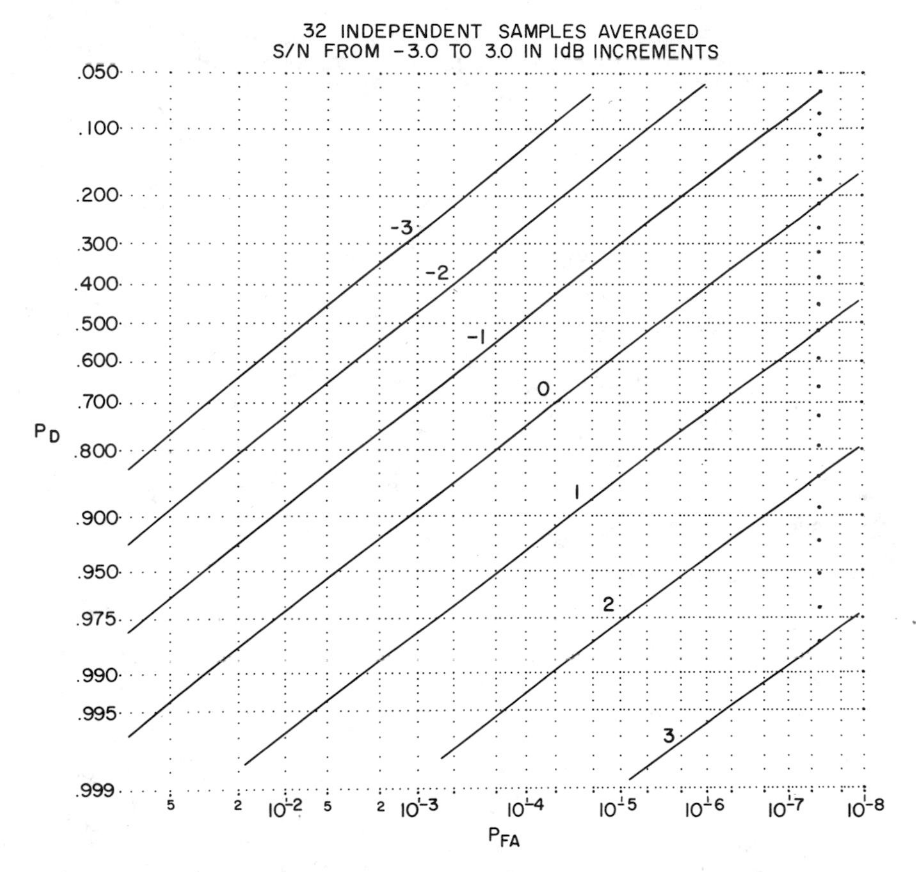

Fig. 8-9 Receiver operating characteristics for detecting sinewaves in white Gaussian noise (32 samples averaged). [From Robertson (*2*); copyright, 1967, The American Telephone and Telegraph Co., reprinted by permission.]

A particularly useful fact is that the Hermite polynomials and the derivatives of the Gaussian function are biorthogonal. That is,

$$\int_{-\infty}^{\infty} H_m(y)\phi^{(n)}(y)\, dy = (-1)^n n!\, \delta_{mn}, \qquad n, m = 0, 1, \ldots$$

where δ_{mn} is the Kronecker delta function ($\delta_{mn} = 1$ for $m = n$ and is zero otherwise). This allows easy determination of the a_n coefficients. Multiplying each side of Eq. (8-34) by $H_n(y)$ and integrating over the real line we find

$$a_n = \frac{(-1)^n}{n!} \int_{-\infty}^{\infty} w(y)H_n(y)\, dy \tag{8-35}$$

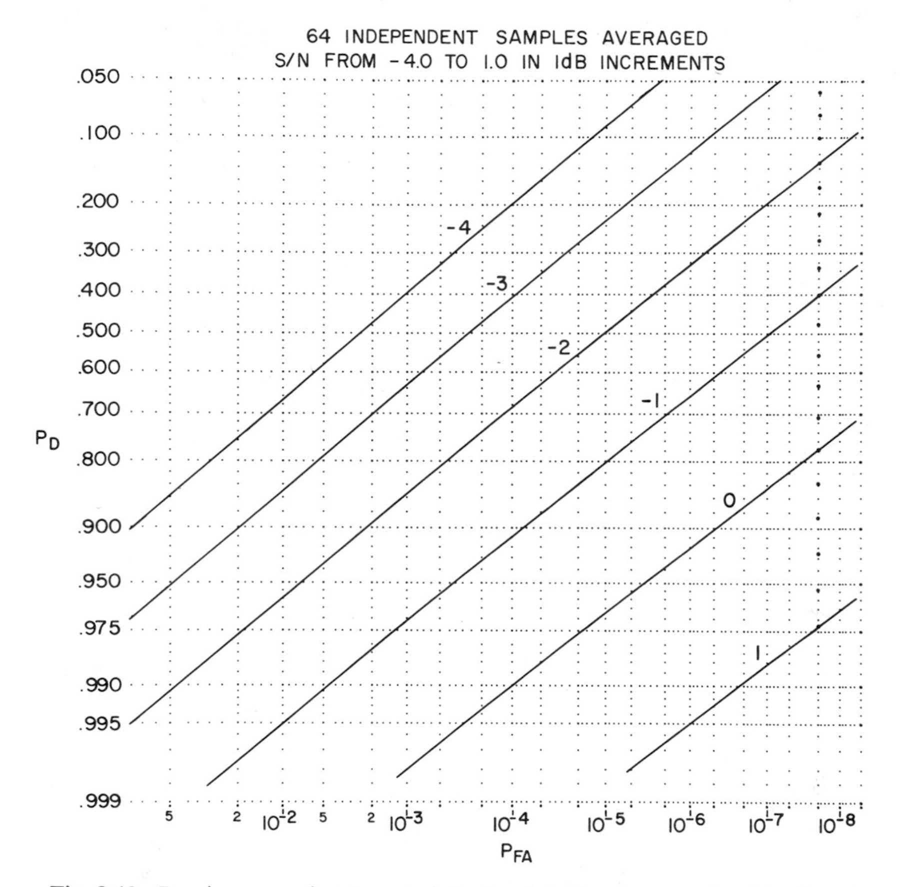

Fig. 8-10 Receiver operating characteristics for detecting sinewaves in white Gaussian noise (64 samples averaged). [From Robertson (*2*); copyright, 1967, The American Telephone and Telegraph Co., reprinted by permission.]

In this form, the coefficients are directly related to the central moments of the probability function being considered.

Substituting the various Hermite polynomials into Eq. (8-35) we have for the coefficients, (recall that $v_1 = 0$ and $v_2 = 1$)

$$a_0 = 1, \qquad a_5 = -(v_5 - 10v_3)/5!$$

$$a_1 = 0, \qquad a_6 = (v_6 - 15v_4 + 30)/6!$$

$$a_2 = 0, \qquad a_7 = -(v_7 - 21v_5 + 105v_3)/7!$$

$$a_3 = -v_3/3!, \qquad a_8 = (v_8 - 28v_6 + 210v_4 - 315)/8!$$

$$a_4 = (v_4 - 3)/4!,$$

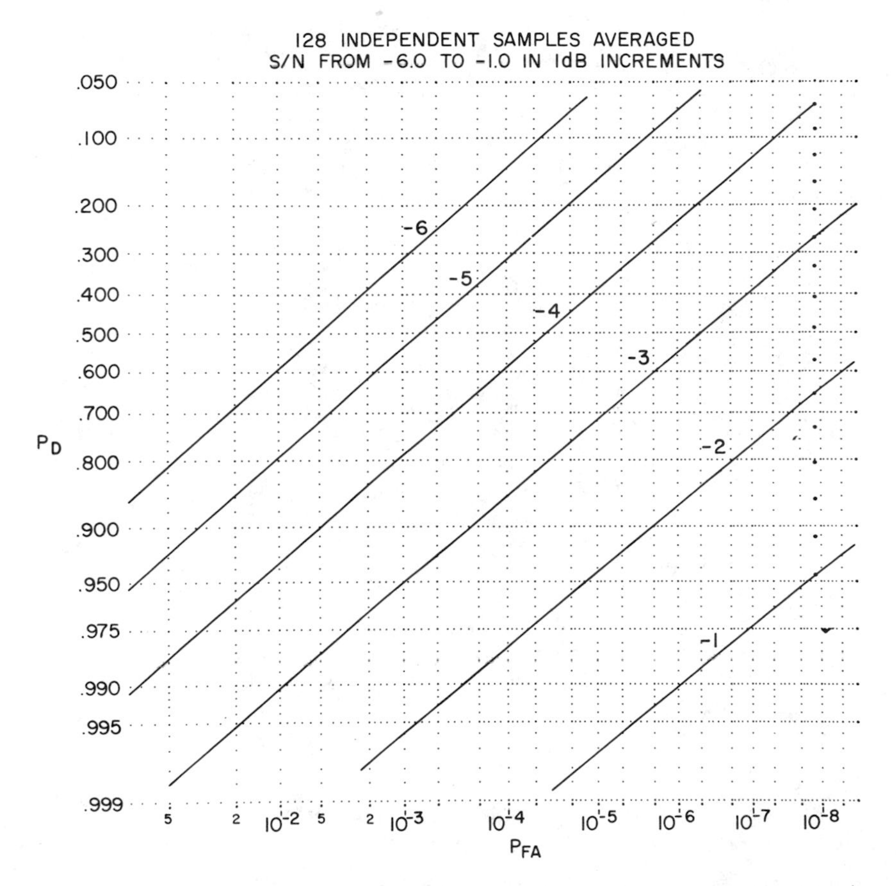

Fig. 8-11 Receiver operating characteristics for detecting sinewaves in white Gaussian noise (128 samples averaged). [From Robertson (*2*); copyright, 1967, The American Telephone and Telegraph Co., reprinted by permission.]

It has been pointed out (*10*) that when a limited number of coefficients are available it may be better to use specific groupings of the coefficients. In particular the grouping of terms

$$0,\ 3$$
$$0,\ 3,\ 4,\ 6$$
$$0,\ 3,\ 4,\ 6,\ 5,\ 7,\ 9$$
$$\cdots$$

is called an Edgeworth series.

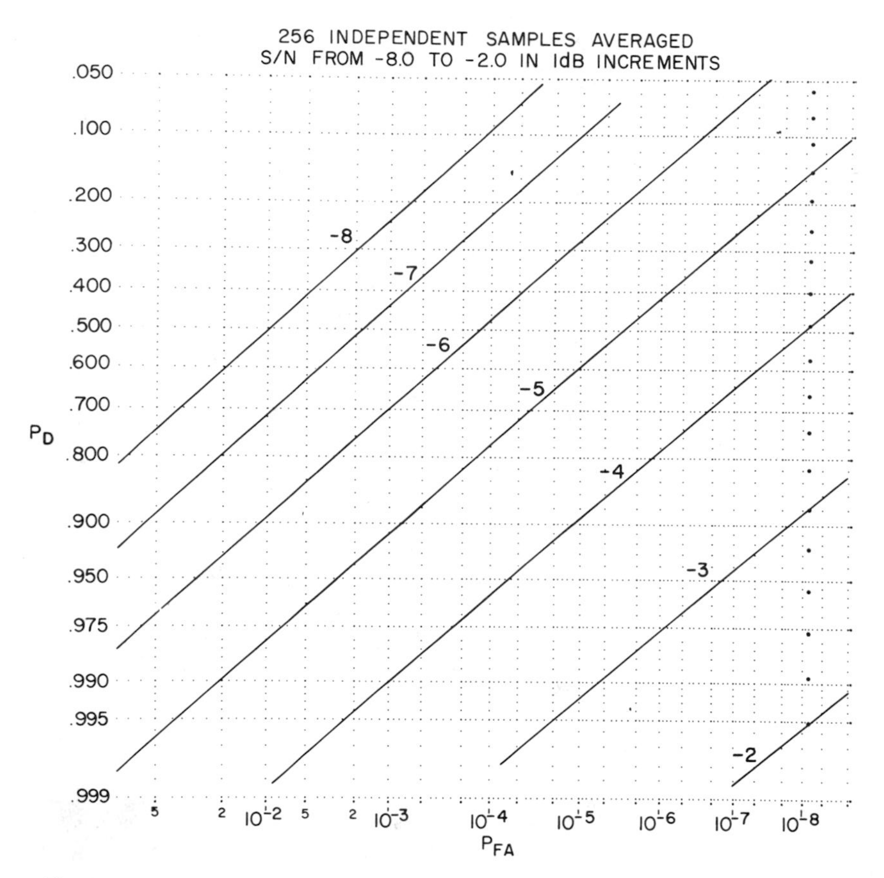

Fig. 8-12 Receiver operating characteristics for detecting sinewaves in white Gaussian noise (256 samples averaged). [From Robertson (*2*); copyright, 1967, The American Telephone and Telegraph Co., reprinted by permission.]

There are at least two areas where a Gram–Charlier series is useful. One is for constructing an empirical density function from experimental measurement of a finite number of moments. A second occurs when the integral over a portion of a known density function cannot be evaluated in closed form. In such cases a series expansion of the function may prove useful. We shall develop such an expression using the Gram–Charlier series. In particular we wish to determine the area in the interval $y \le Y < \infty$ for the series expansion in Eq. (8-34). This is readily determined to be

$$\int_y^\infty w(Y)\,dY = \tfrac{1}{2}\,\mathrm{erfc}\left(\frac{y}{2^{1/2}}\right) - \sum_{n=3}^{\infty} a_n \phi^{(n-1)}(y)$$

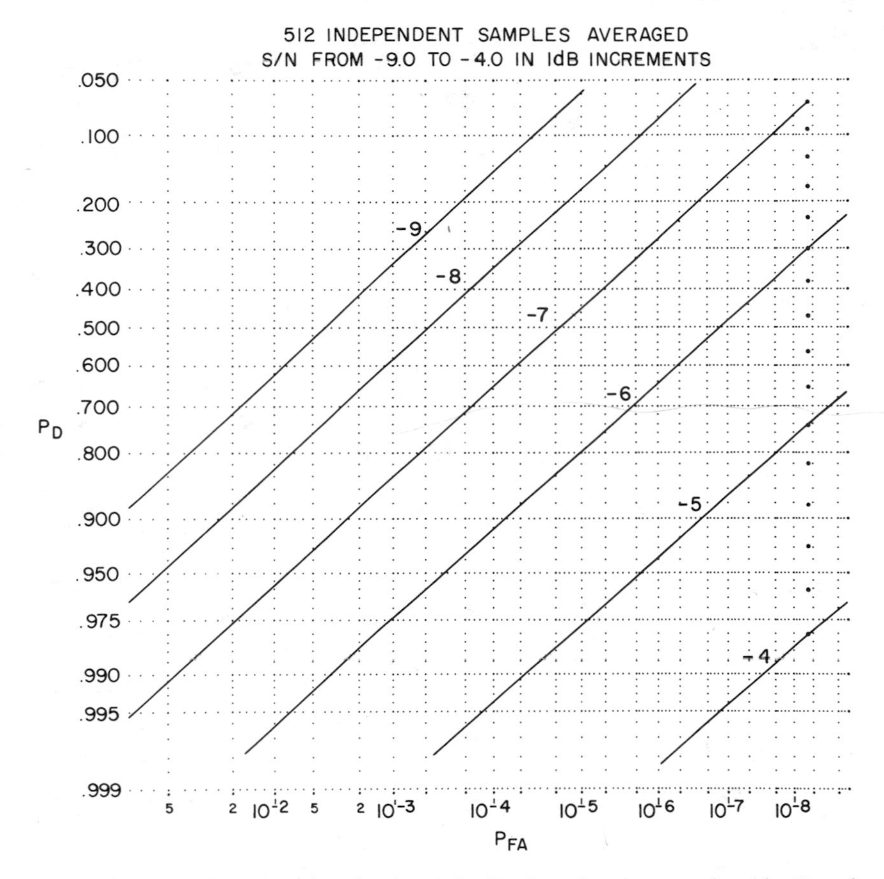

Fig. 8-13 Receiver operating characteristics for detecting sinewaves in white Gaussian noise (512 samples averaged). [From Robertson (*2*); copyright, 1967, The American Telephone and Telegraph Co., reprinted by permission.]

where erfc(x) is the complementary error function

$$\operatorname{erfc}(x) = 1 - \frac{2}{\pi^{1/2}} \int_0^x e^{-z^2}\, dz$$

Since $\phi^{(n-1)}(y) = (-1)^{n-1}\phi(y)H_{n-1}(y)$

$$\int_y^\infty w(Y)\; dY = \tfrac{1}{2}\operatorname{erfc}\left(\frac{y}{2^{1/2}}\right) + \sum_{n=3}^{\infty} (-1)^n a_n\, \phi(y) H_{n-1}(y) \tag{8-36}$$

Therefore, using the central moments to determine the a_n the area may be

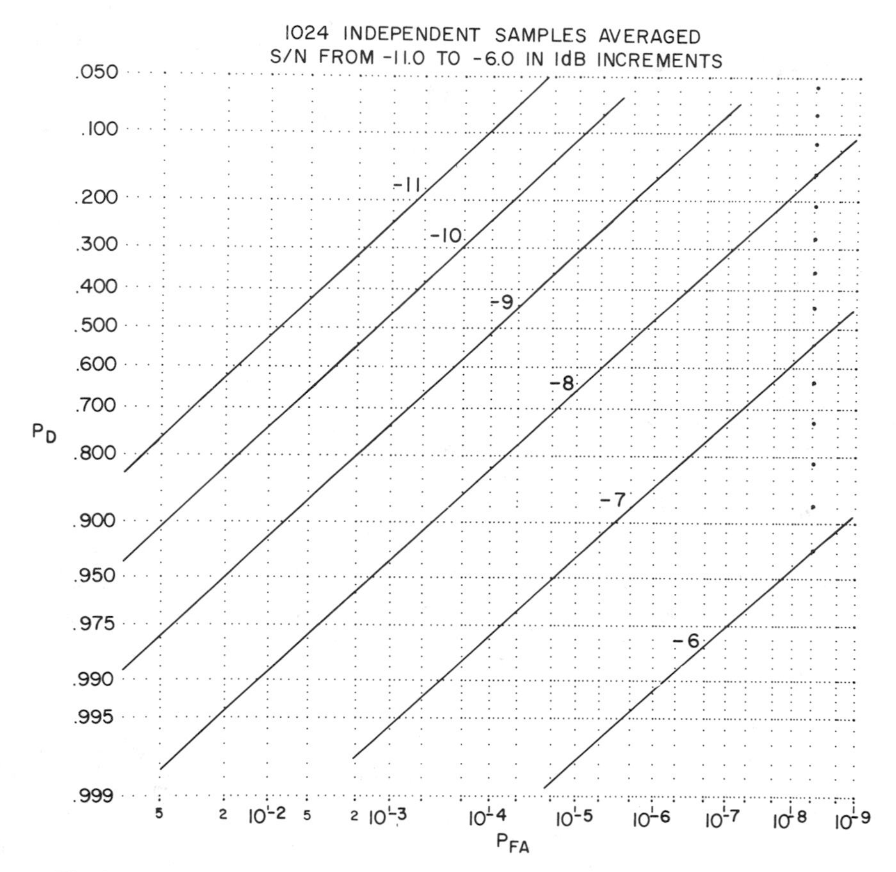

Fig. 8-14 Receiver operating characteristics for detecting sinewaves in white Gaussian noise (1024 samples averaged). [From Robertson (*2*); copyright, 1967, The American Telephone and Telegraph Co., reprinted by permission.]

evaluated. Obviously, a digital computer would be useful since these methods are generally tedious.

Results

We now return to the determination of the receiver operating characteristics for detecting M independent pulses using a linear detector. As before, the output of the linear detector for the ith pulse at time T is denoted q_i. We shall use the normalized variable $z_i = q_i/\sigma_T$. This is distributed as

$$p_1(z_i) = z_i \exp\left(\frac{z_i^2 + \alpha^2}{-2}\right) I_o(z_i \alpha) \qquad \text{or} \qquad p_0(z_i) = z_i e^{-z_i^2/2}$$

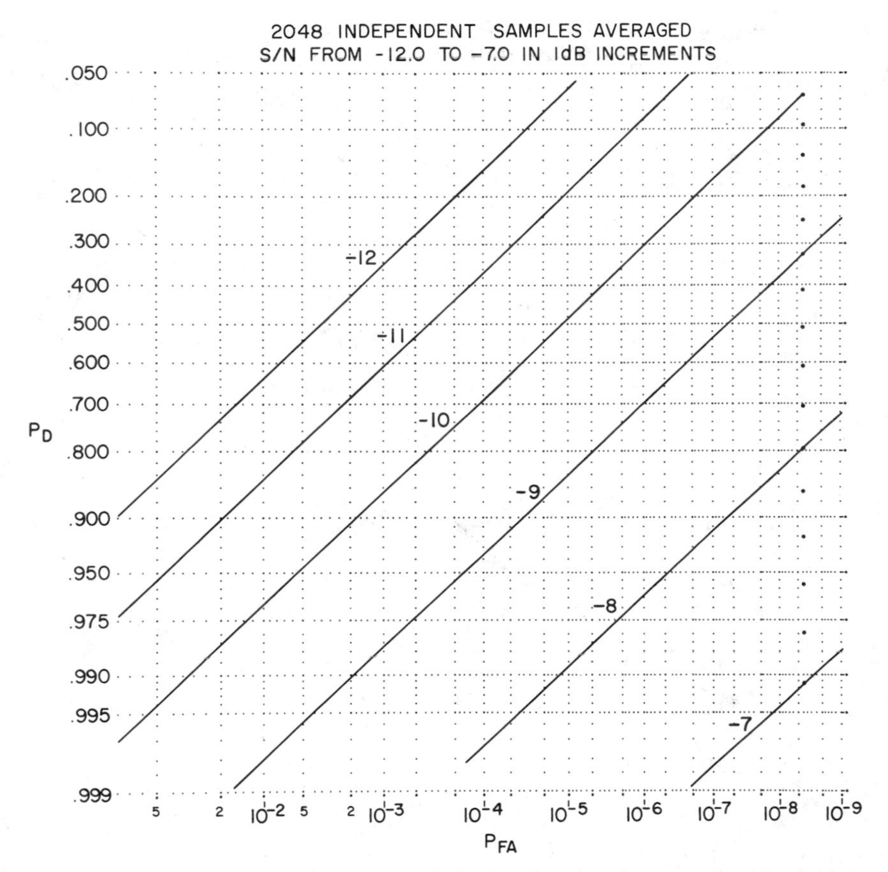

Fig. 8-15 Receiver operating characteristics for detecting sinewaves in white Gaussian noise (2048 samples averaged). [From Robertson (*2*); copyright, 1967, The American Telephone and Telegraph Co., reprinted by permission.]

depending on whether the signal is present or not. The parameter $\alpha^2/2$ is the signal-to-noise ratio at the sampling instant. The moments of z_i for signal present are, cf Eq. (4-54)

$$E_s\{z_i^n\} = 2^{n/2}\Gamma\left(\frac{n}{2}+1\right){}_1F_1\left(-\frac{n}{2}; 1; -\frac{\alpha^2}{2}\right)$$

and for noise only ($\alpha^2/2 = 0$)

$$E_N\{z_i^n\} = 2^{n/2}\Gamma\left(\frac{n}{2}+1\right)$$

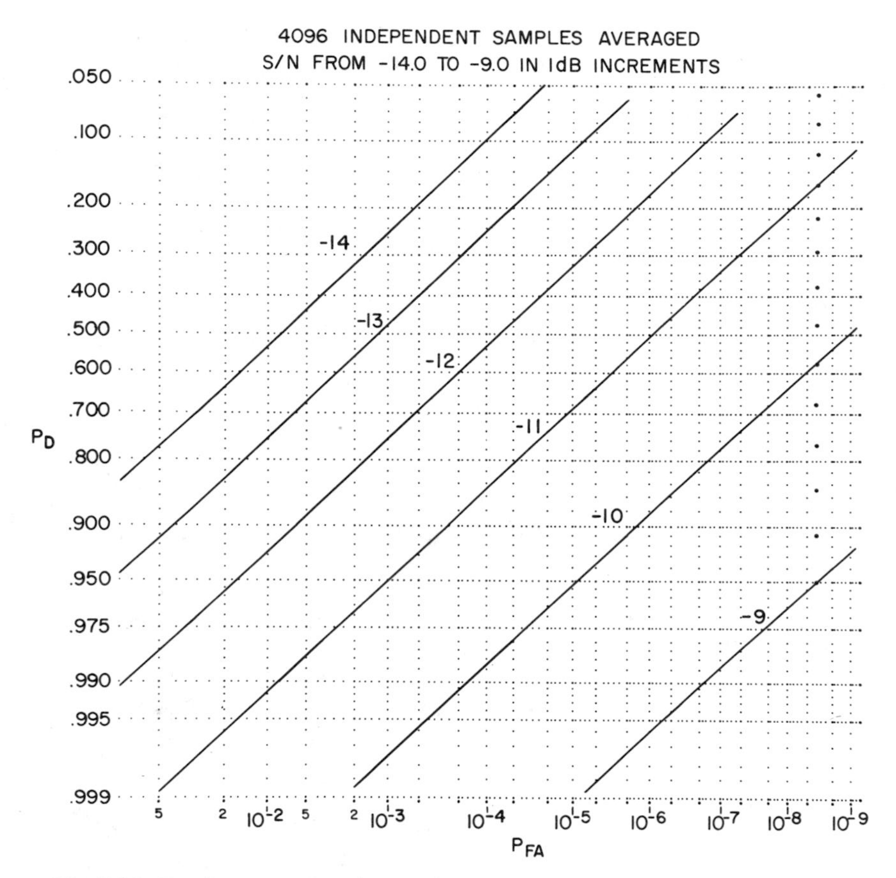

Fig. 8-16 Receiver operating characteristics for detecting sinewaves in white Gaussian noise (4096 samples averaged). [From Robertson (*2*); copyright, 1967, The American Telephone and Telegraph Co., reprinted by permission.]

Having the moments for a single pulse, it is straightforward but tedious to determine the moments for the sum of M such pulses. Having accomplished that, the coefficients a_n in Eq. (8-36) may be evaluated and the probability of detection and false alarm may be determined. This was the approach taken by Robertson† and his results are shown in Figs. 8-4 through 8-17 for up to 8192 independent pulses. In these figures the divisions of each decade of the P_{fa} scale reading left to right are given by

$$10.0\ (1.0)\ 4.0\ (0.5)\ 2.0\ (0.2)\ 1.0$$

† See Robertson (*2*). The same technique was used by Robertson to determine the effects of nonlinearities on post detection integration (*11*).

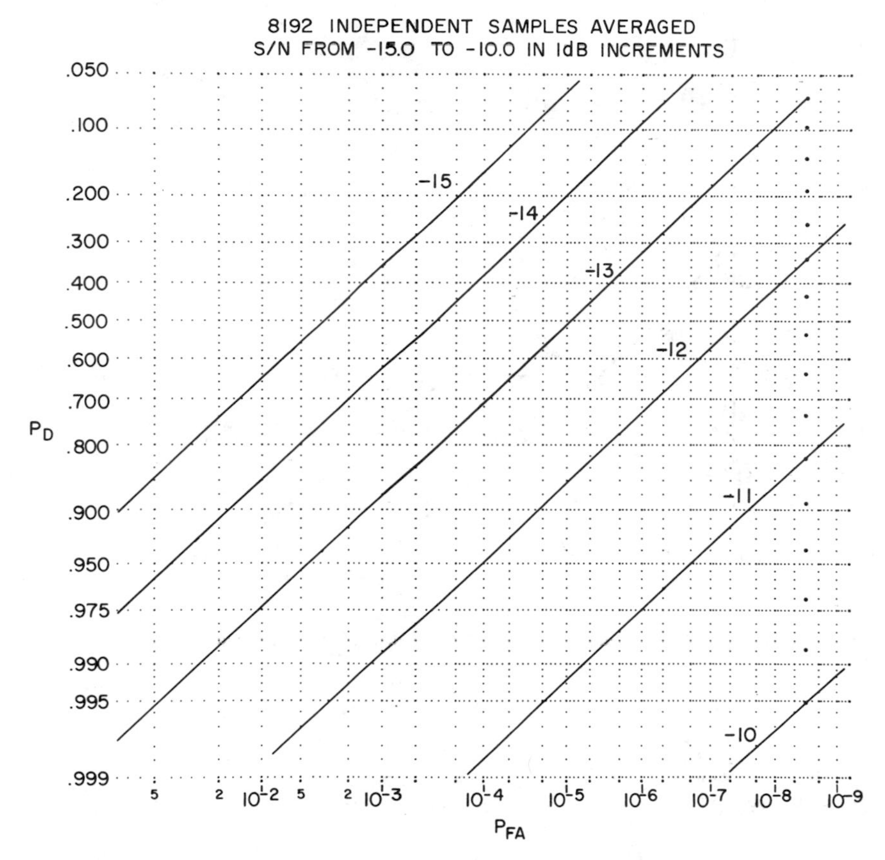

Fig. 8-17 Receiver operating characteristics for detecting sinewaves in white Gaussian noise (8192 samples averaged). [From Robertson (*2*); copyright, 1967, The American Telephone and Telegraph Co., reprinted by permission.]

where the incremental steps are within the parentheses and the applicable range is given by the adjacent numbers. The P_D scale reading top to bottom has divisions

0.05 (0.01) 0.10 (0.02) 0.90 (0.01) 0.950 (0.005) 0.980

0.980 (0.0025) 0.990 (0.001) 0.998 (0.0005) 0.999

The curves are for 1 dB increments. The heavy dots appearing at the right-hand side are spaced 0.2 dB apart and are to aid interpolation

Figure 8-18 shows the behavior for two sets of P_D and P_{fa} operating points as a function of the number of pulses. Note that the abscissa is $\log_2 M$.

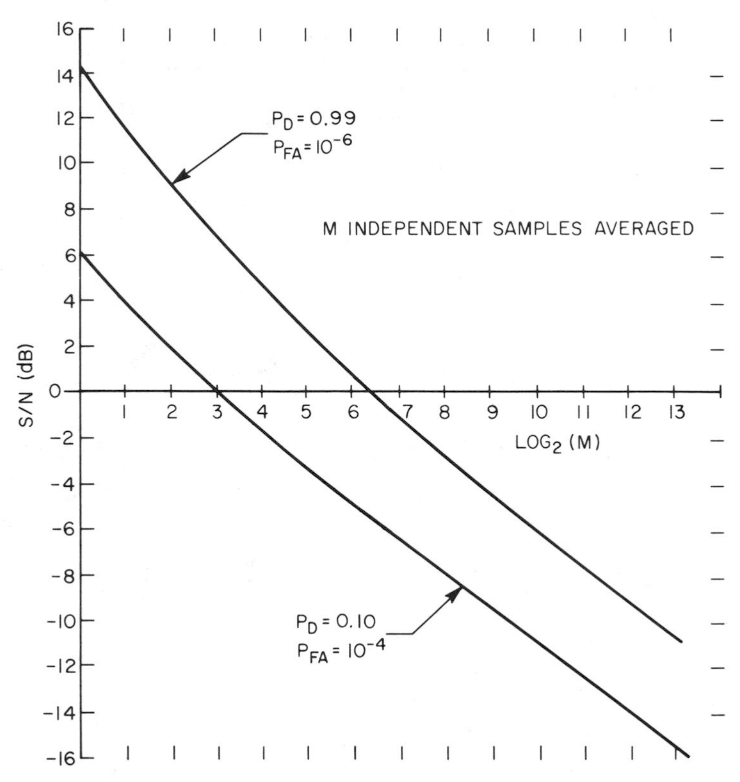

Fig. 8.18 Sensitivity related to number of samples averaged. [From Robertson (*2*); copyright, 1967, The American Telephone and Telegraph Co., reprinted by permission.]

As the number of pulses increases, the required S/N ratio decreases 1.5 dB for each doubling of the number of pulses, a familiar asymptotic result.† Results such as these may be used for interpolating for any number of samples not included in Figs. 8-4 through 8-17. We conclude this section with Fig. 8-19 which is Robertson's comparison of the linear and quadratic detectors (*11*). The difference is seen to be less than 0.2 dB over the entire range. This provides the justification in most cases for considering analytically the quadratic detector even though the receiver in practice will often be implemented with a linear detector [see also Marcum, (*5*)].

Known but Unequal Amplitudes

For this case we assume that the signals are of the form

$$H_1: \quad r_i(t) = A_i \sin(\omega_c t + \theta_i) + n_i(t), \qquad i = 1, \ldots, M$$
$$H_0: \quad r_i(t) = n_i(t), \qquad i = 1, \ldots, M$$

† See Exercises 8.8 and 8.9.

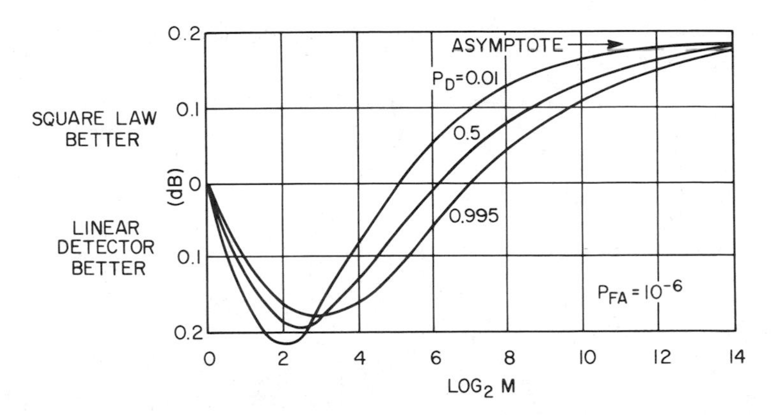

Fig. 8-19 Comparison of square law and linear detectors for detection of M independent pulses. [From Robertson (11); copyright, 1968, The American Telephone and Telegraph Co., reprinted by permission.]

where the amplitudes are known but unequal, and the phases are uniformly distributed. The M signals are assumed statistically independent. Using Eq. (7-10), it is easily seen that the likelihood ratio for the ith pulse is

$$\lambda_i(\mathbf{r}) = \exp\left(-\frac{A_i^2 T}{2N_0}\right) I_0\left(\frac{2A_i q_i}{N_0}\right) \tag{8-37}$$

and therefore, the likelihood ratio for the M pulses is

$$\lambda(\mathbf{r}) = \exp\left(-\sum_{i=1}^{M} \frac{A_i^2 T}{2N_0}\right) \prod_{i=1}^{M} I_0\left(\frac{2A_i q_i}{N_0}\right) \tag{8-38}$$

As in the previous cases, the receiver samples the envelope of the matched filter output of each pulse. Before being combined, the envelope samples are weighted by the respective amplitudes A_i. Consequently, the larger the amplitude, the greater is the weight given to that particular pulse. The implementation of the optimum receiver is therefore dependent upon these amplitudes and it changes if we assume a different set of amplitudes. It follows that the test arising from Eq. (8-38) is not one which is uniformly most powerful with respect to amplitude.

Amplitude and Phase Unknown

We now extend the above case of unequal amplitudes by assuming that they are statistically independent random variables having identical Rayleigh distributions. The slow fading assumption will be made so that for any one pulse the amplitude is constant and independent of the phase. The amplitudes are distributed as

$$w(A_i) = \frac{A_i}{A_o^2} \exp\left[-\frac{A_i^2}{2A_o^2} \right], \qquad i = 1, \ldots, M$$

The averaged likelihood ratio is then

$$\lambda(\mathbf{r}) = \sum_{i=1}^{M} \int_0^\infty \exp\left[-\frac{A_i^2 T}{2N_o} \right] I_o\left(\frac{2A_i q_i}{N_o} \right) \frac{A_i}{A_o^2} \exp\left[-\frac{A_i^2}{2A_o^2} \right] dA_i$$

Applying Eqs. (7-31) and (7-32) this reduces to

$$\lambda(\mathbf{r}) = \prod_{i=1}^{M} \frac{N_o}{N_o + TA_o^2} \exp\left[\frac{2A_o^2 q_i^2}{N_o(N_o + TA_o^2)} \right] \tag{8-39}$$

Using the log-likelihood ratio, and some manipulation, the decision rule may be shown to be: choose H_1 if

$$\sum_{i=1}^{M} q_i^2 \geq V_T$$

The receiver implementation is therefore the same as shown in Fig. 8-3A but for these slow fading signals it is the optimum and not just an approximation to the optimum for low values of signal-to-noise ratio.

To determine the receiver performance we shall use the normalized statistic

$$G = \sum_{i=1}^{M} q_i^2 / \sigma_T^2$$

as was done in Eq. (8-19). It should be clear that the false alarm probability is given by Eq. (8-23). To determine the detection probability we first find the conditional detection probability for a given vector $\mathbf{A}(A_i, i = 1, \ldots, M)$. For a given $\mathbf{A}$, the test statistic G has a noncentral χ^2 distribution with $2M$ degrees-of-freedom and, in analogy to Eq. (8-20), a noncentral parameter given by

$$v = \sum_{i=1}^{M} \left(\frac{A_i^2 T^2}{4\sigma_T^2} \sin^2 \theta_i + \frac{A_i^2 T^2}{4\sigma_T^2} \cos^2 \theta_i \right) = \sum_{i=1}^{M} \frac{A_i^2 T^2}{4\sigma_T^2} = \frac{A_o^2 T}{N_o} \sum_{i=1}^{M} \left(\frac{A_i}{A_o} \right)^2 \tag{8-40}$$

Note that v is itself a random variable since it is a function of the amplitudes A_i which are random variables. We shall therefore determine the performance for a given v and average the result over the probability distribution of v. For a given value of v, the detection performance is found from Eq. (8-24)

$$P_D(v) = \int_{G_T}^\infty \frac{1}{2} \left(\frac{G}{v} \right)^{(M-1)/2} \exp\left(-\frac{G}{2} - \frac{v}{2} \right) I_{M-1}((Gv)^{1/2})\, dG \tag{8-41}$$

Now, it may easily be shown that the probability density function of $u_i \triangleq A_i^2/A_o^2$ is

$$p(u_i) = \tfrac{1}{2}e^{-u_i/2}$$

This is an exponential distribution and is a special case of the central χ^2 distribution with two degrees of freedom (cf Eq. 4-60). It then follows that $Y = \sum_{i=1}^{M} u_i$ is central χ^2 with $2M$ degrees-of-freedom. Finally, the distribution for $v = A_o^2TY/N_o$, Eq. (8-40), is

$$p(v) = \frac{v^{M-1}e^{-v/2\varepsilon}}{(2\varepsilon)^M\Gamma(M)} \tag{8-42}$$

where $\varepsilon \triangleq A_o^2T/N_o$, the ratio of the average signal energy to the noise spectral density. The conditional detection probability, Eq. (8-41), is averaged over this function. We omit the details of this evaluation and present only the result. It is

$$P_D = 1 - \int_0^{G_T/2(1+\varepsilon)} \frac{z^{M-1}e^{-z}}{\Gamma(M)}\,dz \tag{8-43}$$

The performance for detecting up to 1024 pulses is shown in Figs. 8-20 through 8-30.† The integral in Eq. (8-43) is the cumulative distribution of a

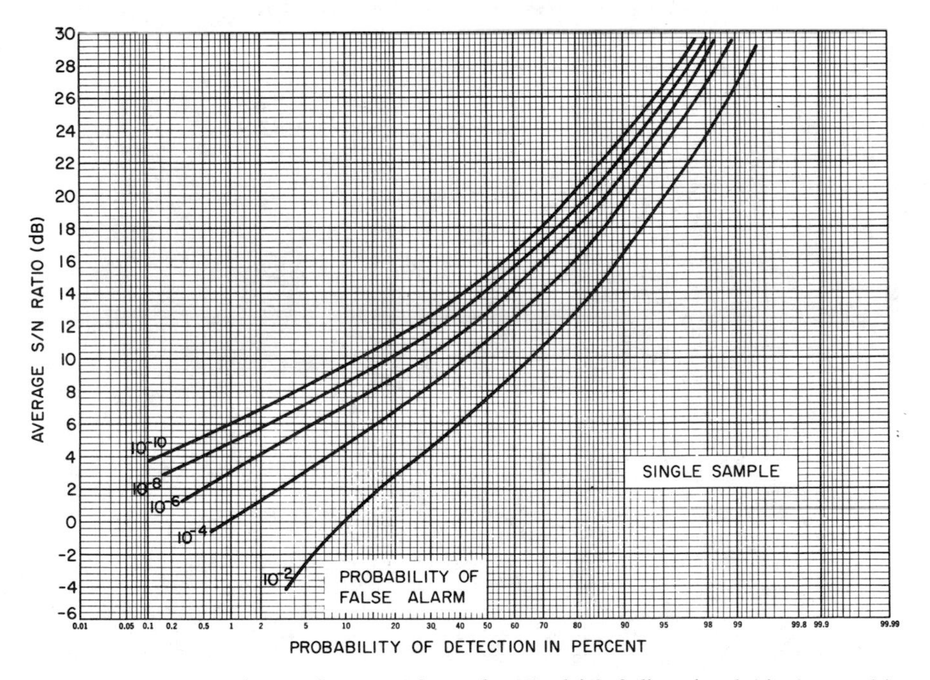

Fig. 8-20 Detection performance for a slow Rayleigh fading signal (single sample).

† These results were obtained using a computer program written by Robertson (*9*) for the noncentral χ^2 distribution.

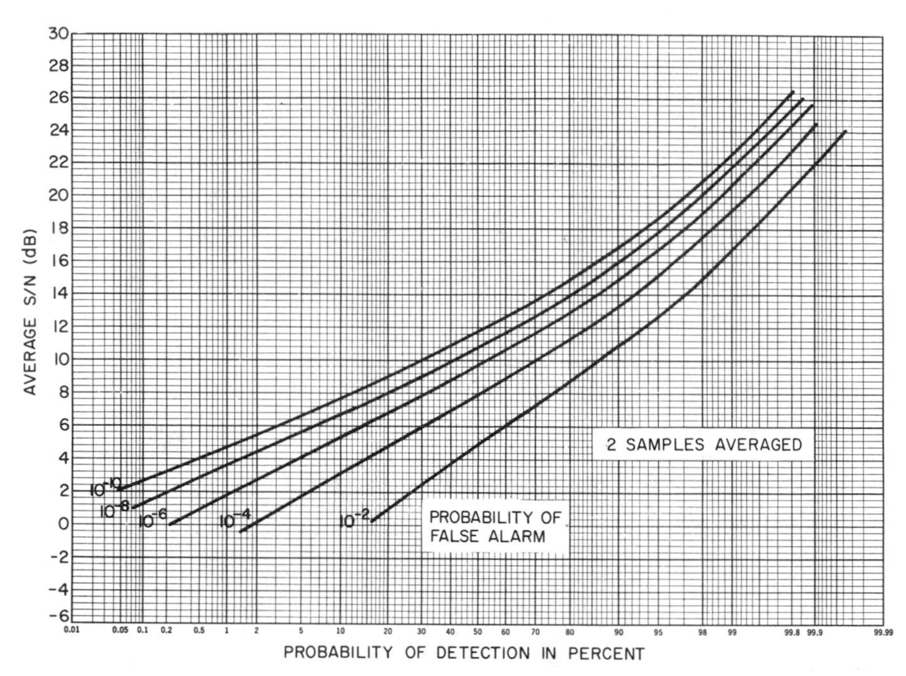

Fig. 8-21 Detection performance for slow Rayleigh fading signals (2 samples averaged).

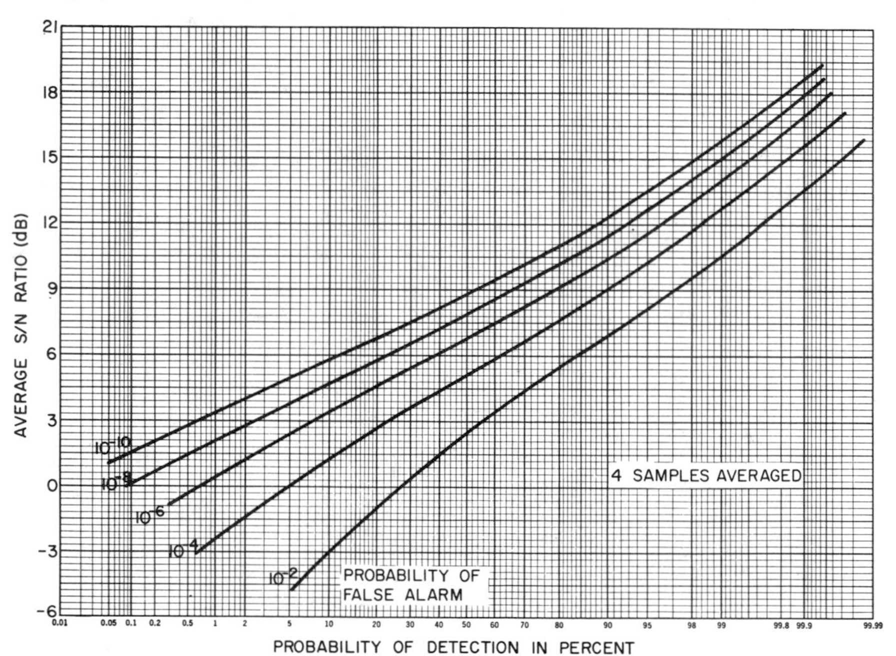

Fig. 8-22 Detection performance for slow Rayleigh fading signals (4 samples averaged).

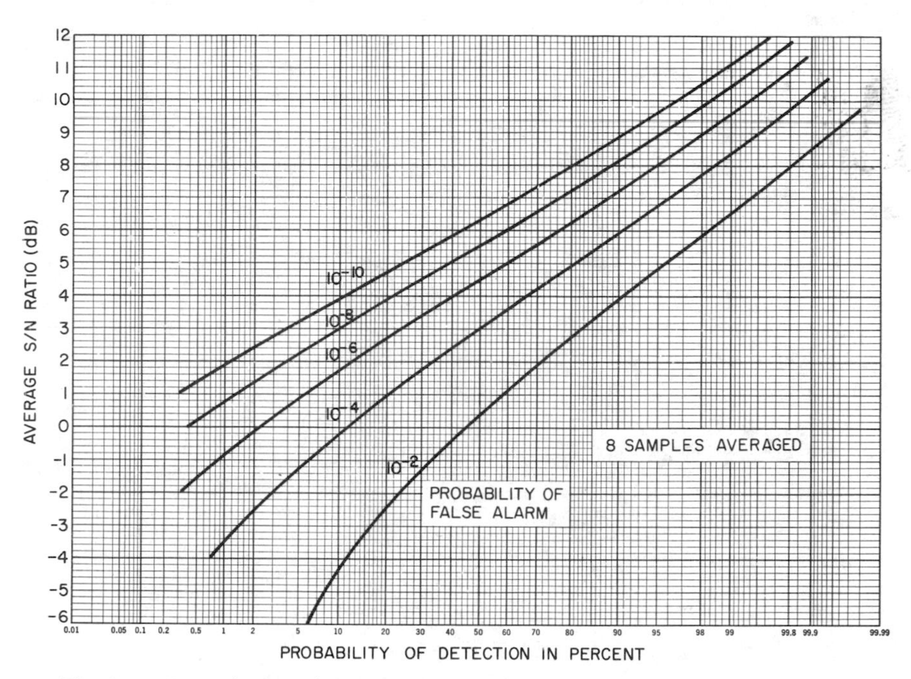

Fig. 8-23 Detection performance for slow Rayleigh fading signals (8 samples averaged).

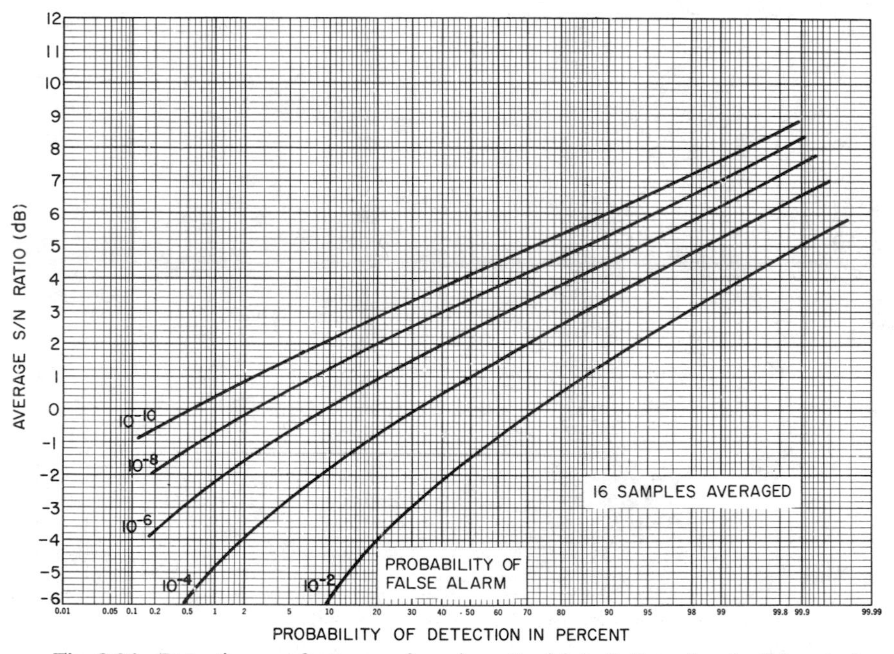

Fig. 8-24 Detection performance for slow Rayleigh fading signals (16 samples averaged).

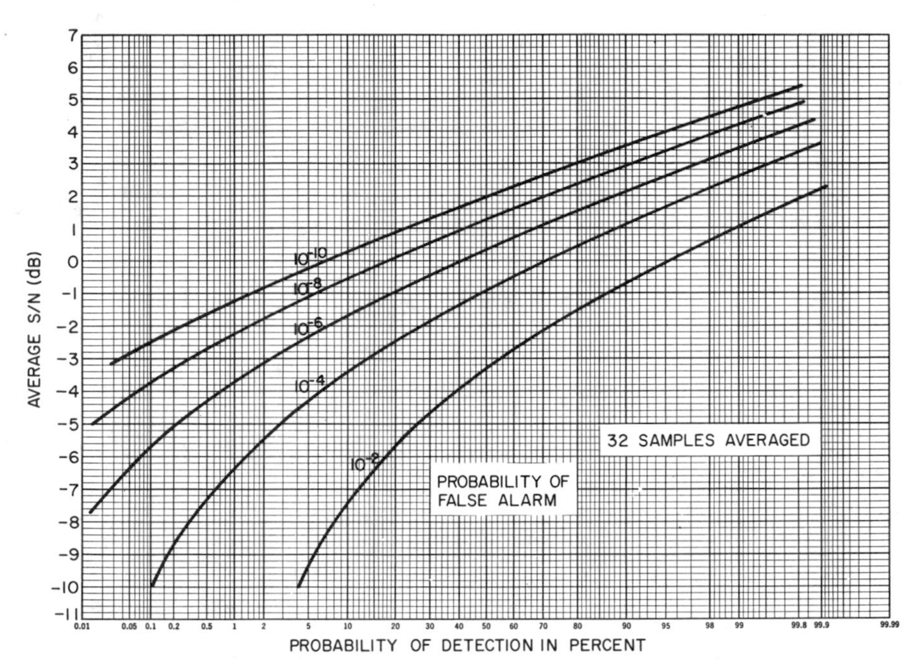

Fig. 8-25 Detection performance for slow Rayleigh fading signals (32 samples averaged).

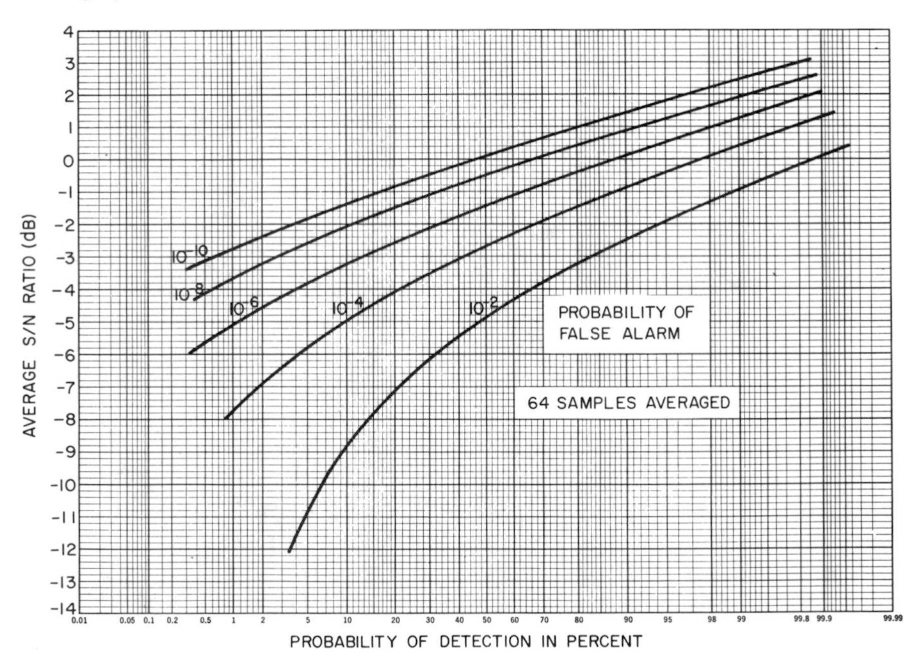

Fig. 8-26 Detection performance for slow Rayleigh fading signals (64 samples averaged).

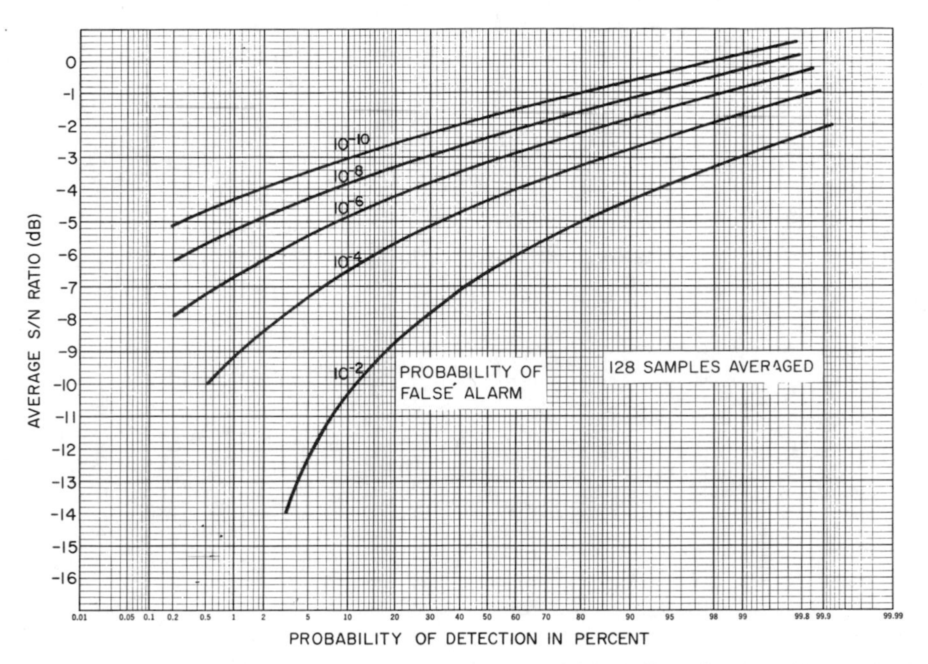

Fig. 8-27 Detection performance for slow Rayleigh fading signals (128 samples averaged).

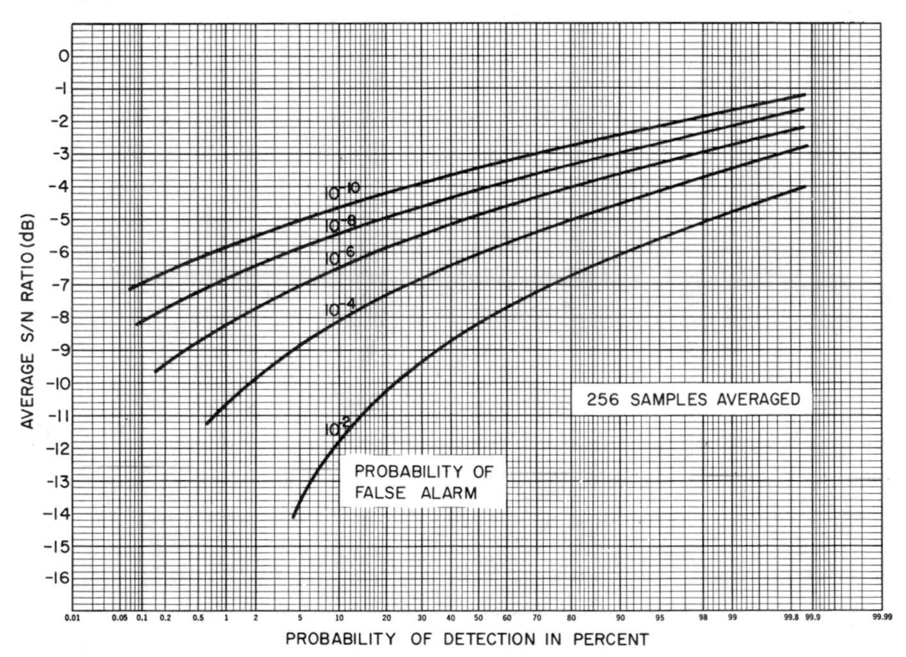

Fig. 8-28 Detection performance for slow Rayleigh fading signals (256 samples averaged).

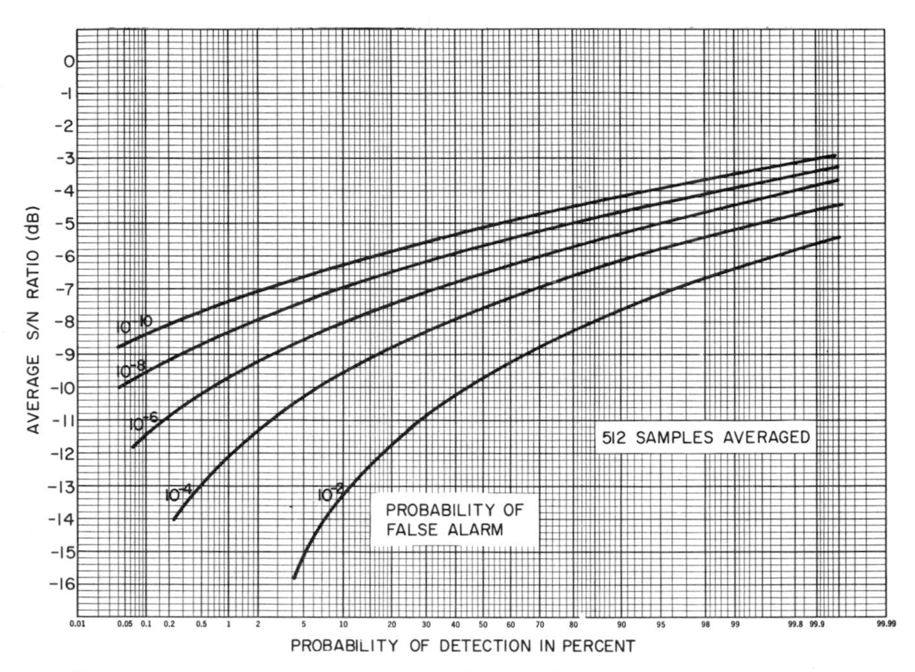

Fig. 8-29 Detection performance for slow Rayleigh fading signals (512 samples averaged).

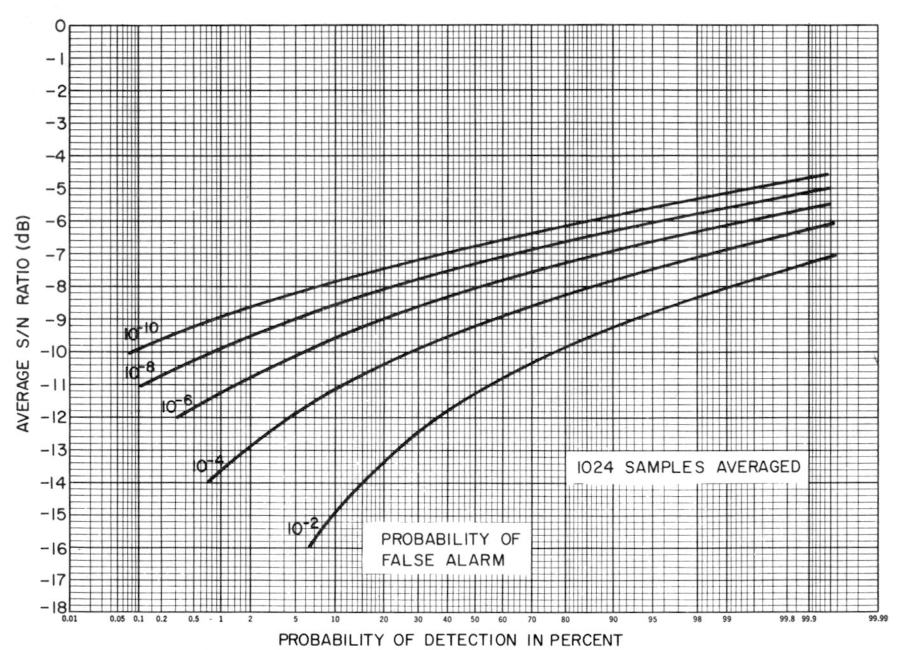

Fig. 8-30 Detection performance for slow Rayleigh fading signals (1024 samples averaged).

central χ^2 variable (cf Eq. 4-66). This result should not be too surprising since the Rayleigh fading signal is a sample function of a narrowband Gaussian process. The sum of such a process and the narrowband Gaussian noise process is itself narrowband Gaussian. The detector sums M statistically independent samples of the square of the envelope and this sum is therefore central χ^2 with $2M$ degrees of freedom. (See Exercise 8.5.)

8.4 Diversity†

The preceding sections on multiple-pulse observations leads quite naturally to the subject of diversity or diversity combining which will be briefly discussed here. The treatment is intended only to acquaint the reader with the existence of such techniques and to draw a parallel to the multiple-pulse detection problem.

We were previously concerned with the problem of receiving a single signal whose amplitude and phase fluctuate in time. This phenomenon is known as fading. In Fig. 7-8, which considered a Rayleigh fading model, the harmful effects of fading for a frequency shift keying communication system are readily apparent. For that case, the error probability decreased linearly with the average signal-to-noise ratio. Because of this linear relationship, even moderate decreases in error probability would require prohibitive increases in the signal power. Other techniques must therefore be used to obtain substantial improvements in performance. One such technique is diversity combining.

If two or more communication channels are sufficiently separated in time, frequency, space, polarization, etc., the fading on these channels may be more or less independent. In that case the probability that the signals on all of these channels fade together is much less than the probability of a fade for any one of the signals. Consequently if these signals were "combined" in a suitable manner, an improvement in the overall performance could be expected.

In the multiple-pulse material in the preceding sections, a similar problem was treated. By combining M statistically independent signals, a decision was to be made which satisfied a given criterion. By determining the likelihood ratio we were able to see how these M pulses should be combined. Methods for combining were determined and illustrated in Figs. 8-1, 8-2, and 8-3. These examples were framed for the "multiple-pulse" radar detection problem. For communications however, the phrase "diversity" is usually employed. It should be apparent that the techniques used in either case are in principle the same, and the results determined for one system apply, with slight variation, to the other.

† A good treatment of this subject is given in Brennan (*12*). See also Baghdady (*13*).

We set up the framework of the diversity problem as follows: Suppose we transmit a signal $s(t)$. We may transmit this signal simultaneously on M "widely" separated carrier frequencies, or in M "widely" separated time intervals. These are frequency and time diversity respectively. The term "widely" separated is meant to imply that the received signals are statistically independent or nearly so. Alternatively, we might have transmitted the signal $s(t)$ in only one time interval and at only one carrier frequency, but received the signal at M different receiver locations which presumably are "widely" separated. This is space diversity and is applicable to ionospheric or tropospheric scatter systems, for example. The received signals are then combined at a central location. Independent of the type of diversity, we say that each signal is received over a diversity channel. For example, if $M = 2$, there are two diversity channels and this is called dual diversity. Four channels combined would be called quadruple diversity, etc.

Although in practice there are a number of methods used to combine diversity channels (*12*), we shall employ only the likelihood ratio receiver to determine how the channels should be combined. The combiner will therefore be optimum but only insofar as the criterion and the signal and noise assumption are valid. Diversity receivers find considerable application for analog communication systems. We shall, however, limit our discussion to a few problems which are encountered in digital communications.

Diversity Improvement for FSK

As an example of diversity performance, consider the case of binary frequency shift keying in the presence of Rayleigh fading. We start the example with dual diversity and then generalize to Mth order diversity. As in the case of multiple-pulse detection the likelihood ratio is used to determine a suitable way for combining the diversity channels.

Under hypothesis H_1 the received signals at the output of the two diversity channels are of the form

$$r_1(t) = A_1 \sin(\omega_1 t + \theta_1) + n_1(t)$$

and (8-44)

$$r_2(t) = A_2 \sin(\omega_1 t + \theta_2) + n_2(t)$$

Under hypothesis H_0, the signals are of the form

$$r_1(t) = B_1 \sin(\omega_0 t + \phi_1) + n_1(t)$$

and (8-45)

$$r_2(t) = B_2 \sin(\omega_0 t + \phi_2) + n_2(t)$$

Over any observation interval $(0, T)$ the amplitude and phase are constant, but may vary from interval to interval (slow fading). The additive noise is assumed to be white and Gaussian with spectral density $N_o/2$. The noise sample functions $n_1(t)$ and $n_2(t)$ are assumed to be independent. The amplitudes are Rayleigh distributed and statistically independent. That is, A_1 fades independently of A_2, and B_1 fades independently of B_2. Their probability density functions are given by

$$w_2(A_i) = \frac{A_i}{A_o^2} \exp\left[-\frac{A_i^2}{2A_o^2}\right] \quad \text{and} \quad w_2(B_i) = \frac{B_i}{A_o^2} \exp\left[-\frac{B_i^2}{2A_o^2}\right] \tag{8-46}$$

This assumes that the statistics of A and B are equal. The phases θ and ϕ are uniformly distributed $(0, 2\pi)$.

Because the signals in each diversity channel are independent, the likelihood function of both channels is the product of the likelihood functions of each channel. That is

$$p_1(\mathbf{r}) = \left[\int_{\{A_1\}} \int_{\{\theta_1\}} p_1(\mathbf{r}_1 \mid A_1, \theta_1) w_1(\theta_1) w_2(A_1)\, d\theta_1\, dA_1\right]$$
$$\times \left[\int_{\{A_2\}} \int_{\{\theta_2\}} p_1(\mathbf{r}_2 \mid A_2, \theta_2) w_1(\theta_2) w_2(A_2)\, d\theta_2\, dA_2\right] \tag{8-47}$$

and

$$p_0(\mathbf{r}) = \left[\int_{\{B_1\}} \int_{\{\phi_1\}} p_0(\mathbf{r}_1 \mid B_1, \phi_1) w_1(\phi_1) w_2(B_1)\, d\phi_1\, dB_1\right]$$
$$\times \left[\int_{\{B_2\}} \int_{\{\phi_2\}} p_0(\mathbf{r}_2 \mid B_2, \phi_2) w_1(\phi_2) w_2(B_2)\, d\phi_2\, dB_2\right] \tag{8-48}$$

These likelihood functions are proportional to the likelihood ratio given by Eq. (8-39) with $M = 2$. The ratio of these likelihood functions gives the desired likelihood ratio, namely,

$$\lambda(\mathbf{r}) = \frac{\displaystyle\prod_{i=1}^{2} \frac{N_o}{N_o + TA_o^2} \exp\left[\frac{2A_o^2 q_{i,1}^2}{N_o(N_o + TA_o^2)}\right]}{\displaystyle\prod_{i=1}^{2} \frac{N_o}{N\ + TA_o^2} \exp\left[\frac{2A_o^2 q_{i,0}^2}{N_o(N_o + TA_o^2)}\right]} \tag{8-49}$$

where, for $i = 1,2$,

$$q_{i,1}^2 = \left[\int_0^T r_i(t) \sin \omega_1 t\, dt\right]^2 + \left[\int_0^T r_i(t) \cos \omega_1 t\, dt\right]^2 \tag{8-50}$$

and

$$q_{i,0}^2 = \left[\int_0^T r_i(t) \sin \omega_0 t\, dt\right]^2 + \left[\int_0^T r_i(t) \cos \omega_0 t\, dt\right]^2 \tag{8-51}$$

The likelihood ratio is then

$$\lambda(\mathbf{r}) = \prod_{i=1}^{2} \exp\left\{\frac{2A_o^2}{N_o(N_o + TA_o^2)} [q_{i,1}^2 - q_{i,o}^2]\right\} \tag{8-52}$$

and finally the log-likelihood ratio is

$$\ln \lambda(\mathbf{r}) = \frac{2A_o^2}{N_o(N_o + TA_o^2)} \sum_{i=1}^{2} (q_{i,1}^2 - q_{i,o}^2)$$

For a minimum probability of error criterion and $P(H_1) = P(H_0)$, $\ln \lambda(\mathbf{r})$ is compared to zero. The decision rule becomes: choose H_1 if

$$\sum_{i=1}^{2} q_{i,1}^2 \geq \sum_{i=1}^{2} q_{i,o}^2 \tag{8-53}$$

That is, the square of the envelope for each center frequency is obtained in each channel and added to the corresponding envelope squared of the other channel. The hypothesis corresponding to the largest sum is chosen. Such a decision rule may be implemented as shown in Fig. 8-31.

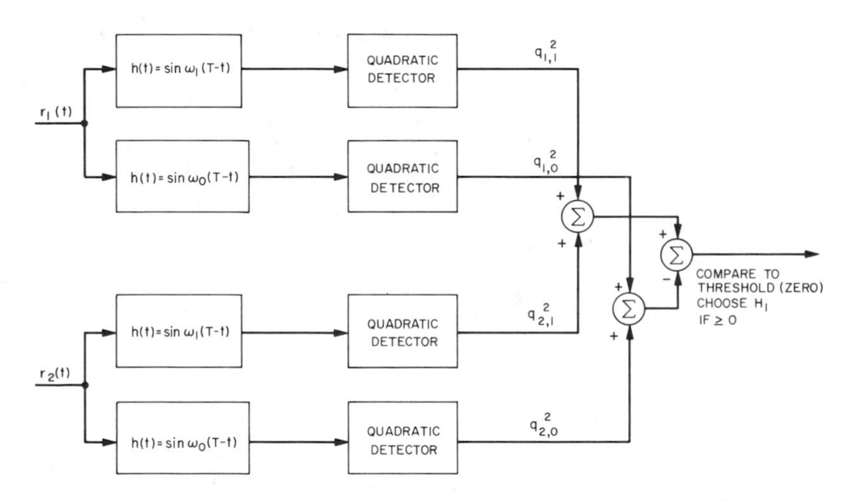

Fig. 8-31 Dual diversity receiver for binary frequency shift keying.

We next generalize to Mth order diversity under a set of assumptions consistent with the dual diversity case. Then, the likelihood ratio is, cf. Eq. (8-52),

$$\lambda(\mathbf{r}) = \prod_{i=1}^{M} \exp\left\{\frac{2A_o^2}{N_o(N_o + TA_o^2)} [q_{i,1}^2 - q_{i,o}^2]\right\} \tag{8-54}$$

and the decision rule may be shown to be, cf. Eq. (8-53): choose H_1 if

$$\sum_{i=1}^{M} q_{i,1}^2 \geq \sum_{i=1}^{M} q_{i,0}^2 \tag{8-55}$$

As before the square of the envelope for each center frequency is extracted and added to the corresponding envelope squared of the other channels.

Diversity Performance

To determine the error performance denote

$$Q_1 = \sum_{i=1}^{M} q_{i,1}^2/\sigma_T{}^2 \qquad \text{and} \qquad Q_0 = \sum_{i=1}^{M} q_{i,0}^2/\sigma_T{}^2$$

where, as before, $\sigma_T{}^2 = N_o T/4$. Assume that hypothesis H_1 (frequency ω_1) is true. Then an error is made if $Q_0 > Q_1$. We assume that Q_0 and Q_1 are statistically independent. As discussed in Sect. 8.3, when H_1 is true Q_0 is due to noise only and it is therefore distributed as central χ^2 with $2M$ degrees of freedom, Eq. (8-22). Further paralleling Sect. 8.3, we shall evaluate the conditional error probability for a given set of the A_i's, that is, determine $P_e(\mathbf{A})$, and then average over the vector $\mathbf{A}$. For a given $\mathbf{A}$, the statistic Q_1 is noncentral χ^2 with $2M$ degrees of freedom and a noncentral parameter given by Eq. (8-40). The noncentral parameter, being a function of $\mathbf{A}$, is a random variable and its distribution is given by Eq. (8-42).

The conditional error probability, in analogy with Eq. (7-43) is

$$P_e(v) = \int_0^\infty dQ_1\, P_1(Q_1 \mid v) \int_{Q_1}^\infty dQ_0\, P(Q_0) \tag{8-56}$$

where $P_1(Q_1 \mid v)$ and $P(Q_0)$ are the probability density functions for Q_1 and Q_0 respectively conditioned on H_1 being true. The average error probability is

$$P_e = \int_0^\infty dQ_1 \int_0^\infty dv\, P_1(Q_1 \mid v) p(v) \int_{Q_1}^\infty dQ_0\, P(Q_0) \tag{8-57}$$

where $p(v)$ is the density function of the noncentral parameter. The integral over v will produce the marginal density function $P_1(Q_1)$. This may be shown to be

$$P_1(Q_1) = \frac{Q_1^{M-1} \exp[-Q_1/2(1+\varepsilon)]}{[2(1+\varepsilon)]^M \Gamma(M)} \tag{8-58}$$

where $\varepsilon = E_{av}/N_o = A_o{}^2 T/N_o$. It may be further shown that

$$\int_{Q_1}^\infty dQ_0\, P(Q_0) = \int_{Q_1}^\infty \frac{Q_0^{M-1} e^{-Q_0/2}}{2^M \Gamma(M)}\, dQ_0$$

$$= e^{-Q_1/2} \sum_{k=0}^{M-1} \frac{Q_1{}^k}{2^k k!} \tag{8-59}$$

Carrying out the operations indicated by Eq. (8-57) will then produce†

$$P_e = \frac{1}{(2+\varepsilon)^M} \sum_{k=0}^{M-1} \binom{M+k-1}{k} \left(\frac{1+\varepsilon}{2+\varepsilon}\right)^k \tag{8-60}$$

In particular for $M = 2$ we have

$$P_e = \frac{4+3\varepsilon}{(2+\varepsilon)^3} \tag{8-61}$$

The error performance for several orders of diversity are shown in Fig. 8-32. The dramatic improvement in performance is readily apparent. Improvements of this order are not generally feasible by merely increasing the transmitted power. The improvement is due almost entirely to the change in the probability distribution of the amplitude of the combined signals. This manifests itself by the change of the number of degrees of freedom of the parameter v of Eq. (8-40). Also shown in Fig. 8-32 for comparison is the performance for the noncoherent FSK case discussed in Chap. 7. The error performance is given in Eq. (7-45).

Exercises

8.1 Verify the probabilities given in Eq. (8-9).

8.2 Consider the detection problem for which the hypotheses are

$$H_1: \quad r_i(t) = s(t) + n_i(t), \qquad i = 1, \ldots, M, \quad 0 \le t \le T$$

$$H_0: \quad r_i(t) = n_i(t), \qquad i = 1, \ldots, M, \quad 0 \le t \le T$$

Assume $s(t) = A \sin(\omega_c t + \theta)$ where θ is uniformly distributed $(0, 2\pi)$, and $n_i(t)$ is white, Gaussian noise with spectral density $N_o/2$. (Note, while the phase is a random variable, it is the same for each signal. This is in contrast to the problem considered in Sect. 8.3). All other parameters are assumed known. What is the form of the likelihood ratio receiver, and what are the expressions for the probabilities of detection and false alarm? How does this compare with the incoherent detection of a single pulse?

8.3 Find an expression to approximate the Rayleigh density function

$$p(x) = xe^{-x^2/2}, \qquad x \ge 0$$

using two terms of the Gram–Charlier series. Would you expect this to be a reasonable approximation?

† See also Pierce (*14*). For higher-order alphabets than binary, see Hahn (*15*).

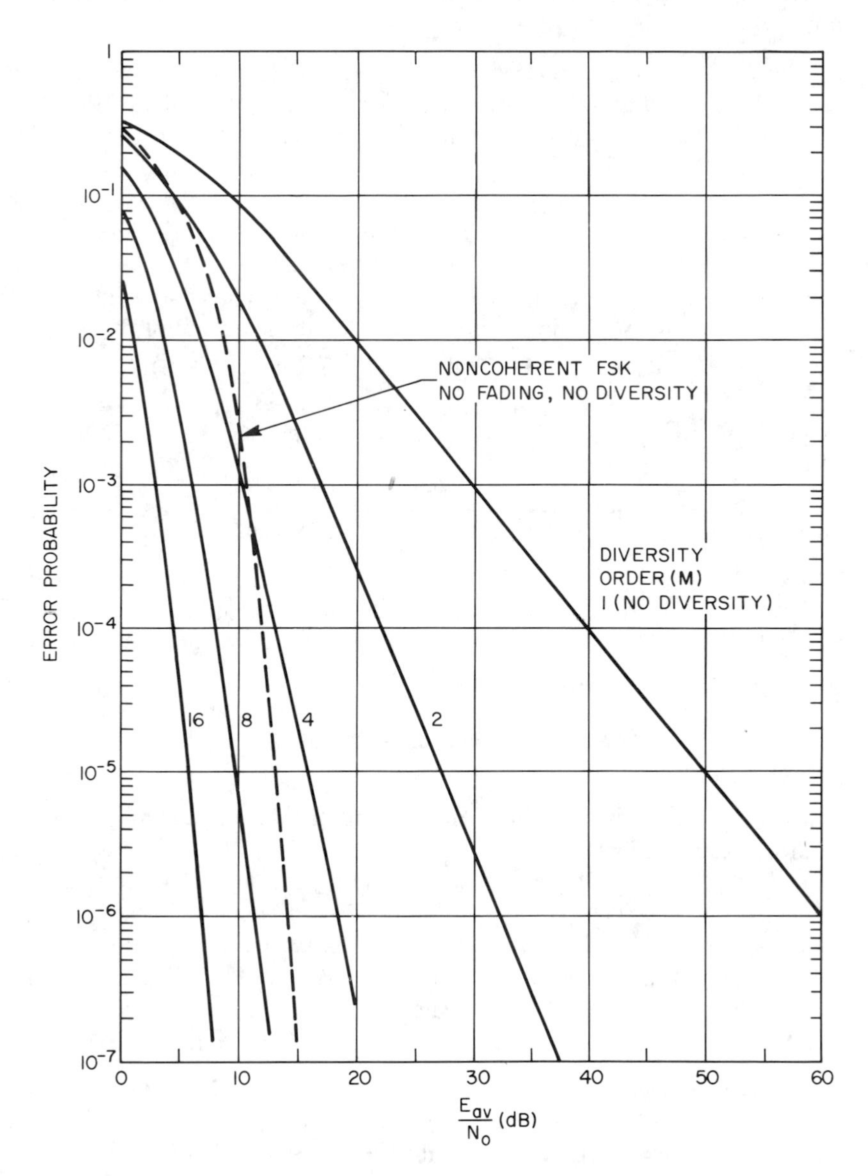

Fig. 8-32 Diversity performance for frequency shift keying and slow Rayleigh fading. E_{av} is the average signal energy in each channel. $N_o/2$ is the two-sided noise spectral density.

8.4 Determine the coefficient a_9 in Eq. (8-35) in terms of the central moments denoted v_n.

8.5 The probability of detection for the case of M pulses having uniform phase and Rayleigh amplitude distribution is given in Eq. (8-43). Rederive that equation by considering the following problem. (See Exercise 7-9 for similar results in the single pulse case.)

A signal $r(t) = a(t)\cos(\omega_c t + \phi(t)) + n(t)$, $-\infty < t < \infty$, is passed through a linear filter, $h(\tau) = \cos \omega_c \tau$, $0 \leq \tau \leq T$ followed by a quadratic detector and a constant multiplier $(1/\sigma_T^2)$. The signal $a(t)\cos(\omega_c t + \phi(t))$ is a Gaussian process, and $R_a(\tau)$ and $\phi(t)$ are substantially constant over any interval T. Assume that M statistically independent samples at the output of the detector are summed. Verify that the probability that this sum exceeds a threshold, G_T, is as given in Eq. (8-43).

8.6 For the coherent combining problem as posed in Fig. 8-33, assume

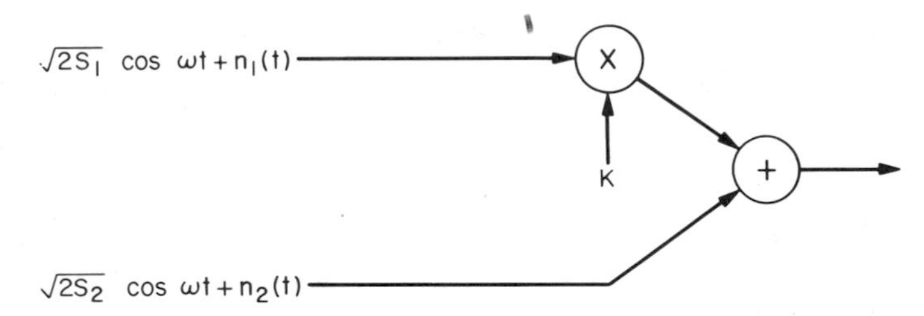

Fig. 8-33 Coherent addition of two signals.

$n_1(t)$ and $n_2(t)$ are statistically independent, and are zero mean with variances N_1 and N_2 respectively.

(a) Find the value of K which will maximize the signal-to-noise ratio of the output. What is the maximum value of signal-to-noise ratio?

(b) Suppose that K is chosen so that the signals are combined on a basis of equal noise powers. That is, $K^2N_1 = N_2$. Find the resulting signal-to-noise ratio. Under what condition is the resulting signal-to-noise ratio greater than the maximum of either S_1/N_1 or S_2/N_2?

8.7 For the incoherent combining problem posed in Fig. 8-34, $n_1(t)$ and $n_2(t)$ are statistically independent, zero mean, with variances N_1 and N_2 respectively. Define the deflection signal-to-noise ratio (*16*) of the output as

$$\text{DSNR} \triangleq \frac{E_{sn}\{z\} - E_n\{z\}}{\sigma_n\{z\}}$$

where $E_{sn}\{z\}$ and $E_n\{z\}$ are expected values with signal plus noise and noise

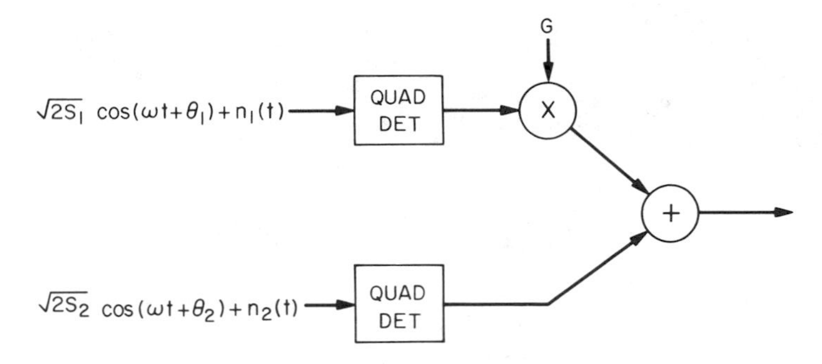

Fig. 8-34 Incoherent addition of two signals.

only respectively, and $\sigma_n\{z\}$ is the standard deviation of z with noise only present.

(a) Show that the maximum value of DSNR is

$$\frac{S_2}{N_2}\left(\frac{S_1}{N_1}+\frac{S_2}{N_2}\right)\Big/\left[\left(\frac{S_1}{N_1}\right)^2+\left(\frac{S_2}{N_2}\right)^2\right]^{1/2}$$

Consider the ratio of DSNR to the maximum input signal-to-noise ratio. What is its maximum value?

(b) Combine the signals on an equal power basis, that is $GN_1 = N_2$. Under what conditions is the DSNR greater than the maximum input signal-to-noise ratio?

8.8 For the receiver shown in Fig. 8-3B, denote the output as

$$Q = \sum_{i=1}^{M} q_i$$

Assume detection of constant amplitude sinewaves in white Gaussian noise. For a given performance level, as $M \to \infty$ the signal-to-noise ratio becomes small.

(a) Show that the asymptotic form (M becoming large and signal-to-noise ratio approaching zero) of the deflection signal-to-noise ratio (*16*) is

$$\text{DSNR} \triangleq \frac{E_1\{Q\} - E_0\{Q\}}{[V_0\{Q\}]^{1/2}} \approx \left(\frac{\pi M}{4-\pi}\right)^{1/2} \frac{\alpha^2}{4}$$

where $\alpha^2/2 = A^2/2\sigma_T^2$, $\sigma_T^2 = N_0T/4$, $E_i\{Q\}$ is the average value of Q for hypothesis i, and $V_0\{Q\}$ is the variance of Q for the noise only case.

(b) Show that the asymptotic form of the detection probability is

$$P_D \approx \int_{\xi}^{\infty} \frac{1}{(2\pi)^{1/2}} e^{-u^2/2}\, du$$

where

$$\xi = [V_T - (\pi/2)^{1/2}\sigma_T(\alpha^2/4)M]/\gamma^{1/2}$$

where

$$\gamma^{1/2} = \left(\frac{4-\pi}{2}\right)^{1/2} M^{1/2}\,\sigma_T$$

and V_T is the detection threshold. (Assume that the variance of Q for signal plus noise is the same as for noise alone.)

(c) Then show that as the number of pulses is doubled, the signal-to-noise ratio required to maintain the same detection probability is reduced by $2^{1/2}$.

8.9 Repeat the preceding problem for the receiver which uses the quadratic detector, Fig. 8-3A. Specifically, show the asymptotic results

(a) $\text{DSNR} = M^{1/2}\alpha^2/2$

(b) $P_D \approx \int_{\xi}^{\infty} (2\pi)^{1/2} e^{-u^2/2}\, du$ where

$$\xi = \frac{V_Q - M{\sigma_T}^2\alpha^2}{2{\sigma_T}^2 M^{1/2}}$$

and V_Q is the threshold for $\sum_{i=1}^{M} {q_i}^2$.

(c) As M is doubled, the signal-to-noise ratio required to maintain the same performance is reduced by $2^{1/2}$.

8.10 Suppose the receiver of Fig. 8-3B is modified as shown in Fig. 8-35 to include the step function nonlinearity. The nonlinearity can be expressed

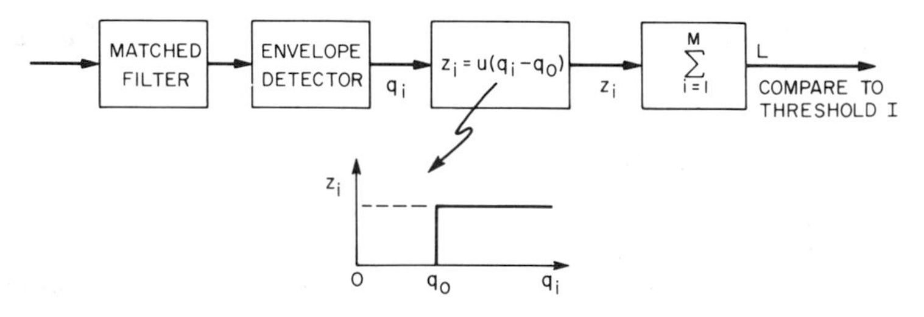

Fig. 8-35 A receiver using binary integration.

mathematically as $z = u(q - q_o)$ where $u(x)$ is 1 for $x \geq 0$, 0 for $x < 0$. The resulting receiver performs what is called binary integration.

Assume that q_o is chosen such that the probability of q_o being exceeded due to noise only is small. Denote this by p_f. Denote the probability of q_o being exceeded by signal plus noise as p_d. Assume that H_1 is chosen if $L \geq I$.

(a) What are the expressions for the probabilities of false alarm and detection in terms of p_d, p_f, I, and M?

(b) Assume that with a signal-to-noise ratio of 6 dB, q_o is chosen such that $p_d = \frac{1}{2}$ and $p_f = 10^{-2}$. Assume $m = 4$. For $I = 1, 2$, and 3. What is the quantitative degradition in detectability, in terms of signal-to-noise ratio, incurred by using the nonlinearity?

8.11 Consider the detection problem for which the hypotheses are

$$H_1: \quad r_i(t) = A_i \sin(\omega_c t + \theta_i) + n_i(t), \qquad i = 1, \ldots, M, \quad 0 \leq t \leq T$$
$$H_0: \quad r_i(t) = n_i(t), \qquad i = 1, \ldots, M, \quad 0 \leq t \leq T$$

The $n_i(t)$ are white Gaussian noise processes with spectral density $N_o/2$, and $n_i(t)$ is uncorrelated with $n_j(t)$, $i \neq j$. The phases θ_i are uniformly distributed $(0, 2\pi)$, and θ_i and θ_j are uncorrelated for $i \neq j$.

(a) Assume that A_i is a discrete random variable such that

$$P(A_i = 0) = 1 - p, \qquad P(A_i = A_o) = p$$

Determine the likelihood ratio. What is the asymptotic form of the likelihood ratio as A_o approaches zero?

8.12 For the preceding problem assume that the probability density function for A_i is

$$p(A_i) = (1 - p)\,\delta(A_i) + p\,\frac{A_i}{A_o^{\,2}} \exp\left(-\frac{A_i^{\,2}}{2A_o^{\,2}}\right)$$

Determine the likelihood ratio, and its form as A_o approaches zero.

References

1. Abramowitz, M., and Stegun, I. A., Handbook of Mathematical Functions, Nat. Bur. of Std. Appl. Math. Ser. 55 (August 1966).
2. Robertson, G. H., Operating characteristics for a linear detector of CW signals in narrow-band gaussian noise, *Bell Syst. Tech. J.* **46,** No. 4, 755–774 (1967).
3. Pearson, K., "Tables of the Incomplete Gamma Function." Cambridge Univ. Press, London and New York, 1965.
4. Pachares, J., A table of bias levels useful in radar detection problems, *IRE Trans. Inform. Theory* (March 1958).
5. Marcum, J. I., and Swerling, P., A statistical theory of target detection by pulsed radar, *IRE Trans. Inform. Theory* (April 1960).

6. Helstrom, C. W., "Statistical Theory of Signal Detection." Pergamon Press, Oxford, 1960.
7. Rubin, W. L., and DiFranco, J. V., Radar Detection, Electro-Technology (1964).
8. Skolnik, M. I., Introduction to Radar Systems." McGraw-Hill, New York, 1962.
9. Robertson, G. H., Computation of the noncentral chi-square distribution, *Bell Syst. Tech. J.* **48,** No. 1, 201–207 (1969).
10. Fry, T. C., "Probability and Its Engineering Uses." 2nd Ed. Van Nostrand, Princeton, New Jersey, 1965.
11. Robertson, G. H., Performance degradation by postdetector nonlinearities, *Bell Syst. Tech. J.* **47,** No. 3, 407–414 (1968).
12. Brennan, D. G. Linear diversity combining techniques, *Proc. IRE,* 1075–1101 (June 1959).
13. Baghdady, E. J., "Communication System Theory." McGraw-Hill, New York, 1961.
14. Pierce, J. N., Theoretical diversity improvement in frequency shift keying, *Proc. IRE,* 903–910 (May 1958).
15. Hahn, P. M. Theoretical diversity improvement in multiple frequency shift keying, *IRE Trans. Commun. Syst.* **CS-10,** No. 2, 177–184 (1962).
16. Lawson, J. L., and Uhlenbeck, G. F., "Threshold Signals, Radiation Laboratory Series," Vol. 24. McGraw-Hill, New York, 1950.

SUPPLEMENTARY BIBLIOGRAPHY

Texts:

Lee, Y. W., "Statistical Theory of Communication." Wiley, New York, 1968.

Schwartz, M., Bennett, W. R., and Stein, S., "Communication Systems and Techniques." McGraw-Hill, New York, 1966.

Van Trees, H. L., "Detection, Estimation, and Modulation Theory," Part I. Wiley, New York, 1968.

Wainstein L. A., and Zubakov, V. D., "Extraction of Signals From Noise" (*transl.* by R. A. Silverman), Prentice Hall, Englewood Cliffs, New Jersey, 1962.

On Communications and Diversity:

Aiken, R. T., Error Probability for binary signaling through a multipath channel, *Bell Syst. Tech. J.* (September 1967).

Bello, P. A., Binary error probabilities over selectively fading channels containing specular components, *IEEE Trans. Commun. Tech.* **COM-14,** No. 4, 400–406 (August 1966).

Brilliant, M. Fading loss in diversity systems, *IRE Trans. Commun. Syst.* (September 1960).

Bullington, K., Radio propagation fundamentals, *Bell Syst. Tech. J.* (May 1957).

Goldman, S., Certain aspects of coherence, modulation, and selectivity in information transmission systems, *IRE Nat. Convention Rec.* Part 4 (1956).

Jacobs, I. M., Probability-of-error bounds for binary transmission on the slowly fading rician channel, *IEEE Trans. Inform. Theory* **IT-2**, No. 4, 431–441 (October 1966).

Kaylor, R. L., A statistical study of selective fading of super-high frequency radio signals, *Bell Syst. Tech. J.* (September 1953).

Kennedy, R. S. and Lebow, I. L., Signal design for dispersive channels, *IEEE Spectrum* (March 1964).

Lindsey, W. C., Error probabilities for rician fading multichannel reception of binary and N-Ary signals, *IEEE Trans. Inform. Theory* **IT-10,** No. 4, 339–350 (October 1964).

Lindsey, W. C., Error probability for incoherent diversity reception, *IEEE Trans. Inform. Theory* **IT-11**, No. 4, 491–499 (October 1965).

Lindsey, W. C., Error probabilities for partially coherent diversity reception, *IEEE Trans. Commun. Tech.* **COM-14,** No. 5, 620–625 (October 1966).
Pierce, J. N., and Stein, S., Multiple diversity with nonindependent fading, *Proc. IRE* (January 1960).
Pierce, J. N., and Stein, S., Ultimate performance of M-Ary transmissions on fading channels, *IEEE Trans. Inform. Theory* **IT-12,** No. 1, 2–5 (January 1966).
Stein, S., and Johansen, D., A statistical description of coincidences among random pulse trains, *Proc. IRE* (May 1958).
Turin, G. L., On optimal diversity reception, *IRE Trans. Inform. Theory* (July 1961).
Vogelman, J. H., Ryerson, J. L., and Bickelhaupt, M. H., Tropospheric scatter system using angle diversity, *Proc. IRE* (May 1959).

On Radar and Detection:

Brennan, L. E., and Hill, F. S. Jr., A two step sequential procedure for improving the cumulative probability of detection in radars, *IEEE Trans. Military Electronics* (July–October 1965).
Bussgang, J. J., Nesbeda, P., and Safran, H., A unified analysis of range performance of CW, pulse, and pulse doppler radar, *Proc. IRE* (October 1959).
Galvin, A. A., A sequential detection system for the processing of radar returns, *Proc. IRE* (September 1961).
Harrington, J. V., An analysis of the detection of repeated signals in noise by binary integration, *IRE Trans. Inform. Theory* (March 1955).
Heatley, A. H., A short table of the Toronto Function, *Trans. Roy. Soc. Can.* Section III (1943).
Heidbreder, G. R., and Mitchell, R. L., Detection Probabilities for Log-Normally Distributed Signals, Aerospace Corporation, TR-669(9990)-6, AD 485827 (April 1966). Also, *IEEE Trans. Aerospace Electron. Syst.* (January 1967).
Helstrom, C. W., A range sampled sequential detection system, *IRE Trans. Inform. Theory* **IT-8,** No. 1, 43–47 (1962).
Kaplan, E. L., Signal-detection studies with application, *Bell Syst. Tech. J.* (March 1955).
Kendall, W. B., and Reed, I. S., A sequential test for radar detection of multiple targets, correspondence, *IEEE Trans. Inform. Theory* **IT-9,** No. 1, 51–53 (1963).
Mallett, J. D., and Brennan, L. E., Cumulative probability of detection for targets approaching a uniformly scanning search radar, *Proc. IEEE* (April 1963).
Marcus, M. B., and Swerling, P., Sequential detection in radar with multiple resolution elements, *IRE Trans. Inform. Theory* **IT-8,** No. 3, 237–245 (1962).
Middleton, D., Error Probabilities for the Detection of Radar Targets by Mismatched Receivers Using Half-Wave νth-law Rectifiers, Lincoln Lab. Tech. Note 1966–22, (14 July 1966).
Schwartz, M., Effects of signal fluctuation on the detection of pulse signals in noise, *IRE Trans. Inform. Theory* **IT-2,** No. 2, 66–71 (1956).
Selin, I., Detection of coherent radar returns of unknown doppler shift, *IEEE Trans. Inform. Theory*, **IT-11,** No. 3, 396–408 (1965).
Siebert, W. M., Some applications of detection theory to radar, *IRE Nat. Convention Rec.* Part 4 (1958).
Steensen, B. O., and Stirling, N. C., The amplitude distribution and false-alarm rate of filtered noise, *Proc. IEEE* (January 1965).
Swerling, P. Detection of fluctuating pulsed signals in the presence of noise, *IRE Trans. Inform. Theory* **IT-3,** No. 3, 175–178 (1957).

Thaler, S., and Meltzer, S. A., The amplitude distribution and false-alarm rate of noise after post-detection filtering, *Proc. IRE* (February 1961).

Worley, R. A., Optimum Binary Integration, Naval Electron. Lab. Center, San Diego, Calif. Rep. 1550, AD 672015 (26 March 1968).

Worley, R. A., Optimum thresholds for binary integration, correspondence, *IEEE Trans. Inform. Theory* (March 1968).

On Phrased Arrays:

Special issue on electronic scanning, *Proc. IEEE* (November 1968).

Cheston, T. C., Phased arrays for radars, *IEEE Spectrum* (November 1968).

Schell, A. C., Sletten, C. J., Blacksmith, P., and Pankiewicz, C. J., Electronic scanning, *Electro-Techn.* (November 1968).

Von Aulock, W. H., Properties of phased arrays, *Proc. IRE* (October 1960).

Allen, J. L., Array antennas, new applications for an old technique, *IEEE Spectrum* (November 1964).

Chapter 9

Detection of Signals in Colored Gaussian Noise

9.1 Introduction

The emphasis of the preceding chapters was on additive white noise. Unfortunately, in practice this assumption is often invalid; we therefore consider methods of detecting signals in colored noise. Recall that for the white noise case we first considered a flat bandlimited spectrum and found that appropriate uniformly spaced amplitude samples were statistically independent. The results for "continuous sampling" were then obtained by allowing the bandwidth to become infinite. For colored noise such uniformly spaced samples are dependent and the limiting form is difficult to evaluate. There is, however, another method which can be used to generate statistically independent "samples." While these samples are not amplitude samples they can be used to reconstruct the signal and provide a statistical description of the signal.

9.2 Karhunen–Loeve Expansion (*1*, *2*)

The approach will be to expand the received signal in a particular kind of series. This is analogous to a Fourier series expansion in terms of sine and cosine functions with appropriate weighting coefficients. Indeed, we could use the Fourier expansion but the weighting coefficients would be correlated.†

† Recall from Chap. 3 that the Fourier coefficients were uncorrelated only in the limit of large T.

This we wish to avoid. In particular we desire an expansion of a received signal in terms of a set of orthonormal functions $f_k(t)$ with uncorrelated weighting coefficients r_k.

A set of functions $\{f_i(t): i = 1, 2, \ldots\}$ defined over the interval $(0, T)$ is orthonormal on this interval if

$$\int_0^T f_i(t) f_j^*(t)\, dt = \begin{cases} 1, & i = j \\ 0, & i \neq j \end{cases}$$

If the orthonormal set is complete (*3, 4*), then an integrable square function $r(t)$, defined on the interval $(0, T)$ may be represented as

$$r(t) = \sum_k r_k f_k(t) \tag{9-1}$$

To determine the coefficients r_k, multiply each side of Eq. (9-1) by $f_j^*(t)$ and integrate over the interval $(0, T)$,

$$\int_0^T r(t) f_j^*(t)\, dt = \sum_k \int_0^T r_k f_k(t) f_j^*(t)\, dt$$

Since the functions are orthonormal, the right-hand side is zero except for $j = k$, and

$$r_k = \int_0^T r(t) f_k^*(t)\, dt \tag{9-2}$$

The coefficients r_k may be generated by inserting $r(t)$ into a linear filter matched to $f_k^*(t)$ and sampling the output at time $t = T$ (cf Fig. 9-1). The linear filter has an impulse response $f_k^*(T - t)$.

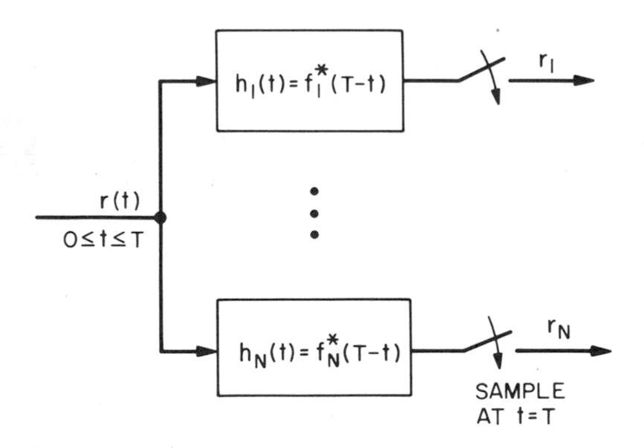

Fig. 9-1 Filters to generate Karhunen–Loeve coefficients.

We also desire the coefficients to be uncorrelated. To see how this may come about, assume that the function is the sum of a completely known signal plus noise, $r(t) = s(t) + n(t)$. (Obviously, the results can be specialized for the case $s(t) = 0$.) Assume the noise is wide-sense stationary with zero mean and autocorrelation function $R_n(\tau)$. Then using Eq. (9-2), the covariance of the coefficients is

$$E\{(r_k - \bar{r}_k)(r_j^* - \bar{r}_j^*)\} = E\left\{\int_0^T \int_0^T f_k^*(t_1) f_j(t_2) n(t_1) n^*(t_2)\, dt_1\, dt_2\right\}$$

$$= \int_0^T \int_0^T f_k^*(t_1) f_j(t_2) R_n(t_1 - t_2)\, dt_1\, dt_2 \tag{9-3}$$

where $\bar{r}_k$ is the average value of r_k.

Now, *suppose* that the following relationship holds

$$\int_0^T f_j(t_2) R_n(t_1 - t_2)\, dt_2 = \lambda_j f_j(t_1) \tag{9-4}$$

Such an equation is called a homogeneous integral equation. $R_n(t_1 - t_2)$ is called the kernel, $f_j(t)$ is an eigenfunction, and λ_j an eigenvalue. The term $f_j(t)$ and λ_j are also known as characteristic function and characteristic value, as well as proper function and proper value, and are analogous to the eigenvectors and eigenvalues discussed for matrices in Chap. 4. Substituting Eq. (9-4) into (9-3) yields

$$E\{(r_k - \bar{r}_k)(r_j^* - \bar{r}_j^*)\} = \int_0^T \lambda_j f_j(t) f_k^*(t)\, dt = \begin{cases} \lambda_j, & j = k \\ 0, & j \neq k \end{cases} \tag{9-5}$$

Consequently, if Eq. (9-4) holds, the coefficients are uncorrelated and their variance is equal to λ_j.

It must be shown that Eq. (9-4) is consistent with the assumed orthonormality of the functions $f_j(t)$. We proceed by multiplying Eq. (9-4) by $f_m^*(t_1)$ and integrating over the interval $(0 \leq t_1 \leq T)$

$$\lambda_j \int_0^T f_j(t_1) f_m^*(t_1)\, dt_1 = \int_0^T \int_0^T f_j(t_2) R_n(t_1 - t_2) f_m^*(t_1)\, dt_1\, dt_2 \tag{9-6}$$

Rewrite Eq. (9-4) by interchanging the variables t_1 and t_2

$$\int_0^T f_j(t_1) R_n(t_2 - t_1)\, dt_1 = \lambda_j f_j(t_2) \tag{9-7}$$

Taking its complex conjugate, and rewriting it for the mth eigenfunction we have

$$\lambda_m^* f_m^*(t_2) = \int_0^T f_m^*(t_1) R_n^*(t_2 - t_1)\, dt_1 \tag{9-8}$$

But $R_n^*(t_2 - t_1) = R_n(t_1 - t_2)$ so that Eq. (9-8) becomes

$$\lambda_m^* f_m^*(t_2) = \int_0^T f_m^*(t_1) R_n(t_1 - t_2)\, dt_1 \tag{9-9}$$

Multiplying this equation by $f_j(t_2)$ and integrating over the interval $(0 \leq t_2 \leq T)$

$$\lambda_m^* \int_0^T f_j(t_2) f_m^*(t_2)\, dt_2 = \int_0^T \int_0^T f_j(t_2) R_n(t_1 - t_2) f_m^*(t_1)\, dt_1\, dt_2 \tag{9-10}$$

Now, the right-hand sides of Eqs. (9-6) and (9-10) are identical so that the difference of their left-hand sides is zero. That is

$$(\lambda_j - \lambda_m^*) \int_0^T f_j(t) f_m^*(t)\, dt = 0 \tag{9-11}$$

The dummy variables t_1 and t_2 have been replaced by t. At this point several properties of the eigenfunctions and eigenvalues can be determined by using Eq. (9-11).

Assume the indices are equal, that is, $j = m$. Then

$$(\lambda_j - \lambda_j^*) \int_0^T |f_j(t)|^2\, dt = 0$$

Since the integral is positive, it follows that $\lambda_j = \lambda_j^*$ so that the *eigenvalues are real*. We therefore drop the complex conjugate notation associated with the eigenvalue.

Assume that $\lambda_j \neq \lambda_m$. Then it follows from Eq. (9-11) that

$$\int_0^T f_j(t) f_m^*(t)\, dt = 0$$

and therefore that eigenfunctions associated with distinct eigenvalues are orthogonal.

Assume $\lambda_j = \lambda_m$. For this case two or more eigenfunctions have the same eigenvalue, and the kernel is said to be degenerate. Such eigenfunctions may nevertheless be made orthonormal by the Gram–Schmidt orthogonalization procedure of forming a weighted linear combination of the degenerate eigenfunctions and choosing the weights to ensure the normalization. (See Chap. 6 for the analogous procedure for decorrelating random variables.)

For a real symmetric kernel, the complex conjugate of Eq. (9-7) may be written

$$\lambda_j f_j^*(t_2) = \int_0^T f_j^*(t_1) R_n(t_2 - t_1)\, dt_1$$

and it follows that $f_j^*(t_2)$ is also an eigenfunction. However, if there is no degeneracy, then it must be that $f_j(t) = f_j^*(t)$ so that the eigenfunctions corresponding to a real symmetric kernel are real.

From Eq. (9-4) we see that the eigenfunctions are arbitrary to within a multiplying constant. This constant may then be chosen to normalize the orthogonal functions, that is, to make them orthonormal.

We next show that nonzero eigenvalues are positive. Consider the inequality

$$E\left\{\left|\int_0^T g(t)n(t)\,dt\right|^2\right\} \geq 0$$

where $g(t)$ is an arbitrary function. The left-hand side may be written

$$\int_0^T\int_0^T g^*(t_1)g(t_2)E\{n(t_1)n^*(t_2)\}\,dt_1\,dt_2 = \int_0^T\int_0^T g^*(t_1)g(t_2)R_n(t_1-t_2)\,dt_1\,dt_2$$

Therefore

$$\int_0^T\int_0^T g^*(t_1)g(t_2)R_n(t_1-t_2)\,dt_1\,dt_2 \geq 0 \tag{9-12}$$

A kernel $R_n(t_1 - t_2)$ which satisfies this equation is said to be positive semidefinite or nonnegative definite. Multiply each side of Eq. (9-4) by $f_j^*(t_1)$ and integrate over the interval $(0 \leq t_1 \leq T)$

$$\lambda_j\int_0^T f_j(t_1)f_j^*(t_1)\,dt_1 = \int_0^T\int_0^T f_j(t_2)f_j^*(t_1)R_n(t_1-t_2)\,dt_1\,dt_2$$

Since the eigenfunctions are orthonormal,

$$\lambda_j = \int_0^T\int_0^T f_j(t_2)f_j^*(t_1)R_n(t_1-t_2)\,dt_1\,dt_2 \tag{9-13}$$

Since $R_n(t_1 - t_2)$ is positive semidefinite, the double integral is nonnegative and, hence, $\lambda_j \geq 0$.

If the inequality in Eq. (9-12) is strict, that is,

$$\int_0^T\int_0^T g^*(t_1)g(t_2)R_n(t_1-t_2)\,dt_1\,dt_2 > 0 \tag{9-14}$$

then the kernel is said to be positive definite. Thus for a positive definite kernel the eigenvalues are strictly positive. Furthermore, if $R_n(t_1, t_2)$ is positive definite, the eigenfunctions form a complete set (*3*). The set of functions $f_i(t)$ is said to be complete if the only function $g(t)$ satisfying

$$\int_0^T g(t)f_i(t)\,dt = 0$$

for all i is the function $g(t) \equiv 0$. In essence this means that if a function $g(t)$ not identically zero, is orthogonal to the set of functions $f_i(t)$, then $g(t)$ is also an eigenfunction.

The significance of the completeness property of the functions $f_i(t)$ is that the series expansion Eq. (9-1) converges in the mean to $r(t)$. The convergence in the mean implies that

$$\lim_{N\to\infty} E\left\{\left[\sum_{k=1}^{N} r_k f_k(t) - r(t)\right]^2\right\} = 0$$

The series expansion of Eq. (9-1) with the functions $f_k(t)$ chosen as the eigenfunctions of the integral equation (9-4) is the Karhunen–Loeve expansion. It is sometimes referred to as a generalized Fourier series.

Some other facts will be of eventual interest. One is Mercer's theorem which states that if $R_n(t_1, t_2)$ is positive semidefinite it can be expanded in terms of the eigenvalues and eigenfunctions as

$$R_n(t_1, t_2) = \sum_{j=1}^{\infty} \lambda_j f_j(t_1) f_j^*(t_2) \tag{9-15}$$

In some circumstances it is convenient to use the inverse kernel $R_n^{-1}(\tau, z)$ defined by

$$\int_0^T R_n^{-1}(t, \tau) R_n(\tau, z)\, d\tau = \delta(t - z), \qquad 0 \le t, z \le T \tag{9-16}$$

The inverse kernel has an expansion in terms of the eigenfunctions $f_k(t)$ and eigenvalues λ_k given by

$$R_n^{-1}(t, \tau) = \sum_{k=1}^{\infty} \frac{1}{\lambda_k} f_k(t) f_k^*(\tau) \tag{9-17}$$

The usefulness of the inverse kernel is mainly analytic. In practice, it may be difficult to determine.

In summary, we may represent an integrable square sample function of a random process over a finite observation interval in a series of orthonormal functions. The coefficients may be made to have the useful property of being uncorrelated.

9.3 Detection of Known Signals

With the above expansion we are now able to handle the problem of detection of known signals in nonwhite Gaussian noise. We again use the likelihood ratio to determine the optimum receiver. The signals will be represented equivalently by the coefficients of the Karhunen–Loeve expansion, and the likelihood functions will be the joint probability density of these coefficients. In that sense, the Karhunen–Loeve coefficients are "samples" of the signal. (Contrast this to the amplitude samples discussed in Chap. 6.)

The objective is to determine a receiver which chooses between two hypotheses

$$H_1: \; r(t) = s_1(t) + n(t)$$
$$H_0: \; r(t) = s_0(t) + n(t)$$

where $r(t)$ is the received signal and the signals $s_1(t)$ and $s_0(t)$ are completely known. The noise is Gaussian with autocorrelation function $R_n(\tau)$. We take as samples of the received signal the first N coefficients, r_k, of the Karhunen–Loeve expansion. Later we allow N to approach infinity. The coefficients r_k are, from Eq. (9-2),

$$r_k = \int_0^T r(t) f_k(t)\, dt \tag{9-18}$$

where the eigenfunctions are the solutions to the integral equation

$$\lambda_j f_j(t) = \int_0^T f_j(s) R_n(t - s)\, ds \tag{9-19}$$

Since $R_n(\tau)$ is real and symmetric, the eigenfunctions are real. The coefficients could be generated in practice by passing the signal $r(t)$ through a bank of filters matched to the eigenfunctions $f_k(t)$ as shown in Fig. 9-1. This is only academic since, as we let $N \to \infty$, the final result will not contain the coefficients explicitly.

By the theory above the coefficients are uncorrelated. Furthermore, the r_k result from a linear operation on the Gaussian process $r(t)$. They are therefore Gaussian and because they are uncorrelated, they are also statistically independent. We need only the mean and variance of the r_k to determine their joint density function.

Since $r(t) = s_i(t) + n(t)$, the coefficients are, for $i = 0, 1$

$$r_k = \int_0^T [s_i(t) + n(t)] f_k(t)\, dt = \int_0^T s_i(t) f_k(t)\, dt + \int_0^T n(t) f_k(t)\, dt \tag{9-20}$$

Denote the first integral by s_{ik}. Then

$$E_1\{r_k\} = s_{1k} \qquad \text{and} \qquad E_0\{r_k\} = s_{0k} \tag{9-21}$$

The variance of r_k is by Eq. (9-5), equal to λ_k and is the same for each hypothesis. The likelihood functions are then

$$p_1(\mathbf{r}) = \prod_{k=1}^{N} \left(\frac{1}{2\pi\lambda_k}\right)^{1/2} \exp\left[\frac{(r_k - s_{1k})^2}{-2\lambda_k}\right] \tag{9-22}$$

and

$$p_0(\mathbf{r}) = \prod_{k=1}^{N} \left(\frac{1}{2\pi\lambda_k}\right)^{1/2} \exp\left[\frac{(r_k - s_{0k})^2}{-2\lambda_k}\right] \tag{9-23}$$

The likelihood ratio is

$$\lambda(\mathbf{r}) = \frac{\exp[\sum_{k=1}^{N} (r_k - s_{1k})^2 / -2\lambda_k{}^2]}{\exp[\sum_{k=1}^{N} (r_k - s_{0k})^2 / -2\lambda_k]}$$

or

$$\lambda(\mathbf{r}) = \exp\left[\tfrac{1}{2}\sum_{k=1}^{N} \frac{s_{1k}}{\lambda_k}(2r_k - s_{1k}) - \tfrac{1}{2}\sum_{k=1}^{N} \frac{s_{0k}}{\lambda_k}(2r_k - s_{0k})\right] \tag{9-24}$$

and the log-likelihood ratio is

$$\ln \lambda(\mathbf{r}) = \tfrac{1}{2}\sum_{k=1}^{N} \frac{s_{1k}}{\lambda_k}(2r_k - s_{1k}) - \tfrac{1}{2}\sum_{k=1}^{N} \frac{s_{0k}}{\lambda_k}(2r_k - s_{0k}) \tag{9-25}$$

We next find an equivalent form for

$$G_1(N) \triangleq \tfrac{1}{2}\sum_{k=1}^{N} \frac{s_{1k}}{\lambda_k}(2r_k - s_{1k})$$

and determine its limit as $N \to \infty$. Observe from Eq. (9-20)

$$\begin{aligned} 2r_k - s_{1k} &= 2\int_0^T r(t) f_k(t)\,dt - \int_0^T s_1(t) f_k(t)\,dt \\ &= \int_0^T [2r(t) - s_1(t)] f_k(t)\,dt \end{aligned} \tag{9-26}$$

and therefore

$$G_1(N) = \int_0^T [r(t) - \tfrac{1}{2}s_1(t)] \sum_{k=1}^{N} \frac{s_{1k} f_k(t)}{\lambda_k}\,dt \tag{9-27}$$

Denote the summation term by $h_{1,N}(t)$, then

$$G_1(N) = \int_0^T [r(t) - \tfrac{1}{2}s_1(t)] h_{1,N}(t)\,dt$$

where

$$h_{1,N}(t) = \sum_{k=1}^{N} \frac{s_{1k} f_k(t)}{\lambda_k}$$

In the limit as $N \to \infty$ we get†

$$G_1 = \int_0^T [r(t) - \tfrac{1}{2}s_1(t)] h_1(t)\,dt \tag{9-28}$$

where

$$h_1(t) = \sum_{k=1}^{\infty} \frac{s_{1k} f_k(t)}{\lambda_k} \tag{9-29}$$

† We shall always assume that the limits exist. For convergence here, it is necessary and sufficient that $\sum_{k=1}^{\infty} |s_{1k}|^2/\lambda_k{}^2 < \infty$, see Grenander (5).

We next determine $h_1(t)$ in terms of more familiar functions. Multiply each side of Eq. (9-29) by $R_n(t-\tau)$ and integrate over the interval $(0 \le \tau \le T)$

$$\int_0^T R_n(t-\tau)h_1(\tau)\,d\tau = \sum_{k=1}^{\infty} \frac{s_{1k}}{\lambda_k} \int_0^T R_n(t-\tau)f_k(\tau)\,d\tau$$

The integral on the right-hand side is $\lambda_k f_k(t)$. Therefore

$$\int_0^T R_n(t-\tau)h_1(\tau)\,d\tau = \sum_{k=1}^{\infty} s_{1k} f_k(t) \tag{9-30}$$

Since

$$s_{1k} = \int_0^T s_1(t)f_k(t)\,dt$$

the summation is an expansion of $s_1(t)$ in terms of the eigenfunctions $f_k(t)$. Therefore Eq. (9-30) becomes

$$s_1(t) = \int_0^T R_n(t-\tau)h_1(\tau)\,d\tau \tag{9-31}$$

so that $h_1(t)$ is the solution to an integral equation.

The second summation in Eq. (9-25) is by analogy

$$G_0 \triangleq \int_0^T [r(t) - \tfrac{1}{2}s_0(t)]h_0(t)\,dt \tag{9-32}$$

where $h_0(t)$ is the solution to

$$s_0(t) = \int_0^T R_n(t-\tau)h_0(\tau)\,d\tau \tag{9-33}$$

The decision rule then becomes: Choose H_1 if $G = G_1 - G_0 \ge \ln \lambda_o$ where λ_o is determined by our criterion.

Thus choose H_1 if

$$G \triangleq \int_0^T [r(t) - \tfrac{1}{2}s_1(t)]h_1(t)\,dt - \int_0^T [r(t) - \tfrac{1}{2}s_0(t)]h_0(t)\,dt \ge \ln \lambda_o \tag{9-34}$$

where $h_1(t)$ and $h_0(t)$ are the solutions to the integral equations (9-31) and (9-33) respectively. An equivalent decision rule is to choose H_1 if

$$\int_0^T r(t)h_1(t)\,dt - \int_0^T r(t)h_0(t)\,dt \ge \ln \lambda_o + \int_0^T \tfrac{1}{2}s_1(t)h_1(t)\,dt - \int_0^T \tfrac{1}{2}s_0(t)h_0(t)\,dt$$

Two forms of this receiver are shown in Fig. 9-2. The first is a correlation receiver while the second receiver uses filters matched to $h_1(t)$ and $h_0(t)$.

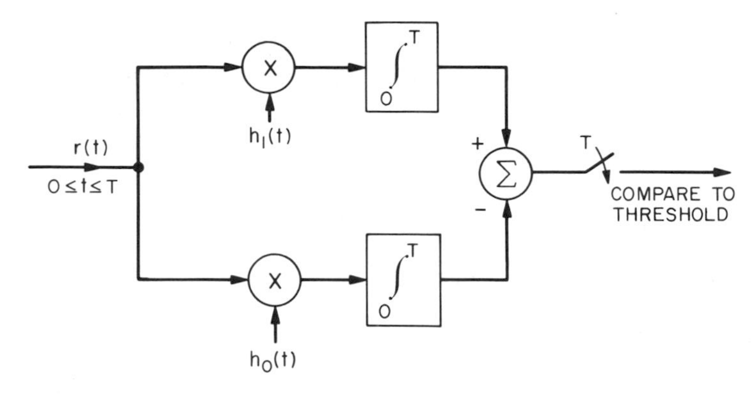

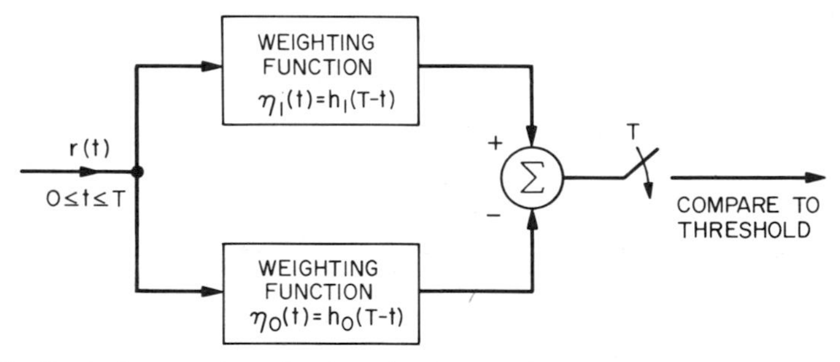

Fig. 9-2 Equivalent receivers for detecting known signals in nonwhite Gaussian noise.

The significance of the latter receiver should be mentioned. If the filter $h_1(T - t)$ is replaced by $\eta_1(t)$, Eq. (9-31) may be written

$$s_1(t) = \int_0^T R_n(t - \tau)\eta_1(T - \tau)\, d\tau$$

Replace the argument t by $T - x$:

$$s_1(T - x) = \int_0^T R_n(T - x - \tau)\eta_1(T - \tau)\, d\tau$$

with a change of variable ($z = T - \tau$) the equation becomes

$$s_1(T - x) = \int_0^T R_n(z - x)\eta_1(z)\, dz$$

This is identical to Eq. (6-63) which gave the solution for the filter which maximized the signal-to-noise ratio when the noise autocorrelation function was given by $R_n(\tau)$. Therefore the receiver of Fig. 9-2 uses generalized matched filters for the colored noise case.

It may be verified for the special case of white noise with $R_n(\tau) = (N_o/2)\,\delta(\tau)$ that the results above reduce to those derived in Sect. 6.2.

9.4 Receiver Performance

The receiver decision rule is to choose H_1 whenever $G \geq \ln \lambda_o$. An error is therefore made if H_1 is true and $G < \ln \lambda_o$, or if H_0 is true and $G \geq \ln \lambda_o$. To determine the probability of these errors we must first determine the density function for G. The statistic G is Gaussian since it results from a linear operation on a Gaussian process $r(t)$. The means of G under each hypothesis are

$$E_1\{G\} = \tfrac{1}{2}\int_0^T s_1(t)h_1(t)\,dt - \tfrac{1}{2}\int_0^T [2s_1(t) - s_0(t)]h_0(t)\,dt \tag{9-35}$$

and

$$E_0\{G\} = -\tfrac{1}{2}\int_0^T s_0(t)h_0(t)\,dt + \tfrac{1}{2}\int_0^T [2s_0(t) - s_1(t)]h_1(t)\,dt \tag{9-36}$$

We shall express these in terms of the signals and the autocorrelation function or its inverse.

Consider the equation for $h_1(t)$

$$s_1(t) = \int_0^T R_n(t-\tau)h_1(\tau)\,d\tau$$

Multiply this equation by $R_n^{-1}(z-t)$, Eq. (9-16), and integrate over the interval $(0 \leq t \leq T)$.

$$\int_0^T d\tau\, h_1(\tau)\int_0^T dt\, R_n(t-\tau)R_n^{-1}(z-t) = \int_0^T s_1(t)R_n^{-1}(z-t)\,dt$$

The inner integral is just $\delta(z-\tau)$ so that the expression on the left is just $h_1(z)$. Therefore

$$h_1(z) = \int_0^T s_1(t)R_n^{-1}(z-t)\,dt \tag{9-37}$$

and by analogy

$$h_0(z) = \int_0^T s_0(t)R_n^{-1}(z-t)\,dt \tag{9-38}$$

Substituting these into Eqs. (9-35) and (9-36) produces after some manipulation

$$E_1\{G\} = \tfrac{1}{2}\int_0^T\int_0^T [s_1(t) - s_0(t)]R_n^{-1}(t-x)[s_1(x) - s_0(x)]\,dt\,dx \tag{9-39}$$

and

$$E_0\{G\} = -\tfrac{1}{2}\int_0^T\int_0^T [s_1(t) - s_0(t)]R_n^{-1}(t-x)[s_1(x) - s_0(x)]\,dt\,dx \tag{9-40}$$

Denote

$$\sigma_G^2 = \int_0^T\int_0^T [s_1(t) - s_0(t)]R_n^{-1}(t-x)[s_1(x) - s_0(x)]\,dt\,dx \tag{9-41}$$

Then

$$E_1\{G\} = -E_0\{G\} = \sigma_G^2/2$$

The variance of G, which is the same under each hypothesis, is

$$V\{G\} = \int_0^T\int_0^T R_n(t-\tau)[h_1(t) - h_0(t)][h_1(\tau) - h_0(\tau)]\,dt\,d\tau \tag{9-42}$$

but, from Eqs. (9-31) and (9-33)

$$s_1(t) - s_0(t) = \int_0^T R_n(t-\tau)[h_1(\tau) - h_0(\tau)]d\tau$$

so that

$$V\{G\} = \int_0^T [s_1(t) - s_0(t)][h_1(t) - h_0(t)]\,dt \tag{9-43}$$

From Eqs. (9-37) and (9-38)

$$h_1(t) - h_0(t) = \int_0^T [s_1(x) - s_0(x)]R_n^{-1}(t-x)\,dx$$

so that

$$V\{G\} = \int_0^T\int_0^T [s_1(t) - s_0(t)]R_n^{-1}(t-x)[s_1(x) - s_0(x)]\,dt\,dx = \sigma_G^2 \tag{9.44}$$

The probability density functions of G may now be expressed.

$$P_0(G) = \frac{1}{(2\pi\sigma_G^2)^{1/2}}\exp\left[\frac{(G + \frac{1}{2}\sigma_G^2)^2}{-2\sigma_G^2}\right] \tag{9-45}$$

and

$$P_1(G) = \frac{1}{(2\pi\sigma_G^2)^{1/2}}\exp\left[\frac{(G - \frac{1}{2}\sigma_G^2)^2}{-2\sigma_G^2}\right] \tag{9-46}$$

Assume that the a priori probabilities are equal, and $\lambda_o = 1$. For such a system, the average probability of error is

$$P_e = P(D_0|H_1) = P(D_1|H_0)$$

In particular, assume H_0 is true. Then

$$P_e = \int_0^\infty \frac{1}{(2\pi\sigma_G^2)^{1/2}} \exp\left[-\frac{(G + \frac{1}{2}\sigma_G^2)^2}{2\sigma_G^2}\right] dG = \int_{\sigma_G/2}^\infty \frac{1}{(2\pi)^{1/2}} e^{-z^2/2}\, dz \quad (9\text{-}47)$$

This result parallels that derived for detection of a known signal in white noise, Eq. (6-32), except that $\sigma_G/2$ replaces $[(1-\rho)E/N_o]^{1/2}$. However, finding σ_G requires solving integral equations such as those given by Eqs. (9-31) and (9-33). This is at best tedious as we show later on.

9.5 Optimum Signal Waveform

The error probability given in Eq. (9-47) decreases monotonically as σ_G^2 increases. Therefore to obtain the best performance a set of signals $s_1(t)$ and $s_0(t)$ must be found which will maximize σ_G^2 subject to a constraint on the signal energy. The constraint is

$$\int_0^T [s_1^2(t) + s_0^2(t)]\, dt = \text{constant} = 2E$$

We therefore set out to maximize

$$Q = \sigma_G^2 - 2\mu E$$

where μ is a Lagrange undetermined multiplier, and Q is given by

$$\int_0^T \int_0^T [s_1(t) - s_0(t)] R_n^{-1}(t-x)[s_1(x) - s_0(x)]\, dt\, dx - \mu \int_0^T [s_0^2(t) + s_1^2(t)]\, dt$$

Let $m_1(t)$ and $m_0(t)$ represent the optimum waveforms for $s_1(t)$ and $s_0(t)$ respectively. Then for $0 \le t \le T$ let

$$s_1(t) = m_1(t) + \alpha_1 \Delta_1(t) \qquad \text{and} \qquad s_0(t) = m_0(t) + \alpha_0 \Delta_0(t)$$

where α_0 and α_1 are arbitrary multipliers and $\Delta_1(t)$ and $\Delta_0(t)$ are arbitrary functions defined over the interval $(0, T)$. We first maximize the equation for $s_1(t)$ holding $s_0(t)$ fixed and then maximize with respect to $s_0(t)$ holding $s_1(t)$ fixed. This results in two simultaneous equations. This problem is similar to the minimization problem which was discussed in Sect. 6.4. Using calculus of variation, the two equations are

$$\left.\frac{\partial Q(\alpha_1)}{\partial \alpha_1}\right|_{\alpha_1=0} = 0 \qquad \text{and} \qquad \left.\frac{\partial Q(\alpha_0)}{\partial \alpha_0}\right|_{\alpha_0=0} = 0$$

These lead to the following equations which $m_0(t)$ and $m_1(t)$ must satisfy

$$\int_0^T [m_1(t) - m_0(t)]R_n^{-1}(t - x)\, dt = 2\mu m_1(x) \tag{9-48}$$

and

$$\int_0^T [m_1(t) - m_0(t)]R_n^{-1}(t - x)\, dt = -2\mu m_0(x) \tag{9-49}$$

Since the left-hand sides of these equations are equal, it follows that the optimum waveforms satisfy the relation

$$m_1(x) = -m_0(x) \tag{9-50}$$

This is the same as the result found for the optimum binary system with white noise and corresponds to a signal correlation $\rho = -1$, cf Eq. (6-25).

While Eq. (9-50) gives the relation between signals, the functional form remains to be determined. Substituting Eq. (9-50) into Eq. (9-48) produces

$$\int_0^T m_1(t)R_n^{-1}(t - x)\, dt = \mu m_1(x) \tag{9-51}$$

Multiply each side by $R_n(\tau - x)$ and integrate over $(0 \le x \le T)$,

$$\int_0^T m_1(x)R_n(\tau - x)\, dx = (1/\mu)m_1(\tau) \tag{9-52}$$

Comparing this to the integral equation of Eq. (9-4), we see that the optimum signal waveform $m_1(x)$ is the eigenfunction of the kernel $R_n(x)$ corresponding to the eigenvalue $1/\mu$. We therefore state formally that the optimum signal waveform is a solution to the integral equation

$$\int_0^T m_1(x)R_n(\tau - x)\, dx = \lambda m_1(\tau) \tag{9-53}$$

To see which particular eigenfunction to choose, substitute $m_1(x) = -m_0(x)$ in the equation for $\sigma_G{}^2$, Eq. (9-41). Then

$$\sigma_G{}^2 = 4 \int_0^T \int_0^T m_1(t)R_n^{-1}(t - x)m_1(x)\, dx\, dt \tag{9-54}$$

Applying Eq. (9-51), this becomes

$$\sigma_G{}^2 = 4\mu \int_0^T m_1{}^2(t)\, dt = 4\mu E \tag{9-55}$$

where E is the signal energy. Therefore

$$\sigma_G{}^2 = 4E/\lambda \tag{9-56}$$

We set out to maximize σ_G^2 keeping E constant. We therefore get for the optimum signal, $m_1(t)$, that particular eigenfunction which has the *smallest* eigenvalue λ. We of course also choose $m_0(t) = -m_1(t)$.

It is interesting to apply this result to the white noise case. With $R_n(\tau) = (N_o/2)\,\delta(\tau)$ Eq. (9-53) reduces to

$$(N_o/2)m_1(t) = \lambda m_1(t)$$

The eigenvalues are therefore all equal and it makes no difference which waveforms we choose, provided that $m_1(t) = -m_0(t)$. This is in agreement with Sect. 6.2.

9.6 The Likelihood Functions

As we have done for the white noise case, we seek a convenient expression for the likelihood function in the limit as $N \to \infty$. The likelihood function for finite N given in Eq. (9-22) is repeated below with slight modification

$$p(\mathbf{r}) = \prod_{k=1}^{N} \left(\frac{1}{2\pi\lambda_k}\right)^{1/2} \exp\left[\frac{(r_k - s_k)^2}{-2\lambda_k}\right]$$

The multiplier $\prod_{k=1}^{N}(1/(2\pi\lambda_k))^{1/2}$ will cancel when a likelihood ratio is used. Therefore we are only interested in determining the limit

$$q \triangleq \lim_{N\to\infty} \prod_{k=1}^{N} \exp\left[\frac{(r_k - s_k)^2}{-2\lambda_k}\right]$$

or what is the same

$$q = \lim_{N\to\infty} \exp\left[\sum_{k=1}^{N} \frac{(r_k - s_k)^2}{-2\lambda_k}\right]$$

As $N \to \infty$

$$q = -\sum_{k=1}^{\infty} \frac{r_k^2}{2\lambda_k} + \tfrac{1}{2}\sum_{k=1}^{\infty} \frac{s_k}{\lambda_k}(2r_k - s_k) \tag{9-57}$$

In analogy with the derivation, Eq. (9-25) through (9-28), the second term is

$$\int_0^T [r(t) - \tfrac{1}{2}s(t)]h(t)\,dt$$

but

$$h(t) = \int_0^T s(\tau)R_n^{-1}(t-\tau)\,d\tau$$

and the second term becomes

$$\int_0^T \int_0^T [r(t) - \tfrac{1}{2}s(t)]s(\tau)R_n^{-1}(t-\tau)\,d\tau\,dt \tag{9-58}$$

We now determine an equivalent expression for the first term of Eq. (9-57). The coefficient r_k is given by Eq. (9-18). Thus

$$-\sum_{k=1}^{\infty} \frac{r_k^2}{2\lambda_k} = -\tfrac{1}{2}\int_0^T \int_0^T r(t)r(\tau) \sum_{k=1}^{\infty} \frac{f_k(t)f_k(\tau)}{\lambda_k}\, dt\, d\tau \tag{9-59}$$

The summation is an expansion for $R_n^{-1}(t-\tau)$ so that

$$-\sum_{k=1}^{\infty} \frac{r_k^2}{2\lambda_k} = -\tfrac{1}{2}\int_0^T \int_0^T r(t)r(\tau)R_n^{-1}(t-\tau)\, dt\, d\tau \tag{9-60}$$

Combining Eqs. (9-58) and (9-60) yields after some manipulation

$$q = -\tfrac{1}{2}\int_0^T \int_0^T [r(t)-s(t)]R_n^{-1}(t-\tau)[r(\tau)-s(\tau)]\, dt\, d\tau \tag{9-61}$$

The likelihood function then becomes

$$p(\mathbf{r}) = C \exp\left\{-\tfrac{1}{2}\int_0^T \int_0^T [r(t)-s(t)]R_n^{-1}(t-\tau)[r(\tau)-s(\tau)]\, dt\, d\tau\right\} \tag{9-62}$$

For noise only

$$p(\mathbf{r}) = C \exp\left[-\tfrac{1}{2}\int_0^T \int_0^T r(t)R_n^{-1}(t-\tau)r(\tau)\, dt\, d\tau\right] \tag{9-63}$$

(We shall not normally be concerned with the multiplier C.)

9.7 Integral Equations†

Integral equations are frequently encountered in detection theory. We found in Sect. 6.4 that the impulse response of the linear filter $h(t)$ which maximizes the signal-to-noise ratio is specified by the integral equation

$$s(T-t) = \int_0^T h(z)R_n(t-z)\, dz, \qquad 0 \le t \le T \tag{9-64}$$

Two integral equations were encountered in connection with the detection of signals in nonwhite noise. These are

$$\int_0^T R_n(t-\tau)f_j(\tau)\, d\tau = \lambda_j f_j(t), \qquad 0 \le t \le T \tag{9-65}$$

where $f_j(t)$ and λ_j are to be determined, and

$$\int_0^T R_n(t-\tau)q(\tau)\, d\tau = s(t), \qquad 0 \le t \le T \tag{9-66}$$

where $q(t)$ is to be determined.

† For the solution of integral equations, consult Davenport and Root (*1*), Helstrom (*2*), and Van Trees (*6*) and the texts and papers listed in the supplementary bibliography.

These equations are defined only over the time interval $0 \leq t \leq T$, and are not, therefore, simple convolution operations. Since we have assumed wide-sense stationary noise the argument of the kernel is simply the difference between t and τ. This is not true for the general case.

Another integral equation is also encountered. Assume that the autocorrelation function is the sum of parts corresponding to white and nonwhite noise. Then

$$R_n(t-\tau) = (N_o/2)\,\delta(t-\tau) + C_n(t-\tau)$$

Substituting this into Eq. (9-66) produces the integral equation

$$\int_0^T C_n(t-\tau)q(\tau)\,d\tau = -(N_o/2)q(t) + s(t), \qquad 0 \leq t \leq T \tag{9-67}$$

The integral equation (9-66) is known as a Fredholm integral equation of the first kind, whereas that of Eq. (9-67) is a Fredholm integral equation of the second kind. The equation (9-65) is called a homogeneous integral equation.

The solution to these integral equations is at best tedious and in many cases they do not appear solvable. After briefly discussing a formal solution we shall discuss a solution of the integral equations for a particular form of the kernel.

Formal Solution

The formal solution of the Fredholm integral equation of the first kind, Eq. (9-66), can be expressed in terms of the eigenvalues and eigenfunctions of the homogeneous integral equation. Assume that the kernel is positive definite so that the eigenfunctions form a complete set. Therefore, if the function $q(t)$ is integrable square it has the expansion

$$q(t) = \sum_{k=1}^{\infty} q_k f_k(t) \tag{9-68}$$

which converges in the mean. The coefficients are given by

$$q_k = \int_0^T q(t) f_k^*(t)\,dt \tag{9-69}$$

The known signal, $s(t)$, may also be expanded in terms of the same eigenfunctions as

$$s(t) = \sum_{k=1}^{\infty} s_k f_k(t) \tag{9-70}$$

where

$$s_k = \int_0^T s(t) f_k^*(t)\,dt \tag{9-71}$$

Substituting Eqs. (9-68) and (9-70) into (9-66), we get

$$\sum_{k=1}^{\infty} s_k f_k(t) = \sum_{k=1}^{\infty} q_k \int_0^T R_n(t-\tau) f_k(\tau)\, d\tau$$

The integral is equal to $\lambda_k f_k(t)$ and therefore

$$\sum_{k=1}^{\infty} s_k f_k(t) = \sum_{k=1}^{\infty} \lambda_k q_k f_k(t)$$

Equating the coefficients term by term we find that

$$q_k = s_k/\lambda_k \tag{9-72}$$

Using these coefficients in Eq. (9-68) we get for the formal solution

$$q(t) = \sum_{k=1}^{\infty} \frac{s_k}{\lambda_k} f_k(t) \tag{9-73}$$

Therefore, if the eigenvalues and eigenfunctions are known, and the s_k coefficients determined, we may formally determine the unknown function $q(t)$. Such an integral square function will exist if and only if (*1*, *5*)

$$\sum_{k=1}^{\infty} \frac{|s_k|^2}{\lambda_k^2} < \infty$$

Solution for Rational Power Spectral Density

If white noise is inserted into any lumped parameter linear filter containing only resistors, inductors, and capacitors, the power spectral density at the output of the filter will be a rational function, that is, a ratio of polynomials in ω^2. An example of this was seen for the low-pass filter of Example 2.7-1 for which the output spectral density was

$$S_n(\omega) = \frac{N_o \omega_o^2/2}{\omega^2 + \omega_o^2}$$

where $\omega_o = 1/RC$. The autocorrelation function is an exponential

$$R_n(\tau) = \tfrac{1}{4} N_o \omega_o e^{-\omega_o|\tau|}$$

For rational spectra the autocorrelation function can be expressed as the sum of exponential terms such as this. It is for such rational spectra that we shall investigate the solution to the integral equations. Some preliminary derivations are required. The objective is to transform the integral equations into differential equations with certain boundary conditions.

We assume the power spectral density can be expressed as

$$S_n(\omega^2) = \frac{N(\omega^2)}{D(\omega^2)} \tag{9-74}$$

where $N(\omega^2)$ is a polynomial in ω^2 of order n. The highest order of ω in the numerator is therefore $2n$. Similarly, $D(\omega^2)$ is a polynomial of order d in ω^2 or $2d$ in ω.

Consider the function

$$g(t) = \int_{-\infty}^{\infty} \frac{N(\omega^2)}{D(\omega^2)} e^{j\omega t} \frac{d\omega}{2\pi} \tag{9-75}$$

The first derivative is

$$\frac{dg(t)}{dt} = \int_{-\infty}^{\infty} j\omega \frac{N(\omega^2)}{D(\omega^2)} e^{j\omega t} \frac{d\omega}{2\pi} \tag{9-76}$$

and the $2d$th derivative is

$$\frac{d^{2d}g(t)}{dt^{2d}} = \int_{-\infty}^{\infty} (-\omega^2)^d \frac{N(\omega^2)}{D(\omega^2)} e^{j\omega t} \frac{d\omega}{2\pi} \tag{9-77}$$

Suppose $D(\omega^2) = a + b\omega^2$. Replace ω^2 by $-d^2/dt^2$ and consider the operator

$$D\left(-\frac{d^2}{dt^2}\right) = a - b\frac{d^2}{dt^2}$$

This operator has the following meaning

$$D\left(-\frac{d^2}{dt^2}\right)g(t) = ag(t) - b\frac{d^2g(t)}{dt^2} \tag{9-78}$$

where $g(t)$ is any differentiable function. In particular, let $g(t)$ be defined as in Eq. (9-75). Then

$$D\left(-\frac{d^2}{dt^2}\right)g(t) = D\left(-\frac{d^2}{dt^2}\right)\int_{-\infty}^{\infty} \frac{N(\omega^2)}{a + b\omega^2} e^{j\omega t} \frac{d\omega}{2\pi} \tag{9-79}$$

Using Eq. (9-78)

$$D\left(-\frac{d^2}{dt^2}\right)g(t) = a\int_{-\infty}^{\infty} \frac{N(\omega^2)}{a + b\omega^2} e^{j\omega t} \frac{d\omega}{2\pi} - b\frac{d^2}{dt^2}\int_{-\infty}^{\infty} \frac{N(\omega^2)}{a + b\omega^2} e^{j\omega t} \frac{d\omega}{2\pi} \tag{9-80}$$

Combining the two integrals, and using Eq. (9-77),

$$D\left(-\frac{d^2}{dt^2}\right)g(t) = \int_{-\infty}^{\infty} N(\omega^2)e^{j\omega t} \frac{d\omega}{2\pi} \tag{9-81}$$

To summarize, we form the operator $D(-d^2/dt^2)$ by replacing ω^2 in $D(\omega^2)$ by $-d^2/dt^2$ wherever it appears. For example, if ω^4 appears, it is replaced by d^4/dt^4, etc. This is equivalent to replacing ω^n by the operator $(-j)^n d^n/dt^n$.

Although Eq. (9-81) was developed for a specific $D(\omega^2)$, a little thought will show it holds for any polynomial $D(\omega^2)$.

We next consider a similar technique to handle $N(\omega^2)$ which appears in the numerator of Eq. (9-75). Observe that

$$\delta(t) = \int_{-\infty}^{\infty} e^{j\omega t} \frac{d\omega}{2\pi}$$

and

$$\frac{d\delta(t)}{dt} = \int_{-\infty}^{\infty} j\omega\, e^{j\omega t} \frac{d\omega}{2\pi}$$

and in general

$$\frac{d^{2n}}{dt^{2n}} \delta(t) = \int_{-\infty}^{\infty} (-\omega^2)^n e^{j\omega t} \frac{d\omega}{2\pi} \tag{9-82}$$

Consider the function

$$u(t) = \int_{-\infty}^{\infty} N(\omega^2) e^{j\omega t} \frac{d\omega}{2\pi} \tag{9-83}$$

As an example, let $N(\omega^2) = \alpha + \beta\omega^2$, and consider the operator

$$N\left(-\frac{d^2}{dt^2}\right) = \alpha - \beta \frac{d^2}{dt^2}$$

Then

$$N\left(-\frac{d^2}{dt^2}\right) \delta(t) = \alpha \int_{-\infty}^{\infty} e^{j\omega t} \frac{d\omega}{2\pi} - \beta \frac{d^2}{dt^2} \int_{-\infty}^{\infty} e^{j\omega t} \frac{d\omega}{2\pi}$$

Taking the second derivative and combining the integrals produces

$$N\left(-\frac{d^2}{dt^2}\right) \delta(t) = \int_{-\infty}^{\infty} N(\omega^2) e^{j\omega t} \frac{d\omega}{2\pi} \tag{9-84}$$

Although this relationship is shown for a specific $N(\omega^2)$ it is easily verified to hold for any polynomial $N(\omega^2)$. The procedure for generating the operator is the same as for $D(\omega^2)$ and is obtained by replacing ω^n by $(-j)^n d^n/dt^n$.

The autocorrelation function appearing as the kernel in the integral equations may be expressed in terms of the power spectral density as

$$R_n(t-\tau) = \int_{-\infty}^{\infty} S_n(\omega^2) e^{j\omega(t-\tau)} \frac{d\omega}{2\pi} = \int_{-\infty}^{\infty} \frac{N(\omega^2)}{D(\omega^2)} e^{j\omega(t-\tau)} \frac{d\omega}{2\pi} \tag{9-85}$$

The Fredholm integral equation of the first kind may now be expressed

$$s(t) = \int_0^T d\tau\, q(\tau) \int_{-\infty}^{\infty} \frac{d\omega}{2\pi} \frac{N(\omega^2)}{D(\omega^2)} e^{j\omega(t-\tau)} \tag{9-86}$$

From Eq. (9-81) this may be replaced by

$$D\left(-\frac{d^2}{dt^2}\right)s(t) = \int_0^T d\tau\, q(\tau) \int_{-\infty}^{\infty} \frac{d\omega}{2\pi} N(\omega^2)e^{j\omega(t-\tau)} \tag{9-87}$$

Applying Eq. (9-84), Eq. (9-87) may now be expressed as

$$D\left(-\frac{d^2}{dt^2}\right)s(t) = N\left(-\frac{d^2}{dt^2}\right) \int_0^T d\tau\, q(\tau)\, \delta(t-\tau)$$

or the equivalent

$$D\left(-\frac{d^2}{dt^2}\right)s(t) = N\left(-\frac{d^2}{dt^2}\right)q(t), \qquad 0 \le t \le T \tag{9-88}$$

Therefore the integral equation has been replaced by a differential equation which must be satisfied over the interval $(0, T)$. Davenport and Root (*1*) investigate the conditions for which the solution $q(t)$ of Eq. (9-88) satisfies the integral equation. They show that this solution satisfies the integral equation if and only if the known function $s(t)$ and its derivatives satisfy certain homogeneous boundary conditions at $t = 0$ and $t = T$. (See Exercises 6.11 and 9.9.) However, by adding delta functions and their derivatives to $q(t)$, a formal solution may always be found. In particular, if a solution to Eq. (9-88) is found without regard to the time interval, then

$$\sum_{i=0}^{2d-2n-2} a_i\, \delta^{(i)}(t) + b_i\, \delta^{(i)}(t-T) \tag{9-89}$$

should be added to that solution to satisfy the integral equation. Here $\delta^{(i)}(t)$ is the ith derivative of the delta function. The coefficients a_i and b_i may be determined by substituting the solution into the integral equation.

The solution will therefore consist of three parts: The complementary and particular solution to Eq. (9-88), and the delta functions.

The complementary solution is determined from the homogenous equation

$$N\left(-\frac{d^2}{dt^2}\right)q(t) = 0 \tag{9-90}$$

It is determined from the $2n$ roots of $N(\omega^2)$ and will consist of exponential terms.

The particular solution can be found by solving Eq. (9-88) directly without regard to the limits $(0, T)$ or, equivalently, using transform techniques (*6*). For example, disregarding the limits in the integral equation (9-66) and taking Fourier transforms would produce

$$Q(j\omega) = \frac{S(j\omega)}{S_n(\omega)}$$

where $Q(j\omega)$ and $S(j\omega)$ are the Fourier transforms of $q(t)$ and $s(t)$. Then the particular solution is

$$q_p(t) = \int_{-\infty}^{\infty} \frac{S(j\omega)}{S_n(\omega)} e^{j\omega t} \frac{d\omega}{2\pi}, \qquad 0 \le t \le T$$

We shall demonstrate this material with an example.

EXAMPLE 9.7-1 Consider the correlation receiver for a known signal in colored noise, Fig. 9-2. The operation performed is

$$\int_0^T r(t)h(t)\,dt$$

where $r(t)$ is the received signal and $h(t)$ is the solution to the Fredholm integral equation of the first kind, Eq. (9-31). Determine the function $h(t)$ and the correlator output for a noise power spectral density and the corresponding autocorrelation function given by

$$S_n(\omega) = \frac{2\alpha\beta}{\omega^2 + \beta^2} \qquad \text{and} \qquad R_n(\tau) = \alpha e^{-\beta|\tau|}; \qquad \alpha, \beta > 0$$

For this,

$$N(\omega^2) = 2\alpha\beta \qquad \text{and} \qquad D(\omega^2) = \omega^2 + \beta^2$$

and the differential equation to be solved is

$$2\alpha\beta h(t) = \beta^2 s(t) - s''(t), \qquad 0 \le t \le T$$

where the double prime denotes second derivative. The complementary solution is determined from $2\alpha\beta h_c(\tau) = 0$ and is therefore zero.

The particular solution is immediately seen to be

$$h_p(t) = \frac{1}{2\alpha\beta}[\beta^2 s(t) - s''(t)]$$

(It is left as an exercise to show the particular solution does not identically satisfy the integral equation.)

Since $2d - 2n - 2 = 0$, delta functions are included at $t = 0$, and $t = T$. However, no derivatives of delta functions are included. The total solution is therefore of the form

$$h(t) = \frac{1}{2\alpha\beta}[\beta^2 s(t) - s''(t)] + a\delta(t) + b\delta(t - T)$$

To determine the coefficients, this solution must be substituted into the integral equation producing

$$s(t) = \int_0^T d\tau\, \alpha e^{-\beta|t-\tau|}\left\{\frac{1}{2\alpha\beta}[\beta^2 s(\tau) - s''(\tau)] + a\delta(t) + b\delta(t-T)\right\}$$

The delta functions are meant to be included within the limits of integration. That is, the limits of integration are really 0_- and T_+. Eliminating the absolute value signs in the exponent, $s(t)$ may be rewritten as

$$s(t) = \int_0^t d\tau\, \alpha e^{-\beta(t-\tau)}\left\{\frac{1}{2\alpha\beta}[\beta^2 s(\tau) - s''(\tau)] + a\delta(\tau)\right\}$$

$$+ \int_t^T d\tau\, \alpha e^{\beta(t-\tau)}\left\{\frac{1}{2\alpha\beta}[\beta^2 s(\tau) - s''(\tau)] + b\delta(\tau - T)\right\}$$

Integrating those terms containing the delta functions

$$s(t) = a\alpha e^{-\beta t} + b\alpha e^{\beta(t-T)}$$

$$+ \int_0^t d\tau\, \alpha e^{-\beta(t-\tau)} \frac{1}{2\alpha\beta}[\beta^2 s(\tau) - s''(\tau)]$$

$$+ \int_t^T d\tau\, \alpha e^{\beta(t-\tau)} \frac{1}{2\alpha\beta}[\beta^2 s(\tau) - s''(\tau)]$$

Consider the integral

$$\int_0^t e^{\beta\tau} s''(\tau)\, d\tau$$

Integrating by parts it may be shown to be equal to

$$e^{\beta\tau} s'(t) - s'(0) - \beta e^{\beta t} s(t) - \beta s(0) + \int_0^t \beta^2 e^{\beta\tau} s(\tau)\, d\tau$$

Similarly, the integral $\int_t^T e^{-\beta\tau} s''(\tau)\, d\tau$ is found to be equal to

$$e^{-\beta T} s'(T) - e^{-\beta t} s'(t) + \mu e^{-\beta T} s(T) - \mu e^{-\beta t} s(t) + \int_t^T \beta^2 e^{-\beta\tau} s(\tau)\, d\tau$$

Substituting these back into the equation for $s(t)$

$$s(t) = s(t) + e^{-\beta t}\left[a\alpha + \frac{s'(0)}{2\beta} - \frac{s(0)}{2}\right] + e^{\beta t} e^{-\beta T}\left[\alpha b - \frac{s'(T)}{2\beta} - \frac{s(T)}{2}\right]$$

Since this equation must be satisfied for all t in the interval $(0, T)$, the coefficients of $e^{-\beta t}$ and $e^{\beta t}$ must be identically zero. Therefore, the values of a and b are

$$a = \frac{1}{2\alpha\beta}[\beta s(0) - s'(0)] \qquad \text{and} \qquad b = \frac{1}{2\alpha\beta}[\beta s(T) + s'(T)]$$

The solution for $h(t)$ is therefore

$$h(t) = \frac{1}{2\alpha\beta}\{[\beta s(0) - s'(0)]\,\delta(t) + [\beta s(T) + s'(T)]\,\delta(t - T) + \beta^2 s(t) - s''(t)\}$$
$$0 \leq t \leq T \qquad (9\text{-}91)$$

The output of the correlator is

$$\int_0^T r(t)h(t)\,dt = \frac{1}{2\alpha\beta}\{[\beta s(0) - s'(0)]r(0) + [\beta s(T) + s'(T)]r(T)\}$$
$$+ \frac{1}{2\alpha\beta}\int_0^T r(t)[\beta^2 s(t) - s''(t)]\,dt \qquad (9\text{-}92)$$

The effect of the delta function at the end points is to sample the received signal $r(t)$ at $t = 0$ and $t = T$. If derivatives of delta functions must be included, their effect is to sample the derivatives of the received signal at times $t = 0$, and $t = T$.

Homogeneous Integral Equation

We next treat the homogeneous integral equation (9-65) for spectral density functions which are rational. The method is substantially the same as for the Fredholm integral equation. One difference is that the eigenfunction solutions will not contain delta functions. This can be reasoned by observing that a delta function within the integral cannot produce a delta function outside the integral provided that the kernel itself is well behaved.

From Eq. (9-88) it is clear that the homogeneous integral equation may be transformed, for $0 \leq t \leq T$, into

$$N\left(-\frac{d^2}{dt^2}\right)f_k(t) = \lambda_k D\left(-\frac{d^2}{dt^2}\right)f_k(t)$$

or the equivalent

$$\left[N\left(-\frac{d^2}{dt^2}\right) - \lambda_k D\left(-\frac{d^2}{dt^2}\right)\right]f_k(t) = 0 \qquad (9\text{-}93)$$

which is a homogeneous differential equation. Its solution may be determined by straightforward but nevertheless tedious techniques as we shall see in the following example.

EXAMPLE 9.7-2 Determine the eigenfunctions and eigenvalues for the RC power spectral density of the preceding example.

Since

$$N\left(-\frac{d^2}{dt^2}\right) = 2\alpha\beta \qquad \text{and} \qquad D\left(-\frac{d^2}{dt^2}\right) = \beta^2 - \frac{d^2}{dt^2}$$

the differential equation to be solved is

$$\frac{1}{\lambda_k}[2\alpha\beta - \lambda_k \beta^2]f_k(t) + \frac{d^2}{dt^2}f_k(t) = 0$$

The roots of this equation are $\pm j\gamma_k$ where

$$\gamma_k{}^2 = \frac{2\alpha\beta}{\lambda_k} - \beta^2 \tag{9-94}$$

The eigenfunctions are therefore of the form

$$f_k(t) = a_k e^{j\gamma_k t} + b_k e^{-j\gamma_k t}$$

But the kernel is real and symmetric so that the eigenfunctions are also real and may be expressed as

$$f_k(t) = a_k \cos \gamma_k t + b_k \sin \gamma_k t \tag{9-95}$$

To determine the constants this solution is substituted into the homogeneous integral equation producing

$$\begin{aligned}\lambda_k a_k \cos \gamma_k t + \lambda_k b_k \sin \gamma_k t &= \int_0^T \alpha e^{-\beta|t-\tau|}(a_k \cos \gamma_k \tau + b_k \sin \gamma_k \tau)\, d\tau \\ &= \int_0^t \alpha e^{-\beta(t-\tau)}(a_k \cos \gamma_k \tau + b_k \sin \gamma_k \tau)\, d\tau \\ &\quad + \int_t^T \alpha e^{\beta(t-\tau)}(a_k \cos \gamma_k \tau + b_k \sin \gamma_k \tau)\, d\tau\end{aligned}$$

Carrying out the integration

$$\begin{aligned}\lambda_k a_k \cos \gamma_k t + \lambda_k b_k \sin \gamma_k t = {} & 2a_k\alpha\beta \cos \gamma_k t + 2b_k\alpha\beta \sin \gamma_k t + (b_k\alpha\gamma_k - a_k\alpha\beta)e^{-\beta t} \\ & + [a_k\alpha(-\beta \cos \gamma_k T + \gamma_k \sin \gamma_k T) \\ & + b_k\alpha(-\beta \sin \gamma_k T - \gamma_k \cos \gamma_k T)]e^{\beta(t-T)}\end{aligned}$$

The coefficients of $e^{-\beta t}$ and $e^{\beta t}$ must vanish identically in the interval $0 \le t \le T$ so that

$$\beta a_k - \gamma_k b_k = 0$$

and

$$(-\beta \cos \gamma_k T + \gamma_k \sin \gamma_k T)a_k + (-\beta \sin \gamma_k T - \gamma_k \cos \gamma_k T)b_k = 0$$

For a nontrivial solution for a and b, the determinant of

$$\begin{vmatrix} \beta & -\gamma_k \\ -\beta \cos \gamma_k T + \gamma_k \sin \gamma_k T & -\beta \sin \gamma_k T - \gamma_k \cos \gamma_k T \end{vmatrix}$$

must be zero. This produces the following transcendental equation to be solved for γ_k:

$$\sin \gamma_k T = \frac{2\gamma_k \beta}{\gamma_k^2 - \beta^2} \cos \gamma_k T$$

Following Helstrom's lead (*2*) and ordering the eigenvalues from $k = 0$ results in the equations

$$\begin{aligned} \xi_k \tan \xi_k &= \beta T/2, \qquad k \text{ even} \\ \xi_k \cot \xi_k &= -\beta T/2, \qquad k \text{ odd} \end{aligned} \tag{9-96}$$

where $\xi_k = \gamma_k T/2$. To determine the eigenvalues we of course use

$$\lambda_k = \frac{2\alpha\beta}{\gamma_k^2 + \beta^2}$$

Fredholm Integral Equation of the Second Kind

The Fredholm integral equation of the second kind is as given in Eq. (9-67). The function $q(t)$ is the only unknown and it may be reasoned that it does not contain any delta functions or derivatives of delta functions.

For the case where the kernel corresponds to the Fourier transform of a rational spectra, the integral equation can be stated in terms of a differential equation as shown in the preceding sections. In particular, if the power spectral density corresponding to the kernel $C_n(\tau)$ in Eq. (9.67) is given by the rational function $N'(\omega^2)/D'(\omega^2)$, the solution of the integral equation is the solution of the differential equation

$$\left[N'\left(-\frac{d^2}{dt^2}\right) + \frac{N_0}{2} D'\left(-\frac{d^2}{dt^2}\right)\right] q(t) = D'\left(-\frac{d^2}{dt^2}\right) s(t), \qquad 0 \le t \le T \tag{9-97}$$

9.8 Detection of Signals With Unknown Phase

We now treat the detection of signals with unknown phase when the additive noise is colored but nevertheless Gaussian. For the random phase case, results are presented in terms of the autocorrelation function of the complex envelope of the noise. For the narrowband representation of noise,

$$n(t) = x(t) \cos \omega_c t - y(t) \sin \omega_c t$$

the preenvelope is $\tilde{z}(t)e^{j\omega_c t}$ and the complex envelope is $\tilde{z}(t) = x(t) + jy(t)$. The autocorrelation function of the complex envelope is

$$E\{\tilde{z}(t)\tilde{z}^*(t-\tau)\} = R_x(\tau) + R_y(\tau) - jR_{xy}(\tau) + jR_{yx}(\tau)$$

For real functions $x(t)$ and $y(t)$, it follows from Eq. (2-18) that $R_{yx}(\tau) = R_{xy}(-\tau)$, and from Eq. (3-55) that $R_{xy}(-\tau) = -R_{xy}(\tau)$. Furthermore, from Eq. (3-51), $R_x(\tau) = R_y(\tau)$. The autocorrelation function of $\tilde{z}(t)$ is therefore

$$E\{\tilde{z}(t)\tilde{z}^*(t-\tau)\} = 2[R_x(\tau) - jR_{xy}(\tau)] = 2\tilde{R}(\tau) \qquad (9\text{-}98)$$

where $\tilde{R}(\tau)$ is the complex envelope of the noise autocorrelation function, Eq. (3-57).

The Likelihood Function

For the derivation of the likelihood function, it is assumed that the complex envelope of the information signal is a known function of time, and the signal phase is uniformly distributed from 0 to 2π. The received signal may be represented as

$$r(t) = \text{Re}\, \tilde{r}(t)e^{j\omega_c t}, \qquad 0 \le t \le T$$

where

$$\tilde{r}(t) = \tilde{A}(t)e^{j\theta} + \tilde{z}(t), \qquad 0 \le t \le T$$

is the complex envelope of the received signal, and $\tilde{A}(t)$ and $\tilde{z}(t)$ are the complex envelopes for the signal and noise respectively. The complex envelope $\tilde{r}(t)$ will be expanded in a Karhunen–Loeve expansion choosing the *complex* eigenfunctions as solutions of the integral equation

$$\lambda_k f_k(t) = \int_0^T \tilde{R}(t-\tau) f_k(\tau)\, d\tau$$

The expansion coefficients are given by

$$r_k = \alpha_k + j\beta_k = \int_0^T \tilde{r}(t) f_k^*(t)\, dt \qquad (9\text{-}99)$$

In a manner similar to that in Sect. 9.2 it can be shown that

$$E\{r_k r_m^*\} = 2\lambda_k \delta_{km}$$

so the complex coefficients are uncorrelated and since they are also Gaussian, they are statistically independent. Note that r_k is a bivariate Gaussian variable since it has both a real and an imaginary part given by α_k and β_k respectively.

It can also be shown that $E\{\tilde{r}(t)\tilde{r}(u)\} = 0$ and from this is follows that $E\{r_k r_m\} = 0$. It then follows for all k and m that

$$E\{\alpha_k \alpha_m\} = E\{\beta_k \beta_m\} = \lambda_k \delta_{km} \qquad \text{and} \qquad E\{\alpha_k \beta_m\} = 0$$

Therefore the real and imaginary parts of r_k are statistically independent and the bivariate density is easily written.

The expected value of r_k is

$$E\{r_k\} = E\left\{\int_0^T [\tilde{A}(t)e^{j\theta} + \tilde{z}(t)]f_k^*(t)\,dt\right\} \triangleq a_k e^{j\theta}$$

Since $r_k = \alpha_k + j\beta_k$

$$E\{\alpha_k\} = \text{Re}\, a_k e^{j\theta} \qquad \text{and} \qquad E\{\beta_k\} = \text{Im}\, a_k e^{j\theta}$$

where Im denotes imaginary part. The likelihood function of $r(t)$ for the first N Karhunen–Loeve coefficients ($2N$ dimension Gaussian density) is

$$p(\mathbf{r} \mid \theta) = \prod_{k=1}^{N} \left(\frac{1}{2\pi\lambda_k}\right) \exp\left[-\frac{(\alpha_k - \text{Re}\, a_k e^{j\theta})^2}{2\lambda_k} - \frac{(\beta_k - \text{Im}\, a_k e^{j\theta})^2}{2\lambda_k}\right]$$

Note however that the exponential term may be written in terms of the absolute value

$$-\frac{1}{2\lambda_k}|\alpha_k - \text{Re}\, a_k e^{j\theta} + j(\beta_k - \text{Im}\, a_k e^{j\theta})|^2$$

or the equivalent

$$-\frac{1}{2\lambda_k}|\alpha_k + j\beta_k - a_k e^{j\theta}|^2 = -\frac{1}{2\lambda_k}|r_k - a_k e^{j\theta}|^2$$

The likelihood function then becomes

$$p(\mathbf{r} \mid \theta) = \left[\prod_{k=1}^{N} \left(\frac{1}{2\pi\lambda_k}\right)\right] \exp\left[-\sum_{k=1}^{N} |r_k - a_k e^{j\theta}|^2 / 2\lambda_k\right]$$

The formal limit as N approaches infinity is

$$p(\mathbf{r} \mid \theta) = C \exp\left[-\sum_{k=1}^{\infty} |r_k - a_k e^{j\theta}|^2 / 2\lambda_k\right] \tag{9-100}$$

(We again ignore the multiplier C.)

In order to carry out the averaging over the phase we shall put the exponential in still another form. Working with just the absolute values term,

$$\begin{aligned} |r_k - a_k e^{j\theta}|^2 &= (r_k - a_k e^{j\theta})(r_k^* - a_k^* e^{-j\theta}) \\ &= |r_k|^2 + |a_k|^2 - 2\,\text{Re}\, r_k a_k^* e^{-j\theta} \end{aligned}$$

Then

$$p(\mathbf{r} \mid \theta) = C \exp\left[\sum_{k=1}^{\infty} \frac{|r_k|^2 + |a_k|^2}{-2\lambda_k}\right] \exp\left[\sum_{k=1}^{\infty} \operatorname{Re} \frac{r_k a_k^* e^{-j\theta}}{\lambda_k}\right] \tag{9-101}$$

Define the real quantities D and η by

$$D = \left|\sum_{k=1}^{\infty} \frac{r_k a_k^*}{\lambda_k}\right| \tag{9-102}$$

and

$$De^{j\eta} = \sum_{k=1}^{\infty} \frac{r_k a_k^*}{\lambda_k}$$

Then the second exponential term in Eq. (9-101) is

$$\operatorname{Re} De^{j(\eta - \theta)} = D \cos(\eta - \theta)$$

and the likelihood function is

$$p(\mathbf{r} \mid \theta) = C \exp\left[\sum_{k=1}^{\infty} \frac{|r_k|^2 + |a_k|^2}{-2\lambda_k}\right] \exp[D \cos(\eta - \theta)]$$

Averaging this function over the density function for the phase produces

$$p(\mathbf{r}) = C \exp\left[\sum_{k=1}^{\infty} \frac{|r_k|^2 + |a_k|^2}{-2\lambda_k}\right] I_0(D) \tag{9-103}$$

To determine an equivalent expression for the term D, define

$$U = \sum_{k=1}^{\infty} \frac{r_k a_k^*}{\lambda_k} = De^{j\eta}$$

and note that $D = |U|$. From Eq. (9-99)

$$U = \int_0^T \tilde{r}(t) \left[\sum_{k=1}^{\infty} \frac{a_k^* f_k^*(t)}{\lambda_k}\right] dt = \int_0^T \tilde{r}(t) \tilde{h}^*(t)\, dt \tag{9-104}$$

where $\tilde{h}^*(t)$ is used to represent the bracketed term. Thus

$$D = \left|\int_0^T \tilde{r}(t) \tilde{h}^*(t)\, dt\right| \tag{9-105}$$

and is the envelope of the function U. Thus the statistic D is generated by inserting the received signal $r(t)$ into a filter whose center frequency is comparable to that of the signal and whose complex impulse response is $\tilde{h}^*(t)$, and following the filter with an envelope detector.

The next step is to find an equivalent expression for the complex envelope of the filter function $\tilde{h}(t)$. Now,

$$\tilde{h}(\tau) = \sum_{k=1}^{\infty} \frac{a_k f_k(\tau)}{\lambda_k}$$

Multiplying each side by $\tilde{R}(t - \tau)$ and integrating over τ

$$\int_0^T \tilde{R}(t - \tau)\tilde{h}(\tau)\, d\tau = \sum_{k=1}^{\infty} \frac{a_k}{\lambda_k} \int_0^T \tilde{R}(t - \tau) f_k(\tau)\, d\tau = \sum_{k=1}^{\infty} a_k f_k(t) \quad (9\text{-}106)$$

But the summation term is just an expansion of the known complex envelope of the signal and is therefore equal to $\tilde{A}(t)$. Thus, the filter function $\tilde{h}(t)$ is the solution of the integral equation

$$\int_0^T \tilde{R}(t - \tau)\tilde{h}(\tau)\, d\tau = \tilde{A}(t), \qquad 0 \le t \le T \quad (9\text{-}107)$$

In summary, the average likelihood function is given by Eq. (9-103) where D is given by (9-105) and $\tilde{h}(\tau)$ is the solution to the integral equation (9-107).

The Binary Communications Case

We shall now apply the preceding results to the binary communications problem. The hypotheses are

$$H_1: \quad r(t) = a(t)\cos(\omega_1 t + \theta) + n(t), \qquad 0 \le t \le T$$
$$H_0: \quad r(t) = b(t)\cos(\omega_0 t + \psi) + n(t), \qquad 0 \le t \le T$$

where $a(t)$ and $b(t)$ are known real functions, and θ and ψ are statistically independent and uniformly distributed over the inverval $(0, 2\pi)$. The signal $r(t)$ is assumed to be a narrowband signal. The signal under hypothesis H_1 may be expressed as

$$\text{Re } a(t)e^{j(\omega_1 - \omega_c)t} e^{j\theta} e^{j\omega_c t} = \text{Re } \tilde{A}(t) e^{j\theta} e^{j\omega_c t}$$

Here ω_c is taken to be the "center frequency" of the narrowband of interest. The frequency difference $\Delta\omega = \omega_1 - \omega_c$ is assumed to be small so that the function $\tilde{A}(t)$ is a slowly changing function of time. The signal for hypothesis H_0 may be similarly expressed

$$\text{Re } b(t)e^{j(\omega_0 - \omega_c)t} e^{j\psi} e^{j\omega_c t} = \text{Re } \tilde{B}(t) e^{j\psi} e^{j\omega_c t}$$

and, as above, assume the frequency difference to be small.

In analogy to the preceding section, in particular Eq. (9-103), the likelihood functions are

$$p_1(\mathbf{r}) = C \exp\left[\sum_{k=1}^{\infty} \frac{|r_k|^2 + |a_k|^2}{-2\lambda_k}\right] I_0(D_1) \quad (9\text{-}108)$$

and

$$p_0(\mathbf{r}) = C \exp\left[\sum_{k=1}^{\infty} \frac{|r_k|^2 + |b_k|^2}{-2\lambda_k}\right] I_0(D_0) \tag{9-109}$$

The coefficients a_k and b_k are the Karhunen–Loeve coefficients for the expansion of $\tilde{A}(t)$ and $\tilde{B}(t)$ in terms of the eigenfunctions of the kernel $\tilde{R}(t)$. From Eq. (9-105)

$$D_1 = \left|\int_0^T \tilde{r}(t)\tilde{h}_1{}^*(t)\,dt\right| \qquad \text{and} \qquad D_0 = \left|\int_0^T \tilde{r}(t)\tilde{h}_0{}^*(t)\,dt\right|$$

Finally, the functions $\tilde{h}_1(t)$ and $\tilde{h}_0(t)$ may be found by solving integral equations such as Eq. (9-107) using $\tilde{A}(t)$ and $\tilde{B}(t)$ respectively. The log-likelihood ratio then becomes

$$\ln \lambda(\mathbf{r}) = \gamma + \ln I_0(D_1) - \ln I_0(D_0) \tag{9-110}$$

where

$$\gamma = \exp\left[\sum_{k=1}^{\infty} \frac{|a_k|^2 - |b_k|^2}{-2\lambda_k}\right]$$

and is a completely known function since $\tilde{A}(t)$ and $\tilde{B}(t)$ are assumed known.

The receiver therefore extracts D_1 and D_0. This may be implemented by passing the received signal $r(t)$ through the bandpass filters whose complex envelopes $\tilde{h}_1(t)$ and $\tilde{h}_0(t)$ are found by solving integral equation such as Eq. (9-107). The significance of the absolute value in the equations for D_i is to imply that the envelope of the filter output is sampled. This is analogous to the results which were obtained in Chap. 7 for white noise.

The Radar Case

The extension of the preceding results to the radar case is immediate and they are included for completeness. The hypotheses of interest are

$$H_1: \quad r(t) = A(t)\cos(\omega_c t + \theta) + n(t)$$
$$H_0: \quad r(t) = n(t)$$

where the preceding assumptions apply for $A(t)$, θ, and $n(t)$. For hypothesis H_1, the likelihood function is given by Eq. (9-103). For the noise only case, Eq. (9-103) can also be used to show that the corresponding likelihood function is

$$p_0(\mathbf{r}) = C \exp\left[\sum_{k=1}^{\infty} \frac{|r_k|^2}{-2\lambda_k}\right]$$

It is then easily seen that the likelihood ratio is

$$\lambda(\mathbf{r}) = \exp\left[\sum_{k=1}^{\infty} \frac{|a_k|^2}{-2\lambda_k}\right] I_0(D)$$

and the test statistic D is sufficient for the optimum receiver.

Exercises

9.1 Assume λ_k are the eigenvalues associated with the kernel $R_m(\tau)$. Recall $R_m(t_1, t_2) \triangleq E\{m(t_1)m(t_2)\}$. Show that

(a)
$$\int_0^T R_m(t_1, t_1)\, dt_1 = \sum_k \lambda_k$$

(b)
$$\int_0^T \int_0^T R_m^2(t_1, t_2)\, dt_1\, dt_2 = \sum_k \lambda_k^2$$

9.2 Consider the signal

$$r(t) = m(t) + n(t), \qquad 0 \leq t \leq T$$

where $m(t)$ is a zero mean colored Gaussian noise process with correlation function $R_m(\tau)$, and $n(t)$ is a zero mean white Gaussian noise process with autocorrelation function $N_o/2\, \delta(\tau)$.

(a) Suppose we expand $r(t)$ in a series

$$r(t) = \sum_{k=0}^{K} r_k \psi_k(t) = \sum_{k=0}^{K} (m_k + n_k)\psi_k(t)$$

where $r_k = \int_0^T r(t)\psi_k(t)\, dt$. (Similar definititions apply for m_k and n_k.) Show that the coefficients r_k will be statistically independent if the $\psi_k(t)$ are chosen to be the eigenfunctions of $R_m(\tau)$.

(b) Determine the covariance of the coefficients.

(c) Using a finite set of coefficients, what is the likelihood function of $r(t)$?

9.3 Find the likelihood function of the preceding problem by first determining the conditional density function of the r_k given the coefficients m_k.

9.4 Consider the detection problem having the hypotheses

$$H_1: \quad r(t) = m(t) + n(t), \qquad 0 \leq t \leq T$$
$$H_0: \quad r(t) = n(t), \qquad 0 \leq t \leq T$$

where $m(t)$ and $n(t)$ are the same as given in Exercise 9.2.

(a) Show that we may use as the statistic

$$\lambda_{\mathrm{T}} = \sum_{k=1}^{K} \frac{\lambda_k r_k^2}{2\lambda_k + N_o}$$

where λ_k are the eigenvalues associated with the kernel $R_m(\tau)$.

(b) Determine the mean and variance of λ_{T} for each hypothesis.

9.5 In the preceding problem, neglect any convergence problems and allow K to approach ∞ so that

$$\lambda_{\mathrm{T}} = \sum_{k=1}^{\infty} \frac{\lambda_k r_k^2}{2\lambda_k + N_o}$$

Show that this may be put in the integral form

$$\lambda_{\mathrm{T}} = \int_0^T \int_0^T r(t)r(v)h(t, v)\, dt\, dv$$

where $h(t, v)$ is the solution to the integral equation

$$h(t, v) + \frac{2}{N_o} \int_0^T R_m(v - u)h(t, u)\, du = \frac{1}{N_o} R_m(t, v), \qquad 0 \leq v, \quad t \leq T$$

9.6 Consider the detection of a known signal in a mixture of white and colored noise. That is,

$$H_1: \quad r(t) = s(t) + m(t) + n(t), \qquad 0 \leq t \leq T$$
$$H_0: \quad r(t) = m(t) + n(t), \qquad 0 \leq t \leq T$$

where $m(t)$ and $n(t)$ are described in Exercise 9.2. Show that we may use as the test statistic

$$\int_0^T [r(t) - s(t)]h(t)\, dt$$

where $h(t)$ is the solution to the integral equation

$$h(t) + \frac{2}{N_o} \int_0^T h(\tau)R_m(t - \tau)\, d\tau = \frac{1}{N_o} s(t), \qquad 0 \leq t \leq T$$

9.7 In Sect. 9.5, the optimum signal for a binary communication system was derived. Discuss qualitatively the significance of the solution.

9.8 Consider the detection of a known signal in colored Gaussian noise

$$H_1: \quad r(t) = s(t) + n(t), \qquad 0 \leq t \leq T$$
$$H_0: \quad r(t) = n(t), \qquad 0 \leq t \leq T$$

Denote the noise correlation function as $R_n(\tau)$. Using a Neyman–Pearson criterion, find the signal $s(t)$ which will maximize the detection probability.

9.9 Consider the integral equation for $h(t)$

$$\int_0^T h(\tau)R_n(t-\tau)\,d\tau = g(t), \qquad 0 \le t \le T$$

where $R_n(\tau)$ and $g(t)$ are known. Assume that $S_n(\omega) = \omega_1{}^2/(\omega^2 + \omega_1{}^2)$ or $R_n(\tau) = \frac{1}{2}\omega_1\, e^{-\omega_1|\tau|}$. Show that

$$h(t) = g(t) - \frac{1}{\omega_1{}^2}\frac{d^2}{dt^2}\,g(t)$$

will be a solution to the integral equation only if

$$\frac{g'(0)}{\omega_1} - g(0) = 0 \qquad \text{and} \qquad \frac{g'(T)}{\omega_1} + g(T) = 0$$

where $g'(0)$ and $g'(T)$ are the derivatives of $g(t)$ evaluated at $t = 0$, T respectively.

9.10 If the boundary values of the signal in the preceding problem are not satisfied, what is the total solution to the integral equation?

9.11 If the kernel to the Fredholm integral equation of the first kind is $R_n(\tau) = \alpha\delta(\tau) + \frac{1}{2}\beta\omega_1\, e^{-\omega_1|\tau|}$,

(a) What is the differential equation to be solved?
(b) What is the form of the homogeneous solution?
(c) If $g(t) = 1 - \cos\omega_0 t$, what is the form of the particular solution?

9.12 Assume a kernel of the form

$$R_n(\tau) = \begin{cases} 1 - \dfrac{|\tau|}{L}, & |\tau| \le L \\ 0, & \text{otherwise} \end{cases}$$

What differential equation is associated with determining the eigenfunctions of the integral equation

$$\int_0^T f(\tau)R_n(t-\tau)\,d\tau = \lambda f(t), \qquad 0 \le t \le T$$

where $T < L$?

References

1. Davenport, W. B. Jr., and Root, W. L., "An Introduction to the Theory of Random Signals and Noise." McGraw-Hill, New York, 1958.
2. Helstrom, C. W., "Statistical Theory of Signal Detection. Pergamon Press, Oxford, 1960.

3. Kolmogorov, A. N., and Fomin, F. V., "Measure, Lebesque Integrals, and Hilbert Space." Academic Press, New York, 1961.
4. Kac, M. and Siegert, J. F., An explicit representation of a stationary gaussian process, *Ann. Math. Statistics* **18,** 438–442 (1947).
5. Grenander, V., Stochastic processes and statistical inference, *Ark. Mat.* Band 1, **17,** (1950).
6. Van Trees, H. L., "Detection, Estimation, and Modulation Theory," Part I. Wiley, New York, 1968.

SUPPLEMENTARY BIBLIOGRAPHY

Texts on Integral Equations:

Laning, J. H., and Battin, R. H., "Random Processes in Automatic Control." McGraw-Hill, New York, 1956.

Middleton, D., "An Introduction to Statistical Communication Theory," McGraw-Hill, New York, 1960.

Mikhlin, S. G., "Integral Equations" (*Transl.* by A. H. Armstrong), A Pergamon Press Book, MacMillan, New York, 1964.

Smithies, F., "Integral Equations." Cambridge Univ. Press, London and New York, 1965

Papers on Integral Equations:

Baggeroer, A. B., A State Variable Approach to the Solution of Fredholm Integral Equations, Research Lab. of Electron. Massachusetts Inst. of Techn. Tech. Rep. 459, AD 666215 (15 November 1967). Also presented at 1967 *IEEE* Int. Conf. on Commun.

Capon, J., Asymptotic eigenfunctions and eigenvalues of a homogeneous integral equation, *IRE Trans. Inform. Theory* **IT-8,** No. 1, 2–4 (1962).

Helstrom, C. W., Solution of the detection integral equation for stationary filtered white noise, *IEEE Trans. Inform. Theory* **IT-11,** No. 3, 335–339 (1965).

Kailath, T., Some integral equations with nonrational kernels, *IEEE Trans. Inform. Theory* **IT-12,** No. 4, 442–447 (1966).

Miller, K. S., and Zadeh, L. A., Solution of an integral equation occurring in the theories of prediction and detection, *IRE Trans. Inform. Theory* (June 1956).

Mittra, R., On the solution of an eigenvalue equation of the Weiner–Hopf type in finite and infinite ranges, *IRE Convention Rec.* **7,** Part 4, 170–173 (1959).

Slepian, D., and Kadota, T. T., Four integral equations of detection theory, *J. Soc. Ind. Appl. Math.* **17**, No. 6, 1102–1117 (1969).

Youla, D. C., The solution of a homogenous Weiner–Hopf integral equation occurring in the expansion of second-order stationary random functions, *IRE Trans. Inform. Theory*, 187–193 (September 1957).

Zadeh, L. A., and Ragazzini, J. R., An extension of Weiner's theory of prediction, *J. Appl. Phys.* **21,** 645–655 (1950).

Zadeh, L. A., and Ragazzini, J. R., Optimum filters for the detection of signals in noise, *Proc. IRE*, 1223–1231 (October 1952).

On Detection:

Jacobs, I., Energy detection of gaussian communication signals, *Proc. Nat. Commun. Symp. 10th*, 440–448 (October 1964).

Kadota, T. T., Optimum reception of M-Ary gaussian signals in gaussian noise, *Bell Syst. Tech. J.* (November 1965).

Kadota, T. T., Simultaneously orthogonal expansion of two stationary gaussian processes—examples, *Bell Syst. Tech. J.* (September 1966).

Kadota, T. T., Simultaneous diagonalization of two covariance kernels and application to second-order stochastic processes, *J. Soc. Ind. Appl. Math.* **15,** (1967).

Kadota, T. T., Differentiation of Karhunen-Loève expansion and application to optimum reception of sure signals in noise, *IEEE Trans. Inform. Theory* **IT-13,** No. 2, 255–260 (1967).

Kadota, T. T., Examples of optimum detection of gaussian signals and interpretation of white noise, *IEEE Trans. Inform. Theory* **IT-14,** No. 5, 725–734 (1968).

Kailath, T., Some results in singular detection, *Inform. Control* **9,** 130–152 (1966). Also Stanford Univ. Tech. Rep. No. 7050–2, AD 468308 (June 1965).

Kailath, T., A projection method for signal detection in colored gaussian noise, *IEEE Trans. Inform. Theory* **IT-13,** No. 3, 441–447 (July 1967).

Landau, H. J., and Pollak, H. O., Prolate spheroidal wave functions, fourier analysis and uncertainty—III: The dimension of the space of essentially time- and band-limited signals, *Bell Syst. Tech. J.* (July 1962).

Mullikin, T. W., and Selin, I., The likelihood ratio filter for the detection of gaussian signals in white noise, *IEEE Trans. Inform. Theory* **IT-11,** No. 4, 513–515 (1965).

Root, W. L., Singular gaussian measures in detection theory, "Time Series Analysis" (M. Rosenblatt, ed.), Chapter 20. Wiley, New York, 1963.

Schwartz, M. I., On the detection of known binary signals in gaussian noise of exponential covariance, *IEEE Trans. Inform. Theory* **IT-11,** No. 3, 330–335 (1965).

Slepian, D., Some comments on the detection of gaussian signals in gaussian noise, *IRE Trans. Inform. Theory* (June 1958).

Slepian, D., and Pollak, H. O., Prolate spheroidal wave functions, fourier analysis and uncertainty—Parts I and II, *Bell Syst. Tech. J.* (January 1961).

Weinstein, S. B., The white noise approximation, concise papers, *IEEE Trans. Commun. Techn.* (February 1969).

Chapter 10

Estimation of Signal Parameters

10.1 Introduction

In the preceding chapters our primary concern was with the detection of signals. We wished to know whether or not a signal was present, or for the communications type problem, which one of several signals was present. Very little emphasis was placed on explicit determination of the signal parameters themselves. In a few cases a separate hypothesis was assigned for each possible parameter. Then having chosen a given hypothesis, we accomplished both detection and estimation of the signal parameter. In many instances knowledge of these parameters is important. For example, in a radar system knowledge of the carrier frequency of the received signal would enable us to calculate the radial velocity of a target. Similarly, the amplitude of the received signal is sometimes related to the size of a target, and time of arrival of a radar signal may be related to the range of a target. Similar problems arise in communication systems.

We are therefore led in many cases to measure the various signal parameters. However, the received signal is invariably corrupted by noise and is therefore random in nature. Under such circumstances only estimates of the parameters can be obtained.

As in the case of detection, we have available a received waveform which consists of the transmitted signal, which may be distorted in the transmission medium, and local additive receiver noise. The signal portion is a function of time and m parameters denoted by the vector $\boldsymbol{\alpha}(= \alpha_1, \ldots, \alpha_m)$. The received signal is expressed as

$$r(t) = s(t, \alpha_1, \ldots, \alpha_m) + n(t) \tag{10-1}$$

Based on either a finite number of samples of $r(t)$, or on a continuous observation of $r(t)$, we wish to estimate the parameters of $\boldsymbol{\alpha}$. There are several estimators which may be used and which depend on our criterion of "goodness." Some of these are discussed below.

10.2 Bayes Estimate

As was down for detection theory, assume that a cost function is associated with the estimation error. Denote the estimate of $\boldsymbol{\alpha}$ by $\hat{\boldsymbol{\alpha}}$, and denote the scalar cost function by $C(\hat{\boldsymbol{\alpha}}, \boldsymbol{\alpha})$. Examples of cost functions for a scalar estimate are

$$C(\hat{\alpha}, \alpha) = (\hat{\alpha} - \alpha), \qquad C(\hat{\alpha}, \alpha) = (\hat{\alpha} - \alpha)^2, \qquad C(\hat{\alpha}, \alpha) = |\hat{\alpha} - \alpha|$$

$$C(\hat{\alpha}, \alpha) = \begin{cases} 1, & |\hat{\alpha} - \alpha| > A \\ 0, & |\hat{\alpha} - \alpha| < A \end{cases}$$

The second of these functions, the quadratic cost function, is quite commonly used in practice. It produces a minimum mean-squared error estimator and is analytically convenient to use.

The estimate $\hat{\boldsymbol{\alpha}}$ is a function of the signal observations, denoted by $\mathbf{r}$. To emphasize this we shall occasionally write the estimate as $\hat{\boldsymbol{\alpha}}(\mathbf{r})$ instead of just $\hat{\boldsymbol{\alpha}}$.

Often the joint probability density function $p(\mathbf{r}, \boldsymbol{\alpha})$ of the observations $\mathbf{r}$ and the parameters $\boldsymbol{\alpha}$ is known. There were many examples of this in previous chapters. For example, with a single sample of Gaussian noise with zero mean and unknown variance $\alpha = \sigma^2$, the density function $p(r|\sigma^2)$ is given by

$$p(r \mid \sigma^2) = \frac{1}{(2\pi)^{1/2}\sigma} \exp(-r^2/2\sigma^2)$$

If the variance is also a random variable with an a priori density function $p(\sigma^2)$, then $p(r, \sigma^2)$ is given by

$$p(r, \sigma^2) = p(r \mid \sigma^2)p(\sigma^2) = \frac{p(\sigma^2)}{(2\pi)^{1/2}\sigma} \exp(-r^2/2\sigma^2)$$

The joint density function of the observations and the random parameters may be written

$$p(\mathbf{r}, \boldsymbol{\alpha}) = p(\mathbf{r}|\boldsymbol{\alpha})p(\boldsymbol{\alpha}) = p(\boldsymbol{\alpha}|\mathbf{r})p(\mathbf{r})$$

so that the conditional density function for $\boldsymbol{\alpha}$ given a set of observations $\mathbf{r}$ is

$$p(\boldsymbol{\alpha} \mid \mathbf{r}) = \frac{p(\mathbf{r} \mid \boldsymbol{\alpha}) p(\boldsymbol{\alpha})}{p(\mathbf{r})} \tag{10-2}$$

The average cost associated with the selection of a set of estimates $\hat{\boldsymbol{\alpha}}$ given the observations $\mathbf{r}$ is

$$C(\hat{\boldsymbol{\alpha}} \mid \mathbf{r}) = \int_{\{\boldsymbol{\alpha}\}} C(\hat{\boldsymbol{\alpha}}, \boldsymbol{\alpha}) p(\boldsymbol{\alpha} \mid \mathbf{r})\, d\boldsymbol{\alpha} \tag{10-3}$$

This is the conditional cost and is averaged over all possible values of the parameters $\boldsymbol{\alpha}$. The total average cost is obtained by averaging over all $\mathbf{r}$

$$C(\hat{\boldsymbol{\alpha}}) = \int_{\{\mathbf{r}\}} C(\hat{\boldsymbol{\alpha}} \mid \mathbf{r}) p(\mathbf{r})\, d\mathbf{r} \tag{10-4}$$

An estimator $\hat{\boldsymbol{\alpha}}$ which minimizes this cost is called a Bayes estimator. Minimizing cost is Bayes criterion.

The total average cost is minimized if, for given $\mathbf{r}$, we choose as our estimate that value for $\hat{\boldsymbol{\alpha}}$ which minimizes the conditional cost $C(\hat{\boldsymbol{\alpha}} \mid \mathbf{r})$. As an example of this minimization, consider the case of a single parameter α and a quadratic cost function

$$C(\hat{\alpha}, \alpha) = (\hat{\alpha} - \alpha)^2$$

The objective is therefore to find the minimum mean-square error estimator. It may be obtained by differentiating

$$C(\hat{\alpha} \mid \mathbf{r}) = \int_{\{\alpha\}} (\hat{\alpha} - \alpha)^2 p(\alpha \mid \mathbf{r})\, d\alpha$$

with respect to $\hat{\alpha}$ and setting the result to zero.

$$\frac{\partial C(\hat{\alpha} \mid \mathbf{r})}{\partial \hat{\alpha}} = 0 = \int_{\{\alpha\}} 2(\hat{\alpha} - \alpha) p(\hat{\alpha} \mid \mathbf{r})\, d\alpha$$

Solving for $\hat{\alpha}$:

$$\hat{\alpha} = \frac{\int_{\{\alpha\}} \alpha p(\alpha \mid \mathbf{r})\, d\alpha}{\int_{\{\alpha\}} p(\alpha \mid \mathbf{r})\, d\alpha}$$

Taking note that $\int_{\{\alpha\}} p(\alpha \mid \mathbf{r})\, d\alpha = 1$ we get the Bayes estimator

$$\hat{\alpha} = \int_{\{\alpha\}} \alpha p(\alpha \mid \mathbf{r})\, d\alpha = E\{\alpha \mid \mathbf{r}\} \tag{10-5}$$

This is the *mean* of the a posteriori probability density function of α given the set of measurements $\mathbf{r}$. This result is independent of the underlying probability distribution. This is quite an important result.

The conditional mean is also the Bayes estimate for a much broader class of cost functions than quadratic. Class I. Assume that the cost is a function of the difference, $\varepsilon \triangleq \hat{\alpha} - \alpha$, between the estimator and the true value of the parameter. Assume that the cost function is symmetric, $C(\varepsilon) = C(-\varepsilon)$, and is a convex function.† Then (*1–3*), the conditional expectation, Eq. (10-5), is the optimum (Bayes) estimator. Class II. Assume that the cost function is symmetric, and nondecreasing, $C(\varepsilon_2) \geq C(\varepsilon_1)$ for values of ε such that $\varepsilon_2 \geq \varepsilon_1 \geq 0$. Assume the a posteriori probability density function $p(\alpha \mid \mathbf{r})$ is unimodal, symmetric about its mean, and such that

$$\lim_{\alpha \to \infty} C(\alpha) p(\alpha \mid \mathbf{r}) = 0$$

Then, the conditional mean Eq. (10-5) is the optimum estimator (*1–3*).

Clearly, if $p(\alpha \mid \mathbf{r})$ is Gaussian, the unimodal and symmetric conditions on the density function are satisfied. Then the mean-square error criterion will produce the same estimator as will a much larger class of criteria.

10.3 Maximum A Posteriori Estimate

If a cost function is not available, a reasonable criterion is to maximize the a posteriori probability (*4*) given by Eq. (10-2). That is, choose as the estimate $\hat{\boldsymbol{\alpha}}$ that value of $\boldsymbol{\alpha}$ which maximizes $p(\boldsymbol{\alpha} \mid \mathbf{r})$. This is equivalent to choosing the most likely value. If $p(\boldsymbol{\alpha} \mid \mathbf{r})$ has only one peak (unimodal density function), then this criterion selects the mode (maximum point) of the a posteriori distribution. If $p(\boldsymbol{\alpha} \mid \mathbf{r})$ is Gaussian, the mode and the mean coincide, in which case the maximum a posteriori estimate is the same as the mean-square error estimate.

Maximizing the a posteriori density, Eq. (10-2), is equivalent to maximizing the product $p(\mathbf{r} \mid \boldsymbol{\alpha})p(\boldsymbol{\alpha})$ since $p(\mathbf{r})$ is independent of the parameter $\boldsymbol{\alpha}$. Now, suppose that the density function of $\boldsymbol{\alpha}$ is very broad and void of peaks. Indeed, in the extreme assume that $p(\boldsymbol{\alpha})$ is constant (uniformly distributed) over the range of interest. Except for the limits on the range of $\boldsymbol{\alpha}$, this implies a lack of any real knowledge of the parameter $\boldsymbol{\alpha}$. In this event, maximizing $p(\boldsymbol{\alpha} \mid \mathbf{r})$ is equivalent to maximizing $p(\mathbf{r} \mid \boldsymbol{\alpha})$. The resulting estimate is very important and is discussed below. It is called a maximum-likelihood estimate.

10.4 Maximum-Likelihood Estimates

In the event that neither cost functions nor a priori knowledge of the parameters is available we may use maximum-likelihood estimates (*5–7*). The

† The function is convex (upward) if a chord connecting any two points of the function is greater than or equal to the function.

maximum-likelihood estimate is that estimate which maximizes the likelihood function $p(\mathbf{r} \mid \boldsymbol{\alpha})$† which was used in previous chapters. As an example suppose we are to estimate the mean of a stationary Gaussian process with known variance σ^2 from N independent samples of the process. Denoting the unknown mean by μ, the likelihood function is

$$p(\mathbf{r} \mid \mu) = \prod_{i=1}^{N} \left(\frac{1}{2\pi\sigma^2}\right)^{1/2} \exp\left[-\frac{(r_i - \mu)^2}{2\sigma^2}\right] \tag{10-6}$$

We take for our estimate, $\hat{\mu}$, that value of μ which maximizes $p(\mathbf{r} \mid \mu)$. In this case, it is easier to use the log-likelihood function

$$\ln p(\mathbf{r} \mid \mu) = K - \sum_{i=1}^{N} \frac{(r_i - \mu)^2}{2\sigma^2} \tag{10-7}$$

where K is a constant independent of μ. The derivative of $\ln p(\mathbf{r} \mid \mu)$ with respect to μ is

$$\frac{\partial}{\partial \mu} \ln p(\mathbf{r} \mid \mu) = \sum_{i=1}^{N} \frac{(r_i - \mu)}{\sigma^2}$$

The estimate is obtained by setting this to zero and solving for μ. However, the solution is $\hat{\mu}$ so that once we set the derivative to zero, we must substitute $\hat{\mu}$ for μ. Then

$$\sum_{i=1}^{N} (r_i - \hat{\mu}) = 0 \qquad \text{or} \qquad \hat{\mu} = (1/N) \sum_{i=1}^{N} r_i$$

Thus, in this case, the maximum-likelihood estimate of the mean is the same as the sample mean, an intuitively satisfying result. We shall see many more examples of maximum-likelihood estimates in later material.

10.5 Properties of Estimators

The value of an estimate is dependent on the observations. Hence, as new sets of measurements are taken, the numerical value of the estimate changes. Since the estimates are themselves random variables, they have a mean and variance. It is obviously a desirable condition for repeated measurements that the values of the estimates tend to be grouped near the true value. If the mean of an estimate is equal to the true value, the estimate is said to be *unbiased.* If this is not so, the difference between the expected value of the estimate and the true value is called the bias.

† Since we do not make use of the a priori distribution of $\boldsymbol{\alpha}$, presumably because it is not known, we could treat this problem as estimation of an unknown but nonrandom parameter. We will continue to use the notation $p(\mathbf{r} \mid \boldsymbol{\alpha})$ even through $\boldsymbol{\alpha}$ may not be a random variable.

As the number of observations used in forming an estimate is increased, it is desirable that the density function for the estimate becomes more and more peaked (variance decreases) near the true value of the parameter. If this occurs, the estimate is said to be *consistent*. To be more precise, denote the estimate by $\hat{\alpha}_n$ where the subscript refers to the number of samples used to form the estimate. Then, an estimate $\hat{\alpha}_n$ is a consistent estimate of α if, for any positive and arbitrarily small ε and η, there is a number N such that the probability

$$P\{|\hat{\alpha}_n - \alpha| < \varepsilon\} > 1 - \eta, \qquad n > N$$

This states that $\hat{\alpha}_n$ converges in probability to α. As a simple example of a consistent estimator, we have the maximum-likelihood estimate of the sample mean discussed above. It is easily shown that $\hat{\mu}$ is Gaussian and is an unbiased estimate of the mean, and has a variance which decreases as $1/n$. It follows that $\hat{\mu}$ is a consistent estimator of the mean.

Another important property of some estimators is sufficiency. The significance of a sufficient estimator or statistic is that no other estimator can yield more information about the parameter than the sufficient estimator itself.

A necessary and sufficient condition (7) for $\hat{\alpha}$ to be a sufficient estimator or statistic is that the likelihood function be factorable into the product

$$p(\mathbf{r} \mid \alpha) = p(\hat{\alpha} \mid \alpha) f(\mathbf{r})$$

where $p(\hat{\alpha} \mid \alpha)$ is the density function of the estimator, and $f(\mathbf{r})$ is independent of α. We demonstrate with an example.

Consider the previous problem of estimating the mean of a Gaussian process. For independent samples, the likelihood function is given by Eq. (10-6). The estimate is the sample mean. Note that $\hat{\mu}$ is Gaussian with mean μ and variance σ^2/N. This gives us a good idea of how to factor the likelihood function. Now, after a bit of work we can show that the likelihood function may be written

$$p(\mathbf{r} \mid \mu) = \left(\frac{N}{2\pi\sigma^2}\right)^{1/2} \exp\left[\frac{(\hat{\mu} - \mu)^2}{-2\sigma^2/N}\right]$$

$$\times \left(\frac{2\pi\sigma^2}{N}\right)^{1/2} \left(\prod_{i=1}^{N} \frac{1}{(2\pi\sigma^2)^{1/2}}\right) \exp\left[\frac{N\hat{\mu}^2 - \sum r_i^2}{2\sigma^2}\right]$$

The first term is the density function of $\hat{\mu}$ for a given value of μ, and the second term is independent of μ. It follows therefore that $\hat{\mu}$ is a sufficient statistic.

An estimator which has a one-to-one correspondence with a sufficient estimator is also sufficient (7). In that sense a sufficient estimator is unique.

This allows us some freedom to choose a function of the sufficient statistic which may be consistent and unbiased.

An estimator is said to be a minimum variance estimator if its variance is less than or equal to the variance of any other estimator. Furthermore, the minimum variance *unbiased* estimator is unique (*7*). In this regard, it may be shown subject to certain conditions that any estimator has a variance which cannot be less than a particular lower bound. This is the Cramer–Rao bound (*1*, *7*, *8*) which we now discuss. Our emphasis will be for estimation of a single constant but unknown parameter. [For the case that the parameter is a random variable, see Gart (*9*) or Van Trees (*1*)].

The likelihood function is a proability density function and thus

$$\int p(\mathbf{r} \mid \alpha)\, d\mathbf{r} = 1$$

Differentiating this with respect to α, and assuming the regularity conditions that permit an interchange of the order of integration and differentiation we get†

$$\int \frac{\partial \ln p(\mathbf{r} \mid \alpha)}{\partial \alpha} p(\mathbf{r} \mid \alpha)\, d\mathbf{r} = E\left\{\frac{\partial \ln p(\mathbf{r} \mid \alpha)}{\partial \alpha}\right\} = 0 \tag{10-8}$$

We have used the fact that $df(x)/dx$ may be expressed as $(d \ln f(x)/dx)\, f(x)$. Since the expectation is with respect to the variables $\mathbf{r}$, then for any function of α, say $\psi(\alpha)$,

$$E\left\{\psi(\alpha) \frac{\partial \ln p(\mathbf{r} \mid \alpha)}{\partial \alpha}\right\} = 0$$

Assume that the expectation of an estimator of α is equal to $\psi(\alpha)$. Then

$$E\{\hat{\alpha}\} = \int \hat{\alpha} p(\mathbf{r} \mid \alpha)\, d\mathbf{r} = \psi(\alpha)$$

Differentiating with respect to α

$$\int \hat{\alpha} \frac{\partial \ln p(\mathbf{r} \mid \alpha)}{\partial \alpha} p(\mathbf{r} \mid \alpha)\, dr = \frac{\partial \psi(\alpha)}{\partial \alpha}$$

Since the expectation of $\psi(\alpha)\,[\partial \ln p(\mathbf{r} \mid \alpha)/\partial \alpha]$ is zero we can include it in the above equation to obtain

$$\int [\hat{\alpha} - \psi(\alpha)]\, \frac{\partial \ln p(\mathbf{r} \mid \alpha)}{\partial \alpha} p(\mathbf{r} \mid \alpha)\, d\mathbf{r} = \frac{\partial \psi(\alpha)}{\partial \alpha}$$

† The limits of integration are assumed independent of the parameter α.

Using Schwartz' inequality† we find

$$\int [\hat{\alpha} - \psi(\alpha)]^2 p(\mathbf{r} \mid \alpha)\, d\mathbf{r} \int \left[\frac{\partial \ln p(\mathbf{r} \mid \alpha)}{\partial \alpha}\right]^2 p(\mathbf{r} \mid \alpha)\, d\mathbf{r} \geq \left[\frac{\partial \psi(\alpha)}{\partial \alpha}\right]^2$$

But, $E\{[\hat{\alpha} - \psi(\alpha)]^2\}$ is the variance of $\hat{\alpha}$ so that

$$\sigma_{\hat{\alpha}}^2 \geq \frac{[\partial \psi(\alpha)/\partial \alpha]^2}{E\{[\partial \ln p(\mathbf{r} \mid \alpha)/\partial \alpha]^2\}} \tag{10-9}$$

This is the Cramer–Rao bound for $\sigma_{\hat{\alpha}}^2$. From the conditions of the Schwartz inequality, the equality will apply when

$$\frac{\partial \ln p(\mathbf{r} \mid \alpha)}{\partial \alpha} = k(\alpha)[\hat{\alpha} - \psi(\alpha)] \tag{10-10}$$

where $k(\alpha)$ is not a function of the data $\mathbf{r}$ or $\hat{\alpha}$ but may be a function of α.

Another form for the inequality results if we substitute

$$E\left\{\left[\frac{\partial \ln p(\mathbf{r} \mid \alpha)}{\partial \alpha}\right]^2\right\} = -E\left\{\frac{\partial^2 \ln p(\mathbf{r} \mid \alpha)}{\partial \alpha^2}\right\} \tag{10-11}$$

This relationship holds when Eq. (10-8) is satisfied (*7*).

If the expectation of the estimator is expressed as $\alpha + b(\alpha)$, where $b(\alpha)$ is the bias, then the numerator of the inequality is $[1 + \partial b(\alpha)/\partial \alpha]^2$. Therefore, for an unbiased estimator the Cramer–Rao inequality becomes

$$\sigma_{\hat{\alpha}}^2 \geq \frac{1}{E\left\{\left[\dfrac{\partial \ln p(\mathbf{r} \mid \alpha)}{\partial \alpha}\right]^2\right\}} = \frac{-1}{E\left\{\dfrac{\partial^2 \ln p(\mathbf{r} \mid \alpha)}{\partial \alpha^2}\right\}} \tag{10-12}$$

An unbiased estimator which attains the minimum variance bound does not always exist. If one does exist, however, it will be a maximum likelihood estimator and it will be unique (*7*). An estimator which attains this lower bound is called a *minimum variance bound* estimator or an *efficient* estimator. We note that if an estimator is efficient it must be a sufficient statistic (*7*). If an efficient estimator does not exist, it may be that a sufficient estimator does exist. Thus, sufficiency is a less restrictive criterion than efficiency.

An efficient estimator can be associated with a maximum likelihood estimator. To see this, assume an unbiased efficient estimator exists so that Eq. (10-10) is satisfied. That is, $\partial \ln p(\mathbf{r} \mid \alpha)/\partial \alpha = k(\alpha)(\hat{\alpha} - \alpha)$. But, the solution

† The form used here is

$$[\textstyle\int g(x)h(x)\, dx]^2 \leq \int g^2(x)\, dx \int h^2(x)\, dx$$

with the equality holding when $g(x)$ and $h(x)$ are linearly related.

for a maximum likelihood estimator is found by setting this equal to zero. It therefore follows that the efficient estimator $\hat{\alpha}$ is also the maximum-likelihood estimator. This is encouraging since we deal primarily with maximum-likelihood estimators. In practice it is often found for the high signal-to-noise case that the maximum likelihood estimator approaches the minimum variance bound, and therefore becomes asymptotically efficient.

If instead of estimating the parameter itself, we are interested in a function of the parameter we have the inequality below. First, define $\hat{f}$ to be an *unbiased* estimate of $f(\alpha)$, a function of α. Then the Cramer–Rao bound applicable to the estimator $\hat{f}$ is

$$\sigma_{\hat{f}}^2 \geq \frac{[\partial f(\alpha)/\partial\alpha]^2}{E\left\{\left[\dfrac{\partial \ln p(\mathbf{r} \mid \alpha)}{\partial\alpha}\right]^2\right\}} \tag{10-13}$$

Expressing this in terms of the minimum variance for the estimator of α we have

$$\sigma_{\hat{f}}^2 \geq \left[\frac{\partial f(\alpha)}{\partial\alpha}\right]^2 \sigma_{\hat{\alpha}}^2$$

The equality will hold if

$$\frac{\partial \ln p(\mathbf{r} \mid \alpha)}{\partial\alpha} = g(\alpha)[\hat{f} - f(\alpha)] \tag{10-14}$$

where $\hat{f}$ is the estimator of $f(\alpha)$ and $g(\alpha)$ is independent of the observations but may be a function of α. Note that if this condition is satisfied, the denominator of the right-hand side of the inequality may be expressed as

$$E\left\{\left[\frac{\partial \ln p(\mathbf{r} \mid \alpha)}{\partial\alpha}\right]^2\right\} = g^2(\alpha)\sigma_{\hat{f}}^2 \tag{10-15}$$

Substituting this into Eq. (10-13) we determine that

$$\sigma_{\hat{f}}^2 = \frac{\partial f(\alpha)}{\partial\alpha}\frac{1}{g(\alpha)} \tag{10-16}$$

if Eq. (10-14) is satisfied.

To demonstrate this with an example, consider the case of estimating the mean of a Gaussian process. The likelihood function is given by Eq. (10-6). The log-likelihood function can be placed in the form

$$\frac{\partial \ln p(\mathbf{r} \mid \mu)}{\partial\mu} = \frac{N}{\sigma^2}(\hat{\mu} - \mu)$$

where $\hat{\mu}$ is the sample mean. In analogy to Eq. (10-14), we known that $\hat{\mu}$ is the minimum variance bound estimator for μ, and $g(\mu) = N/\sigma^2$. Then from Eq. (10-16)

$$\sigma_{\hat{\mu}}^{\ 2} = \sigma^2/N$$

which is the minimum variance.

A fundamental inequality for the variance of simultaneous estimates of parameters also exists (*1*, *8*). For this assume that $\hat{\boldsymbol{\alpha}}$ is an unbiased vector estimate of $\boldsymbol{\alpha}$. For example, $\hat{\alpha}_1$ is an unbiased estimate of α_1, etc. Define the elements of a matrix $\boldsymbol{\gamma}$ as

$$\gamma_{ij} = E\left\{\frac{\partial \ln p(\mathbf{r}\,|\,\boldsymbol{\alpha})}{\partial \alpha_i} \frac{\partial \ln p(\mathbf{r}\,|\,\boldsymbol{\alpha})}{\partial \alpha_j}\right\} = -E\left\{\frac{\partial^2 \ln p(\mathbf{r}\,|\,\boldsymbol{\alpha})}{\partial \alpha_i\, \partial \alpha_j}\right\} \tag{10-17}$$

It is to be understood that the derivatives are evaluated at the true value of the parameter. These are elements of what is sometimes called Fisher's information matrix (*1*, *7*). Denote the inverse of this matrix by $\boldsymbol{\psi}$

$$\boldsymbol{\psi} = \boldsymbol{\gamma}^{-1} \tag{10-18}$$

and its elements by ψ_{ij}. Then the variance of the individual unbiased estimates satisfy the inequality

$$\sigma_{\hat{\alpha}_i}^2 \geq \psi_{ii} \tag{10-19}$$

If the equality holds for the simultaneous estimates, the estimates are said to be jointly efficient. The condition for the equality to hold is (*7*)

$$\hat{\alpha}_i - \alpha_i = \sum_j g_{ij}(\boldsymbol{\alpha}) \frac{\partial \ln p(\mathbf{r}\,|\,\boldsymbol{\alpha})}{\partial \alpha_i} \tag{10-20}$$

where $g_{ij}(\boldsymbol{\alpha})$ is independent of the observation and hence of $\hat{\boldsymbol{\alpha}}$.

By way of illustration, consider the case of only two parameters, α_1 and α_2. The elements of the information matrix are

$$\gamma_{11} = E\left\{\frac{\partial^2 \ln p(\mathbf{r}\,|\,\boldsymbol{\alpha})}{\partial {\alpha_1}^2}\right\}, \qquad \gamma_{22} = E\left\{\frac{\partial^2 \ln p(\mathbf{r}\,|\,\boldsymbol{\alpha})}{\partial {\alpha_2}^2}\right\}$$

$$\gamma_{21} = \gamma_{12} = E\left\{\frac{\partial^2 \ln p(\mathbf{r}\,|\,\boldsymbol{\alpha})}{\partial \alpha_1\, \partial \alpha_2}\right\}$$

Then

$$\boldsymbol{\psi} = \frac{1}{\Delta}\begin{bmatrix} \gamma_{22} & \gamma_{12} \\ \gamma_{12} & \gamma_{11} \end{bmatrix}$$

where

$$\Delta = \gamma_{11}\gamma_{22} - \gamma_{12}^2$$

The variance of jointly efficient estimates may then be expressed as

$$\sigma_{\hat{\alpha}_1}^2 = \frac{\gamma_{22}}{\Delta} = \frac{1}{E\left\{\dfrac{\partial^2 \ln p(\mathbf{r} \mid \boldsymbol{\alpha})}{\partial \alpha_1{}^2}\right\}} \frac{1}{[1 - \rho^2(\hat{\alpha}_1, \hat{\alpha}_2)]} \tag{10-21}$$

and for $\hat{\alpha}_2$

$$\sigma_{\hat{\alpha}_2}^2 = \frac{\gamma_{11}}{\Delta} = \frac{1}{E\left\{\dfrac{\partial^2 \ln p(\mathbf{r} \mid \boldsymbol{\alpha})}{\partial \alpha_2{}^2}\right\}} \frac{1}{[1 - \rho^2(\hat{\alpha}_1, \hat{\alpha}_2)]} \tag{10-22}$$

In these expressions the correlation coefficient of the estimates is

$$\rho(\hat{\alpha}_1, \hat{\alpha}_2) = \gamma_{12}/(\gamma_{11}\gamma_{22})^{1/2} \tag{10-23}$$

It should be noted that the first terms of the right-hand members in the variance expressions are the variances which would apply if only one parameter were unknown. Since $0 \leq \rho^2(\hat{\alpha}_1, \hat{\alpha}_2) \leq 1$, it follows that a nonzero correlation will increase such variances. Consequently, for simultaneous estimation of several parameters the bound given by Eq. (10-12) for a single unknown parameter is still valid. However a better bound will be determined by using the inequality of Eq. (10-19).

10.6 Estimation in Presence of White Noise

We shall now apply the preceding material to the estimation of signal parameters when the additive noise is white and Gaussian. The signal is assumed to be of known form so that the observable signal of interest is

$$r(t) = s(t, \boldsymbol{\alpha}) + n(t), \qquad 0 \leq t \leq T \tag{10-24}$$

where the vector $\boldsymbol{\alpha}$ represents the unknown parameters to be estimated. We shall concentrate on maximum-likelihood estimates, and unless specified otherwise, assume that the time of occurrence of the signal is known.

Using Eq. (6-38), the likelihood function of $r(t)$ is

$$p(\mathbf{r} \mid \boldsymbol{\alpha}) = F \exp\left\{-\frac{1}{N_o}\int_0^T [r(t) - s(t, \boldsymbol{\alpha})]^2 \, dt\right\} \tag{10-25}$$

It is easily shown that

$$\frac{\partial \ln p(\mathbf{r} \mid \boldsymbol{\alpha})}{\partial \alpha_i} = \frac{2}{N_o}\int_0^T [r(t) - s(t, \boldsymbol{\alpha})] \frac{\partial s(t, \boldsymbol{\alpha})}{\partial \alpha_i} \, dt \tag{10-26}$$

Consequently the maximum-likelihood estimates of the parameters of $\boldsymbol{\alpha}$ are the solutions to the set of equations

$$\int_0^T [r(t) - s(t, \boldsymbol{\alpha})] \frac{\partial s(t, \boldsymbol{\alpha})}{\partial \alpha_i} dt = 0, \qquad i = 1, \ldots, M \tag{10-27}$$

where M denotes the number of unknown parameters. Using $n(t) = r(t) - s(t, \boldsymbol{\alpha})$ in Eq. (10-26), the elements of the information matrix, Eq. (10-17), are

$$\gamma_{ij} = \frac{4}{N_o^2} \int_0^T \int_0^T E\{n(t)n(u)\} \frac{\partial s(t, \boldsymbol{\alpha})}{\partial \alpha_i} \frac{\partial s(u, \boldsymbol{\alpha})}{\partial \alpha_j} dt\, du \tag{10-28}$$

The autocorrelation function of the noise is $N_o/2\, \delta(t - u)$ so

$$\gamma_{ij} = \frac{2}{N_o} \int_0^T \frac{\partial s(t, \boldsymbol{\alpha})}{\partial \alpha_i} \frac{\partial s(t, \boldsymbol{\alpha})}{\partial \alpha_j} dt \tag{10-29}$$

In particular, for $i = j$,

$$\gamma_{ii} = \frac{2}{N_o} \int_0^T \left[\frac{\partial s(t, \boldsymbol{\alpha})}{\partial \alpha_i}\right]^2 dt \tag{10-30}$$

Randomly Phased Signals

Of considerable importance in practical application is the sinusoidal signal whose phase is assumed to be uniformly distributed over a $(0, 2\pi)$ interval. We have already discussed in detail the detection of such signals. We shall develop here the ground work necessary for the estimation of their parameters under the assumption that the phase is a stray parameter and is averaged out. Conforming to the notation of Sect. 9.8, the complex envelope of the received signal is

$$\tilde{r}(t) = \tilde{A}(t)e^{j\theta} + \tilde{z}(t)$$

The function $\tilde{A}(t)$ may be complex, allowing for frequency or phase modulation. We are interested here in developing ground work specifically for estimating the time of arrival and frequency of signals with known complex envelope. However, the method is easily extended to estimation of other parameters such as rate of change of frequency.† We shall normalize the signal amplitude so that $\tilde{A}(t) = (2E)^{1/2}\, \tilde{a}(t)$ where

$$\int |\tilde{a}(t)|^2\, dt = 1$$

and E is the signal energy.

† See Exercises 10.10 and 10.11.

For the likelihood function, we shall use the form given by Eq. (9-103) which may be written

$$p_\theta(\mathbf{r}) = KI_o(D) \tag{10-31}$$

where K is a known constant which need not concern us here. The subscript θ is included as a reminder that we have averaged over the phase. The statistic D may be determined from Eqs. (9-105) and (9-107). These are valid for the general case of colored noise. We shall however specialize to the case of white noise for which $\tilde{R}(\tau) = N_o\,\delta(\tau)$.† Using this Eq. (9-107) it follows that

$$\tilde{h}(t) = \frac{1}{N_o}\tilde{A}(t) = \frac{(2E)^{1/2}}{N_o}\tilde{a}(t) \tag{10-32}$$

The statistic D in Eq. (9-105) is then

$$D = \frac{(2E)^{1/2}}{N_o}\left|\int \tilde{r}(t)\tilde{a}^*(t)\,dt\right| \tag{10-33}$$

The limits of integration are understood to be over the range of definition of $\tilde{a}(t)$.

In estimating signal parameters such as time of arrival or frequency, useful results are often obtained only in the case of large ratios of E/N_o. Furthermore, this assumption is often necessary to make the computations tractable. We also find with this assumption that the estimates are asymptotically efficient, that is, the variance approaches the Cramer–Rao bound.

As discussed in Chap. 8 for the high signal-to-noise case, we approximate

$$\ln I_o(D) \approx D$$

and the log-likelihood function by

$$\ln p_\theta(\mathbf{r}) \approx \ln K + D \tag{10-34}$$

For the estimation problem, a receiver is built to implement the statistic D. However, since the various signal parameters are unknown we cannot a priori build a precise matched filter. Thus, while the receiver is matched to a function $\tilde{a}(t)$, the received signal may be somewhat different. To show this explicitly, express the received signal as

$$\tilde{r}(t) = (2E)^{1/2}\,\tilde{a}_r(t)e^{j\theta} + \tilde{z}(t) \tag{10-35}$$

† See Helstrom (*5*), but note that Turin (*10*) and Kailath (*11*) point out that $\tilde{R}(\tau) = (N_o/2)\delta(t) + j(1/\pi t)$ should be used. ($1/\pi t$ is the Hilbert transform of the delta function.) However, they also point out if $\tilde{R}(\tau)$ is used in a convolution such as Eq. (9–107), that the result will be the same as obtained with $\tilde{R}(\tau) = N_o\delta(\tau)$. Since this is our only use for $\tilde{R}(\tau)$, we shall use the simple form.

where $\tilde{a}_r(t)$ differs from $\tilde{a}(t)$ by small differences in the unknown signal parameters. Since the signal-to-noise ratio is large we assume that $\tilde{a}_r(t)$ is very nearly equal to $a(t)$. This will be elaborated in the next section. In the meantime, D may be expressed as

$$D = \frac{2E}{N_o}\left|\int \tilde{a}_r(t)\tilde{a}^*(t)\,dt + \frac{e^{-j\theta}}{(2E)^{1/2}}\int \tilde{z}(t)\tilde{a}^*(t)\,dt\right| \tag{10-36}$$

The first term is a function of only the signal characteristics. Because of the normalization and the assumption that $\tilde{a}_r(t) \approx \tilde{a}(t)$, the term is near unity. The second term relates to the cross correlation of signal and noise. Consistent with the assumption of high signal-to-noise ratio, this term may be neglected and

$$D \approx \frac{2E}{N_o}\left|\int \tilde{a}_r(t)\tilde{a}^*(t)\,dt\right| \tag{10-37}$$

The log-likelihood function is then very nearly equal to

$$\ln p_\theta(\mathbf{r}) \approx \ln K + \frac{2E}{N_o}\left|\int \tilde{a}_r(t)\tilde{a}^*(t)\,dt\right| \tag{10-38}$$

This expression will be useful for the determination of the Cramer–Rao bound.

These assumptions are consistent with the assumptions used in Woodwards' maximum a posteriori approach to the estimation problem.† The above form of the log-likelihood function will be applied in the next section specifically to estimating frequency and time of arrival.

The expression for the cross correlation of the signals in Eq. (10-38) is found very useful in signal design, especially for the radar case. Specifically, the function

$$\chi = \int \tilde{a}_r(t)\tilde{a}^*(t)\,dt$$

is called the ambiguity function (*4*). A plot of $|\chi|^2$ is sometimes referred to as the ambiguity surface‡

10.7 Estimation of Specific Parameters

To show how the preceding material is applied and to provide useful analytic results we shall discuss the estimation of signal amplitude, phase, frequency, and time of arrival. To avoid added complication we first discuss

† Woodward (*4*). See also Kelly *et al.* (*12, 13*).

‡ Consult the bibliography for references relating to the ambiguity function and signal design.

estimation when only one parameter is unknown and later generalize to simultaneous estimation.

For a single unknown parameter α, the maximum-likelihood estimate is the solution of

$$\int_0^T [r(t) - s(t, \hat{\alpha})] \frac{\partial s(t, \hat{\alpha})}{\partial \hat{\alpha}} \, dt = 0 \tag{10-39}$$

and the minimum variance of an unbiased estimate of α is, from Eqs. (10-12) and (10.30),

$$\sigma_{\hat{\alpha}}^2 \geq \left\{ \frac{2}{N_0} \int_0^T \left[\frac{\partial s(t, \alpha)}{\partial \alpha} \right]^2 dt \right\}^{-1} \tag{10-40}$$

Estimation of Amplitude

For amplitude estimation, assume the known signal is of the form $s(t, \alpha) = As(t), 0 \leq t \leq T$. Here $s(t)$ is completely known and we wish to estimate the multiplier A. From Eq. (10-39), the maximum-likelihood estimator is easily seen to be the solution of

$$\int_0^T [r(t) - \hat{A}s(t)]s(t) \, dt = 0$$

Solving for the estimate explicitly

$$\hat{A} = \frac{\int_0^T r(t)s(t) \, dt}{\int_0^T s^2(t) \, dt}$$

Without loss of generality $s(t)$ may be normalized so that $\int_0^T s^2(t) \, dt = 1$. Then

$$\hat{A} = \int_0^T r(t)s(t) \, dt \tag{10-41}$$

The estimate may be obtained by correlating the received signal with the known signal $s(t)$, or by passing the received signal through a matched filter and sampling at time T. (See Chap. 6) We now determine the mean and variance of $\hat{A}$. From Eq. (10-41)

$$E\{\hat{A}\} = E\left\{ \int_0^T [As(t) + n(t)]s(t) \, dt \right\} = A$$

so that the estimator is unbiased. The variance of $\hat{A}$ is

$$E\left\{ \int_0^T \int_0^T n(t)n(u)s(t)s(u) \, dt \, du \right\} = N_0/2$$

Since $\partial s(t, \alpha)/\partial \alpha = s(t)$, it is easily shown using Eq. (10-40) that the Cramer–Rao bound is also $N_o/2$ so that $\hat{A}$ is an efficient estimate of the amplitude. No other unbiased estimate can attain a lower variance.

Estimation of Phase

Consider now the signal of known form $s(t, \alpha) = A \sin(\omega_o t + \theta)$, $0 \leq t \leq T$, where the amplitude and frequency are known and we wish to estimate the phase θ. From Eq. (10-39) the phase estimate is the solution of

$$\int_0^T [r(t) - A \sin(\omega_o t + \hat{\theta})] \cos(\omega_o t + \hat{\theta})\, dt = 0$$

Assume that $\omega_o T = k\pi$ where k is an integer, or $\omega_o T \gg 1$. Then the integral of the second term is zero and the phase estimate is the solution of

$$\int_0^T r(t) \cos(\omega_o t + \hat{\theta})\, dt = 0 \tag{10-42}$$

Expanding the cosine term produces the equation

$$\cos \hat{\theta} \int_0^T r(t) \cos \omega_o t\, dt = \sin \hat{\theta} \int_0^T r(t) \sin \omega_o t\, dt$$

Solving for $\hat{\theta}$:

$$\hat{\theta} = \tan^{-1}\left\{\frac{\int_0^T r(t) \cos \omega_o t\, dt}{\int_0^T r(t) \sin \omega_o t\, dt}\right\} \tag{10-43}$$

These operations may be performed by correlators or filters matched to $\cos \omega_o t$ and $\sin \omega_o t$.

A realization suggested by Eq. (10-42) which is useful in many circumstances is the phase-locked loop (*2*) shown in Fig. 10-1. For example, it may be desired to phase-synchronize an oscillator to a noisy signal. To briefly explain the operation of such a circuit, assume that the signal is noise free. Thus, assume $r(t) \approx A \sin(\omega_o t + \theta)$. The output of the multiplier is then

$$\varepsilon(t) \cong A \sin(\omega_o t + \theta)\cos(\omega_o t + \hat{\hat{\theta}})$$

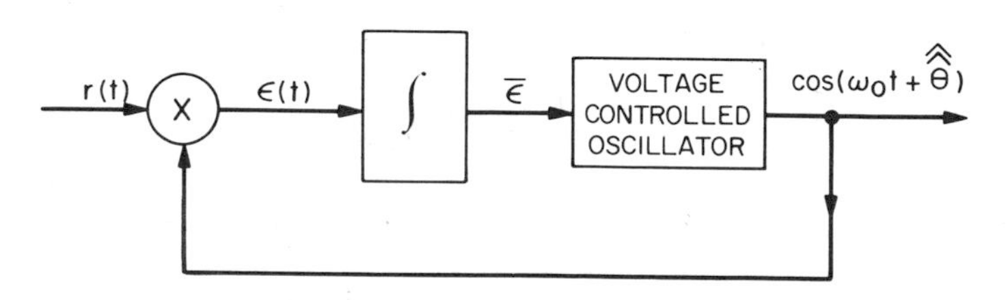

Fig. 10-1 Phase locked loop.

We use the double hat notation "$\hat{\hat{}}$" to distinguish the phase from the maximum likelihood estimate $\hat{\theta}$. Expanding $\varepsilon(t)$ we have

$$\varepsilon(t) = (A/2)\sin(\theta - \hat{\hat{\theta}}) + (A/2)\sin(2\omega_o t + \theta + \hat{\hat{\theta}})$$

The integrator† averages $\varepsilon(t)$, (with respect to time), so that its output is proportional to

$$\bar{\varepsilon} \sim \sin(\theta - \hat{\hat{\theta}})$$

and for small phase differences $\bar{\varepsilon} \sim \theta - \hat{\hat{\theta}}$. The integrator also acts to smooth the variations of $\varepsilon(t)$ due to noise. When the error voltage is applied to the voltage controlled oscillator, the output phase changes in a direction to reduce the average error. Then, $\bar{\varepsilon}$ approaches zero, and $\hat{\hat{\theta}}$ approaches $\hat{\theta}$. This is the desired condition for the phase-locked loop and the condition suggested by Eq. (10-42).

Estimation of Time of Arrival

In Chap. 7 we discussed the problem of simultaneously detecting a signal and determining (estimating) its time of arrival. The approach used a multiple alternative hypothesis test, and the results were substantially independent of signal-to-noise ratio. Here we want to determine the variance of an estimate of time of arrival, and to make the analysis tractable we assume that a high signal-to-noise ratio exists.

Two cases are considered. In the first a low-pass signal is studied (that is, a signal whose Fourier transform is significant only in a region near zero frequency). We do so to avoid a discussion of phase. In the second case a narrowband signal whose phase is uniformly distributed will be treated.

Low-Pass Signal

The signal $s(t, \alpha)$ may be represented as $m(t - \tau)$ where τ is the delay which is to be estimated. Now, if the signal-to-noise ratio is high then we may reasonably assume that we already have a coarse estimate of τ, and without loss of generality, assume that $\tau \approx 0$. Then the problem is to determine how best to process the signal to refine our estimate. Neglecting a slight "edge" effect of the signal, the log-likelihood function may be expressed as

$$\ln p(\mathbf{r} \mid \tau) = \ln F - \frac{1}{N_o}\int_{-T/2}^{T/2} [r(t) - m(t - \tau)]^2 \, dt$$

† The integrator is usually a finite time integrator, or a simple smoothing filter. Since the phase-locked oscillator is a feedback system, care must be exercised to ensure that the system is stable [see Viterbi (*2*)].

The maximum-likelihood estimate is the solution of (*6*)

$$\int_{-T/2}^{T/2} [r(t) - m(t - \hat{\tau})] \frac{\partial}{\partial \hat{\tau}} m(t - \hat{\tau})\, dt = 0$$

The second part of this integral is identically zero so that the estimate of τ is the solution of

$$\int_{-T/2}^{T/2} r(t) \frac{\partial m(t - \tau)}{\partial \tau}\, dt = 0 \qquad (10\text{-}44)$$

The receiver thus correlates the received signal with the derivative of the known signal form. The derivative is like a gating function and has been shown elsewhere (*14*) to be optimum under a different set of conditions which include high signal-to-noise ratio. In practice this gating function is often approximately implemented by two "pulses"—one positive and one negative. To see this consider Fig. 10-2. The function $m(t)$ represents, for example, a typical "video" radar pulse. The derivative of such a signal is a bipolar pulse which is easily approximated by the reactangular gating function shown. The positive pulse is called an "early gate," and the negative pulse a "late gate." In a tracking radar, such a gating function produces a tracking error

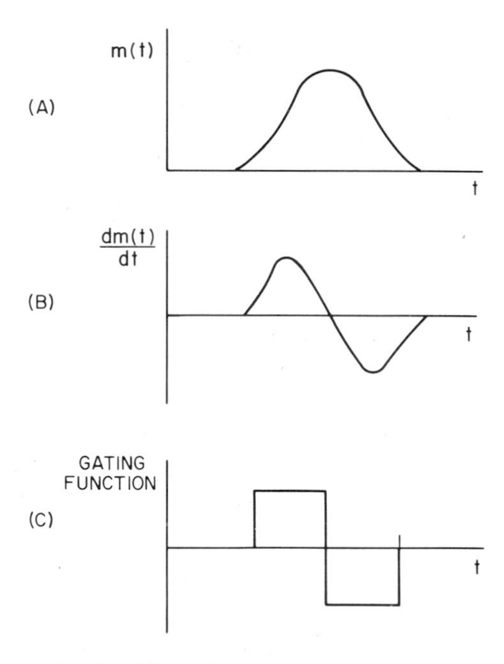

Fig. 10-2 An example of a video pulse (A), and its derivative (B). The gating function (C) may be used to approximate the derivative.

which can be used in a feedback loop analogous to the phase-locked loop. The tracking is accomplished by operating on many sequentially received pulses (*15*).

We now determine the Cramer–Rao bound for an estimate of arrival time. Using Eq. (10-14), for an unbiased estimate

$$\sigma_{\hat{\tau}}^2 \geq \left\{ \frac{2}{N_o} \int_{-T/2}^{T/2} \left[\frac{\partial m(t-\tau)}{\partial \tau} \right]^2 dt \right\}^{-1} \tag{10-45}$$

Since $\tau \approx 0$, we may replace Eq. (10-45) by

$$\sigma_{\hat{\tau}}^2 \geq \left\{ \frac{2}{N_o} \int_{-T/2}^{T/2} \left[\frac{\partial m(t)}{\partial t} \right]^2 dt \right\}^{-1} \tag{10-46}$$

Now assume $m(t)$ has a Fourier transform denoted by $M(j\omega)$ so that

$$m(t) = \int_{-\infty}^{\infty} M(j\omega) e^{j\omega t} \frac{d\omega}{2\pi}$$

Then it is easily seen that $dm(t)/dt$ has a Fourier transform given by $j\omega M(j\omega)$. By Parseval's theorem, it follows that

$$\int_{-T/2}^{T/2} \left[\frac{\partial m(t)}{\partial t} \right]^2 dt = \frac{1}{2\pi} \int_{-\infty}^{\infty} \omega^2 |M(j\omega)|^2 \, d\omega$$

and the minimum variance may then be expressed as

$$\sigma_{\hat{\tau}}^2 \geq \left[\frac{1}{N_o \pi} \int_{-\infty}^{\infty} \omega^2 |M(j\omega)|^2 \, d\omega \right]^{-1} \tag{10-47}$$

Now, the signal energy is

$$E = \int_{-T/2}^{T/2} m^2(t) \, dt = \frac{1}{2\pi} \int_{-\infty}^{\infty} |M(j\omega)|^2 \, d\omega$$

Substituting this into Eq. (10-47)

$$\sigma_{\hat{\tau}}^2 \geq \left(\frac{2E}{N_o} \beta_r^2 \right)^{-1} \tag{10-48}$$

where β_r is a measure of bandwidth (in radians) defined by

$$\beta_r^2 = \frac{\int_{-\infty}^{\infty} \omega^2 |M(j\omega)|^2 \, d\omega}{\int_{-\infty}^{\infty} |M(j\omega)|^2 \, d\omega} \tag{10-49}$$

Note that this is a second moment of the spectrum about $\omega = 0$. In radar estimation problems, this uncertainty in arrival time can be related to uncertainty in range if the velocity of propagation is known.

Narrowband Signal

As discussed in the last section,† the narrowband signal is assumed to have a phase uniformly distributed over the interval $(0, 2\pi)$. The log-likelihood function is found from Eqs. (10-33) and (10-34) and is proportional to

$$\ln p_\theta(\mathbf{r}|\tau) \sim \left| \int \tilde{r}(t-\tau)\tilde{a}^*(t)\, dt \right| \tag{10-50}$$

The estimate of τ is that value which maximizes the expression. The operation indicated is the envelope of the cross correlation function of the complex envelopes of the received signal and the signal of known form. The estimate may be found by inserting the received signal into a narrowband filter whose complex impulse response is $\tilde{a}^*(t)$ followed by an envelope detector, and observing the time at which the output reaches a peak.

We now determine the Cramer–Rao bound for an efficient estimate of time of arrival. Some of the mathematical difficulties are overcome by using the form of the log-likelihood function given in Eq. (10-38) suitably modified for the parameter τ. Since $\tilde{a}_r(t) = \tilde{a}(t-\tau)$

$$\ln p_\theta(\mathbf{r} \mid \tau) \approx \ln K + \frac{2E}{N_o} \left| \int \tilde{a}(t-\tau)\tilde{a}^*(t)\, dt \right| \tag{10-51}$$

For this case, we may write the ambiguity function as

$$\chi(\tau) = \int \tilde{a}(t-\tau)\tilde{a}^*(t)\, dt$$

and the Cramer–Rao bound as‡

$$\sigma_{\hat{\tau}}^{2} \geq \frac{-1}{E\left\{\dfrac{\partial^2 \ln p_\theta(\mathbf{r} \mid \tau)}{\partial \tau^2}\right\}} \approx \frac{-1}{\dfrac{2E}{N_o} \dfrac{\partial^2}{\partial \tau^2} |\chi(\tau)|} \tag{10-52}$$

As a reminder, the derivative is evaluated at the true value of the parameter which we have assumed to be zero without loss of generality. Before performing

† See also the related material in Chap. 7 for simultaneous detection and estimation using multiple alternative hypothesis testing.

‡The assumption of high signal-to-noise ratio permitted the approximation indicated in Eq. (10-51). The noise term does not appear in this expression for $\ln p_\theta(\mathbf{r} \mid \tau)$ so that an approximate form of the Cramer–Rao bound could be written without the expectation operator.

the differentiation, some preliminary derivations will be useful. Denote

$$\chi'(\tau) = \frac{\partial \chi(\tau)}{\partial \tau}, \qquad \chi'(0) = \left.\frac{\partial \chi(\tau)}{\partial \tau}\right|_{\tau=0}$$

$$\chi''(\tau) = \frac{\partial^2 \chi(\tau)}{\partial \tau^2}, \qquad \text{and} \qquad \chi''(0) = \left.\frac{\partial^2 \chi(\tau)}{\partial \tau^2}\right|_{\tau=0}$$

Since

$$|\chi(\tau)| = [\chi(\tau)\chi^*(\tau)]^{1/2}$$

the first derivative is

$$\frac{\partial |\chi(\tau)|}{\partial \tau} = \frac{1}{2|\chi(\tau)|}[\chi(\tau)\chi^{*\prime}(\tau) + \chi^*(\tau)\chi'(\tau)] \tag{10-53}$$

and the second derivative is

$$\frac{\partial^2 |\chi(\tau)|}{\partial \tau^2} = \frac{1}{|\chi(\tau)|}\operatorname{Re}[\chi(\tau)\chi^{*\prime\prime}(\tau) + \chi'(\tau)\chi^{*\prime}(\tau)] - \frac{1}{|\chi(\tau)|^3}[\operatorname{Re}\chi^*(\tau)\chi'(\tau)]^2 \tag{10-54}$$

For the Cramer–Rao bound, we evaluate the second derivative at $\tau = 0$ and obtain (recall $\chi(0) = 1$)

$$\left.\frac{\partial^2 |\chi(\tau)|}{\partial \tau^2}\right|_{\tau=0} = \operatorname{Re}[\chi''(0)] + |\chi'(0)|^2 - [\operatorname{Re}\chi'(0)]^2 \tag{10-55}$$

Evaluating the terms in this equation,

$$\chi'(0) = -\int \tilde{a}^*(t)\tilde{a}'(t)\,dt \tag{10-56}$$

and

$$\chi''(0) = \int \tilde{a}^*(t)\tilde{a}''(t)\,dt = -\int |\tilde{a}'(t)|^2\,dt \tag{10-57}$$

where $\tilde{a}'(t)$ and $\tilde{a}''(t)$ are the first and second derivatives of $\tilde{a}(t)$ with respect to t. The latter form in Eq. (10-57) may be derived by integrating by parts and assuming the function $\tilde{a}(t)$ or its derivative vanishes at the endpoints.

Define the Fourier transform of $\tilde{a}(t)$ by $G(j\omega)$. Then the Fourier transform of $\tilde{a}'(t)$ is $j\omega\, G(j\omega)$, and

$$\chi'(0) = -j\int \omega\,|G(j\omega)|^2\,\frac{d\omega}{2\pi} \tag{10-58}$$

$$\chi''(0) = -\int \omega^2\,|G(j\omega)|^2\,\frac{d\omega}{2\pi} \tag{10-59}$$

Note that $\chi'(0)$ is pure imaginary so Re $\chi'(0) = 0$. Then, using Eqs. (10-51)–(10-59) we have

$$\sigma_{\hat{\tau}}^2 \geq \left(\frac{2E}{N_0}\beta^2\right)^{-1} \tag{10-60}$$

where†

$$\beta^2 = \frac{\int \omega^2 |G(j\omega)|^2(1/2\pi)\, d\omega - [\int \omega |G(j\omega)|^2(1/2\pi)\, d\omega]^2}{\int |G(j\omega)|^2(1/2\pi)\, d\omega} \tag{10-61}$$

is a measure of the signal bandwidth, and is the second moment about the mean of the spectrum $|G(j\omega)|^2$. The second term is the only difference between the result here and that given by Eqs. (10-48) and (10-49). The difference may be specifically attributed to the assumption of random phase for narrowband signals.

Estimation of Frequency‡

We are interested in estimating the frequency of signals having the form $s(t, \alpha) = A(t)\cos(\omega t + \theta)$, $0 \leq t \leq T$, and shall later allow for angle modulated signals. The amplitude and time of arrival are assumed known. As before, phase is treated as a random variable and is integrated out. From Chap. 7, it is easily shown that the likelihood function is

$$p_\theta(\mathbf{r} \mid \omega) = K = e^{-E/N_0} I_0\left(\frac{2q}{N_0}\right)$$

where

$$q^2 = \left[\int_0^T r(t)A(t)\cos\omega t\, dt\right]^2 + \left[\int_0^T r(t)A(t)\sin\omega t\, dt\right]^2 \tag{10-62}$$

The maximum-likelihood estimate of frequency is obtained by maximizing $p_\theta(\mathbf{r} \mid \omega)$ with respect to ω. But, maximizing this is the same as maximizing q. Recall that q may be generated by inserting the received signal $r(t)$ into a filter matched to $A(t)\cos\omega t$ followed by an envelope detector. To determine that value of ω which maximizes q, we can use a parallel bank of filters, each matched to a different frequency. The frequency range of the filters encompasses the entire expected range of frequencies of the received signal. If the filters are closely spaced in frequency, the center frequency of the filter with

† Since $\tilde{a}(t)$ is normalized, the denominator of this expression is unity. We nevertheless include it so that in arbitrary cases the expression is correct.

‡ See also the related material in Chap. 7 which used multiple alternative hypothesis testing.

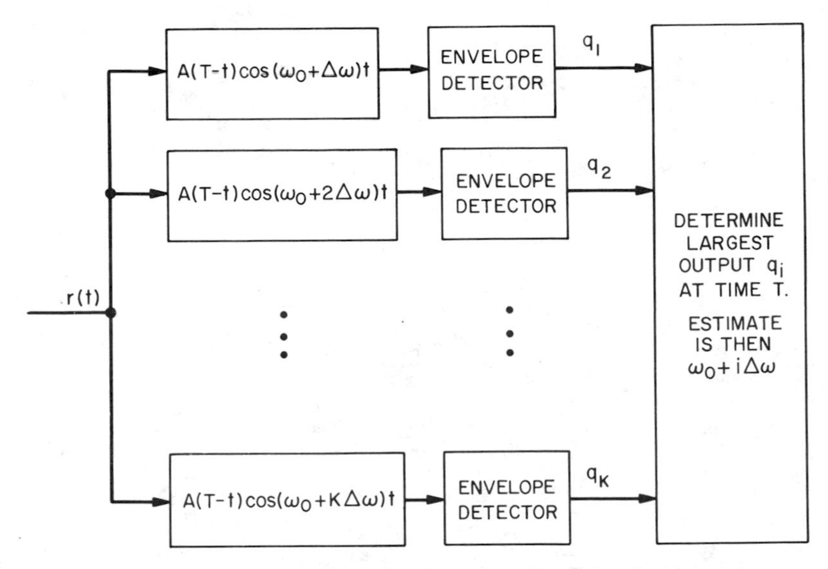

Fig. 10-3 Receiver for estimating frequency (time of arrival known).

the largest output approximates the maximum-likelihood estimate. An example of such an implementation is shown in Fig. 10-3. There is little point in spacing the filters much closer than the standard deviation of the estimate which is calculated below. In practice, the center frequencies of adjacent filters are often chosen to be separated by about $1/T$ or $1/2T$.

The complex envelope of the signal portion of $r(t)$ is given by

$$(2E)^{1/2}\tilde{a}_r(t) = (2E)^{1/2}\tilde{a}(t)\, e^{j\nu t} \tag{10-63}$$

where ν is the difference in frequency between the received signal and the expected signal. This notation allows us to discuss more than signals with just amplitude modulation. Indeed, known frequency or phase modulation may be included in the complex envelope $\tilde{a}(t)$. Then, as in the preceding case where just amplitude modulation was implied, a bank of matched filters and envelope detectors could still be used to extract a frequency estimate. We next determine the Cramer–Rao bound for this general case.

We begin with Eq. (10-38) and the signal form of Eq. (10-63). Thus

$$\ln p_\theta(\mathbf{r}|\omega) \approx \ln K + (2E/N_o)\left|\int |\tilde{a}(t)|^2 e^{j\nu t}\, dt\right| \tag{10-64}$$

In analogy to Eqs. (10-53) through (10-55), define

$$\chi(\nu) = \int |\tilde{a}(t)|^2 e^{j\nu t}\, dt \tag{10-65}$$

where $\chi(0) = 1$. From Eq. (10-55)

$$\left.\frac{\partial^2 |\chi(v)|}{\partial v^2}\right|_{v=0} = \mathrm{Re}[\chi''(0)] + |\chi'(0)|^2 - [\mathrm{Re}\,\chi'(0)]^2 \tag{10-66}$$

For the ambiguity function of Eq. (10-65)

$$\chi'(0) = j\int t\,|\tilde{a}(t)|^2\,dt \tag{10-67}$$

$$\chi''(0) = -\int t^2|\tilde{a}(t)|^2\,dt \tag{10-68}$$

All derivatives are with respect to v. The Cramer–Rao bound is then easily shown to be

$$\sigma_{\hat{\omega}}^2 \geq \left(\frac{2E}{N_o}\,t_d^2\right)^{-1} \tag{10-69}$$

where

$$t_d^2 = \int t^2|\tilde{a}(t)|^2\,dt - \left[\int t|\tilde{a}(t)|^2\,dt\right]^2 \tag{10-70}$$

In analogy to the bandwidth measure of Eq. (10-61), t_d is a measure of the time duration of the signal and is the second moment taken about the mean time of the signal.

Simultaneous Estimation of Parameters (*13, 16*)

We are concerned here with the simultaneous estimation of parameters, particularly time of arrival and frequency.† The results of the previous sections can be easily applied. For the estimation of frequency we can build a bank of matched filters as shown in Fig. 10-3. For the case of a known time of arrival, the signal is gated into the receiver during that time interval in which the signal is known to exist. Except for this time interval, no signal is allowed to enter the receiver. The frequency estimate corresponds to the center frequency of the filter with the largest output. Except for the discrete displacement of the frequency of adjacent filters, this method maximizes the likelihood function. On the other hand, if the frequency is known and the time of arrival is unknown, the estimate of arrival is obtained by finding the maximum of Eq. (10-50). This cross correlation operation may be implemented by inserting the signal into a bandpass filter matched to $\tilde{a}(t)$ with center frequency equal to the known frequency followed by an envelope detector. For the case of high signal-to-noise, the output of the envelope detector will contain a single high peak which can be used for the time of arrival estimate.

† See the related material in Chap. 7.

When neither the frequency nor the time of arrival are known, we can find the peak of the likelihood function by inserting the signal into the bank of matched filters and observing the output over that interval of time in which it is known the signal exists. Again, in the high signal-to-noise case a single high peak will result. The time of the peak, and the filter in which the peak occurred, correspond to the desired simultaneous estimates of time of arrival and frequency respectively. The value of the peak may also be used to estimate the amplitude of the signal.

We now determine the Cramer–Rao bound for joint estimates of frequency and time of arrival. The signal with which we are concerned is of the form $(2E)^{1/2}a(t)\cos(\omega_c t + \phi(t) + \theta)$ where $\phi(t)$ allows for angle modulation. Allowing for a small frequency mismatch v, and a small departure from the estimated time of arrival τ, the signal may be expressed as

$$(2E)^{1/2}a(t-\tau)\cos[(\omega_c + v)(t-\tau) + \phi(t-\tau) + \theta]$$
$$= \mathrm{Re}(2E)^{1/2}\tilde{a}(t-\tau)e^{j[(\omega_c+v)(t-\tau)+\theta]}$$

The angle modulation term has been included in $\tilde{a}(t)$. From this, we see that the signal $\tilde{a}_r(t)$ to be used in Eq. (10-38) to determine the variance is equal to

$$\tilde{a}_r(t) = \tilde{a}(t-\tau)e^{vt}e^{-(\omega_c+v)\tau}$$

The second exponential may be disregarded since it is not a function of t, and has an absolute value of one. Therefore

$$\ln p_\theta(\mathbf{r} \mid \omega, \tau) \approx \ln K + \frac{2E}{N_o}\left|\int \tilde{a}(t-\tau)\tilde{a}^*(t)e^{jvt}\,dt\right| \tag{10-71}$$

It is easily verified from previous results that the diagonal elements of the information matrix, Eq. (10-17), are

$$\gamma_{11} = -\left.\frac{\partial^2 \ln p_\theta(\mathbf{r} \mid v, \tau)}{\partial\tau^2}\right|_{v,\,\tau=0} = \frac{2E}{N_o}\beta^2$$

and

$$\gamma_{22} = -\left.\frac{\partial^2 \ln p_\theta(\mathbf{r} \mid v, \tau)}{\partial v^2}\right|_{v,\,\tau=0} = \frac{2E}{N_o}t_d^{\,2}$$

where β and t_d are given by Eqs. (10-61) and (10-70). The off-diagonal term may be shown to be

$$\gamma_{12} = -\left.\frac{\partial^2 \ln p_\theta(\mathbf{r} \mid v, \tau)}{\partial\tau\,\partial v}\right|_{v,\,\tau=0} = \frac{2E}{N_o}\mathrm{Re}\left[\int jt\,\frac{d\tilde{a}(t)}{dt}\,\tilde{a}^*(t)\,dt - \bar{\omega}\bar{t}\right]. \tag{10-72}$$

where $\bar{\omega}$ and $\bar{t}$ are the mean frequency and mean time of the signal defined by

$$\bar{\omega} = \int \omega |G(\omega)|^2 \frac{d\omega}{2\pi}, \qquad \bar{t} = \int t |\tilde{a}(t)|^2 \, dt$$

To repeat, $G(\omega)$ is the Fourier transform of $\tilde{a}(t)$. Without loss of generality, the frequency and time coordinates may be chosen such that $\bar{\omega} = \bar{t} = 0$.

The interpretation of the right-hand member of Eq. (10-72) becomes clear if we consider a signal such as $f(t)\cos(\omega t + \phi(t))$ where $f(t)$ and $\phi(t)$ are real functions. Then

$$\tilde{a}(t) = f(t)e^{j\phi(t)}$$

In this case

$$\tilde{a}^*(t)\frac{d\tilde{a}(t)}{dt} = jf^2(t)\frac{d\phi(t)}{dt} + f(t)\frac{df(t)}{dt}$$

Therefore, with $\bar{\omega} = \bar{t} = 0$,

$$\begin{aligned} \gamma_{12} &= (2E/N_o)\,\mathrm{Re}\left[\int jt\frac{d\tilde{a}(t)}{dt}\tilde{a}^*(t)\,dt\right] \\ &= -(2E/N_o)\int t\frac{d\phi(t)}{dt}f^2(t)\,dt \end{aligned} \tag{10-73}$$

The term $d\phi(t)/dt$ is equal to the deviation of the instantaneous frequency from the carrier frequency. The integral is then a measure of the average frequency–time product, and we denote it by $\overline{\omega t}$. Then

$$\gamma_{12} = -\frac{2E}{N_o}\overline{\omega t}$$

Clearly, if there is no angle modulation, $\overline{\omega t}$ and γ_{12} are zero. In this case, jointly efficient estimates of frequency and time of arrival are uncorrelated.

For jointly efficient estimates, the covariance matrix of Eq. (10-18) becomes

$$\psi = \frac{\begin{bmatrix} t_d^2 & \overline{\omega t} \\ \overline{\omega t} & \beta^2 \end{bmatrix}}{\left(\frac{2E}{N_o}\right)[\beta^2 t_d^2 - \overline{\omega t}^2]}$$

Specifically

$$\sigma_{\hat{t}}^2 = \frac{t_d^2}{(2E/N_o)[\beta^2 t_d^2 - \overline{\omega t}^2]} \tag{10-74}$$

$$\sigma_{\hat{\omega}}^2 = \frac{\beta^2}{(2E/N_o)(\beta^2 t_d^2 - \overline{\omega t}^2)} \tag{10-75}$$

$$\text{Cov}\{\hat{\omega}, \hat{\tau}\} = \frac{\overline{\omega t}}{(2E/N_o)(\beta^2 t_d^2 - \overline{\omega t}^2)} \tag{10-76}$$

As noted, if $\overline{\omega t}$ is zero these expressions reduce to those values of minimum variance, Eqs. (10-60) and (10-69), which apply when only one parameter is unknown.

An Uncertainty Relation

We have determined the minimum variance of the time of arrival and frequency estimates. Each estimate may be improved by increasing the effective bandwidth and effective time duration respectively. These quantities are not completely independent however. For example, as the effective time duration decreases, the effective bandwidth increases. Consequently, an improvement in the frequency estimate is usually obtained only at a sacrifice of the time of arrival estimate. Conversely, the time of arrival estimate is improved only at a sacrifice of the frequency estimate. Indeed, the product of the minimum standard deviations has a lower bound which is, from Eqs. (10-74) and (10-75)

$$\sigma_{\hat{\tau}}\sigma_{\hat{\omega}} \geq \frac{1}{(2E/N_o)\beta t_d} \tag{10-77}$$

For present purposes we assume that $\overline{\omega t} = 0$. We may therefore improve this bound by either increasing the signal energy, or the effective time-bandwidth product of the signal.

We next determine an inequality, or uncertainty relation, relating to the time–bandwidth product. From Eqs. (10-61) and (10-70)

$$\beta^2 t_d^2 = \int \omega^2 |G(j\omega)|^2 \frac{d\omega}{2\pi} \int t^2 |\tilde{a}(t)|^2 \, dt \tag{10-78}$$

where $\tilde{a}(t)$ and $G(j\omega)$ are a Fourier transform pair, and it is assumed that

$$\overline{\omega} = \int \omega |G(j\omega)|^2 \frac{d\omega}{2\pi} = 0$$

$$\bar{t}_d = \int t |\tilde{a}(t)|^2 \, dt = 0$$

$$\int |G(j\omega)|^2 \frac{d\omega}{2\pi} = \int |\tilde{a}(t)|^2 \, dt = 1$$

If the latter assumption is not used, Eq. (10-78) must be appropriately normalized. Recall that

$$\int \omega^2 |G(j\omega)|^2 \frac{d\omega}{2\pi} = -\int \tilde{a}^*(t)\tilde{a}''(t)\,dt = \int |\tilde{a}'(t)|^2\,dt$$

Then

$$\beta^2 t_d^{\,2} = \int |\tilde{a}'(t)|^2\,dt \int t^2 |\tilde{a}(t)|^2\,dt$$

Applying Schwartz' inequality results in

$$\beta^2 t_d^{\,2} \geq \left| \int t\tilde{a}^*(t)\tilde{a}'(t)\,dt \right|^2 \tag{10-79}$$

In reducing Eq. (10-72) we found it useful to consider the signal in the form

$$\tilde{a}(t) = f(t)e^{j\phi(t)}$$

where $f(t)$ and $\phi(t)$ were real functions of time, and $\phi(t)$ was a measure of the angle modulation of the signal. In this case the inequality (10-79) becomes

$$\beta^2 t_d^{\,2} \geq \left[\int tf^2 \frac{d\phi(t)}{dt}\,dt\right]^2 + \left[\int tf(t)\frac{df(t)}{dt}\,dt\right]^2$$

The first term of this inequality had previously been defined as $\overline{\omega t}$. The second term, after integration by parts, is easily shown to be $\frac{1}{4}$. (Recall $\int f^2(t)\,dt = 1$.) We then have that

$$\beta^2 t_d^{\,2} \geq \overline{\omega t} + \tfrac{1}{4}$$

or, for $\overline{\omega t} = 0$,

$$\beta t_d \geq \tfrac{1}{2} \tag{10-80}$$

This is the uncertainty relation we desired. It shows that a signal cannot simultaneously have an arbitrarily small duration and an arbitrarily small bandwidth.

10.8 Estimation in Nonwhite Gaussian Noise

This section presents some results on estimation of signal parameters when the additive noise is nonwhite Gaussian. The received signal is given by

$$r(t) = s(t, \boldsymbol{\alpha}) + n(t)$$

where $\boldsymbol{\alpha}$ is a parameter which we wish to estimate and $n(t)$ is a sample function of Gaussian noise having an autocorrelation function $R_n(\tau)$.

We shall be concerned with a maximum-likelihood estimation of a single unknown parameter, α. The results are easily generalized to the case of a vector $\boldsymbol{\alpha}$. The log-likelihood function is obtained by appropriately modifying Eq. (9-62). Then,

$$\ln p(\mathbf{r} \mid \alpha) = \ln C - \tfrac{1}{2} \int_0^T \int_0^T [r(t) - s(t, \alpha)] R_n^{-1}(t - \tau) [r(\tau) - s(\tau, \alpha)] \, dt \, d\tau \tag{10-81}$$

where C is a constant independent of α, and where $R_n^{-1}(t, \tau)$ is the solution to

$$\int_0^T R_n^{-1}(t, \tau) R_n(\tau, s) \, d\tau = \delta(t - s)$$

The maximum-likelihood estimate is that value of α which maximizes Eq. (10-81). To show the estimate more explicitly, we require the derivative of the likelihood function:

$$\begin{aligned} \frac{\partial \ln p(\mathbf{r} \mid \alpha)}{\partial \alpha} = &+\tfrac{1}{2} \int_0^T \int_0^T [r(t) - s(t, \alpha)] R_n^{-1}(t - \tau) \frac{\partial s(\tau, \alpha)}{\partial \alpha} \, dt \, d\tau \\ &+ \tfrac{1}{2} \int_0^T \int_0^T [r(\tau) - s(\tau, \alpha)] R_n^{-1}(t - \tau) \frac{\partial s(t, \alpha)}{\partial \alpha} \, dt \, d\tau \end{aligned}$$

Since $R_n^{-1}(t - \tau) = R_n^{-1}(\tau - t)$, both double integrals are identical and therefore

$$\frac{\partial \ln p(\mathbf{r} \mid \alpha)}{\partial \alpha} = \int_0^T \int_0^T \frac{\partial s(t, \alpha)}{\partial \alpha} R_n^{-1}(t - \tau) [r(\tau) - s(\tau, \alpha)] \, dt \, d\tau \tag{10-82}$$

In analogy to Eqs. (9-31) and (9-37) we define a filter $h(t)$ where

$$s(t, \alpha) = \int_0^T R_n(t - \tau) h(\tau) \, d\tau \tag{10-83}$$

or the inverse

$$h(\tau) = \int_0^T s(t, \alpha) R_n^{-1}(\tau - t) \, dt \tag{10-84}$$

It is clear that $h(\tau)$ is a function of the parameter α. For notational convenience we shall not carry through an explicit reminder. Differentiating Eq. (10-84) with respect to α, and using the result in Eq. (10-82) we get

$$\frac{\partial \ln p(\mathbf{r} \mid \alpha)}{\partial \alpha} = \int_0^T \frac{\partial h(\tau)}{\partial \alpha} [r(\tau) - s(\tau, \alpha)] \, d\tau \tag{10-85}$$

The estimate of α is the solution to the equation

$$\int_0^T \frac{\partial h(\tau)}{\partial \hat{\alpha}} [r(\tau) - s(\tau, \hat{\alpha})] \, d\tau = 0 \tag{10-86}$$

To determine the Cramer–Rao bound requires

$$E\left\{\left[\frac{\partial}{\partial \alpha} \ln p(\mathbf{r} \mid \alpha)\right]^2\right\} = E\left\{\int_0^T \int_0^T [r(\tau) - s(\tau, \alpha)][r(t) - s(t, \alpha)] \frac{\partial h(\tau)}{\partial \alpha} \frac{\partial h(t)}{\partial \alpha} \, dt \, d\tau\right\}$$

Substituting $n(t) = r(t) - s(t, \alpha)$;

$$E\left\{\left[\frac{\partial}{\partial \alpha} \ln p(\mathbf{r} \mid \alpha)\right]^2\right\} = \int_0^T \int_0^T R_n(t - \tau) \frac{\partial h(\tau)}{\partial \alpha} \frac{\partial h(t)}{\partial \alpha} \, dt \, d\tau \tag{10-87}$$

Using Eq. (10-83)

$$\frac{\partial s(t, \alpha)}{\partial \alpha} = \int_0^T R_n(t - \tau) \frac{\partial h(\tau)}{\partial \alpha} \, d\tau$$

and therefore

$$E\left\{\left[\frac{\partial}{\partial \alpha} \ln p(\mathbf{r} \mid \alpha)\right]^2\right\} = \int_0^T \frac{\partial s(t, \alpha)}{\partial \alpha} \frac{\partial h(t)}{\partial \alpha} \, dt \tag{10-88}$$

The Cramer–Rao bound is then

$$\sigma_{\hat{\alpha}}^2 \geq \left[\int_0^T \frac{\partial s(t, \alpha)}{\partial \alpha} \frac{\partial h(t)}{\partial \alpha} \, dt\right]^{-1} \tag{10-89}$$

For the general case of a vector $\boldsymbol{\alpha}$, we form an information matrix whose components are given by Eq. (10-17). It is easily shown that the matrix elements are given by

$$\gamma_{ij} = \int_0^T \frac{\partial h(t)}{\partial \alpha_i} \frac{\partial s(t, \boldsymbol{\alpha})}{\partial \alpha_j} \, dt$$

As an illustration for the single variable case, assume we wish to estimate the amplitude A, where

$$r(t) = As(t) + n(t)$$

The estimate for A is found by solving Eq. (10-86). Assume that $k(\tau)$ is the solution of

$$s(t) = \int_0^T R_n(t - \tau) k(\tau) \, d\tau$$

Then, since $As(t) = \int_0^T R_n(t - \tau)h(\tau)\, d\tau$ it follows that $h(\tau) = Ak(\tau)$ and $\partial h(\tau)/\partial A = k(\tau)$. Substituting this into Eq. (10-86) we have

$$\int_0^T k(\tau)[r(\tau) - \hat{A}s(\tau)]\, d\tau = 0$$

or

$$\hat{A} = \frac{\int_0^T k(\tau)r(\tau)\, d\tau}{\int_0^T k(\tau)s(\tau)\, d\tau} = \frac{\int_0^T h(\tau)r(\tau)\, d\tau}{\int_0^T h(\tau)s(\tau)\, d\tau}$$

This is similar to the solution for the white noise case except that $h(\tau)$ replaces $s(\tau)$. This is to be expected since in Eq. (10-83) we use $\delta(t - \tau) = R_n(t - \tau)$, then $s(t) = h(t)$. Consequently, the estimators which were derived using a white Gaussian assumption are similar to those for the nonwhite Gaussian noise case except for a substitution of $h(t)$ for $s(t)$ in appropriate places.

Randomly Phased Signals

For the case where the signal is narrowband and the phase is uniformly distributed, we proceed much the same as for the white noise case (*12*, *17*). Specifically, for a signal of energy E and phase θ, the complex envelope of the received signal is

$$\tilde{r}(t) = (2E)^{1/2}\,\tilde{a}(t)e^{j\theta} + \tilde{z}(t), \qquad 0 \le t \le T$$

The phase averaged log-likelihood function for high signal-to-noise, Eq. (10-34), is

$$\ln p_\theta(\mathbf{r}) \approx \ln K + D$$

where

$$D = \left| \int_0^T \tilde{r}(t)\tilde{h}^*(t)\, dt \right|$$

and $\tilde{h}(t)$ is the solution of

$$\int_0^T \tilde{R}_n(t - \tau)\tilde{h}(\tau)\, d\tau = \tilde{a}(t)$$

Thus, instead of inserting the received signal into a filter matched to $\tilde{a}(t)$, a filter matched to $\tilde{h}^*(t)$ is used (a generalized matched filter). In analogy to the white noise case, if we have unknown parameters such as frequency and time of arrival, we may extract the estimates by building a receiver consisting of a bank of generalized matched filters (as opposed to matched filters for white noise). The generalized matched filters are solutions of

$$\int_0^T \tilde{R}_n(t - \tau)\tilde{h}_{\boldsymbol{\alpha}}(\tau)\, d\tau = \tilde{a}(t, \boldsymbol{\alpha})$$

10.9 Generalized Likelihood Ratio Detection

Except for a brief discussion of the maximum-likelihood principle in Chap. 5, we have not yet discussed the problem of detecting signals with unknown parameters for which there is no a priori information. For those detection problems in which the signal had unknown or random parameters, we assumed an a priori density function for these parameters and generated an averaged likelihood ratio. If the a priori density function was indeed correct, then the resulting receiver was optimum. The question naturally arises as to what methods might be employed if one does not know and is not willing to assume such a priori knowledge. There are a few alternatives. (We will not include among these the adaptive procedures which could be useful in practice.) If a uniformly most powerful test with respect to the unknown parameters is known to exist, then that would be the way to proceed. If no uniformly most powerful test can be found, a minimax criterion might be applied. We would try to accomplish "robust" detection by this method. That is, detection processes which work "well" for whatever value the unknown parameter happens to be.

Still another alternative, and the one considered here, is the maximum likelihood principle. Note, however, that we may no longer be considering "optimum" detection in the sense of minimizing probability of error or maximizing probability of detection.

To use the maximum likelihood principle for two hypotheses, the generalized likelihood ratio is formed

$$\lambda(\mathbf{r}) = \frac{\max_{\boldsymbol{\theta}} p_1(\mathbf{r}|\boldsymbol{\theta})}{\max_{\boldsymbol{\phi}} p_0(\mathbf{r}|\boldsymbol{\phi})} \tag{10-90}$$

and a monotonic function of $\lambda(\mathbf{r})$ is used as the test statistic. The unknown parameters under each hypothesis are denoted $\boldsymbol{\theta}$ and $\boldsymbol{\phi}$. Thus, the maximum-likelihood principle directs us to form maximum-likelihood estimates of the parameters and use these estimates just as though they were known true values. Stated otherwise, the optimum receiver is conceptually the same as the receiver used for the detection of completely known signals, except that estimated values of the parameters are used in place of the true but unknown values. Therefore, for the kinds of signals with which we have been concerned, the receiver would be a correlation receiver. While this is true, we shall derive equivalent forms for these receivers and shall develop some interesting results.

The receiver is shown in Fig. 10-4. A time delay is included since a best estimate cannot be generated until the signal has been received and processed. Once the estimates are formed, the same signal which was used to generate the estimate is then inserted into the receiver. (Using adaptive

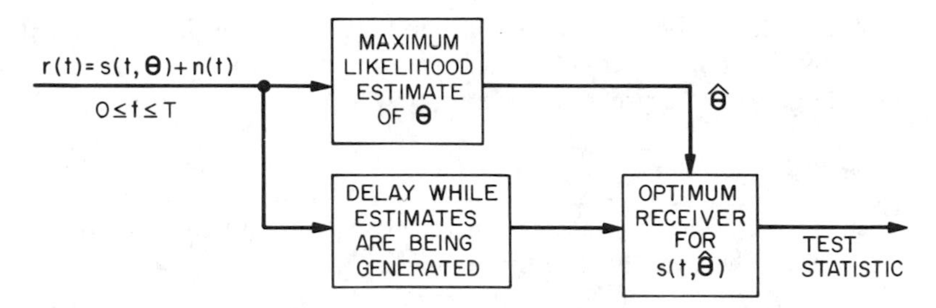

Fig. 10-4 A detector using the maximum likelihood principle.

procedures, a signal other than the one used to generate the estimates might be processed. Indeed, in a real-time system, one might recursively or sequentially update the estimates. Such procedures shall not be considered here even though they may be quite beneficial in practice.)

For simplicity assume that the null hypothesis involves noise only, so that the generalized likelihood ratio is

$$\lambda(\mathbf{r}) = \frac{\max_{\boldsymbol{\theta}} p_1(\mathbf{r}|\boldsymbol{\theta})}{p_0(\mathbf{r})} \tag{10-91}$$

For pedagogical reasons, four examples of increasing complexity are considered. These are

(1) unknown phase
(2) unknown phase and frequency
(3) unknown phase, frequency, and amplitude
(4) unknown phase, frequency, amplitude, and time of arrival.

The signal is assumed to be a pure sinewave in white Gaussian noise. The results may be generalized to other signals and nonwhite noise.

Except for the case of unknown time of arrival the hypotheses will be of the form

$$\begin{aligned} H_1&: \; r(t) = A\sin(\omega t + \theta) + n(t), \qquad 0 \le t \le T \\ H_0&: \; r(t) = n(t) \end{aligned}$$

and the Neyman–Pearson criteria will be used. For such hypotheses with *known* parameters, the likelihood ratio, from Eq. (6-38), is

$$\lambda(\mathbf{r}\,|\,\boldsymbol{\alpha}) = \exp\left[-\frac{A^2T}{2N_0}\right]\exp\left[\frac{2A}{N_0}\int_0^T r(t)\sin(\omega t + \theta)\,dt\right] \tag{10-92}$$

When an unknown time of arrival is considered, the hypotheses must be changed to reflect this uncertainty.

As a reminder, only maximum-likelihood estimates will be considered (as indicated by the generalized likelihood ratio). While other estimates could conceivably be used, they cannot be expected to produce the same results.

Unknown Phase

When only phase is unknown, the receiver which maximizes $\lambda(\mathbf{r}|\hat{\theta})$, Eq. (10-92), is equivalent to one which maximizes

$$\int_0^T r(t) \sin(\omega t + \hat{\theta})\, dt \tag{10-93}$$

where $\hat{\theta}$ is the maximum-likelihood estimate of θ. From Eq. (10-43), this estimate is

$$\hat{\theta} = \tan^{-1}\left\{\frac{\int_0^T r(t) \cos \omega t\, dt}{\int_0^T r(t) \sin \omega t\, dt}\right\}$$

The receiver is shown in Fig. 10-5 along with an equivalent form which is derived next.

Equation (10-93) may be written

$$\cos \hat{\theta} \int_0^T r(t) \sin \omega t\, dt + \sin \hat{\theta} \int_0^T r(t) \cos \omega t\, dt$$

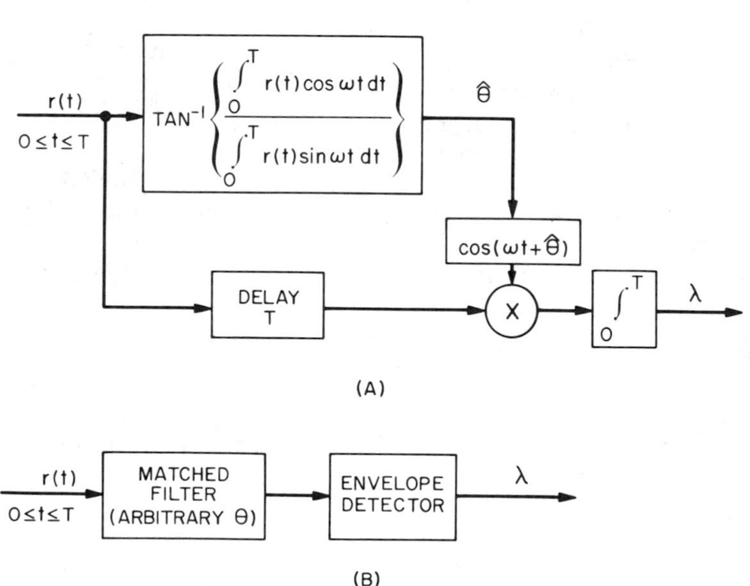

Fig. 10-5 (A) Maximum likelihood receiver for unknown phase, and (B) equivalent receiver (incoherent matched filter).

where

$$\cos \hat{\theta} = \frac{\int_0^T r(t) \sin \omega t \, dt}{q}, \qquad \sin \hat{\theta} = \frac{\int_0^T r(t) \cos \omega t \, dt}{q}$$

$$q^2 = \left[\int_0^T r(t) \sin \omega t \, dt\right]^2 + \left[\int_0^T r(t) \cos \omega t \, dt\right]^2$$

It may then be shown that

$$\int_0^T r(t) \sin(\omega t + \hat{\theta}) \, dt = q$$

The function q may be generated by the incoherent matched filter (filter matched to the signal for arbitrary phase and followed by an envelope detector) discussed in Chap. 7. This receiver is also shown in Fig. 10-5, and we have shown that the maximum-likelihood principle led us to the same receiver derived for a uniform a priori distribution of phase. We shall see more of this as the remaining examples are considered.

Unknown Phase and Frequency

For this case, maximizing $\lambda(\mathbf{r}|\omega, \theta)$ is equivalent to maximizing with respect to ω and θ

$$\lambda(\theta, \omega) \triangleq \int_0^T r(t) \sin(\omega t + \theta) \, dt \tag{10-94}$$

The notation $\lambda(\theta, \omega)$ is used to indicate a monotonic function of $\lambda(\mathbf{r}|\theta, \omega)$. The maximum-likelihood receiver is then indicated by

$$\lambda(\hat{\theta}, \hat{\omega}) = \int_0^T r(t) \sin(\hat{\omega} t + \hat{\theta}) \, dt \tag{10-95}$$

The simultaneous maximization is accomplished if we maximize $\lambda(\hat{\theta}, \omega)$ with respect to ω. But $\lambda(\hat{\theta}, \omega) = q(\omega)$ and

$$q^2(\omega) = \left[\int_0^T r(t) \cos \omega t \, dt\right]^2 + \left[\int_0^T r(t) \sin \omega t \, dt\right]^2 \tag{10-96}$$

We could attempt to find analytically the frequency which maximizes this expression, but a closed form solution does not appear available. This need not deter us however. Conceptually, the maximum may be determined in the same way that frequency was estimated earlier, that is, by constructing a receiver such as shown in Fig. 10-3. (The effects of quantizing the frequency range are being ignored.) This frequency estimate in turn could be used to obtain the phase estimate. Both estimates could then be applied to Eq. (10-95).

But, it has already been shown that Eq. (10-95) with a maximum-likelihood estimate of phase leads to $q(\omega)$. Inserting the frequency estimate produces $q(\hat{\omega})$. This means that the largest output of the filter bank, Fig. 10-3, can also be used as the test statistic—that is, the maximum-likelihood principle receiver. Thus, for this example, the receiver used for estimation can also be used for detection. Of course, $q(\hat{\omega})$ must be greater than some threshold for H_1 to be chosen. It will be shown below that using $q(\hat{\omega})$ as the test statistic is the same as using $\hat{A}$. Accepting the approximation due to quantizing the frequency range, the receiver is the same as that determined in Chap. 7, and shown in Fig. 7-15 for the case where each frequency of interest was equally likely to occur and each was assigned an individual alternative hypothesis.

Unknown Phase, Frequency, and Amplitude

For this case

$$\lambda(\mathbf{r} \mid \omega, \theta, A) = \exp\left[-\frac{A^2T}{2N_0}\right] \exp\left[\frac{2A}{N_0}\int_0^T r(t)\sin(\omega t + \theta)\,dt\right]$$

It may be readily determined that the maximum-likelihood estimate of A is

$$\hat{A} = (2/T)\int_0^T r(t)\sin(\hat{\omega}t + \hat{\theta})\,dt \tag{10-97}$$

But, from the preceding case we already know that the right-hand side is just $2q(\hat{\omega})/T$ so

$$\hat{A} = 2q(\hat{\omega})/T$$

Therefore, the estimate $\hat{A}$ can be obtained by taking the largest output, in Fig. 10-3 or Fig. 7-15, and modifying it by $2/T$, Substituting for $\hat{A}$ we get

$$\lambda(\mathbf{r} \mid \hat{\omega}, \hat{\theta}, \hat{A}) = \exp(2/TN_o)\left[\int_0^T r(t)\sin(\hat{\omega}t + \hat{\theta})\,dt\right]^2$$

The receiver can then be implemented as in the unknown phase and frequency case, that is, $q(\hat{\omega})$ or the equivalent $\hat{A}$ may be used as the test statistic.

Unknown Phase, Frequency, Amplitude, and Time of Arrival

For this case it is assumed that the leading edge of the signal may occur anywhere in an interval of time, say $0 \leq \tau \leq \tau_m$. Then the likelihood function may be written, for $0 \leq \tau \leq \tau_m$,

$$\lambda(\mathbf{r} \mid \omega, \theta, A, \tau) = \exp\left[-\frac{A^2T}{2N_o}\right] \exp\left\{\frac{2A}{N_o}\int_\tau^{\tau+T} r(t)[\sin\omega(t-\tau) + \theta]\,dt\right\}$$

The estimates must maximize this expression. The estimate for τ will be that value which produces a maximum of

$$\int_{\tau}^{\tau+T} r(t) \sin[\hat{\omega}(t-\tau)+\hat{\theta}]\, dt$$

which, in analogy to our previous notation, is denoted $q(\hat{\omega}, \tau)$. Furthermore

$$(2/T)q(\hat{\omega}, \hat{\tau}) = \hat{A} = (2/T)\int_{\hat{\tau}}^{\hat{\tau}+T} r(t) \sin[\hat{\omega}(t-\hat{\tau})+\hat{\theta}]\, dt$$

Therefore, we build a receiver such as shown in Fig. 7-19 (where we assume arbitrarily fine frequency and time granularity if necessary) to do the simultaneous estimation. That is, by finding the maximum $\hat{A}$, we can determine $\hat{\omega}$, $\hat{\tau}$, and $\hat{\theta}$. It may then be easily argued that $\hat{A}$ or $q(\hat{\omega},\hat{\tau})$ may be used as the test statistic.

Exercises

10.1 Denote the set of samples $x_n, x_{n-1}, \ldots, x_1$ by z_n. Show that the a posteriori probability density function of a parameter, say θ, given the set of samples may be expressed

$$p(\theta \mid z_n) = \frac{p(x_n \mid \theta, z_{n-1})p(\theta \mid z_{n-1})}{\int_{\{\theta\}} p(x_n \mid \theta, z_{n-1})p(\theta \mid z_{n-1})\, d\theta}$$

10.2 Based on N statistically independent samples of a Gaussian process of variance σ^2 and unknown mean μ we wish to find MAP estimator for the mean. Assume that the only a priori information about the mean is that it is greater than or equal to zero.

(a) What is the estimator?

(b) What is its probability density function?

10.3 Consider a signal $r(t) = A\sin(\omega_c t + \theta) + n(t)$ where $n(t)$ is white Gaussian noise, θ is uniformly distributed $(0, 2\pi)$. We want a maximum-likelihood estimate for the amplitude A, but since a priori information for the phase is known we first average over θ to obtain $p(\mathbf{r}|A)$. What is the equation which must be solved for the maximum-likelihood estimate of A?

10.4 Using M statistically independent samples x_i, $i = 1, \ldots, M$ of a Gaussian process with mean μ and unknown variance σ^2.

(a) Show that the maximum-likelihood estimate of σ^2 is

$$\hat{\sigma}^2 = (1/M) \sum (x_i - \mu)^2$$

(b) Show that it is a sufficient statistic.

(c) Show that σ^2 is also an efficient estimator.

10.5 Using M statistically independent samples $s_i, i = 1, \ldots, M$ from a gamma distribution with n degrees of freedom so that

$$p(s_i \mid \alpha) = \frac{s_i^{n/2-1} e^{-s_i/2\alpha}}{\alpha^{n/2} 2^{n/2} \Gamma(n/2)}$$

(a) Show that the maximum-likelihood estimate of α is $\hat{\alpha} = \bar{s}/n$ where $\bar{s} = (1/M) \sum_{i=1}^{M} s_i$.
(b) Show that this is a sufficient statistic.
(c) Show that $\hat{\alpha}$ is an efficient estimator for α.

10.6 (a) Determine the Bayes estimator of a parameter α using the cost function

$$C(\hat{\alpha}, \alpha) = \begin{cases} 0, & |\hat{\alpha} - \alpha| < \Delta \\ 1, & \text{otherwise} \end{cases}$$

Assume that the a posteriori density function $p(\alpha|v)$, where v represents data, has definition over the range $-\infty < \alpha < \infty$. Will the estimator generally be a function of Δ?
(b) As Δ approaches 0, what is the estimator?
(c) Suppose $p(\alpha|v)$ is unimodal and symmetric about that mode. What is the Bayes estimator? Is the estimator a function of Δ?

10.7 For a signal $s(t) + n(t)$, where $n(t)$ is white Gaussian noise with power spectral density $N_o/2$, find the minimum variance of an unbiased estimate of the time of arrival of $s(t)$, where $s(t)$ is as shown in Fig. 10-6. Answer: $\sigma_{\hat{t}}{}^2 \geq \Delta(2\Delta + 3L)/(12E/N_o)$.

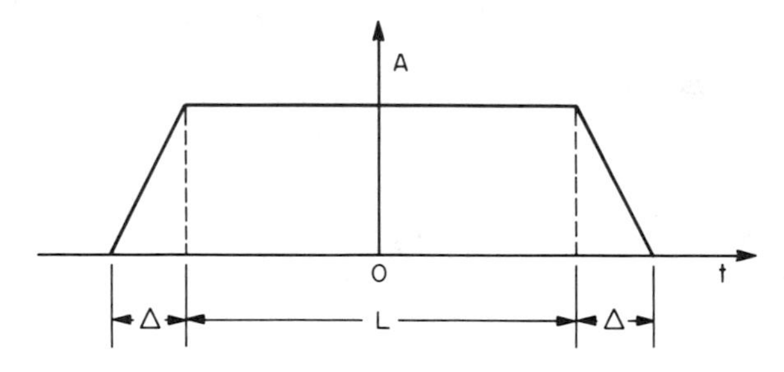

Fig. 10-6 A trapezoidal pulse.

10.8 (a) Repeat the preceding problem for a signal

$$s(t) = \begin{cases} A(1 + \cos \omega_0 t), & -(2m+1)\pi/\omega_0 \leq t \leq -2m\pi/\omega_0 \\ A, & -2m\pi/\omega_0 \leq t \leq 2m\pi/\omega_0 \\ A(1 + \cos \omega_0 t), & 2m\pi/\omega_0 \leq t \leq (2m+1)\pi/\omega_0 \end{cases}$$

to show the minimum variance is

$$\sigma_{\hat{\tau}}^2 \geq \frac{3 + 4m}{(2E/N_o)\omega_0^2}$$

(b) Suppose the signal is of the form $s(t)\cos(\omega_c t + \theta)$ where θ is uniformly distributed $(0, 2\pi)$. Find the minimum variance expressed in terms of A, and in terms of the signal energy E.

(c) For the signal of part (b), show that the minimum variance of an unbiased estimate of frequency (known time of arrival) is

$$\sigma_{\hat{\omega}}^2 \geq \left[\frac{2E}{N_o}\left(\frac{2}{3 + 4m}\right)\left(\frac{mT_0^2}{6\omega_0} + 1\right)\right]^{-1}$$

where $T_0 \triangleq 4m\pi/\omega_0$.

10.9 Repeat problem 10.7 for a signal

$$s(t) = \frac{A}{(2\pi)^{1/2}} e^{-t^2/2\alpha}, \qquad -\infty < t < \infty$$

to show that $\sigma_{\tau}^2 \geq \alpha/(E/N_o)$.

10.10 Consider the signal $A(t)\cos(\omega_c t + \gamma t^2 + \theta) + n(t)$, $0 \leq t \leq T$, where $n(t)$ is white Gaussian noise. The signal envelope and carrier frequency ω_c are known. The phase is uniformly distributed $(0, 2\pi)$. (The parameter γ may be related to the acceleration of a moving source.)

(a) Use the usual narrowband assumptions and show that the minimum variance of an unbiased estimate of γ is given by

$$\sigma_{\hat{\gamma}}^2 \geq \frac{1}{(2E/N_o)(\mu_4 - \mu_2^2)}$$

where

$$\mu_2 = \int t^2|\tilde{a}(t)|^2\,dt, \qquad \mu_4 = \int t^4|\tilde{a}(t)|^2\,dt$$

(d) What is $\sigma_{\hat{\gamma}}^2$ when $A(t)$ is a constant?

10.11 For the signal of the preceding exercise, assume the time of arrival is known, and that simultaneous estimates of frequency and acceleration parameter γ are desired. Show that

$$\sigma_{\hat{\omega}}^2 \geq [(2E/N_o)\,t_d^2\,(1 - \rho^2)]^{-1}$$
$$\sigma_{\hat{\gamma}}^2 \geq [(2E/N_o)\,t_4^2\,(1 - \rho^2)]^{-1}$$

where

$$\mu_i = \int t^i \, |\tilde{a}(t)|^2 \, dt, \qquad t_d{}^2 = \mu_2 - \mu_1{}^2$$

$$t_3 = \mu_3, \qquad t_4{}^2 = \mu_4 - \mu_2{}^2, \qquad \rho = \frac{t_3}{t_d t_4}$$

10.12 Consider the detection problem with hypotheses

$$H_1: \quad r(t) = B\cos(\omega_2 t + \phi) + n(t), \qquad 0 \le t \le T$$
$$H_0: \quad r(t) = n(t)$$

where $n(t)$ is white, Gaussian noise, B is a known constant, ϕ is uniformly distributed $(0, 2\pi)$, and ω_2 is unknown. What is the form of the receiver if the maximum-likelihood principle is used after having averaged over the phase ϕ?

10.13 Consider the signal $r(t) = A\cos\omega_1 t + B\cos(\omega_2 t + \phi) + n(t)$ where $n(t)$ is white, Gaussian noise. Assume A, B, and ω_1 are known, and ϕ is uniformly distributed $(0, 2\pi)$. How might a receiver be implemented to extract a maximum-likelihood estimate of ω_2?

10.14 In the preceding problem, suppose the signal is modified to be $A\cos(\omega_1 t + \theta) + B\cos(\omega_2 t + \phi) + n(t)$ where θ and ϕ are statistically independent and uniformly distributed $(0, 2\pi)$. Assume $\int \cos(\omega_1 t + \theta)\cos(\omega_2 t + \phi)\, dt = 0$. How might a receiver be implemented to simultaneously extract maximum-likelihood estimates of ω_1 and ω_2?

References

1. Van Trees, H. L., "Detection, Estimation, and Modulation Theory," Part I. Wiley, New York, 1968.
2. Viterbi, A. J., "Principles of Coherent Communication." McGraw-Hill, New York, 1966.
3. Sherman, S., Non-mean-square error criteria, *IRE Trans. Inform. Theory* **IT-4,** (1958).
4. Woodward, P. M., "Probability and Information Theory, With Applications to Radar," 2nd Ed. Pergamon Press, Oxford, 1964.
5. Helstrom, C. W., "Statistical Theory of Signal Detection." Pergamon Press, Oxford, 1960.
6. Slepian, D., Estimation of signal parameters in the presence of noise, *IRE Trans. Inform. Theory*, 68–89 (March 1954).
7. Kendall, M. G., and Stuart, A., "Inference and Relationship, The Advanced Theory of Statistics," Vol. 2. Hafner Publ., New York, 1961.
8. Cramer, H., "Mathematical Methods of Statistics," 8th Printing. Princeton Univ. Press, Princeton, New Jersey, 1958.
9. Gart, J. J., An extension of the Cramer-Rao inequality, *Ann. Math. Statistics* **30,** 367–380 (1959).
10. Turin, G. L., On optimal diversity reception, *IRE Trans. Inform. Theory*, 154–166 (July 1961).

11. Kailath, T., The complex envelope of white noise, correspondence, *IEEE Trans. Inform. Theory*, 397–398 (July 1966).
12. Kelly, E. J., Reed, I. S., and Root, W. L., Part II, The Accuracy of Radar Measurements, *J. Soc. Ind. Appl. Math.* **8,** No. 3, Part II (1960).
13. Kelly, E. J., The radar measurement of range, velocity, and acceleration, *IRE Trans. Military Electron.* (April 1961).
14. Mallenckrodt, A. J., and Sollenberger, T. E., Optimum Pulse-Time Determination, *IRE Trans. Inform. Theory*, March 1954.
15. Mityashev, B. N., "The Determination of the Time Position of Pulses in the Presence of Noise" (*Transl.* by D. L. Jones). MacDonald, London, 1965.
16. Bello, P., Joint estimation of delay, doppler, and doppler rate, *IRE Trans. Inform. Theory* **IT-6,** No. 3, 330–341 (1960).
17. Kelly, E. J., Reed, I. S., and Root, W. L., The detection of radar signals in noise, *J. Soc. Ind. Appl. Math.* **8,** No. 2, Part I (1960).

SUPPLEMENTARY BIBLIOGRAPHY

Texts:

Barton, D., "Radar System Analysis." Prentice Hall, Englewood Cliffs, New Jersey, 1964.

Berkowitz, R. S., ed., "Modern Radar, Analysis, Evaluation, and System Design." Wiley, New York, 1965.

Cook, C. E., and Bernfeld, M., "Radar Signals, An Introduction to Theory and Application." Academic Press, New York, 1967.

Deutsch, R., "Estimation Theory." Prentice Hall, Englewood Cliffs, New Jersey, 1965.

Hancock, J. C., and Wintz, P. A., "Signal Detection Theory," McGraw-Hill, New York, 1966.

Lee, Y. W., "Statistical Theory of Communication." Wiley, New York, 1960.

Middleton, D., "Topics in Communication Theory." McGraw-Hill, New York, 1965.

Miller, K. S., "Multidimensional Gaussian Distribution." Wiley, New York, 1964.

Mood, A. M., and Graybill, F. A., "In Introduction to the Theory of Statistics," 2nd Ed. McGraw-Hill, New York, 1963.

Vakman, D. E., "Sophisticated Signals and the Uncertainty Principle in Radar" (E. Jacobs, ed.) (*Transl.* by K. N. Trirogoff). Springer-Verlag, New York, 1968.

Wainstein, L. A., and Zubakov, V. D., "Extraction of Signals From Noise" (*Transl.* by R. A. Silverman). Prentice Hall, Englewood Cliffs, New Jersey, 1962.

Papers:

Balch, H. T., Dale, J. C., Eddy, T. W., and Lauver, R. M., Estimation of the mean of a stationary random process by periodic sampling, The *Bell Syst. Tech. J.* (May-June 1966).

Benedict, T. R., and Soony, T. T., The joint estimation of signal and noise from the sum envelope, *IEEE Trans. Inform. Theory* **IT-13,** No. 3, 447–454 (1967).

Bode, H. W., and Shannon, C. E., A simplified derivation of linear least square smoothing and prediction theory, *Proc. IRE* (April 1950).

Fowle, E. N., The Design of Radar Signals, Mitre Corp. Tech. Rep. ESD-TR-65-97, AD 617711 (June 1965).

Grenander, U., Stochastic processes and statistical inference, *Ark. Mat.* **17,** Band 1 (1950).

Hansen, V. G., The accuracy of radar range measurements using non-optimum filters, *IEEE Trans. Aerospace Navigational Electron.* (June 1965).

Lerner, R. M., Signals with uniform ambiguity functions, *IRE Nat. Convention Rec.* Part 4 (1958).

Marchese, J. F., Precision radar, *Electro-Techn.* (February 1965).

Middleton, D., "The Incoherent Estimation of Signal Amplitudes in Normal Noise Backgrounds," (M. Rosenblatt, ed.), Chap. 24, Time Series Analysis, Wiley, New York, 1963.

Middleton, D., and Esposito, R., Simultaneous optimum detection and estimation of signals in noise, *IEEE Trans. Inform. Theory* **IT-14,** No. 3, 434–444 (1968).

Rao, C. R., Information and the accuracy attainable in the estimation of statistical parameters, *Bull. Calcutta Math. Soc.* **37,** (1945).

Rihaczek, A. W., Radar resolution properties of pulse trains, *Proc. IEEE* (February 1964).

Rihaczek, A. W., Target Resolution: Capabilities of Modern Radar and Fundamental Limits, Electron. Res. Lab., Aerospace Corp., SSD-TDR-64-107, AD 605221 (27 July 1964).

Rihaczek, A. W., Radar signal design for target resolution, *Proc. IEEE* (February 1965).

Rihaczek, A. W., Radar resolution of moving targets, *IEEE Trans. Inform. Theory* **IT-13,** No. 1, 51–56 (1967).

Rihaczek, A. W., Signal energy distribution in time and frequency, *IEEE Trans. Inform. Theory*, **IT-14,** No. 3, 369–374 (1968).

Siebert, W., A radar detection philosophy, *IRE Trans. Inform. Theory* (September 1956).

Stutt, C. A., Some results on real-part/imaginary—part and magnitude—phase relations on ambiguity functions, *IEEE Trans. Inform. Theory* **IT-10,** No. 4, 321–327 (1964).

Swerling, P., Maximum angular accuracy of a pulsed search radar, *Proc. IRE* (September 1956).

Swerling, P., Parameter estimation for waveforms in additive gaussian noise, *J. Soc. Ind. Appl. Math.* **7,** No. 2 (1959).

Swerling, P., Parameter estimation accuracy formulas, *IEEE Trans. Inform. Theory* **IT-10,** No. 4, 302–314 (1964).

Swerling, P. Detection of radar echoes in noise revisited, *IEEE Trans. Inform. Theory* **IT-12,** No. 3, 348–361 (1966). Also, Rand Corp. RM-4586-PR, AD 616700 (May 1965).

Urkowitz, H., The accuracy of maximum likelihood angle estimates in radar and sonar, *IEEE Trans. Military Electron.* **MIL-8,** No. 1, 39–45 (1965).

Urkowitz, H., Hauer, C. A., and Koval, J. F., Generalized resolution in radar systems, *Proc. IRE* (October 1962).

Van Trees, H. L., Bounds on the accuracy attainable in the estimation of continuous random processes, *IEEE Trans. Inform. Theory* **IT-12,** No. 3, 298–305 (1966).

Woodward, P. M., and Davies, I. L., A theory of radar information, *Phil. Mag.* **47,** 7th Ser. (1950).

Youla, D. C., The use of the method of maximum likelihood in estimating continuous-modulated intelligence which has been corrupted by noise, *IRE Trans. Inform. Theory* (March 1954).

Zadeh, L. A., and Ragazzini, J. R., An extension of Wiener's theory of prediction, *J. Appl. Phys.* **21,** (1950).

Zadeh, L. A., and Ragazzini, J. R., Optimum filters for the detection of signals in noise, *Proc. IRE* (October 1952).

Chapter 11

Extensions Using Matrix Formulation

11.1 Introduction

In this chapter we want to emphasize, much more than we have done in Chaps. 6 and 7, the sampled approach to detection and estimation problems. Thus, signals will be represented in terms of samples rather than as continuous functions of time. We shall use vector and matrix notation. This will result in a reasonably compact formulation of the material. (When dealing with signals represented as continuous functions of time, the shorthand notation of abstract vector spaces (*1*, *2*) also results in a compact notation. It is assumed however that the reader is more familiar with matrix operations and we proceed accordingly.)

The fact that samples are discussed should not be construed to mean only time samples. While such a sampling plan is often used in practice, there are others. Two other sampling plans already used are the Fourier coefficients as discussed in Chap. 3, and the Karhunen–Loeve coefficients as discussed in Chap. 9. The fact is that any square integrable signal can be expanded in terms of any complete set of functions. The coefficients of such functions then constitute the "samples." Using time samples implies the set of $(\sin kt)/t$ type of functions as in Shannon's sampling theorem (*3*). Using the Fourier coefficients, the samples are interpreted in the frequency domain. The Karhunen–Loeve coefficients, of course, have the advantage of being uncorrelated. Unless there is a specific need to do so, we shall not stress how the samples are generated. This will keep the results quite general.

If a continuous signal is being approximated by a finite number of samples, then a "sufficient" number of samples must be used. This will not be discussed any further except to say that enough samples are needed to approximate the probability measure of the signal, and this could be fewer samples than that required to accurately reconstruct the signal from the samples (*4–6*).

11.2 Matrix Preliminaries

It will be convenient for later use if a number of properties associated with matrix operation and manipulation are developed.§

Let $\mathbf{\Lambda}$, $\mathbf{A}$, $\mathbf{J}$ be $n \times n$, $m \times m$, and $n \times m$ matrices respectively, and assume that $\mathbf{\Lambda}$ and $\mathbf{A}$ are nonsingular so that their inverses exist. By carrying out the multiplication, the following equation will be seen to be an identity

$$\mathbf{J'\Lambda^{-1}(\Lambda \pm JAJ') = (A^{-1} \pm J'\Lambda^{-1} J)AJ'}$$

Therefore

$$\mathbf{J'\Lambda^{-1} = (A^{-1} \pm J'\Lambda^{-1}J)AJ'(\Lambda \pm JAJ')^{-1}} \tag{11-1}$$

and

$$\mathbf{(A^{-1} \pm J'\Lambda^{-1}J)^{-1}J'\Lambda^{-1} = AJ'(\Lambda \pm JAJ')^{-1}} \tag{11-2}$$

The following equation may also be verified as an identity

$$\mathbf{(A^{-1} \pm J'\Lambda^{-1}J)A \mp J'\Lambda^{-1}JA = I}$$

where $\mathbf{I}$ is the unit matrix with the diagonal elements equal to one, and all other elements equal to zero. Using Eq. (11-1) for $\mathbf{J'\Lambda^{-1}}$ in the second term we can find

$$\mathbf{(A^{-1} \pm J'\Lambda^{-1}J)[A \mp AJ'(\Lambda \pm JAJ')^{-1}JA] = I}$$

and therefore

$$\mathbf{(A^{-1} \pm J'\Lambda^{-1}J)^{-1} = A \mp AJ'(\Lambda \pm JAJ')^{-1}JA} \tag{11-3}$$

or

$$\mathbf{\pm AJ'(\Lambda \pm JAJ')^{-1}JA = A - (A^{-1} \pm J'\Lambda^{-1}J)^{-1}} \tag{11-4}$$

§ Whenever matrix multiplication is indicated, it is of course assumed that the matrices are conformable so that the product is defined. For example, if the matrix product **AB** is given, the number of columns of **A** is equal to the number of rows of **B**.

Next, assume $\mathbf{g}$ is an $m \times 1$ vector§ (analogous to $\mathbf{J}'$), and that $\mathbf{\Lambda}^{-1} = 1$, a scalar. Then Eq. (11-3) becomes

$$(\mathbf{A}^{-1} \pm \mathbf{g}\mathbf{g}')^{-1} = \mathbf{A} \mp \frac{\mathbf{A}\mathbf{g}\mathbf{g}'\mathbf{A}}{1 \pm \mathbf{g}'\mathbf{A}\mathbf{g}} \tag{11-5}$$

The development leading to Eq. (11-3) did not make use of the fact that $\mathbf{J}$ and $\mathbf{J}'$ are related through the transpose operation. Therefore, by substituting $\mathbf{g}$ and $\mathbf{f}'$ for $\mathbf{J}'$ and $\mathbf{J}$ respectively whenever they appear produces a more general identity; namely,

$$(\mathbf{A}^{-1} \pm \mathbf{g}\mathbf{\Lambda}^{-1}\mathbf{f}')^{-1} = \mathbf{A} \mp \mathbf{A}\mathbf{g}(\mathbf{\Lambda} \pm \mathbf{f}'\mathbf{A}\mathbf{g})^{-1}\mathbf{f}'\mathbf{A}$$

If $\mathbf{\Lambda}$ is unity, a scalar, and $\mathbf{g}$ and $\mathbf{f}$ are $m \times 1$ vectors, and $\mathbf{A}$ is $m \times m$ the above identity becomes

$$(\mathbf{A}^{-1} \pm \mathbf{g}\mathbf{f}')^{-1} = \mathbf{A} \mp \frac{\mathbf{A}\mathbf{g}\mathbf{f}'\mathbf{A}}{1 \pm \mathbf{f}'\mathbf{A}\mathbf{g}} \tag{11-6}$$

which is a more general form than Eq. (11-5).

Next assume $\mathbf{\Lambda}^{-1}$ is a nonsingular $n \times n$ matrix, and $\mathbf{B}$ is an $n \times n$ matrix which may be singular. Assume that $\mathbf{B} + \mathbf{\Lambda}$ is nonsingular. Then postmultiplying the identity $\mathbf{B} + \mathbf{\Lambda} - \mathbf{\Lambda} = \mathbf{B}$ by $(\mathbf{B} + \mathbf{\Lambda})^{-1}$ we get

$$\mathbf{I} - \mathbf{\Lambda}(\mathbf{B} + \mathbf{\Lambda})^{-1} = \mathbf{B}(\mathbf{B} + \mathbf{\Lambda})^{-1}$$

Premultiplying by $\mathbf{\Lambda}^{-1}$

$$\mathbf{\Lambda}^{-1} - (\mathbf{B} + \mathbf{\Lambda})^{-1} = \mathbf{\Lambda}^{-1}\mathbf{B}(\mathbf{B} + \mathbf{\Lambda})^{-1} \tag{11-7}$$

Still another matrix identity will be of value. For $\mathbf{J}'\mathbf{\Lambda}^{-1}\mathbf{J}$ nonsingular,

$$\mathbf{\Lambda}^{-1} = \mathbf{\Lambda}^{-1}\mathbf{J}(\mathbf{J}'\mathbf{\Lambda}^{-1}\mathbf{J})^{-1}\mathbf{J}'\mathbf{\Lambda}^{-1} \tag{11-8}$$

We prove this by contradiction. Assume that the relationship is not true. That is

$$\mathbf{\Lambda}^{-1} \neq \mathbf{\Lambda}^{-1}\mathbf{J}(\mathbf{J}'\mathbf{\Lambda}^{-1}\mathbf{J})^{-1}\mathbf{J}'\mathbf{\Lambda}^{-1}$$

Premultiply by $\mathbf{\Lambda}$, then

$$\mathbf{I} \neq \mathbf{J}(\mathbf{J}'\mathbf{\Lambda}^{-1}\mathbf{J})^{-1}\mathbf{J}'\mathbf{\Lambda}^{-1}$$

New postmultiply by $\mathbf{J}$ obtaining

$$\mathbf{J} \neq \mathbf{J}(\mathbf{J}'\mathbf{\Lambda}^{-1}\mathbf{J})^{-1}\mathbf{J}'\mathbf{\Lambda}^{-1}\mathbf{J}$$

and therefore $\mathbf{J} \neq \mathbf{J}$ which is an obvious contradiction. We therefore accept the truth of Eq. (11-8).

§ Normally, a boldface lower case will indicate a column vector, whereas the boldface upper case will denote a matrix of arbitrary size.

Partitioned Matrices

It is convenient at times to work with partitioned matrices. Such matrices are composed of submatrices rather than individual elements, although individual elements may also appear. For example, the following matrix is partitioned into four submatrices

$$\mathbf{A} = \left[\begin{array}{ccc|c} a_{11} & a_{12} & a_{13} & a_{14} \\ a_{21} & a_{22} & a_{23} & a_{24} \\ a_{31} & a_{32} & a_{33} & a_{34} \\ \hline a_{41} & a_{42} & a_{43} & a_{44} \end{array}\right] = \begin{bmatrix} \mathbf{A}_{11} & \mathbf{A}_{12} \\ \mathbf{A}_{21} & \mathbf{A}_{22} \end{bmatrix}$$

where

$$\mathbf{A}_{11} = \begin{bmatrix} a_{11} & a_{12} & a_{13} \\ a_{21} & a_{22} & a_{23} \\ a_{31} & a_{32} & a_{33} \end{bmatrix}, \qquad \mathbf{A}_{12} = \begin{bmatrix} a_{14} \\ a_{24} \\ a_{34} \end{bmatrix}$$

$$\mathbf{A}_{21} = [a_{41} \quad a_{42} \quad a_{43}], \qquad \mathbf{A}_{22} = a_{44}$$

The submatrices in any row have the same number of row vectors; similarly, the submatrices in any column have the same number of column vectors. Such partitions may be generated by drawing horizontal and vertical lines to delineate acceptable submatrices. The number of partitions used is arbitrary, and is usually dictated by the nature of the problem.

Matrix multiplication can be accomplished in partitioned form. Given two partitioned matrices

$$\mathbf{A} = \begin{bmatrix} \mathbf{A}_{11} & \mathbf{A}_{12} \\ \mathbf{A}_{21} & \mathbf{A}_{22} \end{bmatrix} \quad \text{and} \quad \mathbf{B} = \begin{bmatrix} \mathbf{B}_{11} & \mathbf{B}_{12} \\ \mathbf{B}_{21} & \mathbf{B}_{22} \end{bmatrix}$$

their matrix product is

$$\begin{bmatrix} \mathbf{A}_{11}\mathbf{B}_{11} + \mathbf{A}_{12}\mathbf{B}_{21} & \mathbf{A}_{11}\mathbf{B}_{12} + \mathbf{A}_{12}\mathbf{B}_{22} \\ \mathbf{A}_{21}\mathbf{B}_{11} + \mathbf{A}_{22}\mathbf{B}_{21} & \mathbf{A}_{21}\mathbf{B}_{12} + \mathbf{A}_{22}\mathbf{B}_{22} \end{bmatrix}$$

provided that the matrix multiplications are defined. Thus, $\mathbf{A}_{11}$ must have as many columns as $\mathbf{B}_{11}$ has rows, and so forth for the other matrices.

Consider a matrix partitioned into four submatrices such as

$$\mathbf{A} = \begin{bmatrix} \mathbf{A}_{11} & \mathbf{A}_{12} \\ \mathbf{A}_{21} & \mathbf{A}_{22} \end{bmatrix} \tag{11-9}$$

Assume that $\mathbf{A}_{11}$ and $\mathbf{A}_{22}$ are square and nonsingular, then the inverse of the matrix $\mathbf{A}$ is (7)

$$\mathbf{A}^{-1} = \left[\begin{array}{c|c} (\mathbf{A}_{11} - \mathbf{A}_{12}\mathbf{A}_{22}^{-1}\mathbf{A}_{21})^{-1} & -(\mathbf{A}_{11} - \mathbf{A}_{12}\mathbf{A}_{22}^{-1}\mathbf{A}_{21})^{-1}\mathbf{A}_{12}\mathbf{A}_{22}^{-1} \\ \hline -(\mathbf{A}_{22} - \mathbf{A}_{21}\mathbf{A}_{11}^{-1}\mathbf{A}_{12})^{-1}\mathbf{A}_{21}\mathbf{A}_{11}^{-1} & (\mathbf{A}_{22} - \mathbf{A}_{21}\mathbf{A}_{11}^{-1}\mathbf{A}_{12})^{-1} \end{array}\right] \tag{11-10}$$

We next express the determinant of $\mathbf{A}$ in terms of the submatrices. For this, assume $\mathbf{A}_{11}$ and $\mathbf{A}_{22}$ are square matrices, and that $\mathbf{A}_{22}$ is nonsingular. Consider the matrix product

$$\begin{bmatrix} \mathbf{A}_{11} & \mathbf{A}_{12} \\ \mathbf{A}_{21} & \mathbf{A}_{22} \end{bmatrix} \begin{bmatrix} \mathbf{I} & \mathbf{0} \\ -\mathbf{A}_{22}^{-1}\mathbf{A}_{21} & \mathbf{A}_{22}^{-1} \end{bmatrix} = \begin{bmatrix} (\mathbf{A}_{11} - \mathbf{A}_{12}\mathbf{A}_{22}^{-1}\mathbf{A}_{21}) & \mathbf{A}_{12}\mathbf{A}_{22}^{-1} \\ \mathbf{0} & \mathbf{I} \end{bmatrix}$$

The second and third matrices are "block" triangular and their determinants are easily verified to be $|\mathbf{A}_{22}^{-1}| = 1/|\mathbf{A}_{22}|$ and $|\mathbf{A}_{11} - \mathbf{A}_{12}\mathbf{A}_{22}^{-1}\mathbf{A}_{21}|$ respectively. Therefore

$$|\mathbf{A}| = |\mathbf{A}_{22}| \cdot |\mathbf{A}_{11} - \mathbf{A}_{12}\mathbf{A}_{22}^{-1}\mathbf{A}_{21}| \tag{11-11}$$

Expectations

The expectation of a matrix is a matrix with elements equal to the expected values of the original matrix elements. There are a few matrix expectations which shall be periodically encountered so they will be presented here for easy reference.

We have already made use of the fact that

$$E\{\mathbf{a}'\mathbf{x}\mathbf{y}'\mathbf{b}\} = \mathbf{a}'E\{\mathbf{x}\mathbf{y}'\}\mathbf{b} = \mathbf{a}'\mathbf{R}_{xy}\mathbf{b} \tag{11-12}$$

were $\mathbf{a}$ and $\mathbf{b}$ are given column vectors and $\mathbf{x}$ and $\mathbf{y}$ are the vector random variables. To show this result, put the matrix product in the summation form and carry out the averaging on the elements

$$E\{\mathbf{a}'\mathbf{x}\mathbf{y}'\mathbf{b}\} = E\left\{\sum_{i,j} a_i x_i y_j b_j\right\} = \sum_{i,j} a_i R_{x_i y_j} b_j = \mathbf{a}'\mathbf{R}_{xy}\mathbf{b}$$

Also of interest is the form $E\{\mathbf{x}'\mathbf{A}\mathbf{y}\}$ where $\mathbf{A}$ is a given square matrix and $\mathbf{x}$ and $\mathbf{y}$ are the random vectors. Thus

$$E\{\mathbf{x}'\mathbf{A}\mathbf{y}\} = E\left\{\sum_{i,j} x_i A_{ij} y_j\right\}$$

$$= \sum_{i,j} A_{ij} R_{y_j x_i} = \operatorname{tr} \mathbf{A}\mathbf{R}_{yx} \tag{11-13}$$

where tr denotes the trace of a square matrix and it is equal to the sum of the elements along the main diagonal. Since the trace of a matrix is equal to the trace of the transposed matrix, it also follows that

$$E\{\mathbf{x}'\mathbf{A}\mathbf{y}\} = \operatorname{tr} \mathbf{R}_{yx}'\mathbf{A}' = \operatorname{tr} \mathbf{R}_{xy}\mathbf{A}'$$

Another useful relation involving the trace of a matrix is $\operatorname{tr} \mathbf{AB} = \operatorname{tr} \mathbf{BA}$. Of course, for $\mathbf{A}$ and $\mathbf{B}$ to be comformable in each combination, they must be square matrices.

We can also derive an expression for the expectation of the product of four Gaussian vectors, in analogy to that given in Eq. (4-40). Hence, we wish to determine the scalar quantity $E\{\mathbf{w}'\mathbf{x}\mathbf{y}'\mathbf{z}\}$ where the column vectors $\mathbf{w}$, $\mathbf{x}$, $\mathbf{y}$, and $\mathbf{z}$ are zero mean jointly Gaussian vectors. Now,

$$E\{\mathbf{w}'\mathbf{x}\mathbf{y}'\mathbf{z}\} = \sum_i \sum_k E\{w_i x_i y_k z_k\}$$

Using Eq. (4-40), this is equal to

$$\sum_i E\{w_i x_i\} \sum_k E\{y_k z_k\} + \sum_{i,k} E\{w_i y_k\} E\{z_k x_i\} + \sum_{i,k} E\{w_i z_k\} E\{y_k x_i\}$$

With the notations, $E\{\mathbf{w}\mathbf{x}'\} \triangleq \mathbf{R}_{wx}$ etc., we have

$$E\{\mathbf{w}'\mathbf{x}\mathbf{y}'\mathbf{z}\} = \operatorname{tr} \mathbf{R}_{wx} \operatorname{tr} \mathbf{R}_{yz} + \operatorname{tr}(\mathbf{R}_{wy}\mathbf{R}_{zx}) + \operatorname{tr}(\mathbf{R}_{wz}\mathbf{R}_{yx}) \tag{11-14a}$$

From this relationship, a number of special cases of interest may be generated. For example, it follows that

$$E\{\mathbf{w}'\mathbf{x}\mathbf{x}'\mathbf{w}\} = \operatorname{tr}^2 \mathbf{R}_{wx} + \operatorname{tr} \mathbf{R}_{wx}^2 + \operatorname{tr} \mathbf{R}_w \mathbf{R}_x \tag{11-14b}$$

where $\mathbf{R}_w = E\{\mathbf{w}\mathbf{w}'\}$, etc. Since the product $\mathbf{x}'\mathbf{w}$ is a scalar, it is also equal to $\mathbf{w}'\mathbf{x}$ so that

$$E\{\mathbf{w}'\mathbf{x}\mathbf{w}'\mathbf{x}\} = E\{(\mathbf{w}'\mathbf{x})^2\} = \operatorname{tr}^2 \mathbf{R}_{wx} + \operatorname{tr} \mathbf{R}_{wx}^2 + \operatorname{tr} \mathbf{R}_w \mathbf{R}_x \tag{11-14c}$$

For the case $\mathbf{x} = \mathbf{A}\mathbf{w}$, where $\mathbf{A}$ is a square matrix conformable with $\mathbf{w}$,

$$E\{(\mathbf{w}'\mathbf{A}\mathbf{w})^2\} = \operatorname{tr}^2 \mathbf{A}\mathbf{R}_w + \operatorname{tr}(\mathbf{A}\mathbf{R}_w)^2 + \operatorname{tr} \mathbf{A}\mathbf{R}_w \mathbf{A}'\mathbf{R}_w \tag{11-14d}$$

If, in addition, $\mathbf{A}$ is symmetric

$$E\{(\mathbf{w}'\mathbf{A}\mathbf{w})^2\} = \operatorname{tr}^2 \mathbf{A}\mathbf{R}_w + 2\operatorname{tr}(\mathbf{A}\mathbf{R}_w)^2 \tag{11-14e}$$

We finally note that

$$\operatorname{tr} E\{\mathbf{w}\mathbf{x}'\} = \operatorname{tr} \mathbf{R}_{wx} = E\{\mathbf{w}'\mathbf{x}\} = E\{\mathbf{x}'\mathbf{w}\} \tag{11-15}$$

Isomorphic Matrices

In preparation for material in this chapter dealing with complex Gaussian variables whose covariance matrix satisfies a certain relation, we shall discuss isomorphism (*8, 9*) with regard to matrices. The advantage will become evident.

For the simplest case, a 2×2 matrix of the form

$$\mathbf{C} = \begin{bmatrix} a & -b \\ b & a \end{bmatrix} \tag{11-16}$$

is isomorphic to the complex number $a + jb$, where a and b are real. For such an isomorphism, instead of multiplying matrices having the form of **C**, the same answer can be obtained by multiplying the complex numbers which are isomorphic, and then taking the matrix which is isomorphic to that product. For example, the matrix product

$$\begin{bmatrix} a & -b \\ b & a \end{bmatrix} \begin{bmatrix} c & -d \\ d & c \end{bmatrix} = \begin{bmatrix} ac - bd & -(bc + ad) \\ bc + ad & ac - bd \end{bmatrix}$$

should be isomorphic to the complex number $(ac - bd) + j(bc + ad)$. By direct multiplication we can verify that $(a + jb)(c + jd) = (ac - bd) + j(bc + ad)$. Thus our assertion is correct.

We can extend this kind of result to higher order matrices by proper partitioning into 2×2 submatrices. For example, the matrix

$$\left[\begin{array}{cc:cc:cc} a_{11} & -b_{11} & a_{12} & -b_{12} & a_{13} & -b_{13} \\ b_{11} & a_{11} & b_{12} & a_{12} & b_{13} & a_{13} \end{array}\right]$$

is isomorphic to the matrix of complex numbers

$$[a_{11} + jb_{11} \quad a_{12} + jb_{12} \quad a_{13} + jb_{13}]$$

Thus, providing each 2×2 submatrix has the form given in Eq. (11-16), each real submatrix is replaced by its isomorphic complex number. This of course reduces the dimensionality of the matrix by a factor of 2.

We see that an isomorphism is a one-to-one correspondence between a matrix of a given form and a complex number, or matrix of complex numbers. If the matrix is denoted by **C** and it has the form for which an isomorphism may be defined, its isomorphic counterpart will be denoted by $\vec{\mathbf{C}}$, and the transformation from one to the other as $\mathbf{C} \to \vec{\mathbf{C}}$. Using this notation, we then have the relations

$$\mathbf{A} + \mathbf{B} \to \vec{\mathbf{A}} + \vec{\mathbf{B}}, \qquad \mathbf{AB} \to \vec{\mathbf{A}}\vec{\mathbf{B}}, \qquad k\mathbf{A} \to k\vec{\mathbf{A}} \tag{11-17}$$

where k is a scalar.

There are a number of other relations concerning isomorphisms that will be of later use. They are stated without proof which may be found in Goodman (*8*).

Property 1. If **R** is symmetric,§ then $\vec{\mathbf{R}}$ is Hermitian and conversely. A matrix is Hermitian if it is equal to its complex conjugate transpose, which we denote by a superscript (†). Thus if the matrix **A** is Hermitian

$$\mathbf{A} = \mathbf{A}^{\dagger} \qquad \text{or} \qquad \mathbf{A}^{*} = \mathbf{A}'$$

where * indicates complex conjugate.

§ If **R** is symmetric, and it has the form for which the isomorphism may be defined then its elements will have the form given in Eq. (11–20) which will be discussed shortly.

Property 2. If $\mathbf{R}$ is nonsingular, $\vec{\mathbf{R}}$ is also nonsingular and conversely. Also, $\mathbf{R}^{-1} \rightarrow (\vec{\mathbf{R}})^{-1}$.

Property 3. If $\mathbf{R}$ is orthogonal, $\vec{\mathbf{R}}$ is unitary and conversely. A matrix is orthogonal (Chap. 4) if its transpose is also its inverse. That is, for an orthogonal matrix

$$\mathbf{A}^{-1} = \mathbf{A}' \qquad \text{and} \qquad \mathbf{AA}' = \mathbf{I}$$

If a matrix is complex and the complex conjugate transpose is equal to its inverse, it is unitary. That is

$$\mathbf{A}^{-1} = \mathbf{A}^{\dagger} \qquad \text{and} \qquad \mathbf{AA}^{\dagger} = \mathbf{I}$$

Property 4. The determinant $|\mathbf{R}|$ is equal to the square of the absolute value of $|\vec{\mathbf{R}}|$. That is, $|\mathbf{R}| = [abs\, |\vec{\mathbf{R}}|]^2$ where abs indicates absolute value, and $|\cdot|$ indicates the determinant.

Property 5. If $\mathbf{R}$ is symmetric, then $|\mathbf{R}| = |\vec{\mathbf{R}}|^2$. This follows from Property 4 since the determinant of a Hermitian matrix is real.

Property 6. Consider the vectors

$$\mathbf{w} = \begin{bmatrix} x_1 \\ y_1 \\ x_2 \\ y_2 \\ \vdots \\ x_n \\ y_n \end{bmatrix}, \qquad \mathbf{z} = \begin{bmatrix} z_1 \\ \vdots \\ z_n \end{bmatrix} = \begin{bmatrix} x_1 + jy_1 \\ \vdots \\ x_n + jy_n \end{bmatrix} \tag{11-18}$$

Then, if $\mathbf{R}$ is symmetric,

$$\mathbf{w}'\mathbf{R}\mathbf{w} = \mathbf{z}^{\dagger}\vec{\mathbf{R}}\mathbf{z} \tag{11-19}$$

Note that if $\mathbf{R}$ is symmetric, then the 2×2 submatrices along the diagonal of $\mathbf{R}$ must be of the form $k\left[\begin{smallmatrix} 1 & 0 \\ 0 & 1 \end{smallmatrix}\right]$. See Eq. (11-20).

Property 7. If $\mathbf{R}$ is positive definite, $\vec{\mathbf{R}}$ is Hermitian positive definite. A Hermitian matrix $\vec{\mathbf{R}}$ is Hermitian positive definite if $\mathbf{b}^{\dagger}\vec{\mathbf{R}}\mathbf{b} > 0$ for all nonzero vectors $\mathbf{b}$.

11.3 Multivariate Complex Gaussian Distribution

In Sect. 3.7, the Fourier series representation of a finite time segment of narrowband processes was discussed. It was found, Eq. (3-76), that the covariance of Fourier components of different processes has the form given in Eq. (11-16) for which the isomorphism discussed above can be applied. It is for such Gaussian random variables that we can discuss the idea of a complex Gaussian distribution (*8*). Using the notation of Eq. (11-18), assume the $2n$ components of $\mathbf{w}$ have a $2n$-variate Gaussian distribution with zero mean. In analogy to Eq. (3-76), assume that the covariance matrix $\mathbf{R} = E\{\mathbf{w}\mathbf{w}'\}$ has 2×2 matrix partitions of the form

$$E\begin{Bmatrix} x_i x_k & x_i y_k \\ y_i x_k & y_i y_k \end{Bmatrix} = \begin{cases} \frac{1}{2}\begin{bmatrix} \alpha_{ii} & 0 \\ 0 & \alpha_{ii} \end{bmatrix}, & i = k \\ \frac{1}{2}\begin{bmatrix} \alpha_{ik} & -\beta_{ik} \\ \beta_{ik} & \alpha_{ik} \end{bmatrix}, & i \neq k \end{cases} \tag{11-20}$$

In analogy to Eq. (3-77), the covariance matrix of the complex samples $\mathbf{C} = E\{\mathbf{z}\mathbf{z}^\dagger\}$ has elements

$$E\{z_i z_k{}^*\} = \begin{cases} \alpha_{ii}, & i = k \\ \alpha_{ik} + j\beta_{ik}, & i \neq k \end{cases} \tag{11-21}$$

It is then easily seen that $2\mathbf{R} \rightarrow \mathbf{C}$. Since $\mathbf{R}$ is symmetric, $\mathbf{C}$ is Hermitian. From Property 2 it follows that $\mathbf{C}^{-1} \rightarrow \frac{1}{2}\mathbf{R}^{-1}$, and from Property 5, $|2\mathbf{R}| = |\mathbf{C}|^2$. From the latter we also have $2^{2n}|\mathbf{R}| = |\mathbf{C}|^2$ and therefore $2^n|\mathbf{R}|^{1/2} = |\mathbf{C}|$. We are now prepared to apply these relations to the Gaussian distribution

$$p(\mathbf{w}) = \frac{1}{(2\pi)^n\,|\mathbf{R}|^{1/2}} \exp(-\tfrac{1}{2}\mathbf{w}'\mathbf{R}^{-1}\mathbf{w})$$

Since $\frac{1}{2}\mathbf{w}'\mathbf{R}^{-1}\mathbf{w} = \mathbf{z}^\dagger\mathbf{C}^{-1}\mathbf{z}$ (Property 6), and $|\mathbf{R}|^{1/2} = 2^{-n}|\mathbf{C}|$, we have

$$p(\mathbf{w}) \triangleq p(\mathbf{z}) = \frac{1}{\pi^n\,|\mathbf{C}|} \exp(-\mathbf{z}^\dagger\mathbf{C}^{-1}\mathbf{z}) \tag{11-22}$$

which Goodman calls the multivariate complex Gaussian distribution.

11.4 Estimation

Unlike the organization of the earlier chapters of this book, we shall discuss signal estimation before signal detection. Whereas in the preceding chapter, estimation of signal parameters was the primary concern, in this chapter the emphasis is on the estimation of the signal waveform. Nominally, the estimation cases will be taken in order of increasing a priori information. Starting with only knowledge of covariance functions, we then progress to maximum-likelihood estimation of an unknown but nonrandom signal which of course presumes a probability density function for the noise, and finally to Bayes estimation where the signal is a vector random variable with a known a priori density function.

11.5 Best Linear Estimator

Consider the problem posed in Fig. 11-1. The received vector **r** is to be processed by a linear filter **W** in such a way that its output **y** is a minimum mean square error estimate of a desired signal **d**. That is, we wish to minimize

Fig. 11-1 Best linear filter to estimate **d**.

$\mathscr{E} = E\{(\mathbf{y} - \mathbf{d})'(\mathbf{y} - \mathbf{d})\}$ where the error **e** is $\mathbf{y} - \mathbf{d}$. Here **r** is $N \times 1$, **W** is $M \times N$, and **y** and **d** are $M \times 1$. Since **y** is **Wr**, minimize

$$\mathscr{E} = E\{(\mathbf{Wr} - \mathbf{d})'(\mathbf{Wr} - \mathbf{d})\} = E\{\mathbf{r}'\mathbf{W}'\mathbf{Wr} - 2\mathbf{r}'\mathbf{W}'\mathbf{d} + \mathbf{d}'\mathbf{d}\} \qquad (11\text{-}23)$$

We should emphasize that the filter is constrained to be linear. In general, a lower value of mean squared error might be obtained with a nonlinear filter. It is shown in Sect. 11.7 for the Gaussian case that the best filter is indeed linear.

Determination of the filter is a problem in calculus of variation similar to that encountered in Sect. 6.4. Substitute $\mathbf{W} = \mathbf{W}_o + \varepsilon\mathbf{V}$ where $\mathbf{W}_o$ is the optimum filter to be determined. The equation to be satisfied is

$$\left.\frac{d\mathscr{E}}{d\varepsilon}\right|_{\varepsilon=0} = 0$$

Substituting for **W** in Eq. (11-23) and carrying out the differentiation produces

$$0 = E\{\mathbf{d}'\mathbf{Vr} - \mathbf{r}'\mathbf{W}_o{}'\mathbf{Vr}\} \qquad (11\text{-}24)$$

where we have made use of the facts that $\mathbf{r}'\mathbf{V}'\mathbf{d} = \mathbf{d}'\mathbf{Vr}$ and $\mathbf{r}'\mathbf{W}_o{}'\mathbf{Vr} = \mathbf{r}'\mathbf{V}'\mathbf{W}_o\mathbf{r}$. Equation (11-24) can also be written in the form

$$0 = E\{(\mathbf{d} - \mathbf{W}_o\mathbf{r})'\mathbf{Vr}\} \qquad (11\text{-}25)$$

Since this holds for arbitrary **V**, it will also hold for a particular value of **V**, say $\mathbf{W}_o$. Then

$$0 = E\{(\mathbf{d} - \mathbf{W}_o\mathbf{r})'\mathbf{W}_o\mathbf{r}\} \qquad (11\text{-}26)$$

The first part $(\mathbf{d} - \mathbf{W}_o\mathbf{r})$ is the error term, and $\mathbf{W}_o\mathbf{r}$ is the filter output. Thus the error and the optimum filter output are orthogonal. Papoulis (*10*) refers to this as the orthogonality principle, and it is a necessary condition for the optimum filter. To continue with the derivation, using Eq. (11-15), Eq. (11-25) may be expressed as

$$0 = \text{tr}\, \mathbf{V}E\{\mathbf{r}(\mathbf{d} - \mathbf{W}_o\mathbf{r})'\}$$

or

$$0 = \text{tr}\, \mathbf{V}(\mathbf{R}_{rd} - \mathbf{R}_r\mathbf{W}_o')$$

Since $\mathbf{V}$ is arbitrary, we get for the optimum filter (see Fig. 11.1)§

$$\mathbf{W}_o = \mathbf{R}_{dr}\mathbf{R}_r^{-1} \tag{11-27}$$

and therefore

$$\hat{\mathbf{d}} \triangleq \mathbf{y} = \mathbf{R}_{dr}\mathbf{R}_r^{-1}\mathbf{r} \tag{11-28}$$

The correlation matrix of $\hat{\mathbf{d}}$ is

$$E\{\hat{\mathbf{d}}\hat{\mathbf{d}}'\} = \mathbf{R}_{dr}\mathbf{R}_r^{-1}E\{\mathbf{r}\mathbf{r}'\}\mathbf{R}_r^{-1}\mathbf{R}_{dr}' = \mathbf{R}_{dr}\mathbf{R}_r^{-1}\mathbf{R}_{dr}' \tag{11-29}$$

The mean square error matrix of the error vector $\mathbf{e}$ is

$$\begin{aligned} E\{(\hat{\mathbf{d}} - \mathbf{d})(\hat{\mathbf{d}} - \mathbf{d})'\} &= E\{\mathbf{e}\mathbf{e}'\} = E\{(\mathbf{W}_o\mathbf{r} - \mathbf{d})(\mathbf{W}_o\mathbf{r} - \mathbf{d})'\} \\ &= E\{\mathbf{W}_o\mathbf{r}\mathbf{r}'\mathbf{W}_o' - \mathbf{d}\mathbf{r}'\mathbf{W}_o' - \mathbf{W}_o\mathbf{r}\mathbf{d}' + \mathbf{d}\mathbf{d}'\} \\ &= \mathbf{W}_o\mathbf{R}_r\mathbf{W}_o' - \mathbf{R}_{dr}\mathbf{W}_o' - \mathbf{W}_o\mathbf{R}_{rd} + \mathbf{R}_d \end{aligned}$$

Substituting for the optimum filter, Eq. (11-27), we find

$$E\{(\hat{\mathbf{d}} - \mathbf{d})(\hat{\mathbf{d}} - \mathbf{d})'\} = \mathbf{R}_d - \mathbf{R}_{dr}\mathbf{R}_r^{-1}\mathbf{R}_{rd} \tag{11-30}$$

The total mean square error is

$$\begin{aligned} \mathscr{E} &= E\{(\mathbf{W}_o\mathbf{r} - \mathbf{d})'(\mathbf{W}_o\mathbf{r} - \mathbf{d})\} = \text{tr}\, E\{(\mathbf{W}_o\mathbf{r} - \mathbf{d})(\mathbf{W}_o\mathbf{r} - \mathbf{d})'\} \\ &= \text{tr}\, E\{(\hat{\mathbf{d}} - \mathbf{d})(\hat{\mathbf{d}} - \mathbf{d})'\} \end{aligned}$$

so that

$$\mathscr{E} = \text{tr}(\mathbf{R}_d - \mathbf{R}_{dr}\mathbf{R}_r^{-1}\mathbf{R}_{rd}) \tag{11-31}$$

As a point of interest, the mean square error may be expressed in the form

$$\mathscr{E} = E\{(\mathbf{W}_o\mathbf{r} - \mathbf{d})'\mathbf{W}_o\mathbf{r}\} - E\{(\mathbf{W}_o\mathbf{r} - \mathbf{d})'\mathbf{d}\}$$

§ The linear filter which minimizes the mean square error is often called the Wiener filter.

The first term, due to the orthogonality of the error vector and $\mathbf{W}_o\mathbf{r}$, Eq. (11-26), is zero. Hence

$$\mathscr{E} = -E\{(\mathbf{W}_o\mathbf{r} - \mathbf{d})'\mathbf{d}\} \tag{11-32}$$

which is the correlation of the error vector and the desired signal.

Special Case

Suppose the data $\mathbf{r}$ contains signal and noise in the form (*11*)

$$\mathbf{r} = \mathbf{Hs} + \mathbf{n}$$

and we desire the best linear filter to estimate the signal vector $\mathbf{s}$. Thus $\mathbf{s}$ is the desired signal $\mathbf{d}$. The signal and noise have covariance matrices $\mathbf{R}_s$ and $\mathbf{R}_n$ respectively, and they are assumed to be uncorrelated with zero mean. The covariance of $\mathbf{d}$ and $\mathbf{r}$ is

$$\mathbf{R}_{dr} = E\{\mathbf{s}(\mathbf{Hs} + \mathbf{n})'\} = E\{\mathbf{ss}'\mathbf{H}' + \mathbf{sn}'\} = \mathbf{R}_s\mathbf{H}'$$

Similarly

$$\mathbf{R}_r = E\{(\mathbf{Hs} + \mathbf{n})(\mathbf{Hs} + \mathbf{n})'\} = E\{\mathbf{Hss}'\mathbf{H}' + \mathbf{nn}'\} = \mathbf{HR}_s\mathbf{H}' + \mathbf{R}_n$$

Thus the estimate $\hat{\mathbf{s}} = \mathbf{R}_{dr}\mathbf{R}_r^{-1}\mathbf{r}$ is

$$\hat{\mathbf{s}} = \mathbf{R}_s\mathbf{H}'(\mathbf{HR}_s\mathbf{H}' + \mathbf{R}_n)^{-1}\mathbf{r} \tag{11-33}$$

For any given value of $\mathbf{s}$,

$$E\{\hat{\mathbf{s}}|\mathbf{s}\} = \mathbf{R}_s\mathbf{H}'(\mathbf{HR}_s\mathbf{H}' + \mathbf{R}_n)^{-1}\mathbf{Hs} \tag{11-34}$$

so that the estimate is biased. For the correlation matrix of the error, from Eq. (11-30), we have

$$E\{(\hat{\mathbf{s}} - \mathbf{s})(\hat{\mathbf{s}} - \mathbf{s})'\} = \mathbf{R}_s - \mathbf{R}_s\mathbf{H}'(\mathbf{HR}_s\mathbf{H}' + \mathbf{R}_n)^{-1}\mathbf{HR}_s \tag{11-35}$$

and the total mean square error $\mathscr{E}$ is the trace of this result.

11.6 Maximum Likelihood Estimation

We now add a bit of information to our a priori knowledge about the noise, and assume that the received data vector is Gaussian distributed. Consider the case where the received signal is of the form (*11*)

$$\mathbf{r} = \mathbf{Hs} + \mathbf{n}$$

The additive noise $\mathbf{n}$ is a zero mean multivariate Gaussian random vector with covariance $\mathbf{R}_n$. We wish to estimate $\mathbf{s}$, which for the case at hand is not assumed to be random. (The dimensionality of $\mathbf{H}$, $\mathbf{s}$, and $\mathbf{n}$ are $N \times M$,

$M \times 1$, and $N \times 1$ respectively.) The appropriate estimator to use is the maximum likelihood estimator which is that value of $\mathbf{s}$ which maximizes

$$p(\mathbf{r} \mid \mathbf{s}) = \frac{1}{(2\pi)^{N/2} |\mathbf{R}_n|^{1/2}} \exp[-\tfrac{1}{2}(\mathbf{r} - \mathbf{Hs})'\mathbf{R}_n^{-1}(\mathbf{r} - \mathbf{Hs})]$$

This is equivalent to finding the value of $\mathbf{s}$ which minimizes $-2\mathbf{r}'\mathbf{R}_n^{-1}\mathbf{Hs} + \mathbf{s}'\mathbf{H}'\mathbf{R}_n^{-1}\mathbf{Hs}$. Thus, solve for $\mathbf{s}$ such that§

$$\frac{\partial}{\partial \mathbf{s}}(-2\mathbf{r}'\mathbf{R}_n^{-1}\mathbf{Hs} + \mathbf{s}'\mathbf{H}'\mathbf{R}_n^{-1}\mathbf{Hs}) = 0 \tag{11-36}$$

Now, the matrix $\mathbf{r}'\mathbf{R}_n^{-1}\mathbf{H}$ is a row vector. Denote its elements by v_i. Then,

$$\frac{\partial}{\partial s_i} \mathbf{r}'\mathbf{R}_n^{-1}\mathbf{Hs} = v_i$$

Therefore, forming the column vector

$$\frac{\partial}{\partial \mathbf{s}} \triangleq \begin{bmatrix} \partial/\partial s_1 \\ \vdots \\ \partial/\partial s_M \end{bmatrix}$$

we can easily see that

$$\frac{\partial}{\partial \mathbf{s}} \mathbf{r}'\mathbf{R}_n^{-1}\mathbf{Hs} = (\mathbf{r}'\mathbf{R}_n^{-1}\mathbf{H})' = \mathbf{H}'\mathbf{R}_n^{-1}\mathbf{r} \tag{11-37}$$

We also wish to determine the derivative with respect to the s_i of the second term in Eq. (11-36) which is a quadratic form. Define the square matrix $\mathbf{P}$ with elements p_{ij}, and consider the quadratic form $\mathbf{s}'\mathbf{Ps} = \sum_{i,j} s_i p_{ij} s_j$. Then

$$\frac{\partial}{\partial s_1} \mathbf{s}'\mathbf{Ps} = \sum_i s_i p_{i1} + \sum_j p_{1j} s_j$$

$$\vdots$$

$$\frac{\partial}{\partial s_M} \mathbf{s}'\mathbf{Ps} = \sum_i s_i p_{iM} + \sum_j p_{Mj} s_j$$

and it may be reasoned that

$$\frac{\partial}{\partial \mathbf{s}} \mathbf{s}'\mathbf{Ps} = \mathbf{P}'\mathbf{s} + \mathbf{Ps} \tag{11-38}$$

§ The solution may be found without recourse to differentiation. Since $\mathbf{R}_n^{-1}$ is positive definite, $p(\mathbf{r} \mid \mathbf{s})$ is maximized when $\mathbf{r} = \mathbf{H}\hat{\mathbf{s}}$. After some manipulation, and using Eq. (11–8), the desired result Eq. (11–40) may be found. See Exercise 11.9.

For the special case when $\mathbf{P}$ is symmetric, then

$$\frac{\partial}{\partial \mathbf{s}} \mathbf{s}'\mathbf{P}\mathbf{s} = 2\mathbf{P}\mathbf{s} \tag{11-39}$$

Applying these to Eq. (11-36) to determine the minimum we obtain

$$0 = \frac{\partial}{\partial \mathbf{s}}(-2\mathbf{r}'\mathbf{R}_n^{-1}\mathbf{H}\mathbf{s} + \mathbf{s}'\mathbf{H}'\mathbf{R}_n^{-1}\mathbf{H}\mathbf{s}) = -2\mathbf{H}'\mathbf{R}_n^{-1}\mathbf{r} + 2\mathbf{H}'\mathbf{R}_n^{-1}\mathbf{H}\mathbf{s}$$

Providing that the matrix $\mathbf{H}'\mathbf{R}_n^{-1}\mathbf{H}$ is nonsingular, the solution of this is

$$\hat{\mathbf{s}} = (\mathbf{H}'\mathbf{R}_n^{-1}\mathbf{H})^{-1}\mathbf{H}'\mathbf{R}_n^{-1}\mathbf{r} \tag{11-40}$$

and it is the maximum-likelihood estimator. The estimate is a linear function of the data and can be implemented as shown in Fig. 11.1 with $\mathbf{W} = (\mathbf{H}'\mathbf{R}_n^{-1}\mathbf{H})^{-1}\mathbf{H}'\mathbf{R}_n^{-1}$. Note that for a given $\mathbf{s}$, the average value of the estimator is

$$E\{\hat{\mathbf{s}}\} = (\mathbf{H}'\mathbf{R}_n^{-1}\mathbf{H})^{-1}\mathbf{H}'\mathbf{R}_n^{-1}\mathbf{H}\mathbf{s} = \mathbf{s} \tag{11-41}$$

so that the estimator is unbiased. To determine the covariance matrix of $\mathbf{s}$,

$$\hat{\mathbf{s}} - E\{\hat{\mathbf{s}}\} = (\mathbf{H}'\mathbf{R}_n^{-1}\mathbf{H})^{-1}\mathbf{H}'\mathbf{R}_n^{-1}\mathbf{n}$$

Then

$$E\{(\hat{\mathbf{s}} - \mathbf{s})(\hat{\mathbf{s}} - \mathbf{s})'\} = (\mathbf{H}'\mathbf{R}_n^{-1}\mathbf{H})^{-1}\mathbf{H}'\mathbf{R}_n^{-1}\mathbf{H}(\mathbf{H}'\mathbf{R}_n^{-1}\mathbf{H})^{-1} = (\mathbf{H}'\mathbf{R}_n^{-1}\mathbf{H})^{-1} \tag{11-42}$$

As a final note, the maximum-likelihood estimator in the Gaussian case minimizes the quadratic form

$$(\mathbf{r} - \mathbf{H}\mathbf{s})'\mathbf{R}_n^{-1}(\mathbf{r} - \mathbf{H}\mathbf{s})$$

In the event that the additive noise cannot be assumed to be Gaussian, and $\mathbf{s}$ is an unknown but nonrandom function, minimizing the above quadratic form may be a reasonable criterion for the determination of an estimate. Of course, the solution will be given by Eq. (11-40).

11.7 Maximum A Posteriori Estimation

We now include in our a priori knowledge the probability density of the unknown signal, say $\mathbf{s}$. We therefore use the maximum a posteriori estimator (MAP) discussed in Chap. 10. The a posteriori probability density for $\mathbf{s}$ given the received data, say $\mathbf{r}$, is equal to

$$p(\mathbf{s} \mid \mathbf{r}) = \frac{p(\mathbf{s}, \mathbf{r})}{p(\mathbf{r})} = \frac{p(\mathbf{r} \mid \mathbf{s})p(\mathbf{s})}{p(\mathbf{r})} \tag{11-43}$$

and the MAP estimate is that value of $\mathbf{s}$ which produces a maximum.

We determine that a posteriori density function under the assumption that **s** and **r** are individually and jointly zero mean multivariate Gaussian variables. Denote the vector $\mathbf{z} = \begin{bmatrix}\mathbf{s}\\ \mathbf{r}\end{bmatrix}$. Then $p(\mathbf{s}, \mathbf{r}) = p(\mathbf{z})$ is multivariate Gaussian. The dimensions of **s** and **r** are $M \times 1$ and $N \times 1$ respectively. The partitioned covariance matrix of **z** is

$$\mathbf{R}_z = E\left\{\begin{bmatrix}\mathbf{s}\\ \mathbf{r}\end{bmatrix}[\mathbf{s}'\ \mathbf{r}']\right\} = \begin{bmatrix}\mathbf{R}_s & \mathbf{R}_{sr}\\ \mathbf{R}'_{sr} & \mathbf{R}_r\end{bmatrix} \tag{11-44}$$

Using Eq. (11-10), the inverse is

$$\mathbf{R}_z^{-1} = \left[\begin{array}{c|c}(\mathbf{R}_s - \mathbf{R}_{sr}\mathbf{R}_r^{-1}\mathbf{R}'_{sr})^{-1} & -(\mathbf{R}_s - \mathbf{R}_{sr}\mathbf{R}_r^{-1}\mathbf{R}'_{sr})^{-1}\mathbf{R}_{sr}\mathbf{R}_r^{-1}\\ \hline -(\mathbf{R}_r - \mathbf{R}'_{sr}\mathbf{R}_s^{-1}\mathbf{R}_{sr})^{-1}\mathbf{R}'_{sr}\mathbf{R}_s^{-1} & (\mathbf{R}_r - \mathbf{R}'_{sr}\mathbf{R}_s^{-1}\mathbf{R}_{sr})^{-1}\end{array}\right] \tag{11-45}$$

and from Eq. (11-11) the determinant is

$$|\mathbf{R}_z| = |\mathbf{R}_r| \cdot |\mathbf{R}_s - \mathbf{R}_{sr}\mathbf{R}_r^{-1}\mathbf{R}'_{sr}| \tag{11-46}$$

Therefore

$$p(\mathbf{s}\,|\,\mathbf{r}) = \frac{(2\pi)^{N/2}|\mathbf{R}_r|^{1/2}\exp(-\frac{1}{2}\mathbf{z}'\mathbf{R}_z^{-1}\mathbf{z})}{(2\pi)^{(N+M)/2}|\mathbf{R}_z|^{1/2}\exp(-\frac{1}{2}\mathbf{r}'\mathbf{R}_r^{-1}\mathbf{r})}$$

$$= \frac{|\mathbf{R}_r|^{1/2}}{(2\pi)^{M/2}|\mathbf{R}_z|^{1/2}}\exp\left[-\tfrac{1}{2}(\mathbf{z}'\mathbf{R}_z^{-1}\mathbf{z} - \mathbf{r}'\mathbf{R}_r^{-1}\mathbf{r})\right]$$

We have explicit expressions of all terms in this equation as functions of **s**, **r**, $\mathbf{R}_s$, $\mathbf{R}_{sr}$, and $\mathbf{R}_r$. If these are used along with some of the matrix identities of Sect. 11.2, it may be verified after some labor that the a posteriori density function may be expressed as

$$p(\mathbf{s}\,|\,\mathbf{r}) = \frac{\exp[-\frac{1}{2}(\mathbf{s} - \mathbf{R}_{sr}\mathbf{R}_r^{-1}\mathbf{r})'(\mathbf{R}_s - \mathbf{R}_{sr}\mathbf{R}_r^{-1}\mathbf{R}'_{sr})^{-1}(\mathbf{s} - \mathbf{R}_{sr}\mathbf{R}_r^{-1}\mathbf{r})]}{(2\pi)^{M/2}|\mathbf{R}_s - \mathbf{R}_{sr}\mathbf{R}_r^{-1}\mathbf{R}'_{sr}|^{1/2}} \tag{11-47}$$

From this equation it follows directly that

$$E\{\mathbf{s}|\mathbf{r}\} = \mathbf{R}_{sr}\mathbf{R}_r^{-1}\mathbf{r} \tag{11-48}$$

From Eq. (10-5), we know that this is also the expression for the Bayes estimator which minimize the mean square error.

The covariance of **s** given **r**, denoted $\text{Cov}\{\mathbf{s}|\mathbf{r}\}$, is

$$\text{Cov}\{\mathbf{s}|\mathbf{r}\} = \mathbf{R}_s - \mathbf{R}_{sr}\mathbf{R}_r^{-1}\mathbf{R}'_{sr} \tag{11-49}$$

Since the expression $p(\mathbf{s}|\mathbf{r})$ is Gaussian, its mean value and maximum value coincide so that the MAP estimator is

$$\hat{\mathbf{s}} = \mathbf{R}_{sr}\mathbf{R}_r^{-1}\mathbf{r} \tag{11-50}$$

and is for this case also the minimum mean square error estimator. Since $\hat{\mathbf{s}}$ is a linear function of the data $\mathbf{r}$, we conclude in the Gaussian case, that the minimum mean square estimator is a linear function of the data. The estimate can be obtained as shown in Fig. 11.1 with $\mathbf{W} = \mathbf{R}_{sr}\mathbf{R}_r^{-1}$.

For the special case of $\mathbf{r} = \mathbf{Hs} + \mathbf{n}$, the estimator is

$$\hat{\mathbf{s}} = \mathbf{R}_s\mathbf{H}'(\mathbf{H}\mathbf{R}_s\mathbf{H}' + \mathbf{R}_n)^{-1}\mathbf{r}$$

11.8 Detection

The vector-matrix formulation of detection of completely known signals was discussed in Chap. 6. Detection of signals of known form but with random parameters was treated in Chap. 7. In the following material we discuss the detection of nonrandom signals of unknown form, and the detection of Gaussian signals in Gaussian noise.

Consider the hypotheses

$$H_1: \quad \mathbf{r} = \mathbf{Hs} + \mathbf{n}$$
$$H_0: \quad \mathbf{r} = \mathbf{n}$$

where $\mathbf{n}$ is zero mean Gaussian, and $\mathbf{s}$ is considered to be unknown but nonrandom. Because of these assumptions it is reasonable, although not necessarily optimum, to apply the maximum likelihood principle, Sect. 10.9. Thus we set out to determine the ratio

$$\ln \lambda = \ln \frac{\max_{\mathbf{s}} p_1(\mathbf{r}|\mathbf{s})}{p_0(\mathbf{r})}$$

From Eq. (6-88), this can be directly written as

$$\ln \lambda = \mathbf{r}'\mathbf{R}_n^{-1}\mathbf{H}\hat{\mathbf{s}} - \tfrac{1}{2}\hat{\mathbf{s}}'\mathbf{H}'\mathbf{R}_n^{-1}\mathbf{H}\hat{\mathbf{s}} \tag{11-51}$$

If the signal were known completely, the first part of this equation, $\mathbf{r}'\mathbf{R}_n^{-1}\mathbf{Hs}$, would be used as the test statistic. It follows that in the case at hand this form with $\mathbf{s}$ replaced by $\hat{\mathbf{s}}$ is to be used as the test statistic.

The estimation part of this problem is, of course, that estimation problem discussed in Sect. 11.6 where it was determined that the maximum-likelihood estimate of the signal is, from Eq. (11-40) and providing $\mathbf{H}'\mathbf{R}_n^{-1}\mathbf{H}$ is nonsingular,

$$\hat{\mathbf{s}} = (\mathbf{H}'\mathbf{R}_n^{-1}\mathbf{H})^{-1}\mathbf{H}'\mathbf{R}_n^{-1}\mathbf{r}$$

Substituting this into $\mathbf{r}'\mathbf{R}_n^{-1}\mathbf{H}\hat{\mathbf{s}}$ produces

$$\ln \lambda = \mathbf{r}'\mathbf{R}_n^{-1}\mathbf{H}(\mathbf{H}'\mathbf{R}_n^{-1}\mathbf{H})^{-1}\mathbf{H}'\mathbf{R}_n^{-1}\mathbf{r}$$

Using the matrix identity (11-8), this equation may be expressed as $\mathbf{r}'\mathbf{R}_n^{-1}\mathbf{r}$ providing that $\mathbf{H}'\mathbf{R}_n^{-1}\mathbf{H}$ is nonsingular. We may therefore use as the test statistic

$$\lambda_T = \mathbf{r}'\mathbf{R}_n^{-1}\mathbf{H}\hat{\mathbf{s}} \qquad \text{or} \qquad \mathbf{r}'\mathbf{R}_n^{-1}\mathbf{r} \tag{11-52}$$

If the matrix $\mathbf{R}_n^{-1}$ is factored into the product of triangular matrices $\mathbf{C}'\mathbf{C}$ (see Chap. 6) then λ_T may be expressed

$$\lambda_T = (\mathbf{Cr})'\mathbf{CH}\hat{\mathbf{s}} = (\mathbf{Cr})'(\mathbf{Cr})$$

The operator $\mathbf{C}$ "whitens" the data, and the receiver correlates the result with the estimated signal $\mathbf{H}\hat{\mathbf{s}}$ after it too is operated on by $\mathbf{C}$.

We can view the receiver in still another way. The fact that the test statistic is a quadratic form in the data suggests a weighted energy or power detector. For the uncorrelated noise case, the test statistic would be proportional to $\mathbf{r}'\mathbf{r} = \sum r_i^2$ which is a power or energy detector with all samples equally weighted.

The fact that the test statistic is also equivalent to $\mathbf{r}'\mathbf{R}_n^{-1}\mathbf{r}$ should not be too surprising. Indeed, consider the case where the received signal is represented as $\mathbf{r} = \mathbf{y} + \mathbf{n}$ and $\mathbf{y}$ is known. It may be verified that the test statistic is proportional to $\mathbf{r}'\mathbf{R}_n^{-1}\mathbf{y}$. Now, if $\mathbf{y}$ is nonrandom but unknown the maximum likelihood estimate would be $\hat{\mathbf{y}} = \mathbf{r}$. Hence the estimate of the signal $\mathbf{y}$ is the data itself. Now, applying the maximum-likelihood principle to the detection problem would result in the test statistic $\mathbf{r}'\mathbf{R}_n^{-1}\hat{\mathbf{y}} = \mathbf{r}'\mathbf{R}_n^{-1}\mathbf{r}$.

We now determine the moments of the test statistic. These may be used to determine the deflection signal-to-noise ratio (*12*). Denote the expectations when H_0 and H_1 are true by $E_0\{\cdot\}$ and $E_1\{\cdot\}$ respectively. Then

$$E_0\{\lambda_T\} = E\{\mathbf{n}'\mathbf{R}_n^{-1}\mathbf{n}\} = \operatorname{tr} \mathbf{R}_n^{-1}\mathbf{R}_n = N$$

where N is the dimension of the column vector $\mathbf{n}$. For H_1,

$$\begin{aligned} E_1\{\lambda_T\} &= E\{(\mathbf{Hs} + \mathbf{n})'\mathbf{R}_n^{-1}(\mathbf{Hs} + \mathbf{n})\} = (\mathbf{Hs})'\mathbf{R}_n^{-1}\mathbf{Hs} + E\{\mathbf{n}'\mathbf{R}_n^{-1}\mathbf{n}\} \\ &= (\mathbf{Hs})'\mathbf{R}_n^{-1}\mathbf{Hs} + N \end{aligned}$$

It then follows that the difference in the means is

$$\Delta\mu = (\mathbf{Hs})'\mathbf{R}_n^{-1}\mathbf{Hs}$$

Now, $\lambda_T^2 = \mathbf{r}'\mathbf{R}_n^{-1}\mathbf{r}\mathbf{r}'\mathbf{R}_n^{-1}\mathbf{r}$. Using Eq. (11-14) we find

$$E_0\{\lambda_T^2\} = N^2 + 2N, \qquad \sigma_0^2(\lambda_T) = 2N$$
$$\sigma_1^2(\lambda_T) = 2\{[(\mathbf{Hs})'\mathbf{R}_n^{-1}\mathbf{Hs}]^2 + 2(\mathbf{Hs})'\mathbf{R}_n^{-1}\mathbf{Hs} + N\}$$

For the no signal case, it is worthwhile noting that the test statistic is $(\mathbf{Cr})'\mathbf{Cr} = \sum_{i=1}^N d_i^2$ where d_i is the ith element of the column vector $\mathbf{Cr}$. It has already been pointed out in Sect. 6.6 that $\mathbf{C}$ is a linear transformation on the data and such that the transformed variables are uncorrelated with unit variance. Since $\mathbf{r}$ is Gaussian, the d_i are Gaussian. Therefore, for H_0, the test statistic is central chi square with N degrees of freedom. From our computations, $2E_0{}^2(\lambda_T)/\sigma_0{}^2(\lambda_T) = N$ which is in agreement with the degrees of freedom.

11.9 Gaussian Signal in Gaussian Noise

We now include in our a priori knowledge the fact that the signal is governed by a multivariate Gaussian density function of zero mean and covariance matrix $\mathbf{R}_s$. As before, consider hypotheses of the form

$$H_1: \quad \mathbf{r} = \mathbf{H}\mathbf{s} + \mathbf{n}$$
$$H_0: \quad \mathbf{r} = \mathbf{n}$$

where $\mathbf{H}$ is $N \times M$, $\mathbf{s}$ is $M \times 1$, and $\mathbf{n}$ is $N \times 1$. The signal and noise are independent of one another. The likelihood functions are

$$p_1(\mathbf{r}) = \frac{\exp[-\frac{1}{2}\mathbf{r}'(\mathbf{H}\mathbf{R}_s\mathbf{H}' + \mathbf{R}_n)^{-1}\mathbf{r}]}{(2\pi)^{N/2}\,|\mathbf{H}\mathbf{R}_s\mathbf{H}' + \mathbf{R}_n|^{1/2}}$$

and

$$p_0(\mathbf{r}) = \frac{\exp[-\frac{1}{2}\mathbf{r}'\mathbf{R}_n^{-1}\mathbf{r}]}{(2\pi)^{N/2}\,|\mathbf{R}_n|^{1/2}}$$

The likelihood ratio is therefore

$$\lambda = \frac{|\mathbf{R}_n|^{1/2}}{|\mathbf{H}\mathbf{R}_s\mathbf{H}' + \mathbf{R}_n|^{1/2}} \exp \tfrac{1}{2}\mathbf{r}'[\mathbf{R}_n^{-1} - (\mathbf{H}\mathbf{R}_s\mathbf{H}' + \mathbf{R}_n)^{-1}]\mathbf{r}$$

Since $\mathbf{R}_s$, $\mathbf{R}_n$, and $\mathbf{H}$ are assumed known, we may use for the test statistic

$$\lambda_T = \mathbf{r}'[\mathbf{R}_n^{-1} - (\mathbf{H}\mathbf{R}_s\mathbf{H}' + \mathbf{R}_n)^{-1}]\mathbf{r} \tag{11-53}$$

Using Eq. (11-7) this may be expressed in the form

$$\lambda_T = \mathbf{r}'\mathbf{R}_n^{-1}\mathbf{H}\mathbf{R}_s\mathbf{H}'(\mathbf{R}_n + \mathbf{H}\mathbf{R}_s\mathbf{H}')^{-1}\mathbf{r} \tag{11-54}$$

provided that $\mathbf{R}_n$ and $\mathbf{R}_n + \mathbf{H}\mathbf{R}_s\mathbf{H}'$ are nonsingular. It is not required that $\mathbf{H}\mathbf{R}_s\mathbf{H}'$ by itself be nonsingular. Now, according to Sect. 11.7, the maximum a posteriori estimator of $\mathbf{s}$ is

$$\hat{\mathbf{s}} = \mathbf{R}_s\mathbf{H}'(\mathbf{R}_n + \mathbf{H}\mathbf{R}_s\mathbf{H}')^{-1}\mathbf{r}$$

so that the test statistic can again be expressed§

$$\lambda_T = \mathbf{r}'\mathbf{R}_n^{-1}\mathbf{H}\hat{\mathbf{s}}$$

Although this is of the same form as that derived in the last section, there is a difference in that here $\hat{\mathbf{s}}$ is a maximum a posteriori estimate, whereas in Sect. 11.8 the estimate is a maximum-likelihood estimate. Nevertheless, the point is clear that the " best " estimate, consistent with the a priori information, is used.

Special Case

As a special case assume that the data is of the form

$$\mathbf{r} = \mathbf{s} + \mathbf{n}$$

The results can be generated directly from the preceding results by making $\mathbf{H} = \mathbf{I}$, the unit matrix. For the matrices to be conformable, the dimensions must be $N \times N$ for $\mathbf{I}$, and $N \times 1$ for $\mathbf{s}$ and $\mathbf{n}$. The test statistic is then expressed as

$$\lambda_T = \mathbf{r}'\mathbf{R}_n^{-1}\mathbf{R}_s(\mathbf{R}_n + \mathbf{R}_s)^{-1}\mathbf{r} \tag{11-55}$$

or $\lambda_T = \mathbf{r}'\mathbf{R}_n^{-1}\hat{\mathbf{s}}$, where

$$\hat{\mathbf{s}} = \mathbf{R}_s(\mathbf{R}_n + \mathbf{R}_s)^{-1}\mathbf{r} \tag{11-56}$$

We point out in passing that for the low signal-to-noise case (*14*) the tess statistic is expressible as¶

$$\lambda_T \approx \mathbf{r}'\mathbf{R}_n^{-1}\mathbf{R}_s\mathbf{R}_n^{-1}\mathbf{r} \tag{11-57}$$

When convenient to do so, this approximation will be used.

The moments of λ_T are generated next

$$E_0\{\lambda_T\} = E_0\{\mathbf{n}'\mathbf{R}_n^{-1}\mathbf{R}_s(\mathbf{R}_n + \mathbf{R}_s)^{-1}\mathbf{n}\} = \operatorname{tr} \mathbf{R}_n^{-1}\mathbf{R}_s(\mathbf{R}_n + \mathbf{R}_s)^{-1}\mathbf{R}_n \tag{11-58}$$

Using the fact that tr $\mathbf{AB}$ = tr $\mathbf{BA}$, and associating $\mathbf{A}$ with $\mathbf{R}_n^{-1}\mathbf{R}_s$, and $\mathbf{B}$ with $(\mathbf{R}_n + \mathbf{R}_s)^{-1}\mathbf{R}_n$, Eq. (11-58) may be expressed as

$$E_0\{\lambda_T\} = \operatorname{tr}(\mathbf{R}_n + \mathbf{R}_s)^{-1}\mathbf{R}_s = \operatorname{tr} \mathbf{R}_s(\mathbf{R}_n + \mathbf{R}_s)^{-1} \tag{11-59}$$

For hypothesis H_1,

$$\begin{aligned} E_1\{\lambda_T\} &= E_1\{(\mathbf{s} + \mathbf{n})'\mathbf{R}_n^{-1}\mathbf{R}_s(\mathbf{R}_n + \mathbf{R}_s)^{-1}(\mathbf{s} + \mathbf{n})\} \\ &= \operatorname{tr} \mathbf{R}_n^{-1}\mathbf{R}_s(\mathbf{R}_n + \mathbf{R}_s)^{-1}(\mathbf{R}_n + \mathbf{R}_s) \\ &= \operatorname{tr} \mathbf{R}_n^{-1}\mathbf{R}_s \end{aligned} \tag{11-60}$$

§ See also Kailath (*13*).

¶ This result may be derived by assuming that the largest eigenvalue of $\mathbf{R}_n^{-1}\mathbf{R}_s$ is much less than unity. See Exercise 11.18.

For the difference in the mean values,

$$\begin{aligned}\Delta\mu &\triangleq E_1\{\lambda_T\} - E_0\{\lambda_T\}\\ &= \text{tr}[\mathbf{R}_n^{-1}\mathbf{R}_s - (\mathbf{R}_n + \mathbf{R}_s)^{-1}\mathbf{R}_s]\\ &= \text{tr}[\mathbf{R}_n^{-1} - (\mathbf{R}_n + \mathbf{R}_s)^{-1}]\mathbf{R}_s\end{aligned}$$

Again using Eq. (11-7) we get an alternative expression

$$\Delta\mu = \text{tr}[\mathbf{R}_n^{-1}\mathbf{R}_s(\mathbf{R}_n + \mathbf{R}_s)^{-1}\mathbf{R}_s] \tag{11-61}$$

In the small signal case

$$\Delta\mu \approx \text{tr}(\mathbf{R}_n^{-1}\mathbf{R}_s)^2$$

To determine second moments, denote the kernel in the test statistic by $\mathbf{A} = \mathbf{R}_n^{-1}\mathbf{R}_s(\mathbf{R}_n + \mathbf{R}_s)^{-1}$. By comparing this to Eq. (11-7) it may be shown that $\mathbf{A}$ is symmetric. Then $\lambda_T = \mathbf{r}'\mathbf{A}\mathbf{r}$, and from Eq. (11-14e) we know

$$E\{\lambda_T{}^2\} = \text{tr}^2\, \mathbf{A}E\{\mathbf{r}\mathbf{r}'\} + 2\text{tr}(\mathbf{A}E\{\mathbf{r}\mathbf{r}'\})^2$$

Now

$$E_0\{\mathbf{r}\mathbf{r}'\} = \mathbf{R}_n \qquad \text{and} \qquad E_1\{\mathbf{r}\mathbf{r}'\} = \mathbf{R}_s + \mathbf{R}_n$$

Therefore

$$\begin{aligned}E_0\{\lambda_T{}^2\} &= \text{tr}^2\, \mathbf{R}_n^{-1}\mathbf{R}_s(\mathbf{R}_n + \mathbf{R}_s)^{-1}\mathbf{R}_n + 2\text{tr}[\mathbf{R}_n^{-1}\mathbf{R}_s(\mathbf{R}_n + \mathbf{R}_s)^{-1}\mathbf{R}_n]^2\\ &= \text{tr}^2(\mathbf{R}_n + \mathbf{R}_s)^{-1}\mathbf{R}_s + 2\text{tr}[(\mathbf{R}_n + \mathbf{R}_s)^{-1}\mathbf{R}_s]^2\end{aligned} \tag{11-62}$$

and the variance is

$$\sigma_0{}^2(\lambda_T) = 2\text{tr}[(\mathbf{R}_n + \mathbf{R}_s)^{-1}\mathbf{R}_s]^2 \tag{11-63}$$

Similarly

$$\begin{aligned}E_1\{\lambda_T{}^2\} &= \text{tr}^2\, \mathbf{R}_n^{-1}\mathbf{R}_s(\mathbf{R}_n + \mathbf{R}_s)^{-1}(\mathbf{R}_n + \mathbf{R}_s)\\ &\quad + 2\text{tr}[\mathbf{R}_n^{-1}\mathbf{R}_s(\mathbf{R}_n + \mathbf{R}_s)^{-1}(\mathbf{R}_n + \mathbf{R}_s)]^2\\ &= \text{tr}^2\, \mathbf{R}_n^{-1}\mathbf{R}_s + 2\text{tr}(\mathbf{R}_n^{-1}\mathbf{R}_s)^2\end{aligned} \tag{11-64}$$

and the variance is

$$\sigma_1{}^2(\lambda_T) = 2\text{tr}(\mathbf{R}_n^{-1}\mathbf{R}_s)^2 \tag{11-65}$$

For the small signal case $\sigma_0{}^2(\lambda_T) \simeq \sigma_1{}^2(\lambda_T)$ and the deflection signal-to-noise ratio (*12*) is

$$\left(\frac{\Delta\mu}{\sigma(\lambda_T)}\right)^2 \approx \frac{[\text{tr}(\mathbf{R}_n^{-1}\mathbf{R}_s)^2]^2}{2\,\text{tr}(\mathbf{R}_n^{-1}\mathbf{R}_s)^2} = \tfrac{1}{2}\,\text{tr}(\mathbf{R}_n^{-1}\mathbf{R}_s)^2$$

11.10 Space–Time Processing

In this section we want to emphasize multichannel or space–time signal processing brought about by signals being received at spacially separated sensors. Applications of this include detection of seismic signals (*15*) with geophone sensors distributed over a wide geographic region, detection of acoustic signals in the ocean by means of hydrophone arrays (*16*, *17*), detection of electromagnetic signals for communication purposes using space diversity, and phased arrays used for radar detection. In these cases the object of signal processing is to combine the signal received at each sensor in some "optimum" manner.

In the simplest case, as we shall see, where the signal is identical at each of the sensors except for a known time delay, and the noise at each sensor is statistically independent of that at all other sensors, the best way to combine the signals is to adjust for the known time delays and add the resulting signals coherently. This is the concept of modern phased array antennas used primarily in radar applications. When these assumptions are no longer valid, there is a strong need for more sophisticated methods of combining.

We should point out that the multichannel case has really been included in our previous material in this chapter. More specifically, the preceding material can be applied directly to the multichannel case. This is so because the origin of the samples **r**, **s**, and **n** etc. was not specified. Thus, in a single channel case they may have been time samples in that one channel. For the multichannel case, however, these samples could still be used to represent the data. For example, we could denote

$$\mathbf{s} = \begin{bmatrix} \mathbf{s}_1 \\ \vdots \\ \mathbf{s}_M \end{bmatrix}$$

where $\mathbf{s}_1$ denotes the time vector samples in channel 1, $\mathbf{s}_2$ represents the time samples in channel 2, and so on for all M channels. We could also have $\mathbf{s}_1$ represent a vector of samples taken one from each channel at time t_1, $\mathbf{s}_2$ at time t_2, etc. Furthermore, the samples need not be time samples, but could represent Fourier coefficients, or Karhunen–Loeve coefficients, etc. As the samples are rearranged in the data vectors, or as the sample definition changes, the covariance functions must follow suit. Thus, by proper definition of the vector samples, the preceding material is still applicable here.

Nevertheless, the space–time processing will be pursued to show explicitly the multichannel character and implementation of the receiver. This opportunity will also be taken to work in the frequency domain with Fourier coefficients, and to work with complex Gaussian random variables.

Before proceeding, a caution should be sounded. We shall assume that the signal and noise covariances are completely known, and that signal and noise are Gaussian. Some difficulty arises when these covariances are unknown. In the communications case where control of the transmitted signal is possible, one could presumably measure the sample covariance functions and use these as estimates of the true covariance function. If one has no control over the transmitted signal, and furthermore does not know when a "signal" is present or not, then it is not known whether $\mathbf{R}_s + \mathbf{R}_n$ or $\mathbf{R}_n$ alone is being estimated. This can lead to some surprising and interesting results. A discussion of this is beyond the scope of this text. It suffices to say that much work remains to be done on the subject of space–time processing.

Sampling Plan

As shown in Fig. 11-2, consider an array of M elements whose input signals are $r_i(t) = v_i(t) + n_i(t)$, $i = 1, \ldots, M$. We shall later specialize to the case where $v_i(t)$ is a convolution of a channel filter $h_i(t)$ and a signal $s(t)$.

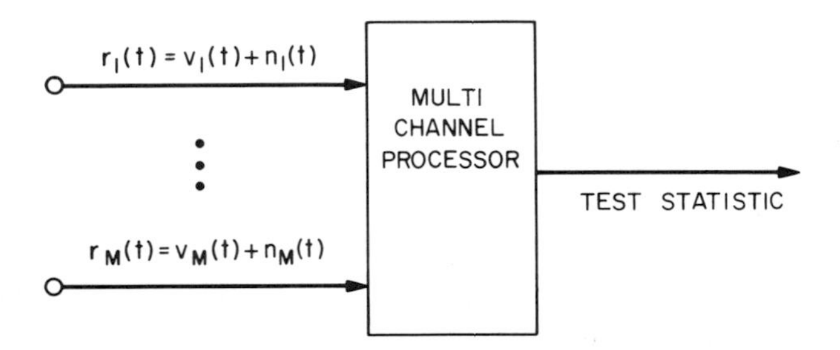

Fig. 11-2 Configuration for space–time processing.

Assume that the signals are decomposed into their Fourier components. Using the tilde to denote the Fourier coefficients we have for $0 \le t \le T$ and $i = 1, \ldots, M$

$$r_i(t) = \frac{1}{T^{1/2}} \sum_k \tilde{r}_i(k) e^{j\omega_k t} \quad \text{and} \quad \tilde{r}_i(k) = \frac{1}{T^{1/2}} \int_0^T r_i(t) e^{-j\omega_k t}\, dt$$

$$v_i(t) = \frac{1}{T^{1/2}} \sum_k \tilde{v}_i(k) e^{j\omega_k t} \quad \text{and} \quad \tilde{v}_i(k) = \frac{1}{T^{1/2}} \int_0^T v_i(t) e^{-j\omega_k t}\, dt$$

$$n_i(t) = \frac{1}{T^{1/2}} \sum_k \tilde{n}_i(k) e^{j\omega_k t} \quad \text{and} \quad \tilde{n}_i(k) = \frac{1}{T^{1/2}} \int_0^T n_i(t) e^{-j\omega_k t}\, dt$$

If we consider only the significant frequency region we can label the lowest frequency by ω_1 and the highest by ω_K. For real functions, the Fourier coefficient for negative frequencies can be expressed in terms of the corresponding positive frequency coefficients. For example, $\tilde{n}(-k) = \tilde{n}^*(k)$. Thus as far as the probability measure of the signals is concerned, only the Fourier coefficients corresponding to positive frequencies need to be considered. The time length T of the data analyzed is assumed to be sufficiently long so that the Fourier components in any one channel are assumed uncorrelated, and so for the Gaussian case statistically independent. Thus relations such as those in Eq. (3-75) are assumed to hold.

The samples will be the complex Fourier coefficients. The same results could be derived using the inphase and quadrature components, both of which are real, as the samples. However, the notation becomes much too cumbersome. Because of the special form of the covariance matrices in these problems, the material on isomorphic matrices and the multivariate complex Gaussian distribution permit some notation simplification.

Arrange the samples in the following manner

$$\tilde{\mathbf{r}} = \begin{bmatrix} \tilde{r}_1(1) \\ \vdots \\ \tilde{r}_M(1) \\ \hdashline \tilde{r}_1(2) \\ \vdots \\ \tilde{r}_M(2) \\ \hdashline \vdots \\ \hdashline \tilde{r}_1(K) \\ \vdots \\ \tilde{r}_M(K) \end{bmatrix}, \qquad \tilde{\mathbf{v}} = \begin{bmatrix} \tilde{v}_1(1) \\ \vdots \\ \tilde{v}_M(1) \\ \hdashline \tilde{v}_1(2) \\ \vdots \\ \tilde{v}_M(2) \\ \hdashline \vdots \\ \hdashline \tilde{v}_1(K) \\ \vdots \\ \tilde{v}_M(K) \end{bmatrix}, \qquad \tilde{\mathbf{n}} = \begin{bmatrix} \tilde{n}_1(1) \\ \vdots \\ \tilde{n}_M(1) \\ \hdashline \tilde{n}_1(2) \\ \vdots \\ \tilde{n}_M(2) \\ \hdashline \vdots \\ \hdashline \tilde{n}_1(K) \\ \vdots \\ \tilde{n}_M(K) \end{bmatrix} \tag{11-66}$$

The subscript refers to the channel, and the argument refers to the frequency index. The vectors have dimension MK. To condense the notation, define column vectors such as

$$\tilde{\mathbf{r}}(i) = \begin{bmatrix} \tilde{r}_1(i) \\ \vdots \\ \tilde{r}_M(i) \end{bmatrix}, \qquad \tilde{\mathbf{v}}(i) = \begin{bmatrix} \tilde{v}_1(i) \\ \vdots \\ \tilde{v}_M(i) \end{bmatrix}, \qquad \tilde{\mathbf{n}}(i) = \begin{bmatrix} \tilde{n}_1(i) \\ \vdots \\ \tilde{n}_M(i) \end{bmatrix} \tag{11-67}$$

Then we have

$$\tilde{\mathbf{r}} = \begin{bmatrix} \tilde{\mathbf{r}}(1) \\ \hdashline \vdots \\ \hdashline \tilde{\mathbf{r}}(K) \end{bmatrix}, \qquad \tilde{\mathbf{v}} = \begin{bmatrix} \tilde{\mathbf{v}}(1) \\ \hdashline \vdots \\ \hdashline \tilde{\mathbf{v}}(K) \end{bmatrix}, \qquad \tilde{\mathbf{n}} = \begin{bmatrix} \tilde{\mathbf{n}}(1) \\ \hdashline \vdots \\ \hdashline \tilde{\mathbf{n}}(K) \end{bmatrix} \tag{11-68}$$

The sample vectors $\tilde{\mathbf{r}}$, $\tilde{\mathbf{v}}$, and $\tilde{\mathbf{n}}$ are therefore in partition form, each subvector of which refers to the samples at a given frequency. It was pointed out in

Chap. 3 that the covariance matrices of the Fourier components at a given frequency were really the cross spectral density matrices at the frequency of interest. Any 2×2 submatrix of the covariance matrix will have the form given in Eq. (3-77). Denote the covariance matrix of $\tilde{\mathbf{r}}(i)$, $\tilde{\mathbf{v}}(i)$, and $\tilde{\mathbf{n}}(i)$ by $\tilde{\mathbf{S}}_r(i)$, $\tilde{\mathbf{S}}_v(i)$, $\tilde{\mathbf{S}}_n(i)$ respectively. Then

$$E\{\tilde{\mathbf{r}}(i)\tilde{\mathbf{r}}^\dagger(i)\} = \tilde{\mathbf{S}}_r(i) = \begin{bmatrix} S_{r_1}(i) & S_{r_1r_2}(i) & \cdots & S_{r_1r_M}(i) \\ S^*_{r_1r_2}(i) & S_{r_2}(i) & \cdots & S_{r_2r_M}(i) \\ \vdots & & & \\ S^*_{r_1r_M}(i) & & \cdots & S_{r_M}(i) \end{bmatrix} \tag{11-69}$$

where $S_{r_i}(k)$ is the power spectral density of $r_i(t)$ evaluated at frequency ω_k, and $S_{r_ir_j}(k)$ is the power cross spectral density of $r_i(t)$ amd $r_j(t)$ also evaluated at frequency ω_k. The † denotes complex conjugate transpose. The matrix $\tilde{\mathbf{S}}_r(i)$ is Hermitian. The form is of course similar for $\tilde{\mathbf{S}}_v(i)$ and $\tilde{\mathbf{S}}_n(i)$.

From the assumption of sufficiently long T, which allows the Fourier components at different frequencies to be uncorrelated, we know for $i \neq j$ that

$$E\{\tilde{\mathbf{r}}(i)\tilde{\mathbf{r}}^\dagger(j)\} = 0, \qquad E\{\tilde{\mathbf{v}}(i)\tilde{\mathbf{v}}^\dagger(j)\} = 0, \qquad E\{\tilde{\mathbf{n}}(i)\tilde{\mathbf{n}}^\dagger(j)\} = 0$$

Therefore

$$\tilde{\mathbf{S}}_r = E\{\tilde{\mathbf{r}}\tilde{\mathbf{r}}^\dagger\} = \begin{bmatrix} \tilde{\mathbf{S}}_r(1) & 0 & \cdots & 0 \\ 0 & \tilde{\mathbf{S}}_r(2) & \cdots & 0 \\ \vdots & & & \\ 0 & \cdots & & \tilde{\mathbf{S}}_r(K) \end{bmatrix} \tag{11-70}$$

$$\tilde{\mathbf{S}}_v = E\{\tilde{\mathbf{v}}\tilde{\mathbf{v}}^\dagger\} = \begin{bmatrix} \tilde{\mathbf{S}}_v(1) & 0 & \cdots & 0 \\ 0 & \tilde{\mathbf{S}}_v(2) & \cdots & 0 \\ \vdots & & & \\ 0 & \cdots & & \tilde{\mathbf{S}}_v(K) \end{bmatrix} \tag{11-71}$$

$$\tilde{\mathbf{S}}_n = E\{\tilde{\mathbf{n}}\tilde{\mathbf{n}}^\dagger\} = \begin{bmatrix} \tilde{\mathbf{S}}_n(1) & 0 & \cdots & 0 \\ 0 & \tilde{\mathbf{S}}_n(2) & \cdots & 0 \\ \vdots & & & \\ 0 & \cdots & & \tilde{\mathbf{S}}_n(K) \end{bmatrix} \tag{11-72}$$

Such matrices are referred to as block diagonal.

It may be verified, by using Eq. (11-45) for example, that the inverse matrices are

$$\tilde{\mathbf{S}}_r^{-1} = \begin{bmatrix} \tilde{\mathbf{S}}_r^{-1}(1) & & \cdots & 0 \\ 0 & \tilde{\mathbf{S}}_r^{-1}(2) & \cdots & 0 \\ \vdots & & & \\ 0 & \cdots & & \tilde{\mathbf{S}}_r^{-1}(K) \end{bmatrix} \tag{11-73}$$

$$\tilde{\mathbf{S}}_v^{-1} = \begin{bmatrix} \tilde{\mathbf{S}}_v^{-1}(1) & & \cdots & 0 \\ 0 & \tilde{\mathbf{S}}_v^{-1}(2) & \cdots & 0 \\ \vdots & & & \\ 0 & \cdots & & \tilde{\mathbf{S}}_v^{-1}(K) \end{bmatrix} \tag{11-74}$$

$$\tilde{\mathbf{S}}_n^{-1} = \begin{bmatrix} \tilde{\mathbf{S}}_n^{-1}(1) & & \cdots & 0 \\ 0 & \tilde{\mathbf{S}}_n^{-1}(2) & \cdots & 0 \\ \vdots & & & \\ 0 & \cdots & & \tilde{\mathbf{S}}_n^{-1}(K) \end{bmatrix} \tag{11-75}$$

The Likelihood Ratio

For the hypotheses

H_1: $\tilde{\mathbf{r}} = \tilde{\mathbf{v}} + \tilde{\mathbf{n}}$
H_0: $\tilde{\mathbf{r}} = \tilde{\mathbf{n}}$

we are in a position to write the likelihood ratio. According to the discussion in Sects. 11.2 and 11.3, the distribution of $\tilde{\mathbf{r}}$ can be expressed as a multivariate complex Gaussian distribution such as given in Eq. (11-22). Thus,

$$p_1(\tilde{\mathbf{r}}) = \frac{1}{\pi^{MK}|\tilde{\mathbf{S}}_v + \tilde{\mathbf{S}}_n|} \exp(-\tilde{\mathbf{r}}^\dagger(\tilde{\mathbf{S}}_v + \tilde{\mathbf{S}}_n)^{-1}\tilde{\mathbf{r}})$$

$$p_0(\tilde{\mathbf{r}}) = \frac{1}{\pi^{MK}|\tilde{\mathbf{S}}_n|} \exp(-\tilde{\mathbf{r}}^\dagger\tilde{\mathbf{S}}_n^{-1}\tilde{\mathbf{r}})$$

and the likelihood ratio is

$$\lambda(\tilde{\mathbf{r}}) = \frac{|\tilde{\mathbf{S}}_n|}{|\tilde{\mathbf{S}}_v + \tilde{\mathbf{S}}_n|} \exp \tilde{\mathbf{r}}^\dagger[\tilde{\mathbf{S}}_n^{-1} - (\tilde{\mathbf{S}}_v + \tilde{\mathbf{S}}_n)^{-1}]\tilde{\mathbf{r}} \tag{11-76}$$

We can therefore use for the test statistic

$$\lambda_T = \tilde{\mathbf{r}}^\dagger[\tilde{\mathbf{S}}_n^{-1} - (\tilde{\mathbf{S}}_v + \tilde{\mathbf{S}}_n)^{-1}]\tilde{\mathbf{r}} \tag{11-77}$$

or the equivalent

$$\lambda_T = \tilde{\mathbf{r}}^\dagger\tilde{\mathbf{S}}_n^{-1}\tilde{\mathbf{S}}_v(\tilde{\mathbf{S}}_n + \tilde{\mathbf{S}}_v)^{-1}\tilde{\mathbf{r}} \tag{11-78}$$

As expected, this is of the same form as Eq. (11-55). We point out that the test statistic is a real scalar quantity.

Using the matrix inverses such as given in Eq. (11-75), the kernel of Eq. (11-78) is

$$\tilde{\mathbf{S}}_n^{-1}\tilde{\mathbf{S}}_v(\tilde{\mathbf{S}}_n + \tilde{\mathbf{S}}_v)^{-1}$$

$$= \begin{bmatrix} \tilde{\mathbf{S}}_n^{-1}(1)\tilde{\mathbf{S}}_v(1)[\tilde{\mathbf{S}}_n(1) + \tilde{\mathbf{S}}_v(1)]^{-1} & \cdots & 0 \\ \vdots & & \vdots \\ 0 & & \tilde{\mathbf{S}}_n^{-1}(K)\tilde{\mathbf{S}}_v(K)[\tilde{\mathbf{S}}_n(K) + \tilde{\mathbf{S}}_v(K)]^{-1} \end{bmatrix}$$

The test statistic may then be expressed as

$$\lambda_{\mathrm{T}} = \sum_{k=1}^{K} \tilde{\mathbf{r}}^{\dagger}(k)\tilde{\mathbf{S}}_n^{-1}(k)\tilde{\mathbf{S}}_v(k)[\tilde{\mathbf{S}}_n(k) + \tilde{\mathbf{S}}_v(k)]^{-1}\tilde{\mathbf{r}}(k) \tag{11-79}$$

Consequently, the signal processing can be done at each Fourier component and the results summed across the frequency band. This is a direct consequence of the assumption that different frequency components are uncorrelated. At each frequency the processor generates

$$\lambda_{\mathrm{T}}(k) = \tilde{\mathbf{r}}^{\dagger}(k)\tilde{\mathbf{S}}_n^{-1}(k)\tilde{\mathbf{S}}_v(k)[\tilde{\mathbf{S}}_n(k) + \tilde{\mathbf{S}}_v(k)]^{-1}\tilde{\mathbf{r}}(k)$$

and then

$$\lambda_{\mathrm{T}} = \sum_{k=1}^{K} \lambda_{\mathrm{T}}(k)$$

The moments of $\lambda_{\mathrm{T}}(k)$ may be shown to be

$$E_0\{\lambda_{\mathrm{T}}(k)\} = \mathrm{tr}[\tilde{\mathbf{S}}_n(k) + \tilde{\mathbf{S}}_v(k)]^{-1}\tilde{\mathbf{S}}_v(k) \tag{11-80a}$$

$$E_1\{\lambda_{\mathrm{T}}(k)\} = \mathrm{tr}\,\tilde{\mathbf{S}}_n^{-1}(k)\tilde{\mathbf{S}}_v(k) \tag{11-80b}$$

$$\sigma_0{}^2(\lambda_{\mathrm{T}}) = 2\mathrm{tr}[(\tilde{\mathbf{S}}_n(k) + \tilde{\mathbf{S}}_v(k))^{-1}\tilde{\mathbf{S}}_v(k)]^2 \tag{11-80c}$$

$$\sigma_1{}^2(\lambda_{\mathrm{T}}) = 2\mathrm{tr}[\tilde{\mathbf{S}}_n^{-1}(k)\tilde{\mathbf{S}}_v(k)]^2 \tag{11-80d}$$

Consult Eqs. (11-59), (11-60), (11-63), and (11-65) for analogous expressions.

Digression

Although Eqs. (11-80a) and (11-80b) are easily found, Eqs. (11-80c) and (11-80d) need a bit of justification since we have not discussed fourth mixed moments of complex Gaussian vectors.

In one of the steps leading to Eq. (11-55) the test statistic was multiplied by 2 to remove the factor $\frac{1}{2}$. (No modification was used in the complex sample case.) Keeping this factor in the equation the test statistic would have been $G_{\mathrm{T}} = \lambda_{\mathrm{T}}/2$. Then we would have had $E_1\{G_{\mathrm{T}}\} = \frac{1}{2}\,\mathrm{tr}\,\mathbf{R}_n^{-1}\mathbf{R}_s$, for example. Just this moment will be considered and the results extrapolated to the other moments in Eq. (11-80). For convenience the frequency index will be suppressed.

If instead of working with complex samples we used the inphase and quadrature components individually, the covariance matrix $\mathbf{R}_n^{-1}$ would be isomorphic to $\tilde{\mathbf{S}}_n/2$, and $\mathbf{R}_s$ would be isomorphic to $\tilde{\mathbf{S}}_v/2$. The matrices $\tilde{\mathbf{S}}_v$ and $\tilde{\mathbf{S}}_n$ are Hermitian and the trace of a Hermitian matrix is real. If a matrix $\mathbf{Q}$ is isomorphic to $\vec{\mathbf{Q}}$, then $\mathrm{tr}\,\mathbf{Q} = 2\,\mathrm{Re}\,\mathrm{tr}\,\vec{\mathbf{Q}}$. If $\vec{\mathbf{Q}}$ is the product of Hermitian matrices then $\mathrm{tr}\,\vec{\mathbf{Q}}$ is real so that $\frac{1}{2}\,\mathrm{tr}\,\mathbf{Q} = \mathrm{tr}\,\vec{\mathbf{Q}}$. Therefore

$$E_1\{G_{\mathrm{T}}\} = \tfrac{1}{2}\,\mathrm{tr}\,\mathbf{R}_n^{-1}\mathbf{R}_s = \mathrm{tr}(\tilde{\mathbf{S}}_n/2)^{-1}(\tilde{\mathbf{S}}_v/2) = \mathrm{tr}\,\tilde{\mathbf{S}}_n^{-1}\tilde{\mathbf{S}}_v$$

in agreement with Eq. (11-80b). Using the statistic G_T, Eqs. (11-63) and (11-65), and the relations just discussed, then Eqs. (11-80c) and (11-80d) can be derived.

Channel Model

We now make the signal model more specific and consider the case shown in Fig. 11-3. The signal source is $s(t)$ which is assumed to be a Gaussian process. The transmission medium is characterized by a set of filters $h_i(t)$, $i = 1, \ldots, M$. This model is a particularly simple one but one which can

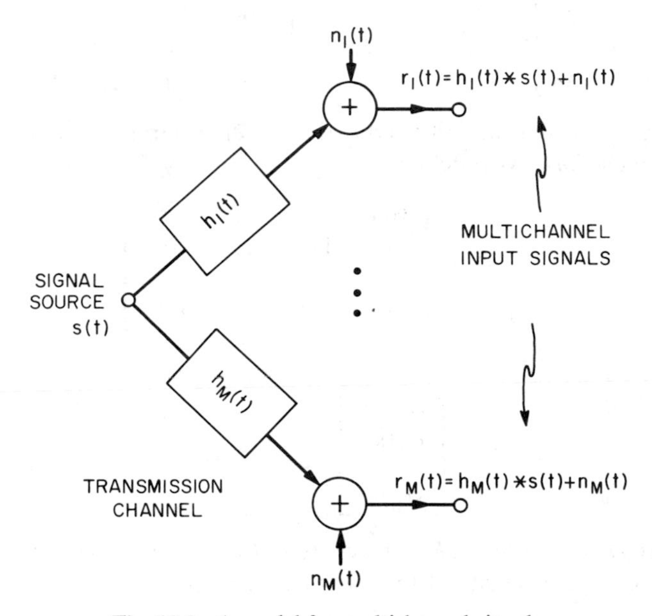

Fig. 11-3 A model for multichannel signals.

yield further insight into the space–time processing scheme. In a less constrained case, the filters could be time varying, and furthermore could be stochastic filters. Also, instead of showing just one path from the signal source to each receiving element, there could be many. In many practical cases of interest, the proper channel characterization is indeed a difficult problem. We shall avoid this problem and deal only with the model of Fig. 11-3.

Denote the Fourier transform of the channel filters by

$$H_i(\omega) = \int_{-\infty}^{\infty} h_i(t)e^{-j\omega t}\, dt$$

The signal $s(t)$ is represented in terms of its Fourier coefficients

$$s(t) = \frac{1}{T^{1/2}} \sum_k \tilde{s}(k) e^{j\omega_k t}$$

The signal at the sensors therefore has Fourier components of the form

$$\tilde{v}_i(k) = H_i(k)\tilde{s}(k), \qquad i = 1, \ldots, M, \quad k = 1, \ldots, K \tag{11-81}$$

where $H_i(k) \triangleq H_i(\omega_k)$. To put this in vector form, define

$$\tilde{\mathbf{s}} = \begin{bmatrix} \tilde{s}(1) \\ \tilde{s}(2) \\ \vdots \\ \tilde{s}(K) \end{bmatrix}, \qquad \tilde{\mathbf{S}}_s = E\{\tilde{\mathbf{s}}\tilde{\mathbf{s}}^\dagger\} = \begin{bmatrix} S_s(1) & 0 & \ldots & 0 \\ 0 & S_s(2) & \ldots & 0 \\ \vdots & \vdots & & \\ 0 & 0 & \ldots & S_s(K) \end{bmatrix}$$

where $S_s(k)$ is a scalar quantity and is the power spectral density of the signal $s(t)$ at frequency ω_k. Also define

$$\tilde{\mathbf{H}} = \begin{bmatrix} \tilde{\mathbf{H}}(1) & 0 & \ldots & 0 \\ 0 & \tilde{\mathbf{H}}(2) & \ldots & 0 \\ \vdots & \vdots & & \vdots \\ 0 & 0 & \ldots & \tilde{\mathbf{H}}(K) \end{bmatrix} \tag{11-82}$$

where

$$\tilde{\mathbf{H}}(k) \triangleq \begin{bmatrix} H_1(k) \\ H_2(k) \\ \vdots \\ H_M(k) \end{bmatrix}, \qquad k = 1, \ldots, K \tag{11-83}$$

The matrix $\tilde{\mathbf{H}}$ is of order $MK \times K$, while $\tilde{\mathbf{H}}(k)$ is of order $M \times 1$.

The Fourier coefficients of the received signal may now be expressed

$$\tilde{\mathbf{v}} = \tilde{\mathbf{H}}\tilde{\mathbf{s}} \tag{11-84}$$

This may now be applied to previous results which were expressed in terms of $\tilde{\mathbf{v}}$.

Processor Realization

The correlation function for $\tilde{\mathbf{v}}$ is equal to

$$\tilde{\mathbf{S}}_v = E\{\tilde{\mathbf{v}}\tilde{\mathbf{v}}^\dagger\} = \tilde{\mathbf{H}}\tilde{\mathbf{S}}_s\tilde{\mathbf{H}}^\dagger \tag{11-85}$$

Because of the block diagonal form of the matrices $\tilde{\mathbf{H}}$ and $\tilde{\mathbf{S}}_s$, it is easily shown that

$$\tilde{\mathbf{S}}_v(k) = \tilde{\mathbf{H}}(k)S_s(k)\tilde{\mathbf{H}}^\dagger(k), \qquad k = 1, \ldots, K \tag{11-86}$$

Inserting this into the equation of the test statistic (11-79) produces

$$\lambda_T = \sum_{k=1}^{K} \tilde{\mathbf{r}}^\dagger(k)\tilde{\mathbf{S}}_n^{-1}(k)\tilde{\mathbf{H}}(k)S_s(k)\tilde{\mathbf{H}}^\dagger(k)[\tilde{\mathbf{S}}_n(k) + \tilde{\mathbf{H}}(k)S_s(k)\tilde{\mathbf{H}}^\dagger(k)]^{-1}\tilde{\mathbf{r}}(k) \quad (11\text{-}87)$$

As a reminder, $S_s(k)$ is a real scalar quantity. By associating $\mathbf{J} = S_s^{1/2}(k)\tilde{\mathbf{H}}(k)$, $\mathbf{\Lambda} = \tilde{\mathbf{S}}_n(k)$, and $\mathbf{A} = 1$, a scalar, in Eq. (11-2),§ we can show that

$$\begin{aligned} &S_s^{1/2}(k)\tilde{\mathbf{H}}^\dagger(k)[\tilde{\mathbf{S}}_n(k) + S_s(k)\tilde{\mathbf{H}}(k)\tilde{\mathbf{H}}^\dagger(k)]^{-1} \\ &\quad = [1 + S_s(k)\tilde{\mathbf{H}}^\dagger(k)\tilde{\mathbf{S}}_n^{-1}(k)\tilde{\mathbf{H}}(k)]^{-1}S_s^{1/2}(k)\tilde{\mathbf{H}}^\dagger(k)\tilde{\mathbf{S}}_n^{-1}(k) \end{aligned}$$

The quantity in the bracket is a real scalar. We may therefore write

$$\lambda_T = \sum_{k=1}^{K} \frac{S_s(k)\tilde{\mathbf{r}}^\dagger(k)\tilde{\mathbf{S}}_n^{-1}(k)\tilde{\mathbf{H}}(k)\tilde{\mathbf{H}}^\dagger(k)\tilde{\mathbf{S}}_n^{-1}(k)\tilde{\mathbf{r}}(k)}{1 + S_s(k)\tilde{\mathbf{H}}^\dagger(k)\tilde{\mathbf{S}}_n^{-1}(k)\tilde{\mathbf{H}}(k)}$$

or

$$\lambda_T = \sum_{k=1}^{K} \frac{S_s(k)\,|\tilde{\mathbf{H}}^\dagger(k)\tilde{\mathbf{S}}_n^{-1}(k)\tilde{\mathbf{r}}(k)|^2}{1 + S_s(k)\tilde{\mathbf{H}}^\dagger(k)\tilde{\mathbf{S}}_n^{-1}(k)\tilde{\mathbf{H}}(k)} \quad (11\text{-}88)$$

The term in the absolute value brackets is also a scalar quantity. Denoting the multiplier

$$|f(k)|^2 = \frac{S_s(k)}{1 + S_s(k)\tilde{\mathbf{H}}^\dagger(k)\tilde{\mathbf{S}}_n^{-1}(k)\tilde{\mathbf{H}}(k)} \quad (11\text{-}89)$$

the test statistic is finally expressed

$$\lambda_T = \sum_{k=1}^{K} |f(k)\tilde{\mathbf{H}}^\dagger(k)\tilde{\mathbf{S}}_n^{-1}(k)\tilde{\mathbf{r}}(k)|^2 \quad (11\text{-}90)$$

Using Parseval's theorem with our particular normalization for the Fourier series, the sum may be expressed

$$\lambda_T = \sum_{k=1}^{\infty} |\tilde{g}(k)|^2 = \tfrac{1}{2}\int_0^T g^2(t)\,dt$$

Here, $g(t)$ is a real function and its Fourier coefficients are the $\tilde{g}(k)$. (The $\tilde{g}(0)$ term is assumed zero.) Thus we may associate the coefficients in Eq. (11-90) with some function, say $g(t)$, where

$$g(t) = \frac{1}{T^{1/2}} \sum_k f(k)\tilde{\mathbf{H}}^\dagger(k)\tilde{\mathbf{S}}_n^{-1}(k)\tilde{\mathbf{r}}(k)e^{j\omega_k t} \quad (11\text{-}91)$$

§ It is easily verified that Eq. (11–2) also holds if the complex conjugate transpose is used in place of the transpose.

Now, define

$$\mathbf{L}^{\dagger}(k) = \tilde{\mathbf{H}}^{\dagger}(k)\tilde{\mathbf{S}}_n^{-1}(k) \tag{11-92}$$

Of course, $\mathbf{L}(k)$ is a column vector with components $l_i(k)$. Then

$$g(t) = \frac{1}{T^{1/2}} \sum_k f(k)[l_1{}^*(k)\tilde{r}_1(k) + \cdots + l_M{}^*(k)\tilde{r}_M(k)]e^{j\omega_k t}$$

$$= \sum_i g_i(t)$$

The signal $g_i(t)$ is seen to be the response of the filters $f(\omega)$ and $l_i(\omega)$ to the signal $r_i(t)$. Thus, $g(t)$ and the test statistic λ_T may be generated as shown in Fig. 11-4. The filter unique to each branch is dependent on the channel model through both $\tilde{\mathbf{H}}(k)$ and $\tilde{\mathbf{S}}_n(k)$, but not on the properties of $s(t)$, whereas the common filter $f(\omega)$ also includes dependence on the signal.

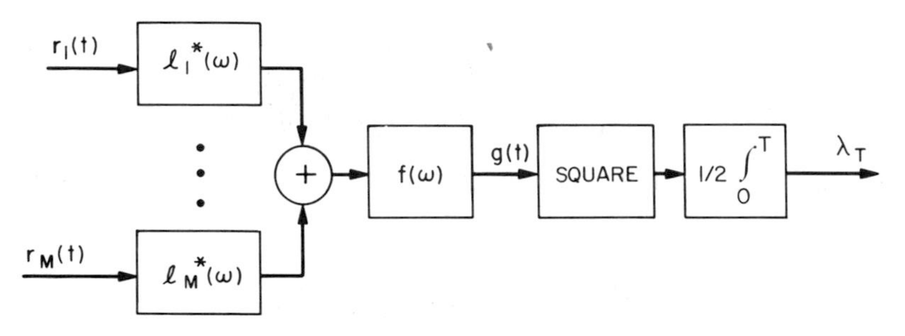

Fig. 11-4 A realization of a space–time processor.

Spatial Prewhitening

Another implementation of the space–time processor may be realized. This version brings out what is called spatial prewhitening. As clearly indicated in the model of the detection problem, the noise is assumed correlated at each sensor. A spatial prewhitener operates on such noise signals to make them uncorrelated. This processor does in the spatial domain what our prewhiteners were shown to do in the time domain.

As before, factor the inverse noise covariance matrix into the product of triangular matrices

$$\tilde{\mathbf{S}}_n^{-1}(k) = \mathbf{C}^{\dagger}(k)\mathbf{C}(k)$$

where $\mathbf{C}(k)$ is lower triangular. The function, $g(t)$ Eq. (11-91) may then be written

$$g(t) = \frac{1}{T^{1/2}} \sum_k f(k)[\mathbf{C}(k)\tilde{\mathbf{H}}(k)]^{\dagger}[\mathbf{C}(k)\tilde{\mathbf{r}}(k)]e^{j\omega_k t}$$

The column vector is

$$\mathbf{C}(k)\tilde{\mathbf{r}}(k) = \begin{bmatrix} c_{11}(k) & 0 & \cdots & 0 \\ c_{21}(k) & c_{22}(k) & \cdots & \\ \vdots & & \ddots & 0 \\ c_{M1}(k) & c_{M2}(k) & \cdots & c_{MM}(k) \end{bmatrix} \begin{bmatrix} \tilde{r}_1(k) \\ \vdots \\ \tilde{r}_M(k) \end{bmatrix}$$

$$= \begin{bmatrix} c_{11}(k)\tilde{r}_1(k) \\ c_{21}(k)\tilde{r}_1(k) + c_{22}(k)\tilde{r}_2(k) \\ \vdots \\ c_{M1}(k)\tilde{r}_1(k) + c_{M2}(k)\tilde{r}_2(k) + \cdots + c_{MM}(k)\tilde{r}_M(k) \end{bmatrix}$$

A similar expression holds for $\mathbf{C}(k)\tilde{\mathbf{H}}(k)$ with obvious modifications. Then $g(t)$ is

$$\begin{aligned} g(t) = \frac{1}{T^{1/2}} \sum_k f(k) e^{j\omega_k t} & \\ \{c_{11}^*(k)H_1{}^*(k)c_{11}(k)\tilde{r}_1(k) & \\ + [c_{21}^*(k)H_1{}^*(k) + c_{22}^*(k)H_2{}^*(k)][c_{21}(k)\tilde{r}_1(k) + c_{22}(k)\tilde{r}_2(k)] & \\ \vdots & \\ + [c_{M1}^*(k)H_1{}^*(k) + \cdots + c_{MM}^*(k)H_M{}^*(k)] & \\ \times [c_{M1}(k)\tilde{r}_1(k) + \cdots + c_{MM}(k)\tilde{r}_M(k)]\} & \qquad \text{(11-93)} \end{aligned}$$

Denote the following filters based on terms in this expression (we are suppressing the ω dependence of the c_{ij} for notational convenience)

$$\begin{aligned} m_{11}(\omega) &= c_{11}^* H_1{}^*(\omega) c_{11} \\ m_{12}(\omega) &= [c_{21}^* H_1{}^*(\omega) + c_{22}^* H_2{}^*(\omega)] c_{21} \\ m_{22}(\omega) &= [c_{21}^* H_1{}^*(\omega) + c_{22}^* H_2{}^*(\omega)] c_{22} \\ &\vdots \\ m_{1M}(\omega) &= [c_{M1}^* H_1{}^*(\omega) + \cdots + c_{MM}^* H_M{}^*(\omega)] c_{M1} \\ m_{MM}(\omega) &= [c_{M1}^* H_1{}^*(\omega) + \cdots + c_{MM}^* H_M^*(\omega)] c_{MM} \end{aligned}$$

Using these definitions

$$\begin{aligned} g(t) = \frac{1}{T^{1/2}} \sum_k f(k) e^{j\omega_k t} [& m_{11}(\omega_k)\tilde{r}_1(k) \\ &+ m_{12}(\omega_k)\tilde{r}_1(k) + m_{22}(\omega_k)\tilde{r}_2(k) \\ &\vdots \\ &+ m_{1M}(\omega_k)\tilde{r}_1(k) + \cdots + m_{MM}(\omega_k)\tilde{r}_M(k)] \end{aligned}$$

The receiver may now be implemented as shown in Fig. 11-5 where the set of signals $r_i(t)$ is transformed into the set $\rho_i(t)$. This implementation is suggestive of the Gram–Schmidt orthogonalization procedure. We shall show that the $\rho_i(t)$ are uncorrelated by showing that their Fourier components, denoted $\tilde{\rho}_i(k)$, are uncorrelated when noise alone is present at the input. The $\tilde{\rho}_i(k)$

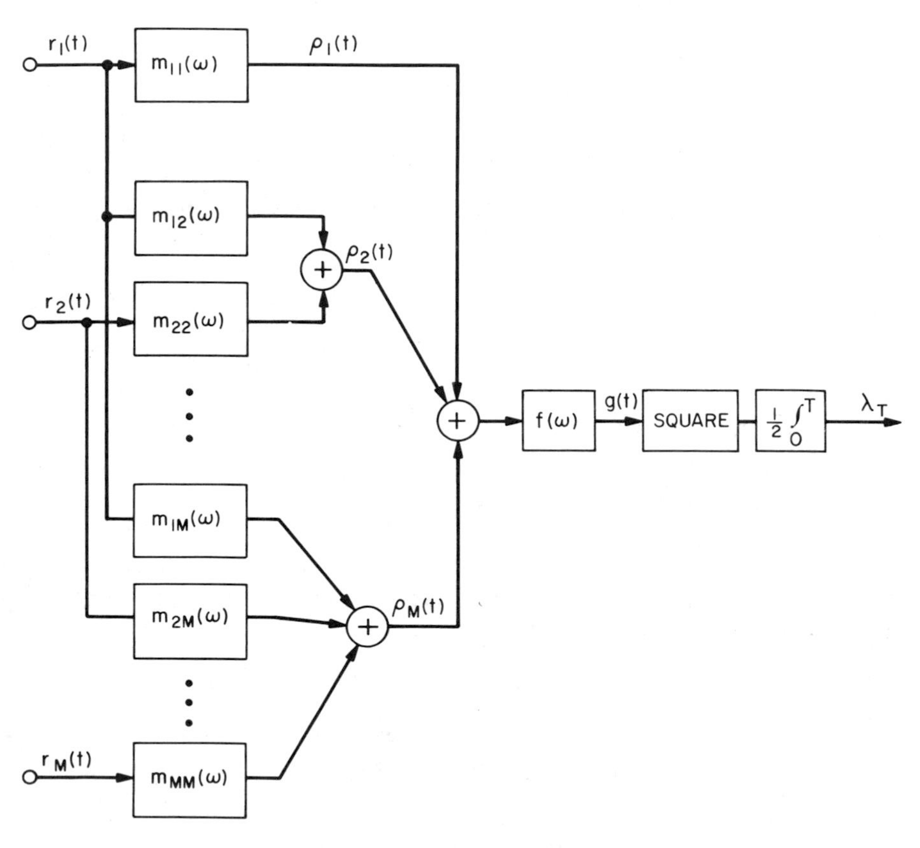

Fig. 11-5 A realization of a space–time processor which points out spatial prewhitening.

are linearly related to the elements of the column vector $\mathbf{C}(k)\tilde{\mathbf{r}}(k)$ which has a covariance matrix

$$E_0\{\mathbf{C}(k)\tilde{\mathbf{r}}(k)\tilde{\mathbf{r}}^\dagger(k)\mathbf{C}^\dagger(k)\} = \mathbf{C}(k)\tilde{\mathbf{S}}_n(k)\mathbf{C}^\dagger(k) = \mathbf{I}$$

Since the terms of this vector are uncorrelated, it also follows that the $\tilde{\rho}_i(k)$ are similarly uncorrelated.

Exercises

11.1 Verify Eq. (11-6).

11.2 Show that the trace of a matrix is equal to the sum of its eigenvalues.

11.3 If $\mathbf{R}$ is a real symmetric matrix, show that

$$\mathbf{x}'\mathbf{R}\mathbf{x} = \mathbf{x}'\mathbf{D}\mathbf{x}$$

where $\mathbf{x}$ is a real column vector, and $\mathbf{D}$ is a triangular matrix with elements

$$d_{ij} = \begin{cases} r_{ii}, & i = j \\ 2r_{ij}, & i > j \\ 0, & i < j \end{cases}$$

11.4 Assume that $\mathbf{H}$ is a Hermitian matrix, and $\mathbf{y}$ is a complex column vector. Show that

$$\mathbf{y}^\dagger\mathbf{H}\mathbf{y} = \text{Re}\ \mathbf{y}^\dagger\mathbf{D}\mathbf{y}$$

where $\mathbf{D}$ is a triangular matrix with elements

$$d_{ij} = \begin{cases} h_{ii}, & i = j \\ 2h_{ij}, & i > j \\ 0, & i < j \end{cases}$$

11.5 Invert the matrix

$$\begin{bmatrix} 1 & \rho & \rho^2 & \rho^3 \\ \rho & 1 & \rho & \rho^2 \\ \rho^2 & \rho & 1 & \rho \\ \rho^3 & \rho^2 & \rho & 1 \end{bmatrix}$$

by inverting only matrices of order two.

11.6 Invert the matrix

$$\begin{bmatrix} 1 & \rho & \rho^2 \\ \rho & 1 & \rho \\ \rho^2 & \rho & 1 \end{bmatrix}$$

by first factoring the matrix into a product of upper and lower triangular matrices.

11.7 Verify Property 6 for isomorphic matrices. That is, for the stated conditions and definitions, show

$$\mathbf{w}'\mathbf{R}\mathbf{w} = \mathbf{z}^\dagger\vec{\mathbf{R}}\mathbf{z}$$

Hint: Carry out the multiplication indicated on the left-hand side using partitioned matrices.

11.8 (Goodman) Consider the complex sample vector

$$\mathbf{z} = \begin{bmatrix} z_1 \\ z_2 \end{bmatrix} = \begin{bmatrix} x_1 + jy_1 \\ x_2 + jy_2 \end{bmatrix}$$

Assume that

$$E\left\{\begin{bmatrix} x_1 \\ y_1 \\ x_2 \\ y_2 \end{bmatrix} [x_1\ y_1\ x_2\ y_2]\right\} = \tfrac{1}{2}\begin{bmatrix} \alpha_{11} & 0 & \alpha_{12} & -\beta_{12} \\ 0 & \alpha_{11} & \beta_{12} & \alpha_{12} \\ \alpha_{12} & \beta_{12} & \alpha_{22} & 0 \\ -\beta_{12} & \alpha_{12} & 0 & \alpha_{22} \end{bmatrix}$$

Show that

$$p(z_1) = \frac{1}{\pi\alpha_{11}} \exp\left\{-\frac{|z_1|^2}{\alpha_{11}}\right\}$$

$$p(\mathbf{z}) = \frac{1}{\pi^2\gamma} \exp\left\{\frac{|z_1|^2\alpha_{22} - 2\,\mathrm{Re}\, z_2{}^* z_1(\alpha_{12} - j\beta_{12}) + |z_2|^2\alpha_{11}}{-\gamma}\right\}$$

where $\gamma = \alpha_{11}\,\alpha_{22} - (\alpha_{12}^2 + \beta_{12}^2)$.

11.9 Derive the maximum likelihood estimator of Eq. (11-40) without recourse to differentiation as was done in Sect. 11.6.

11.10 Consider the signal $\mathbf{r} = \mathbf{Hs} + \mathbf{n}$ where $\mathbf{r}$ and $\mathbf{n}$ are $M \times 1$ column vectors, $\mathbf{s}$ is a $p \times 1$ column vector with elements $s_1, s_2, \ldots, s_p$, and $\mathbf{H}$ is an $M \times p$ matrix. $\mathbf{s}$ and $\mathbf{n}$ are zero mean vectors and are not necessarily Gaussian. Their covariance matrices are $\mathbf{R}_s$ and $\mathbf{R}_n$ respectively. Determine the best linear estimator of s_{p+k}. (That is, we wish to predict "ahead" to time $p + k$.)

11.11 Consider the signal $\mathbf{r} = \mathbf{v} + \mathbf{n}$ where each vector is a zero mean vector sample of dimension 3×1. Denote

$$\mathbf{r} = \begin{bmatrix} r_{p-1} \\ r_{p-2} \\ r_{p-3} \end{bmatrix}, \qquad \mathbf{v} = \begin{bmatrix} v_{p-1} \\ v_{p-2} \\ v_{p-3} \end{bmatrix}, \qquad \mathbf{n} = \begin{bmatrix} n_{p-1} \\ n_{p-2} \\ n_{p-3} \end{bmatrix}$$

Consider these as being consecutive and equally spaced time samples of the processes. Assume $E\{n_{p-i}n_{p-j}\} = \rho^{|j-i|}$, and denote $E\{v_{p-i}v_{p-j}\} = R_v(j - i)$. Derive an explicit expression for the best linear estimate of v_p corresponding to the next consecutive time sample.

11.12 Consider the signal x where

$$\mathbf{x} = \begin{bmatrix} x_{p-1} \\ x_{p-2} \\ x_{p-3} \end{bmatrix} \qquad \text{and} \qquad E\{x_{p-i}x_{p-j}\} = \rho^{|j-i|}$$

The samples represent equally spaced time samples of the signal. Show that the best linear estimator (predictor) of the next consecutive sample x_p is $\hat{x}_p = \rho x_{p-1}$.

11.13 Determine the eigenvectors and eigenvalues of $\mathbf{N} = \sigma_w^2\mathbf{I} + \mathbf{R}_n$ in terms of the eigenvectors and eigenvalues of $\mathbf{R}_n$.

11.14 Consider an $M \times 1$ vector of random samples $\mathbf{n}$, and the corresponding $M \times M$ covariance matrix $\mathbf{R}_n$. Denote the matrix of eigenvectors and eigenvalues of $\mathbf{R}_n$ by $\mathbf{A}$ and $\boldsymbol{\lambda}$ respectively. Find a matrix transformation of $\mathbf{n}$ in terms of $\mathbf{A}$ and $\boldsymbol{\lambda}$ such that the transformed variables have the identity covariance matrix $\mathbf{I}$. The solution is $\boldsymbol{\lambda}^{-1/2}\mathbf{A}'$ where

$$\boldsymbol{\lambda}^{1/2} = \begin{bmatrix} \lambda_1^{1/2} & 0 & \cdots & 0 \\ 0 & \lambda_2^{1/2} & & \vdots \\ \vdots & & & \\ 0 & \cdots & & \lambda_M^{1/2} \end{bmatrix}$$

11.15 (a) Show that an orthogonal transformation of a random vector with a unity covariance matrix produces a new sample vector also having a unity covariance matrix.

(b) Denote the matrices of eigenvectors and eigenvalues of $\mathbf{R}_n$ by $\mathbf{A}$ and $\boldsymbol{\lambda}$ respectively. Denote the eigenvectors and eigenvalues matrices of $\boldsymbol{\lambda}^{-1/2}\mathbf{A}'\mathbf{R}_s\mathbf{A}\boldsymbol{\lambda}^{-1/2}$ as $\mathbf{B}$ and $\boldsymbol{\phi}$ respectively. Consider the random vector $\mathbf{r} = \mathbf{s} + \mathbf{n}$ with covariance matrix $\mathbf{R}_s + \mathbf{R}_n$. In terms of $\mathbf{A}$, $\mathbf{B}$, and $\boldsymbol{\lambda}$, find a transformation of $\mathbf{r}$ which results in a diagonal covariance matrix. What is the resulting covariance matrix? (This is a problem of simultaneous diagonalization.)

11.16 Consider the detection of a zero mean Gaussian signal vector in zero mean Gaussian noise. That is,

$$H_1: \quad \mathbf{r} = \mathbf{s} + \mathbf{n}$$
$$H_0: \quad \mathbf{r} = \mathbf{n}$$

What is the likelihood ratio receiver in terms of the elements of a vector $\mathbf{x}$, where

$$\mathbf{x} = \mathbf{B}'\boldsymbol{\lambda}^{-1/2}\mathbf{A}'\mathbf{r}$$

and $\mathbf{A}$, $\mathbf{B}$, and $\boldsymbol{\lambda}$ are as defined in the preceding exercise?

11.17 (a) Using the matrix definitions of the preceding exercises show that the likelihood ratio receiver for detecting a zero mean Gaussian signal vector in zero mean Gaussian noise can be implemented by generating the test statistic

$$\lambda_T = \mathbf{r}'\mathbf{A}\boldsymbol{\lambda}^{-1/2}\mathbf{B}[\mathbf{I} - (\boldsymbol{\phi} + \mathbf{I})^{-1}]\mathbf{B}'\boldsymbol{\lambda}^{-1/2}\mathbf{A}'\mathbf{r}$$

(McDonough (*18*) refers to this as the canonical receiver.)

(b) Show that this reduces to the test statistic of the preceding exercise.

11.18 (a) Use the canonical form of the test statistic for detecting a Gaussian signal in Gaussian noise. That is

$$\lambda_T = \mathbf{r}'\mathbf{A}\boldsymbol{\lambda}^{-1/2}\mathbf{B}[\mathbf{I} - (\boldsymbol{\phi} + \mathbf{I})^{-1}]\mathbf{B}'\boldsymbol{\lambda}^{-1/2}\mathbf{A}'\mathbf{r}$$

If the largest eigenvalue of $\boldsymbol{\phi}$ is $\ll 1$, show that the test statistic is approximately

$$\lambda_T = \mathbf{r}'\mathbf{R}_n^{-1}\mathbf{R}_s\mathbf{R}_n^{-1}\mathbf{r}$$

(b) Show that the eigenvalues of $\boldsymbol{\lambda}^{-1/2}\mathbf{A}'\mathbf{R}_s\mathbf{A}\boldsymbol{\lambda}^{-1/2}$ are also the eigenvalues of $\mathbf{R}_s\mathbf{R}_n^{-1}$.

(c) Suppose the smallest eigenvalue of $\boldsymbol{\phi}$ is $\gg 1$, show that the test statistic is approximately

$$\lambda_T = \mathbf{r}'\mathbf{R}_n^{-1}\mathbf{r}$$

11.19 Consider the space–time problem as discussed in Sect. 11.10 using the Fourier coefficients as samples. We wish to detect a signal of the form $\tilde{\mathbf{r}} = \tilde{\mathbf{H}}\tilde{\mathbf{s}} + \tilde{\mathbf{n}}$. The vectors are defined in the manner of Eqs. (11-68) and (11-84). The signal and noise are assumed to be Gaussian. Consider a linear array of M equally spaced receiving elements as shown in Fig. 11-6. Assume the incident signal wavefront is known to be planar. Then the channel model is strictly one of pure delay so that at any frequency ω_k

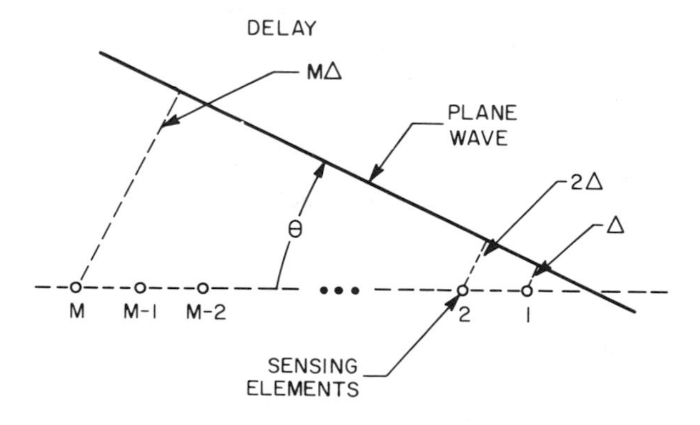

Fig. 11-6 A linear array of M equally spaced sensors in a planewave environment.

$$\tilde{\mathbf{H}}(k) = \begin{bmatrix} e^{-j\omega_k \Delta} \\ e^{-j2\omega_k \Delta} \\ \vdots \\ e^{-jM\omega_k \Delta} \end{bmatrix}$$

Assume that the noise in each channel has the same spectral density, and is statistically independent of the noise in any other channel. Show that the optimum receiver may be implemented as shown in Fig. 11-4 with the filters $l_i^*(\omega)$ representing pure delay compensation to align the signals at each element, and the filter $f(\omega)$ is given by

$$f(\omega) = \left\{ \frac{S_s(\omega)}{S_n(\omega)[S_n(\omega) + MS_s(\omega)]} \right\}^{1/2}$$

where $S_s(\omega)$ and $S_n(\omega)$ are the power spectral density of the signal and noise. (The compensation is the standard method employed in "phased array" antennas.)

11.20 As in the previous example, assume we have a linear array of M equally spaced receiving elements. We are interested in detecting a Gaussian signal $\tilde{\mathbf{s}}$ in the presence of a Gaussian interfering signal $\tilde{\mathbf{m}}$ and Gaussian noise $\tilde{\mathbf{n}}$. All are zero mean. Thus

$$H_1: \quad \tilde{\mathbf{r}} = \tilde{\mathbf{H}}\tilde{\mathbf{s}} + \tilde{\mathbf{W}}\tilde{\mathbf{m}} + \tilde{\mathbf{n}}$$
$$H_0: \quad \tilde{\mathbf{r}} = \tilde{\mathbf{W}}\tilde{\mathbf{m}} + \tilde{\mathbf{n}}$$

The vectors are defined by relations such as those given in Eq. (11-68). Assume that $\tilde{\mathbf{H}}$ and $\tilde{\mathbf{W}}$ correspond to plane waves (pure delay). Denote the incremental delays of each wave by Δ and δ respectively. Assume the noise $\tilde{\mathbf{n}}$ is of equal power at each receiving element and is statistically independent of the noise in any other element. Show that the optimum receiver can be implemented as shown in Fig. 11-7. Here the set $\tau_i(H)$ represents delay compensation to align the signal $\mathbf{s}$ at each element, and $\tau_i(W)$ represents the delay compensation to align the signal $\tilde{\mathbf{m}}$ at each element. (These represent "beams" steered to the signal and interfering noise respectively.) Also

$$f_1(\omega) = \frac{S_m(\omega)}{S_n(\omega) + MS_m(\omega)}$$

$$f_2(\omega) = \frac{1}{S_n(\omega)^{1/2}} \left[\frac{S_s(\omega)}{S_n(\omega) + \dfrac{S_m(\omega)S_s(\omega)\,|D(\omega)|^2}{S_n(\omega) + MS_m(\omega)}} \right]^{1/2}$$

$$D(\omega) = e^{j\omega(\Delta - \delta)} \frac{(1 - e^{jM\omega(\Delta - \delta)})}{1 - e^{j\omega(\Delta - \delta)}}$$

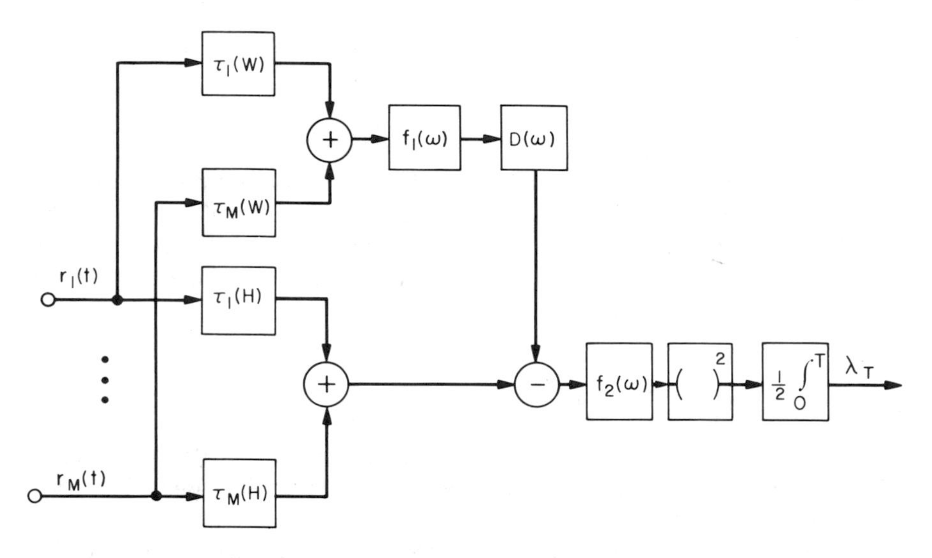

Fig. 11-7 Signal detection in presence of interfering plane wave.

It may be verified that $|D(\omega)|^2$ represents the power response of an array to a plane wave arriving at an angle θ_Δ, when the array is steered to an arrival angle θ_δ.

11.21 Design a space–time processor to detect the signal $\tilde{\mathbf{r}} = \tilde{\mathbf{v}} + \tilde{\mathbf{n}}$ where $\tilde{\mathbf{v}}$ and $\tilde{\mathbf{n}}$ are Gaussian vectors. Assume, as in Fig. 11-4, that there are M receiving elements. Assume that the noise at each element is of equal power, say $S_n(\omega_k)$, and is statistically independent of the noise at any other element. Furthermore, assume the same conditions for the signal except for the level of the signal power which we denote as $S_v(\omega_k)$. Show that the optimum

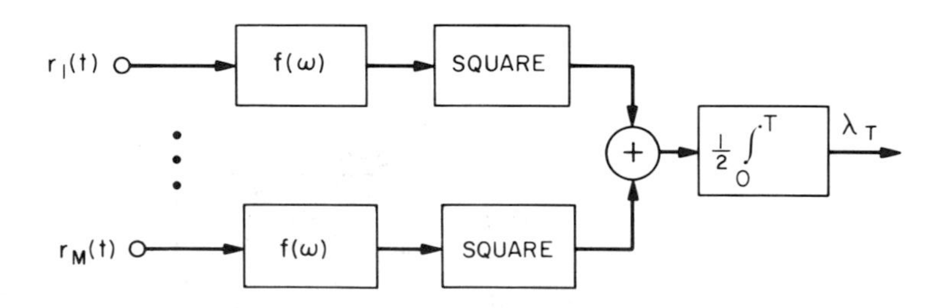

Fig. 11-8 Receiver to detect independent signals assuming independent noise.

receiver may be implemented as shown in Fig. 11-8 with

$$f(\omega) = \left\{ \frac{S_v(\omega)}{S_n(\omega)[S_v(\omega) + S_n(\omega)]} \right\}^{1/2}$$

$S_v(\omega)$ and $S_n(\omega)$ are the power spectral densities of the signal and noise respectively. Thus the channels are incoherently combined after the quadratic detectors.

References

1. Schwartz, M., Abstract vector spaces applied to problems in detection and estimation theory, *IEEE Trans. Inform. Theory* **IT-2,** No. 3, (1966).
2. Schwartz, M., Optimum demodulation of signals through randomly fading media, Presented at the Symposium on System Theory, Polytechnic Institute of Brooklyn, April 1965.
3. Shannon, C. E., and Weaver, W., "The Mathematical Theory of Communication." Univ. of Illinois Press, Urbana, Illinois, 1963.
4. Linden, D. A., A discussion of sampling theorems, *Proc. IRE* (July 1959).
5. Zakai, M., Band-limited functions and the sampling theorem, *Inform. Control* (April 1965).
6. Balakrishnan, A. V., A note on the sampling principle for continuous signals, *IRE Trans. Inform. Theory* **IT-3,** No. 2, 143–146 (1957).
7. Guillemin, E. A., "The Mathematics of Circuit Analysis," Wiley, New York, 1949.
8. Goodman, N. R., Statistical analysis based on a certain multivariate complex gaussian distribution, *Ann. Math. Statistics* **34,** (1963).
9. Birkhoff, G., and MacLane, S., "A Survey of Modern Algebra," 3rd ed., Macmillan, New York, 1965.
10. Papoulis, A., "Probability, Random Variables, and Stochastic Processes." McGraw-Hill, New York, 1965.
11. Cox, H., Interrelated problems in estimation and detection I & II, *Proc. NATO Advan. Study Inst. Signal Proc. Emphasis Underwater Acoust., Enschede, Netherlands* (August 1968).
12. Lawson, J. L., and Uhlenbeck, G. E., "Threshold Signals, Radiation Laboratory Series," Vol. 24. McGraw-Hill, New York, 1950.
13. Kailath, T., Correlation detection of signals perturbed by a random channel, *IRE Trans. Inform. Theory* **IT-6,** No. 3, 361–366 (1960).
14. Bryn, F., Optimum signal processing of three-dimensional arrays operating on gaussian signals and noise, *J. Acoust. Soc. Amer.* **34,** No. 3 (1962).
15. Capon, J., Greenfield, R. J., and Kolker, R. J., Multidimensional maximum likelihood processing of a large aperture seismic array, *Proc. IEEE* (February 1967).
16. Ide, J. M., Development of underwater acoustic arrays for passive detection of sound sources, *Proc. IRE* (May 1959).
17. Schultheiss, P. M., and Tuteur, F. B., Optimum and suboptimum detection of directional gaussian signals in an isotropic gaussian noise field, *IEEE Trans. Military Electron.* 197–208 (July-October 1965).
18. McDonough, R. N., A canonical form of the likelihood detector for gaussian random vectors, *J. Acoust. Soc. Amer.* **49,** 402–406 (1971).

SUPPLEMENTARY BIBLIOGRAPHY

Texts:

Aitken, A. C., "Determinants and Matrices," 9th Ed. Oliver and Boyd, Edinburgh, Scotland, 1958.

Anderson, T. W., "An Introduction to Multivariate Statistical Analysis." Wiley, New York, 1958.

Baghdady, E. J., Ed., "Lectures on Communication System Theory." McGraw-Hill, New York, 1961.

Bellman, R., "Introduction to Matrix Analysis." McGraw-Hill, New York, 1960.

Bodewig, E., "Matrix Calculus," 2nd Ed. North-Holland Publ., Amsterdam, 1959.

Deutsch, R., "Estimation Theory." Prentice Hall, Englewood Cliffs, New Jersey, 1965.

Franklin, J. N., "Matrix Theory." Prentice Hall, Englewood Cliffs, New Jersey, 1968.

Grenander, U., and Rosenblatt, M., "Statistical Analysis of Stationary Time Series." Wiley, New York, 1957.

Hancock, J. C., and Wintz, P. A., "Signal Detection Theory." McGraw-Hill, New York, 1966.

Kullback, S., "Information Theory and Statistics." Wiley, New York, 1959.

Marcus, M., Basic Theorems in Matrix Theory, National Bureau of Standards, Applied Math. Series 57 (January 1960).

Miller, K. S., "Multidimensional Gaussian Distribution." Wiley, New York, 1964.

Perlis, S., "Theory of Matrices." Addison-Wesley, Reading, Massachusetts, 1952.

Westlake, J. R., "A Handbook of Numerical Matrix Inversion and Solution of Linear Equations." Wiley, New York, 1968.

Papers:

Bryn, F., Optimum Theoretical Structures of Sonar Systems Employing Spatially-Distributed Receiving Elements, SACLANT ASW Research Center, Tech. Rep. No. 119, AD 839817 (1 September 1968).

Burg, J. P., Three-dimensional filtering with an array of seismometers, *Geophysics* **29,** No. 5 (1964).

Capon, J., Investigation of Long Period Noise at LASA, Lincoln Lab. Tech. Note 1968–15, AD 671509 (3 June 1968).

Claerbout, J. F., Detection of p-waves from weak sources at great distances, *Geophysics* **29,** No. 2 (1964).

Cox, H., Optimum arrays and the schwartz inequality, *J. Acoust. Soc. Amer.* **45,** No. 1, 228–232 (1969).

Edelblute, D. J., Fisk, J. M., and Kinnison, G. L., Criteria for optimum-signal-detection theory for arrays, *J. Acoust. Soc. Amer.* **41,** No. 1, 199–205 (1967).

Faran, J. J., and Hills, R., NR-384-903, Tech. Rep. No. 28 (1 November 1952).

Griffiths, L. J., A comparison of multidimensional wiener and maximum-likelihood filters for antenna arrays, proceedings letters, *Proc. IEEE* (November 1967).

Griffiths, L. J., Signal Extraction Using Real-Time Adaptation of Linear Multichannel Filter, Stanford Univ. Tech. Rep. 6788–1, AD 836761 (February 1968).

Kadota, T. T., Optimum estimation of nonstationary gaussian signals in noise, *IEEE Trans. Inform. Theory* (March 1969).

Kailath, T., Correlation detection of signals perturbed by a random channel, *IRE Trans. Inform. Theory* (June 1960).

Kennedy, R. M., and O'Neill, R. Jr., Adaptive Multichannel Filtering for Sonar Arrays, U.S. Navy Underwater Sound Lab. USL Rep. No. 809, AD 819096 (23 May 1967).

Middleton, D., and Groginsky, H. L., Detection of random acoustic signals by receivers with distributed elements: optimum receiver structures for normal signal and noise fields, *J. Acoust. Soc. Amer.* **38,** No. 5, 727–737 (1965).

Nuclear Test Detection Issue, *Proc. IEEE* **53,** No. 12 (1965).

Schultheiss, P., Optimal Detection of Directional Gaussian Signals in an Isotropic Gaussian Noise Field, Appendix 3 of: Processing of Data From Sonar, by R. A. McDonald, P. M. Schultheiss, F. B. Tuteur, and T. Usher, Jr., Yale Univ. Rep. No. U417-63-045, AD 420575 (1 September 1963).

Stocklin, P. L., Space-Time Sampling and Likelihood Ratio Processing in Acoustic Pressure Fields, Journal, British IRE (July 1963).

Tuteur, F. B., Some Aspects of the Detectability of Broadband Sonar Signals by Non-directional Passive Hydrophones, Advan. Res. Projects Agency, RM-4578-ARPA, AD 618500 (June 1964).

Tuteur, F. B., Detectability of directional amplitude-modulated noise signals in an isotropic noise background of unknown power level, correspondence, *IEEE Trans. Inform. Theory* **IT-11,** No. 4, 591–593 (1965).

Vanderkulk, W., Optimum processing for acoustic arrays, the radio and electronic engineer, *J. Brit. IRE* **26,** No. 4, 285–292 (1963).

Widrow, B., Mantey, P. E., Griffiths, L. J., and Goode, B. B., Adaptive antenna systems, *Proc. IEEE* (December 1967).

Index

N

O

P

Q

R

S

T

U

V

W

ELECTRICAL SCIENCE

A Series of Monographs and Texts

Editors

Joseph E. Rowe. Nonlinear Electron-Wave Interaction Phenomena. 1965

Max J. O. Strutt. Semiconductor Devices: Volume I. Semiconductors and Semiconductor Diodes. 1966

Austin Blaquiere. Nonlinear System Analysis. 1966

Victor Rumsey. Frequency Independent Antennas. 1966

Charles K. Birdsall and William B. Bridges. Electron Dynamics of Diode Regions. 1966

A. D. Kuz'min and A. E. Salomonovich. Radioastronomical Methods of Antenna Measurements. 1966

Charles Cook and Marvin Bernfeld. Radar Signals: An Introduction to Theory and Application. 1967

J. W. Crispin, Jr., and K. M. Siegel (eds.). Methods of Radar Cross Section Analysis. 1968

Giuseppe Biorci (ed.). Network and Switching Theory. 1968

Ernest C. Okress (ed.). Microwave Power Engineering:
Volume 1. Generation, Transmission, Rectification. 1968
Volume 2. Applications. 1968

T. R. Bashkow (ed.). Engineering Applications of Digital Computers. 1968

Julius T. Tou (ed.). Applied Automata Theory. 1968

Robert Lyon-Caen. Diodes, Transistors, and Integrated Circuits for Switching Systems. 1969

M. Ronald Wohlers. Lumped and Distributed Passive Networks. 1969

Michel Cuenod and Allen E. Durling. A Discrete-Time Approach for System Analysis. 1969

K. Kurokawa. An Introduction to the Theory of Microwave Circuits. 1969

H. K. Messerle. Energy Conversion Statics. 1969

George Tyras. Radiation and Propagation of Electromagnetic Waves. 1969

Georges Metzger and Jean-Paul Vabre. Transmission Lines with Pulse Excitation. 1969

C. L. Sheng. Threshold Logic. 1969

Dale M. Grimes. Electromagnetism and Quantum Theory. 1969

Robert O. Harger. Synthetic Aperture Radar Systems: Theory and Design. 1970

M. A. Lampert and P. Mark. Current Injection in Solids. 1970

W. V. T. Rusch and P. D. Potter. Analysis of Reflector Antennas. 1970

Amar Mukhopadhyay. Recent Developments in Switching Theory. 1971

A. D. Whalen. Detection of Signals in Noise. 1971

J. E. Rubio. The Theory of Linear Systems. 1971

Keinosuke Fukunaga. Introduction To Statistical Pattern Recognition. 1972

Jacob Klapper and John T. Frankle. Phase-Locked and Frequency-Feedback Systems: Principles and Techniques. 1972

Kumpati S. Narendra and James H. Taylor. Frequency Domain Criteria for Absolute Stability. 1973

Daniel P. Meyer and Herbert A. Mayer. Radar Target Detection: Handbook of Theory and Practice. 1973

T. R. N. Rao. Error Coding for Arithmetic Processors. 1974

C. A. Desoer and M. Vidyasagar. Feedback Systems: Input-Output Properties. 1975